Biomass Preprocessing and Pretreatments for Production of Biofuels

Mechanical, Chemical and Thermal Methods

Editor

Jaya Shankar Tumuluru
Biofuel and Renewable Energy Technologies
Idaho National Laboratory
Idaho Fall, USA

CRC Press
Taylor & Francis Group
Boca Raton London New York

CRC Press is an imprint of the
Taylor & Francis Group, an **informa** business
A SCIENCE PUBLISHERS BOOK

CRC Press
Taylor & Francis Group
6000 Broken Sound Parkway NW, Suite 300
Boca Raton, FL 33487-2742

First issued in paperback 2021

© 2018 by Taylor & Francis Group, LLC
CRC Press is an imprint of Taylor & Francis Group, an Informa business

No claim to original U.S. Government works

Version Date: 20180427

ISBN-13: 978-0-367-78100-2 (pbk)
ISBN-13: 978-1-4987-6547-3 (hbk)

Library of Congress Cataloging-in-Publication Data
Names: Tumuluru, Jaya Shankar, editor.
Title: Biomass preprocessing and pretreatments for production of biofuels : mechanical, chemical and thermal methods / editor Jaya Shankar Tumuluru, Biofuel and Renewable Energy Technologies, Idaho National Laboratory, Idaho Falls, USA.
Description: Boca Raton, FL : CRC Press, Taylor & Francis Group, [2018]	"A science publishers book."	Includes bibliographical references and index.
Identifiers: LCCN 2018019173	ISBN 9781498765473 (hardback)
Subjects: LCSH: Biomass energy.	Chemical processes.
Classification: LCC TP339 .B5737 2018	DDC 662/.88--dc23
LC record available at https://lccn.loc.gov/2018019173

Visit the Taylor & Francis Web site at
http://www.taylorandfrancis.com

and the CRC Press Web site at
http://www.crcpress.com

Preface

Biomass—both plant and animal waste—is the oldest source of renewable energy. When compared to fossil fuels, which took hundreds of millions of years to form, biomass is considered a form of renewable energy because it can be regrown in a short period of time. In addition, biomass is considered carbon-neutral because the carbon dioxide released during the conversion process is still part of the carbon cycle. The major advantage of using biomass is that it helps to reduce carbon dioxide emissions, thereby reducing its negative impact on the environment. There are various ways biomass can be used for generating energy. The simplest form is by burning. When biomass is burned, the chemical energy in it converts to heat. In addition, biomass can further be converted to liquid fuels or biogas, which can be utilized as a form of fuel.

Burning of biomass to generate power is only one way to release the energy in biomass. It can also be converted to other useable forms of energy, such as methane gas or transportation fuels like ethanol and biodiesel. Methane gas is a component of landfill gas or biogas that forms when garbage, agricultural waste, and human waste decompose in landfills or in special containers called digesters. Crops such as corn and sugar cane are fermented to produce fuel ethanol for use in vehicles. Currently dedicated energy crops such as switchgrass, miscanthus, bluestem, elephant grass, wheatgrass, and others, as well as short rotation woody crops like willow, cottonwood, silver maple, black locust, and poplar are grown for bioenergy production. In 2016, biomass fuels provided about 5% of the primary energy in the United States. Out of that 5%, 48% is from biofuels (mainly ethanol), 41% from wood and wood-derived biomass, and about 11% from the biomass in municipal waste (EIA, 2018).

The Inspiration for Book

Currently, many biorefineries are facing challenges in terms of both feedstock handling and quality. Due to this, the biorefineries are not able to run their plants at their designed capacity. Physical attributes (i.e., moisture, particle size, and density), rheological properties (i.e., elastic and cohesive), and chemical characteristics (i.e., proximate, ultimate, and energy properties) of raw biomass limits its use at a scale necessary for biofuels applications. This book is designed to understand some of the mechanical, chemical, and thermal preprocessing and pretreatment methods that are available that can help to overcome some of these limitations in terms of physical, chemical, and biochemical composition. There is currently no written material that introduces all of the preprocessing and pretreatment methods into a single volume. As such, it is our hope that this book will help to fill the void of presenting this much-needed information to industry in terms of how biomass feedstock properties can be effectively managed and utilized with the help of these pretreatment and preprocessing technologies.

The Organization

The subject matter is logically divided into 16 chapters. This introduction chapter explains why biomass needs to be preprocessed and pretreated. Chapters 2 through 6 focus on the current state-of-the-art mechanical preprocessing methods where the focus is on how different mechanical methods like size reduction, fractionation, and densification can help to meet physical specifications for use in industry, such as particle size distribution, density ash specification, etc. Chapters 7 through 11 discuss the thermal pretreatments that are available, such as steam explosion, hydrothermal carbonization, torrefaction, and hydrothermal liquefaction, which help to modify the chemical and physical characteristics and energy density of the biomass for thermochemical conversion applications. Finally, Chapters 12 to 16 provide more detail on current chemical pretreatment methods, such as acid, alkaline and ionic, and ammonia fiber explosion pretreatments, which impact the composition of the biomass and its subsequent effect on the conversion process. Many of the chapters also discuss how integrating the thermal methods with the mechanical and chemical methods further help to obtain the physical properties and compositional requirements for biofuels production.

Who will Benefit from This Book?

This book will assist feedstock preprocess engineers, biorefineries managers, and academic researchers to understand how mechanical, chemical, and thermal preprocessing methods can help to overcome some of the limitations in using biomass in terms of its physical properties, chemical and biochemical composition, and morphological characteristics. In addition, this book can serve as a textbook for engineering and science students as it stresses the scientific understanding, analysis, and applications of different pretreatment and preprocessing methods for biofuels production. The book may also serve to create awareness among energy planners, policy makers, and other users about these technologies in general.

Reference

https://www.eia.gov/energyexplained/index.cfm?page=biomass_home

Disclaimer

U.S. Department of Energy Disclaimer

This information was prepared as an account of work sponsored by an agency of the U.S. Government. Neither the U.S. Government nor any agency thereof, nor any of their employees, makes any warranty, express or implied, or assumes any legal liability or responsibility for the accuracy, completeness, or usefulness of any information, apparatus, product, or process disclosed, or represents that its use would not infringe privately owned rights. References herein to any specific commercial product, process, or service by trade name, trademark, manufacturer, or otherwise, does not necessarily constitute or imply its endorsement, recommendation, or favoring by the U.S. Government or any agency thereof. The views and opinions of authors and editor expressed herein do not necessarily state or reflect those of the U.S. Government or any agency thereof.

This work was supported by the U.S. Department of Energy, Office of Energy Efficiency and Renewable Energy, under DOE Idaho Operations Office Contract DE-AC07-05ID14517.

Acknowledgements

The editor would like to thank Dr. Erin Searcy, INL Biofuels Department Manager, for her outstanding support as a reviewer on this project; Dr. Richard Hess, INL Energy and Environment Science Director, for his helpful and timely suggestions; and Mr. Kevin Kenny, INL Bioenergy Program Relationship Manager, for his assistance in getting this book published. In addition, the editor would like to sincerely thank the individual reviewers listed below who assisted greatly in the publication of this book. Sincere thanks is expressed as well to Mr. David Combs, INL Art Director and Branding Specialist, and Mr. Gordon Holt, INL Senior Editor, for their invaluable support. The editor would also like to thank all of the authors who contributed in any way to this book, without whom it would have been impossible to get it published. Finally, the editor would like to express extreme gratitude and thanks to his wife NagaSri Valli Gali and two daughters Priya Lasya Tumuluru and Siva Lalasa Tumuluru for their help and support as well.

List of Reviewers

Dr. Erin Searcy, Biofuels Department Manager, Idaho National Laboratory, Idaho Falls, Idaho, USA.

Mr. John Gardner, P.E., Senior Project Engineer, Environmental Segment, Vermeer, Iowa, USA.

Dr. Nikhil M. Patel, Lead, Distributed Energy Technologies, Energy and Environmental Research Center (EERC), University of North Dakota, Grand Forks, North Dakota, USA.

Dr. Suchithra Thangalazhy Gopakumar, Assistant Professor, Department of Chemical and Environmental Engineering, University of Nottingham Malaysia Campus, Jalan Broga, Semenyih, Malaysia.

Mr. Christopher Lanning, P.E., Design Engineer, Forest Concepts, LLC, Auburn, Washington, USA.

Dr. Deepak Kumar, Postdoctoral Research Associate, Urbana-Champaign, University of Illinois, Chicago, USA.

Dr. Jyoti Prasad Chakraborty, Assistant Professor, Department of Chemical Engineering and Technology, India.

Dr. Nourredine (Nour) Abdoulmoumine, Assistant Professor, Biosystems Engineering and Soil Science, University of Tennessee, Knoxville, Tennessee, USA.

Dr. Oladiran Fasina, Biosystems Engineering Department Head, Samuel Ginn College of Engineering, Auburn University, Auburn, Alabama, USA.

Dr. Wilson Lam, Managing Director, LAMPS Consulting Inc., Vancouver, British Columbia, Canada.

Dr. Erica L. Belmont, Assistant Professor of Mechanical Engineering, College of Engineering and Applied Science, Department of Mechanical Engineering, University of Wyoming, Laramie, Wyoming, USA.

Dr. Campabadal, Carlos, Instructor/IGP Outreach Specialist, Kansas State University, Manhattan, Kansas, USA.

Dr. Lalitendu Das, Post-Doctoral Scholar, University of Kentucky, Lexington, Kentucky, USA.

Dr. Tom Fletcher, Professor and Chair, Chemical Engineering Department, Brigham Young University, Provo, Utah, USA.

Dr. Lijun Wang, Professor of Biological Engineering, North Carolina Agricultural and Technical State University, Greensboro, North Carolina, USA.

Dr. Igathinathane (Igathi) Cannayen, Associate Professor, Agricultural and Biosystems Engineering, Northern Great Plains Research Lab, North Dakota State University, Fargo, North Dakota, USA.

Dr. Michael Tai, ADM Biotech Project Leader, Decatur, Illinois, USA.

Dr. Chenlin Li, Staff Scientist, Energy and Environment Science and Technology, Biofuels Department, Idaho National Laboratory, Idaho Falls, Idaho, USA.

Dr. Joan Lynam, Assistant Professor, Chemical Engineering Department, Louisiana Tech University, Ruston, Louisiana, USA.

Dr. Rajeev Kumar, Assistant Research Engineer, Center for Environmental Research and Technology (CECERT), Bourns College of Engineering, University of California, Riverside, California, USA.

Dr. Joy Agnew, P.Eng., Project Manager, Agricultural Research Services, Prairie Agricultural Machinery Institute, Humboldt, Saskatchewan, Canada.

Dr. David Hodge, Associate Professor, Department of Chemical & Biological Engineering, Montana State University, Bozeman, Montana, USA.

Dr. Karunanithy Chinnadurai (CK), Assistant Professor, Food Engineering, Food and Nutrition Department, University of Wisconsin-Stout, Menomonie, Wisconsin, USA.

Contents

THERMAL PREPROCESSING

CHEMICAL PREPROCESSING

CHAPTER 1

Why Biomass Preprocessing and Pretreatments?

Jaya Shankar Tumuluru

1. Introduction

The 2016 United Nations Framework Convention on Climate Change (UNFCCC) challenged the global reduction of greenhouse gas annual emissions to less than 2°C by 2020 (Tumuluru, 2016). According to Arias et al. (2008), biomass is considered carbon-neutral as the carbon dioxide released during its conversion is still part of the carbon cycle. The use of biomass helps to reduce carbon dioxide emissions and minimize negative impacts on the environment. Physical attributes (i.e., moisture, particle size, and density), rheological properties (i.e., elastic and cohesive), and chemical characteristics (i.e., proximate, ultimate, and energy properties) of raw biomass limits its use at a scale necessary for biofuels applications (Tumuluru, 2016).

2. Biomass Physical and Chemical Properties Limitations for Solid and Liquid Fuels

Biomass energy is produced using thermochemical (i.e., direct combustion, gasification, and pyrolysis), biological (i.e., anaerobic digestion and

750 University Blvd, Energy Systems Laboratory, Idaho National Laboratory, Idaho Falls, Idaho, 83415-3570.
Email: JayaShankar.Tumuluru@inl.gov

fermentation), and chemical (i.e., esterification) technologies. Biomass as harvested lacks both the bulk density and energy density necessary for cost-efficient bioenergy production. Also, biomass lacks flowability characteristics that limit its' ability to move from location to location in existing transportation and handling infrastructures. Compared to fossil fuels, raw biomass has a low bulk density, a high moisture content, a hydrophilic nature, and a low calorific value, which all contribute negatively to logistics and final energy efficiency. The low energy density of biomass, as compared to fossil fuels, results in the requirement of high volumes of biomass to generate a comparable amount of fuels production. These high volumes, in turn, lead to storage, transportation, and feed handling issues at cogeneration, thermochemical, and biochemical conversion plants.

High moisture content in the biomass creates uncertainty in its physical, chemical, and microbiological properties. In general, high moisture raw biomass results in preprocessing, storage, feeding and handling, and conversion challenges. Biomass moisture influences the grinding behavior. According to Tumuluru et al. (2014), it is very challenging to grind high moisture biomass due to its fibrous nature. Also, with the increase in moisture content, the amount of energy needed for grinding increases exponentially, as well as negatively impacting particle size distribution. Higher moisture in the biomass also results in feeding and handling issues due to elastic and cohesive properties, which lead to the plugging of grinder screens, the bridging of particles in the conveyors, and the plugging of the reactors themselves.

High moisture content in the biomass results in variable particle sizes (especially when the particles are less than 2 mm) during grinding. These inconsistent particle sizes may not react consistently, thereby reducing efficiency and increasing the costs of the biofuels' conversion process. Also, raw biomass is thermally unstable due to high moisture, which results in low calorific values when used in thermochemical processes such as gasification. In terms of chemical composition compared to fossil fuels, biomass has more oxygen than carbon and hydrogen. The changes in chemical composition can result in producing inconsistent products and the formation of condensable tars, which further lead to problems like gas-line blockage. As far as storage is concerned, high moisture in biomass results in loss of quality due to chemical oxidation and microbial degradation.

The variable physical, chemical, and rheological properties of the raw biomass are therefore limiting biorefineries to operate at their designed capacities. Variations in physical properties—moisture content, foreign matter (soil), particle size, and particle size distribution—result in reductions in grinding throughput, equipment wear, plugging during conveyance, and reactor feeding problems. These problems all contribute to a decrease in production capacity, an interruption of normal operations, an increase in preprocessing and conversion costs, and a major reduction in yield. Solving the inherent biomass physical, chemical, and rheological issues, whether within a single-plant dedicated supply chain or via a depot system of distributed commercial feedstock suppliers, are crucial in enabling biorefineries to operate at their designed capacities.

3. Preprocessing or Pretreatments

Whether producing biofuels, biopower, or other bioproducts, all bioenergy industries depend on on-spec biomass feedstocks. Biomass cannot be fed into conversion infeed systems until it undergoes some level of preprocessing, such as size reduction or others. The degree of preprocessing depends on the type of conversion, such as biochemical and thermochemical. Pretreatment or preprocessing helps to alter the physical, rheological, and chemical properties of biomass, making it more suitable for conversion. Figure 1 shows how pretreatment impacts lignocellulosic biomass. Recent studies conducted by Tumuluru et al. (2016) suggest that feedstock supply system unit operations have significant impacts on feedstock quality and cost. The aforementioned authors also suggest that novel harvesting methods and mechanical, chemical, and thermal pretreatment technologies can improve the physical, chemical, and rheological properties of biomass, thereby making them better suited to meet the specification in terms of density, particle size, ash composition, and carbohydrate content for both biochemical and thermochemical conversion applications. These mechanical, chemical, and thermal preprocessing and pretreatment methods help to alter the amorphous and crystalline regions of the biomass and bring significant changes in structural and chemical compositions.

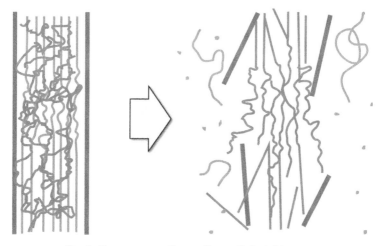

Fig. 1. Pretreatment effect on lignocellulosic biomass.

3.1 Mechanical methods

Currently, biorefineries are experiencing challenges in feeding, handling, transportation, and storage of raw biomass with variable physical properties (e.g., high and low moisture, less flowable formats). Some of the common problems faced by biorefineries in handling the variable moisture of biomass are bridging of the particles, which leads to uneven discharge from silos and the subsequent jamming of conveyors, resulting in inconsistent feeding to reactors and storage issues (e.g., dry matter loss and spontaneous combustion). These challenges need to be solved for smooth biorefinery operation. Mechanical preprocessing technologies such as size reduction and densification are critical to meet the particle size specification and density. These specifications directly impact the front end operations of the biorefineries such as transportation, storage, handling, and feeding of the reactors for the conversion. Developing energy efficient grinding and densification technologies will help to address the storage, feeding and handling issues, and overall mechanical preprocessing costs. In general, grinding energy and particle size distribution are sensitive to moisture content of the biomass. Developing grinding technologies that are less sensitive to biomass moisture content will have a major impact on grinding energy. Optimizing densification technologies for different feedstock moisture contents to produce densified products that can meet the density requirement for different transportation scenarios and conversion requirements will also have an impact on mechanical preprocessing costs. In addition to physical properties such as density and particle size, quality specifications such as ash content are also important. Fractional milling and air classification are promising and relatively low-cost methods for ash reduction of non-traditional, high-ash, feedstocks. Air classification is also suitable for ash reduction and anatomical sorting of herbaceous feedstocks.

3.2 Thermal methods

Thermal methods are typically used to dry biomass and make it aerobically stable. Biomass is usually dried in rotary driers to a lower moisture content of 10% (w.b.) (Yancey et al., 2013; Tumuluru, 2016). High-temperature drying technologies, such as rotary driers, are energy- and capital-intensive and not environmental friendly. They are known to release volatile organic compounds (VOCs) and extractives that are harmful to humans, forests, and crops (Granström, 2005; Johansson and Rasmuson, 1998; Lamers et al., 2015). Developing low temperature drying technologies are critical as they not only help to reduce moisture content but they do not emit VOCs and extractives into the environment as well. Researchers are currently investigating high-temperature thermal methods, such as torrefaction, to improve the physical, chemical, and energy properties of biomass. During torrefaction, biomass is slowly heated in between temperatures of 200–300°C in specialized reactors in an inert or reduced environment. This process results in a solid uniform product with a lower moisture and higher energy content when compared to raw biomass. This technique helps biomass to meet most of the physical, chemical, and energy properties for thermochemical and cofiring applications. Common biomass reactions during torrefaction include the devolatilization, depolymerization, and carbonization of hemicellulose, lignin, and cellulose. In addition to producing a brown to black uniform solid product, torrefaction also results in condensables (e.g., water, organics, and lipids) and non-condensable gases (e.g., CO_2, CO, and CH_4). Typically during torrefaction, 70% of the mass is retained as a solid product containing 90% of the initial energy content, while 30% of the lost mass is converted into condensable and non-condensable products. Advanced torrefaction systems should be developed that can recapture the volatiles and use them as a heat source, thereby making the torrefaction process much more sustainable.

Torrefaction makes biomass brittle and decreases grinding energy by about 70%. This ground torrefied biomass has improved sphericity, particle surface area, and particle size distribution (Phanphanich and Mani, 2011). During torrefaction, carbon content and calorific value of biomass increases by 15–25% wt while the moisture content decreases to < 3% (w.b.) (Tumuluru et al., 2010). The major limitation of torrefaction is the loss of mass density that can be overcome by densifying the torrefied biomass. The Energy Research Centre of the Netherlands (ECN) developed a torrefaction and pelletization process called TOP where the hot torrefied material is pelleted (Bergman, 2005). Their results also indicate that pelleting the torrefied biomass at 225°C increases the throughput of the mill by two times whereas at the same time specific energy consumption of the densification process is lowered by two times. Another major advantage of torrefied material is its hydrophobic nature. Loss of OH functional group during torrefaction make biomass hydrophobic (i.e., loses the ability to attract water molecules), making biomass more stable against chemical oxidation and microbial degradation. In their studies on the storage of woody, herbaceous, and torrified biomass, Tumuluru et al. (2016) indicated that the off-gas (CO and CO_2) emissions from torrefied woody biomass are less compared to non-torrefied woody biomass. These improved properties make torrefied biomass particularly suitable for cofiring in power plants and as an upgraded feedstock for gasification. Currently, researchers are working on developing wet torrefaction techniques because they operate at lower temperatures and can accommodate wet biomasses as well. Bach and Tran (2015) discovered that dry torrefaction removes more hemicellulose from Norway spruce wood than wet torrefaction. They also concluded that dry-torrefied woody is less reactive in devolatilization but more reactive in char combustion. More studies in understanding the techno-economics of these two processes are needed to realize their technical and economic feasibility.

3.3 Chemical methods

Ash in biomass is inorganic and derived from soil physiological sources. Ash in the biomass acts as an inhibitor during conversion and also results in equipment erosion during preprocessing. Various constituents in the ash have multiple implications; therefore, ash removal must be designed to meet the target end use. Also, ash removal is cost- and energy-intensive so selecting the right ash removal technology will have a significant impact on the cost. Ash in the biomass during biochemical conversion can reduce (a) dilute acid pretreatment effectiveness, (b) reduce cell growth, and (c) decrease ethanol yields (Carpenter et al., 2014; Zhang et al., 2013; Palmqvist and Hahn-Hägerdal, 2000; Casey et al., 2013). According to Carpenter et al. (2014), Bridgewater (2012), and Liden et al. (1988), even trace quantities of chlorine, calcium, silica, potassium, and phosphorus can result in corrosion and reduce ash-fusion temperature, which can significantly affect thermochemical conversion product yields.

According to Chen et al. (2011), Deng et al. (2013), and Aston et al. (2016), low-severity hydrothermal or dilute acid leaching is excellent for removing alkaline earth metals and alkali metals for thermochemical conversions. Washing and leaching of raw biomass can also help to minimize ash content, which further reduces corrosion, slagging, fouling (ash deposition), and sintering and agglomeration of the bed. Washing and leaching helps to remove most of the alkalis, which is mainly problematic for herbaceous materials. Recent studies by researchers have indicated that washing and leaching herbaceous biomass efficiently modifies the ash composition to produce clean biomass, which could be further used for firing/co-firing with coal. Also during chemical pretreatment, the biomass undergoes structural changes that can further help the mechanical preprocessing such as grinding and densification and can reduce the wear of the equipment. The major challenge in using chemical pretreatment is the cost. One way to reduce chemical pre-conversion costs is to treat only the biomass fractions, which are high in ash content. Thompson et al. (2016) have suggested that chemically preprocessing only a select, high-ash fraction of mechanically generated feedstock fractions can result in sufficient feedstock quality while minimizing chemical, disposal, and drying costs. It is very critical to develop chemical preprocessing technologies (e.g., washing, leaching, acid and alkali pretreatment, and ammonia fiber explosion) to make biomass to meet the specifications desirable for different conversion pathway. Making these technologies sustainable is critical for their success.

4. Integrating the Different Preprocessing and Pretreatment Technologies

The challenge with the existing conventional systems is that they do not supply quality biomass for integrated biorefineries. Integrating mechanical, chemical, and thermal preprocessing and pretreatment methods helps to overcome the physical, chemical, and energy property limitations of biomass and supply high quality biomass to biorefineries.

Figure 2 portrays possible combinations of formulation, mechanical, and thermal pretreatment methods, which can improve the physical and chemical properties of biomass (Tumuluru et al., 2012). Individually these preprocessing methods can be expensive but combining them creates new opportunities and can reduce costs. For example, chemical and thermal preprocessing technologies combined with mechanical methods such as densification will help to save costs and minimize product losses. This integrated approach helps to develop uniform feedstocks with minimum physical and compositional issues for biochemical, thermochemical, and combustion applications, including co-firing with coal at high percentages. Chemical pretreatments can also be integrated with other mechanical preprocessing methods to meet the feedstock specifications necessary for biochemical conversion.

Some of the new mechanical preprocessing technologies, such as rotary shear grinding and high moisture pelleting, should be explored further. Recent studies conducted by Lamers et al. (2015) indicated that high moisture pelleting combined with low temperature drying reduced pelleting costs by about 40%. New grinding technologies will also have a significant impact on the quality of the grind (i.e., particle size distribution, mean particle size, and overall grinding energy). Other pretreatment options—such as washing, leaching, acid and alkaline, hydrothermal carbonization, and steam explosion—result in biomass with a high moisture

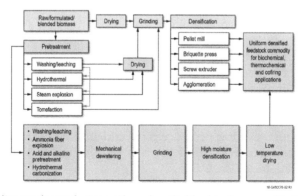

Fig. 2. Flow diagram for producing uniform densified feedstock commodity for biochemical, thermochemical conversion and combustion applications (Tumuluru et al., 2011).

content; however, using these new novel technologies (i.e., high moisture pelleting and low temperature drying) will help to produce a uniform densified feedstock (see Fig. 2) at a reduced cost. Future research should address how these individual operations impact one another and how they can be combined. The focus of the research should be on meeting the desired specifications such as low moisture, ash content, and high mass, and energy densities.

In order to make biorefineries able to operate at their designed capacities, new feedstock supply systems should be designed that can address the existing biomass limitations in terms of physical properties and chemical and biochemical composition. Analysis by Argo et al. (2013), Hess et al. (2009), and Muth et al. (2014) have shown that the conventional system has limitations in meeting the feedstock supply targets outside of highly productive regions due to weather conditions and due to floods or droughts in these areas. Hess et al. (2009) and Searcy and Hess (2010) also reported on an advanced preprocessing system that can reduce the feedstock price and quality uncertainties. They state that this new method, known as a biomass depot, effectively addresses supply system issues because it decentralizes biomass processing facilities and can help to achieve the feedstock cost, quantity, and quality required by the biorefineries. A major motivation of the depot concept is that the feedstock preprocessing can be performed off-site to ensure that all feedstocks that are supplied to the biorefinery are already densified and meet the quality requirements. Currently, the depot concept is gaining wide acceptance because it helps to overcome the supply limitations of the conventional system.

In their studies, Eranki et al. (2011) pointed out that the network of distributed biomass processing centers use one or more types of biomass to generate uniform feedstock 'commodities,' which are intermediate products that have consistent physical properties and chemical composition that meet conversion quality targets with improved flowability, transportability (bulk density), and stability/storability (dry matter loss reduction) properties. Consequently, an integrated approach that weighs the relative importance of different feedstock specifications is necessary. Demonstration of preprocessing technologies (e.g., chemical, mechanical, and thermal) or integrated preprocessing systems in depots will help to produce continuous and consistent feedstocks in terms of chemical and biochemical specifications and physical properties of the biorefineries, thereby reducing the quality and supply risk. Industry engagement is also critical to implement the preprocessing and pretreatment technologies to mobilize the more than a billion ton of woody, herbaceous, and others biomasses available in the U.S.

5. Conclusions

Mechanical, chemical, and thermal preprocessing technologies will have a significant impact on biomass physical properties and chemical composition. Mechanical preprocessing densification systems reduce the bulk density of biomass by five to six times as compared to raw biomass. Thermal processing techniques using torrefaction can reduce grinding energy by 70%, and the torrefied ground biomass has improved sphericity, particle surface area, and particle size distribution. The major limitation of torrefaction is loss of mass density. However, this can be overcome by mechanical densification of the torrefied biomass. Another biomass specification that has a significant impact on the biochemical and thermochemical conversion is ash. In the case of thermochemical conversion, alkaline earth and alkali metals such as calcium magnesium, sodium, and potassium reduce product yield during thermochemical and biochemical conversion. Chemical preprocessing methods such as washing, leaching, acid, alkali, and ammonia fiber explosion pretreatments reduce the recalcitrance in biomass and improve product yields. To meet the specification of the biochemical and thermochemical and cofiring applications, developing advanced preprocessing systems by integrating the mechanical, thermal, and chemical preprocessing is going to be critical to utilize more than a billion ton of biomass available in the U.S. Combining mechanical, thermal, and chemical preprocessing methods can help the biomass to meet not only the ash and carbohydrate specification but also particle size and density.

References

Argo, A. M., E. C. D. Tan, D. Inman, M. H. Langholtz, L. M. Eaton, J. J. Jacobson, C. T. Wright, D. J. Muth, M. M. Wu, Y.-W. Chiu and R. L. Graham. 2013. Investigation of biochemical biorefinery sizing and environmental sustainability impacts for conventional bale system and advanced uniform biomass logistics designs. Biofuels, Bioprod. Biorefin. 7: 282–302.

Arias, B. R., C. G. Pevida, J. D. Fermoso, M. G. Plaza, F. G. Rubiera and J. J. Pis-Martinez. 2008. Influence of torrefaction on the grindability and reactivity of woody biomass. Fuel Process. Technol. 89: 169–175.

Aston, J. E., D. N. Thompson and T. L. Westover. 2016. Performance assessment of dilute-acid leaching to improve corn stover quality for thermochemical conversion. Fuel 186: 311–319.

Bach, S.-V. and K.-Q. Tran. 2015. Dry and wet torrefaction of woody biomass—A comparative study on combustion kinetics. Energy Procedia 75: 150–155.

Bergman, P. C. A. 2005. Combined torrefaction and pelletization: The TOP process. Report ECN-C-05-073, Energy Research Centre of the Netherlands (ECN), Petten.

Bridgewater, T. 2012. Review of fast pyrolysis of biomass and product upgrading. Biomass Bioenergy 38: 68–94.

Carpenter, D., T. L. Westover, S. Czernik and W. Jablonski. 2014. Biomass feedstocks for renewable fuel production: A review of the impacts of feedstock and pretreatment 170 on the yield and product distribution of fast pyrolysis bio-oils and vapors. Green Chem. 16(2): 384–406.

Casey, E., N. S. Mosier, J. Adamec, Z. Stockdale, N. Ho and M. Sedlak. 2013. Effect of salts on the co-fermentation of glucose and xylose by a genetically engineered strain of *Saccharomyces cerevisiae*. Biotechnol. Biofuels 6: 83.

Chen, W., H. Yu, Y. Liu, P. Chen, M. Zhang and Y. Hai. 2011. Individualization of cellulose nanofibers from wood using high-intensity ultrasonication combined with chemical pretreatments. Carbohydr. Polym. 83: 1804–1811.

Deng, Y., D. G. Olson, J. Zhou, C. D. Herring, A. Joe Shaw and L. R. Lynd. 2013. Redirecting carbon flux through exogenous pyruvate kinase to achieve high ethanol yields in *Clostridium thermocellum*. Metab. Eng. 15: 151–158.

Eranki, P. L., B. D. Bals and B. E. Dale. 2011. Advanced regional biomass processing depots: A key to the logistical challenges of the cellulosic biofuel industry. Biofuels, Bioprod. Biorefin. 5: 621–630.

Granström, K. 2005. Emissions of volatile organic compounds from wood. Ph. D. Dissertation, Karlstad University, Karlstad, Sweden. ISBN 91-85335-46-0.

Hess, J. R., K. L. Kenney, L. P. Ovard, E. M. Searcy and C. T. Wright. 2009. Commodity Scale Production of an Infrastructure-Compatible Bulk Solid from Herbaceous Lignocellulosic Biomass. INL/EXT-09-17527, Idaho National Laboratory (INL), Idaho Falls, ID, USA.

Johansson, A. and A. Rasmuson. 1998. The release of monoterpenes during convective drying of wood chips. Dry. Technol. 16(7): 1395–1428.

Lamers, P., M. S. Roni, J. S. Tumuluru, J. J. Jacobson, K. G. Cafferty, J. K. Hansen, K. Kenney, F. Teymouri and B. Bals. 2015. Techno-economic analysis of decentralized biomass processing depots. Bioresour. Technol. 194: 205–213.

Liden, G., F. Berruti and D. S. Scott. 1988. A kinetic model for the production of liquids from the flash pyrolysis of biomass. Chem. Eng. Commun. 65(1): 207–221.

Muth, D. J., M. H. Langholtz, E. C. D. Tan, J. J. Jacobson, A. Schwab, M. M. Wu, A. Argo, C. C. Brandt, K. G. Cafferty, Y.-W. Chiu, A. Dutta, L. M. Eaton and E. M. Searcy. 2014. Investigation of thermochemical biorefinery sizing and environmental sustainability impacts for conventional supply system and distributed preprocessing supply system designs. Biofuels, Bioprod. Biorefin. 8: 545–567.

Palmqvist, E. and B. Hahn-Hägerdal. 2000. Fermentation of lignocellulosic hydrolysates. II: Inhibitors and mechanisms of inhibition. Bioresour. Technol. 74: 25–33.

Phanphanich, M. and S. Mani. 2011. Impact of torrefaction on the grindability and fuel characteristics of forest biomass. Bioresour. Technol. 102: 1246–1253.

Searcy, E. and R. Hess. 2010. Uniform-Format Feedstock Supply System: A Commodity Scale Design to Produce an Infrastructure-Compatible Biocrude from Lignocellulosic Biomass. INL/EXT-10-20372, INL, Idaho Falls, ID, USA.

Thompson, D. N., V. S. Thompson, J. A. Lacey, D. S. Hartley, M. A. Jindra and J. E. Aston. 2016. Strategies for post-harvest reduction of ash content in biomass feedstocks. Oral presentation at the Agricultural and Equipment Technology Conference, February 8–10, 2016, Louisville, KY, USA.

Tumuluru, J. S., S. Sokhansanj, J. R. Hess, C. T. Wright and R. D. Boardman. 2011. A review on biomass torrefaction process and product properties for energy applications. Ind. Biotechnol. 7: 384401.

Tumuluru, J. S., J. R. Hess, R. D. Boardman, C. T. Wright and T. L. Westover. 2012. Formulation, pretreatment, and densification options to improve biomass specifications for cofiring high percentages with coal. Ind. Biotechnol. 8: 113–132.

Tumuluru, J. S., L. G. Tabil, Y. Song, K. L. Iroba and V. Meda. 2014. Grinding energy and physical properties of chopped and hammer-milled barley, wheat, oat, and canola straws. Biomass Bioenergy 60: 5867.

Tumuluru, J. S. 2016. Effect of deep drying and torrefaction temperature on proximate, ultimate composition, and heating value of 2 mm lodgepole pine (*Pinus contorta*) grind. J. Bioeng. 3: 16.

Tumuluru, J. S., E. Searcy, K. L. Kenney, W. A. Smith, G. L. Gresham and N. A. Yancey. 2016. Impact of feedstock supply systems unit operations on feedstock cost and quality for bioenergy applications. *In*: Kumar, R., S. Singh and V. Balan (eds.). Valorization of Lignocellulosic Biomass in a Biorefinery: From Logistic to Environmental and Performance Impact. Nova Science Publishers: New York, NY, USA.

Yancey, N. A., J. S. Tumuluru and C. T. Wright. 2013. Drying, grinding, and pelletization studies on raw and formulated biomass feedstock's for bioenergy applications. J. Biobased Mater. Bioenergy 7: 549–558.

Zhang, W., Z. Yi, J. Huang, F. Li, B. Hao, M. Li, S. Hong, Y. Lv, W. Sun, A. Ragauskas, F. Hu, J. Peng and L. Peng. 2013. Three lignocellulose features that distinctively affect biomass enzymatic digestibility under NaOH and H_2SO_4 pretreatments in *Miscanthus*. Bioresour. Technol. 130: 30–37.

Mechanical Preprocessing

CHAPTER 2

Conventional and Advanced Mechanical Preprocessing Methods for Biomass

Performance Quality Attributes and Cost Analysis

Jaya Shankar Tumuluru* and Neal Yancey

1. Introduction

Biomass preprocessing refers to all of the processing steps that take biomass, whether woody or herbaceous, from its condition after harvest (i.e., bales, logs, chips, loose bulk material, etc.) to the point where it is introduced into a conversion process. In general, these processing steps include size reduction (i.e., chipping, grinding, chopping, knife mill, hammer mill, etc.); in some cases, the drying of the biomass; and densification. Size reduction and densification are the mechanical preprocessing operations used to convert the biomass from its harvested condition to a particle size and moisture content that meets the specifications needed for conversion applications. Preprocessing may include drying, which is a thermal pretreatment technique used to make biomass aerobically stable or to prepare the biomass for thermal conversion.

750 University Blvd, Energy Systems Laboratory, Idaho National Laboratory, Idaho Falls, Idaho, 83415-3570.
* Corresponding author: JayaShankar.Tumuluru@inl.gov

1.1 Conventional Preprocessing

Biomass in the field seldom possesses the right physical characteristics in terms of moisture, particle size, flowability, and density to make it ready for conversion, whether it be biochemical or thermochemical conversion. The biomass must meet physical and chemical specifications required by the conversion process. The physical and chemical properties that are typically considered are moisture, total ash, hemicellulose, cellulose, lignin, and elemental ash content (Lee et al., 2007), as well as format characteristics (i.e., grind size, particle size distribution, fines content, flowability, and durability). Conventional preprocessing converts biomass into a more flowable feedstock with the desired physical and chemical requirements. Conventional preprocessing includes size reduction, drying, and densification. The specifications determined by the conversion process will dictate specific control parameters within each of the preprocessing steps. The major challenge of conventional mechanical preprocessing is the cost.

Searcy et al. (2015) indicated that one of the major limitations of using high-moisture woody and herbaceous biomass by biorefineries for biofuels production is high preprocessing (i.e., size reduction, drying, and densification) costs. The conventional pelleting process includes a Stage-1 grinding, drying, Stage-2 grinding, and pelleting (see Fig. 1). In the conventional pelleting process, the high-moisture woody and herbaceous biomass, at about 30% (w.b.) are conveyed through a stage-1 grinder, which is usually a hammer mill or shear mill fitted with a screen size in the range of 25.4 to 50.8 mm (Yancey et al., 2013). Next, the ground biomass is dried to about 10% (w.b.) in a rotary dryer. The dried biomass is further passed through a stage-2 grinder, which is a hammer mill fitted with a 6.4-mm screen (Yancey et al., 2013). Finally, the ground biomass is further steam-conditioned to soften the lignin and is finally pelleted (Yancey et al., 2013). Figure 2 shows the energy consumption associated with each unit operation in the biomass preprocessing scenario shown in Fig. 1. The cost of pelleting is mainly dependent on the energy (electrical and heat) consumption, production rate, capital, and maintenance costs associated with these unit operations.

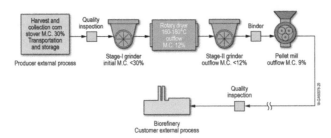

Fig. 1. Various unit operation in the conventional pelleting process (Lamers et al., 2015).

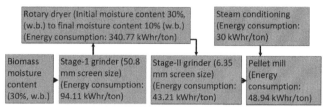

Fig. 2. Energy consumption of various unit operations in the conventional pelleting process (Tumuluru, 2016).

Techno-economic analysis indicates that efficient moisture management is critical for reducing the preprocessing cost of biomass (Tumuluru et al., 2014a). Many researchers such as Sakkampang and Wongwuttanasatian (2014), Yancey et al. (2013), and Pirraglia et al. (2010) indicated that drying is the major energy consumer (65–70%) in the densification process. Also, another major limitation of high-temperature drying is the emission of volatile organic compounds (VOCs) and extractives. According to Granström (2005) and Johansson and Rasmuson (1998), drying of woody biomass at a higher temperature of 160–180°C results in the emission of wood extractives and VOCs. These emissions pose human health and environmental concerns and result in the formation of photo-oxidants, which are harmful to humans if inhaled and disturb photosynthesis causing damage to forests and crops.

1.1.1 Size Reduction

In general, harvested biomass is baled for ease in transportation. Typical formats include bales, logs, or chips. Size reduction is required to convert the biomass from a non-flowable, packaged state to a more flowable feedstock that meets the particle size specifications for the conversion process. Most size reduction processes involve a two-stage grinding process. The first stage grinder breaks the bale apart or chips the log, thereby creating a more flowable feedstock. The second stage includes grinding or milling, which reduces the particle size to meet the specifications of the conversion process. For biochemical conversion, the biomass is typically ground using a grinder fitted with a 25.4 mm screen. For thermochemical conversions, such as pyrolysis and gasification, a 6.35 mm screen produces smaller particle sizes (2 mm), which are desirable to control the reaction kinetics and achieve the desired conversion efficiency (Dibble et al., 2011; van Walsum et al., 1996). In general, herbaceous biomass is transported in bale format to biorefineries or satellite storage points, where it is stored until needed. Normally, bales are stored outside, which can result in the accumulation of moisture. Moisture in biomass has a significant effect on grinding efficiency and product quality.

An experiment conducted at Idaho National Laboratory's (INL) Biomass Feedstock National User Facility (BFNUF) determined the effect of biomass moisture and grinder screen size (25.4, 50.8, 76.2, 101.6, and 152.6 mm) on the energy consumption for corn stover. The Stage-1 grinding tests were carried out using a Vermeer Corporation prototype horizontal hammer mill bale grinder. The corn stover was harvested from Palo Alto County, Iowa. The ground material produced in the Stage-1 grinder was subsequently used to conduct experiments in a Stage-2 grinder fitted with a 25.4 mm or a 6.35 mm screen size. For Stage-2 grinding studies, a Bliss hammer mill (model number E-4424-TF) was used.

1.1.1.1 Stage-1 and Stage-2 Grinding

At each screen size tested in the Stage-1 grinder, the grinding energy increased with biomass moisture content (see Fig. 3). The higher moisture content of 30% w.b. had the least effect on the grinding energy when the grinder was fitted with a bigger screen size (152.6-mm), whereas the smaller screen sizes (< 60 mm) increased the grinding energy exponentially. At a lower biomass moisture (10% or less), the screen size had a very marginal effect on the grinding energy. The trends for Stage-2 grinding were similar; as the moisture content increases, the grinding energy also increases (see Fig. 4). Tests conducted on the Stage-2 grinder with 6.35 and 25.4 mm screens

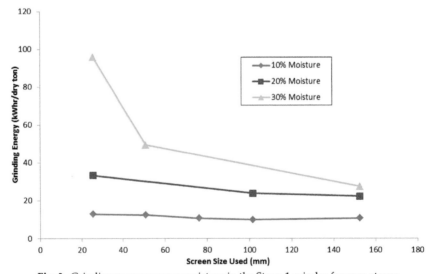

Fig. 3. Grinding energy versus moisture in the Stage-1 grinder for corn stover.

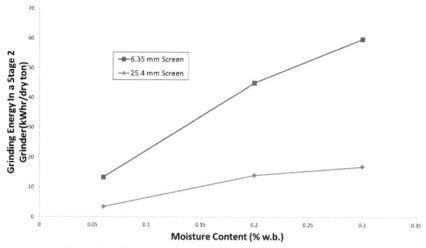

Fig. 4. Grinding energy of corn stover in a Stage-2 hammer mill.[1]

showed that increasing the moisture from low to high moisture (10–30% wt basis) significantly increased the grinding energy (see Fig. 4). Biomass moisture has a much higher impact on grinding energy as the screen size is reduced as seen by the slope of the two lines shown in Fig. 4, which illustrates that both biomass moisture content and screen size have a significant effect on grinding energy. These studies indicated that grinding energy increases exponentially with an increase in moisture content and a decrease in screen size.

1.1.2 Effect of Moisture on Particle Size

As a general rule, the mean particle size generated during hammer milling or other Stage-2 milling process increases with increasing biomass moisture content, as observed in Fig. 5. However, there are many other factors that can influence particle size distribution. As biomass moisture increases, grinder energy increases, which usually results in a decrease in feed rate and can, in turn, decrease the particle size as the material stays in the mill for longer times. Also, if the moisture is too high for a given milling process, it can result in the plugging of screens. If the screens get plugged, the particle size decreases dramatically because the material cannot exit the mill quickly enough and is extruded from the screen instead of being hammered through as it does under drier conditions.

---

[1] Unpublished INL BFNUF data.

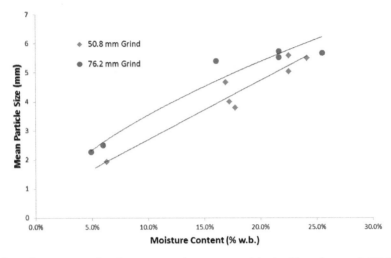

Fig. 5. Screen size and moisture content impact on particle size (Tumuluru et al., 2016).

2. Comparison of Different Grinding Technologies for Herbaceous and Woody Biomass and Mill Types

These studies have indicated that commercial hammer mills, currently used in the biomass industry, are suitable for low moisture biomass; however, at higher moisture content, these mills consume excessive energy and can produce non uniform particle sizes, which may reduce conversion efficiency in downstream biochemical processes. Comminution methods, relying on different modes of failure within biomass materials, have different sensitivities to biomass moisture content and produce significantly different particle sizes. A comparison of various comminution technologies from literature is nearly impossible to compare due to differences in material properties (i.e., biomass type/source, particle size/distribution, moisture content, etc.), methods for measuring power consumption, and other test variables. Studies were conducted at INL using a rotary shear mill, collision mill, and hammer mill to understand their suitability for the grinding applications tested. The specific objectives of these studies were to understand the effect of moisture on grinding energy and particle size and assisting in the comparison of three comminution technologies using the same materials and methods.

Conventional hammer milling is best suited for low-moisture feedstocks but the industry continues to use the hammer milling technology to process high-moisture biomass. Forest Concepts, LLC has developed a new size reduction technology that is based on shearing (Crumbler model M24) to grind biomass at higher moisture while using less energy. The pilot scale unit has a nominal operating capacity of 1 ton per hour and is driven by a 14.19 kW electric motor. The shear uses a 70-cm-wide row of interlocking disks. The width of the interlocking disks has a direct impact on the resulting particle size distribution. For these tests, the disk width was 4.76 mm. The M24 was designed as a Crumbler-style rotary shear, which was used to produce "WoodStraw." In early tests of the machine by Forest Concepts LLC, increased grinding efficiencies are observed for high moisture wood chips. This trend is opposite from most other grinding methods (hammer mill), so this technology was of great interest as a possible substitute for processing herbaceous or woody biomass with moisture levels up to 30%.

A Vortex, Inc., Model VP-2400 collision mill was also tested for grinding biomass. This unit is also a pilot scale mill with a nominal operating capacity of one ton per hour and is driven by a 22.4 kW electric motor. The machine has a rotor inside a specially shaped housing. The rotor acts like a blower, moving air and materials to be processed in through the inlet. The interaction between the rotor and the housing causes random vortices that cause collisions between product particles in the airstream, resulting in particle size reduction. The technology was developed to extract precious metals from loose solid rocks by processing the rock into a powder where the target materials can be separated. This unit could take a softball-sized rock and turn it into a fine powder in less than a second. The technology was successfully tested with wood chips, but not herbaceous materials. The processed materials were collected from cyclones and a vacuum filter bag. Figure 6 shows a flow diagram of the tests conducted using the different grinders. Three crop types were tested: corn stover, switchgrass, and ponderosa pine, with high moisture and low moisture scenarios for each crop type.

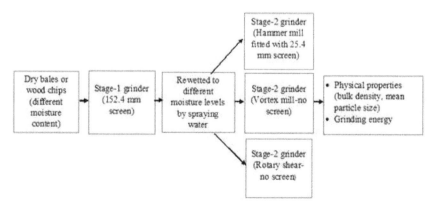

Fig. 6. Flow diagram of the tests done using different size reduction technologies.

For the herbaceous feedstocks, the baled biomass was first processed through the Stage-1 grinder with a 152.4 mm screen and was further fed into each of the three types of mills tested. For the woody biomass, the wood was received as chips and processed directly into each of the three milling processes. The moisture for each of the biomass and milling process are shown in Table 1.

Table 1. Moisture content tested for the comparison of three different mills.

Material	Moisture	Process	Actual moisture
Corn Stover	Low	Hammer Mill	6.04%
	High	(Bliss eliminator hammer mill, Model E-4424-TF, Bliss Industries)	28.02%
Switchgrass	Low		6.10%
	High		14.81%
Ponderosa Pine	Low		11.63%
	High		51.48%
Corn Stover	Low	Rotary Shear	6.18%
	High	(Crumbler model M24, Forest Concepts LLC)	31.02%
Switchgrass	Low		6.27%
	High		16.95%
Ponderosa Pine	Low		14.70%
	High		44.06%
Corn Stover	Low	Vortex Mill	7.51%
	High	(Model VP-2400, Vortex, Inc.)	27.07%
Switchgrass	Low		9.33%
	High		13.95%
Ponderosa Pine	Low		14.97%
	High		45.44%

2.1 Grinding energy

The specific grinding energy was compared for each technology tested for the three crops and at the two moisture contents (see Fig. 7). The specific grinding energy removes the energy required to operate the equipment while it is idling (and no load is applied) and only compares the energy that is required to grind or mill the biomass. One can see that the vortex or collision mill required more energy to process biomass, especially high moisture biomass, as compared to the other two technologies. The shear mill performed the best on the high-moisture corn stover. The hammer mill performed better than the other two technologies for woody material.

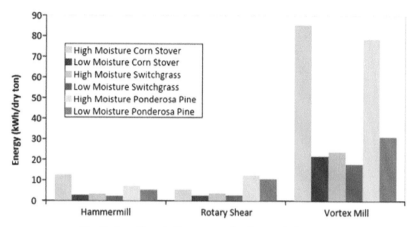

Fig. 7. Specific grinding energy for three grinder types.

2.2 Physical properties

While woody biomass can be processed to a fairly uniform particle size, herbaceous material, when ground or milled, has a much broader particle size distribution including fine, medium, and coarse particles. The particles tend to have much larger length-to-width ratios than wood particles, resulting in long thin particles. Soil and rocks that are brought into the process with the feedstock tend to end up in the fines during the grinding process. As a result, the fines (less than 500 µm) generated in the grinding process typically have the highest concentration of ash (Tumuluru et al., 2016). This inorganic matter (e.g., soil and rock) will increase erosion of the conveying and handling systems (Bell, 2005). The general consensus of the many industrial partners is that fines in the ground material limit the success of downstream processes.

The bulk densities of the processed materials are a function of particle size, particle size distribution, and particle aspect ratio. Small particles result in higher bulk densities than larger particles. Bulk density is also affected by particle aspect ratios with rounder particles, producing somewhat higher bulk densities than nonsymmetrical particles. Figures 8, 9, and 10 provide a comparison between the three technologies for the moistures for all three materials. As expected, the Vortex mill with the smaller particle size and rounder aspect ratio produces the highest loose and tapped bulk densities. The hammer mill has slightly higher densities over the rotary shear. The low moisture produced smaller particle sizes and higher bulk densities. Both the switchgrass and ponderosa pine show little difference over this wide range of processed moisture levels, even with significantly different resulting particle sizes.

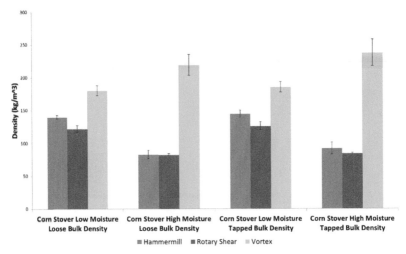

Fig. 8. Comparison of ground corn stover bulk density when using three different mills.

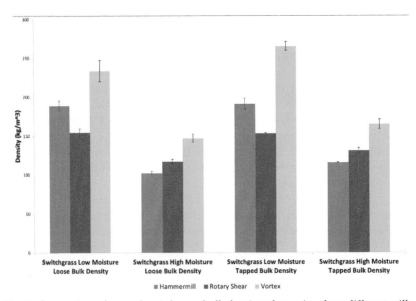

Fig. 9. Comparison of ground switchgrass bulk density when using three different mills.

For particle size analysis, a standard sieve analysis was completed utilizing a Ro-Tap machine with nine sieves (ASABE, 2007). The distribution of the observed particle sizes for the different material is strongly dependent on the way the material deconstructs. Corn stover is very fibrous and tends to shred into thin stringy fibers, whereas switchgrass does not have as much of a fibrous nature so it tends to shatter into small straight pieces. The

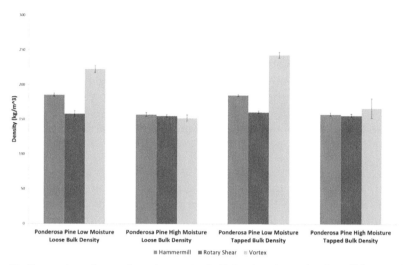

Fig. 10. Comparison of ground ponderosa pine bulk density when using three different mills.

ponderosa pine, being a woody feedstock, tends to shatter along the grain lines producing larger and more symmetrical particles. The moisture levels of the materials also affect how they are deconstructed. Dry materials tend to shatter more easily, resulting in smaller particles, than more flexible high-moisture materials. Figure 11 shows a comparison of the generated mean particle sizes between the three technologies over a range of materials. The hammer mill was fitted with a 25.4 mm screen as this is commonly used for biochemical conversion applications. For the rotary shear, a 4.76-mm

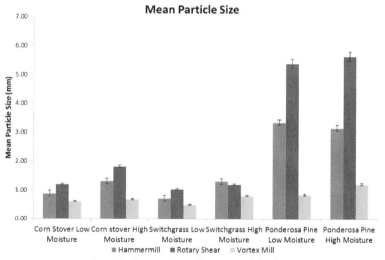

Fig. 11. Comparison of mean particle size.

disk width was selected as the closest match to the hammer mill's 25.4-mm screen. The vortex mill does not provide any changeable configuration that affects particle size, but the shape of the grinding chamber can be designed to generate different levels of vortices, which do affect the resulting particle size. Examining the results, the rotary shear seems to generate somewhat larger particles than the hammer mill in nearly all the cases, except for high moisture switchgrass. In all the cases, the Vortex mill produced the smallest particles that are fairly independent of moisture level and material type.

2.3 Drying

In many cases, the biomass is harvested in the field at higher moisture content that results in the biomass being biologically unstable (up to 30% moisture content). Drying to <10% (w.b.) helps biomass to become aerobically stable and not only stabilizes the biomass for storage, but decreases grinding energy in the process.

2.3.1 Open Air-Drying (Passive Drying)

Open air-drying uses ambient air to dry the biomass left on the field. The final moisture of open air-drying depends on the size and characteristics of the material and its ambient condition and is typically in the range of 15–35% w.b. Open air-drying is slow and is impacted by weather conditions. The pile or windrows may need stirring or turning to facilitate drying. Long drying times in open air-drying of high moisture biomass may result in biomass quality loss. Open air or passive drying is a slow process; it does not dry the material sufficiently to meet the moisture specifications for thermochemical applications such as pyrolysis and gasification (i.e., less than 15% [Jahirul et al., 2012]).

2.3.2 Low-Temperature Drying of Biomass

When compared to high-temperature drying, low-temperature drying methods are in general less expensive, require less equipment, and require less external energy input. Most of the low-temperature dryers such as conveyor or belt dryers fans blow the drying medium either upward or downward. Low-temperature dryers such as belt or conveyor dryers are very convenient as they can handle a variety of biomass feedstocks. In biomass industry, these dryers are used commonly for drying of hog fuel. Another significant advantage of the conveyor or belt dryers is that low-quality heat from various heat sources can be used. Typically, waste heat that is available in the biorefineries can be used. Currently, low-temperature drying methods are gaining importance in Europe to avoid high-drying costs and environmental issues. One example is a forced air convective method where lower drying temperatures in the range of 60 to 80°C and low airflow rates are used for the drying of biomass materials. Some forced air convective dryers include cabinet dryers, grain driers, and belt dryers, which operate at lower temperatures and are not capital intensive (Lamers et al., 2015). Tumuluru et al. (2016) indicated that low-temperature drying methods have several advantages: (a) higher efficiency, (b) a reduced fire hazard, (c) no need of high-quality heat, (d) reduced VOC emissions, (e) reduced particulate emissions, (f) no agglomeration of high clay or sticky biomass, and (g) easy access for maintenance. Several companies are selling commercial-scale conveyors or belt dryers with a throughput of greater than 5 ton/hr. Perry of Oakley Limited of United Kingdom has introduced a belt dryer which can dry various products such as woodchips, cereals, granular products, and recycled waste (see Fig. 12). According to the company, this dryer can handle woodchips at a rate of 10 t/hour with a 30% moisture reduction. The belt for this dryer is made using stainless steel, and it comes with a wide range of perforations to suit different products (http://www.perryofoakley.co.uk/). The dryer is capable of drying with product depths ranging from 50 to 250 mm to allow for optimum moisture content removal and power consumption.

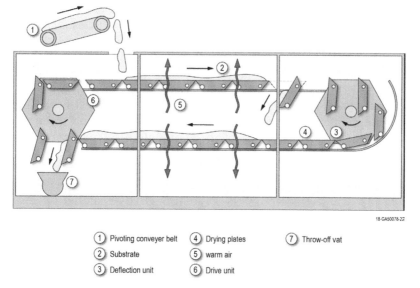

18-GA50078-22

(1) Pivoting conveyer belt	(4) Drying plates	(7) Throw-off vat
(2) Substrate	(5) warm air	
(3) Deflection unit	(6) Drive unit	

Fig. 12. Conveyor or belt dryer for drying of biomass (http://www.qalovis.com/en/dryer-livestock-owner/).

2.3.3 Medium-Temperature Dryers

Cascade, flash, and super-heated dryers are medium temperature dryers, which are widely used for the drying of biomass. Cascade dryers operate at intermediate temperatures between those of low-temperature and high-temperature dryers. Flash dryers (or pneumatic dryers) operate where the feedstock is suspended in an upward flow of the drying medium, usually flue gas. Flash dryers require a small particle size. Superheated steam dryers are very similar to flash dryers, except the drying medium is steam from the boiler. In superheated steam dryers, the drying temperature stays above the saturation temperature to avoid steam condensation.

2.3.4 High-Temperature Dryers

High-temperature drying methods, which are also called active drying methods, are performed using commercial-scale dryers to meet the moisture specifications of biomass conversion processes. Rotary dryers are commonly used to dry biomass to less than 10% (w.b.) moisture content. High drying temperature removes all the surface as well as most of the chemically bound (i.e., moisture that may be caught in capillaries or fibers or held onto via chemical reactions) moisture in the biomass. Among the different types of biomass, woody biomass takes more energy for drying compared to herbaceous biomass. The major reasons for lower drying energy for herbaceous biomass is due to thinner particles and lower particle density that allow for faster transfer of moisture out of the particles. Among the woody biomass, hardwood takes more drying energy compared to softwood (Yancey et al., 2013). According to Yancey et al. (2013), the drying energy for woody biomass is in the range of 340 to 400 kWhr/ton, whereas herbaceous biomasses are in the range of 200 to 300 kWh/ton.

These dryers are classified as either indirect- or direct-fired. In the case of direct-fired dryers, flue gas or hot air is passed directly through the material to be dried. These dryers are commonly used for drying of hog fuel, sawdust, and bark, as well as many other materials. For indirect-fired dryers, the heat is passed to the material through tubes or heat exchangers, which are placed inside the dryer. Indirect-fired dryers are less efficient as they have to transfer heat from the steam tubes to the material.

2.3.5 Densification

Ground agricultural biomass has low-bulk density, which reduces transportation efficiency. Low-bulk densities also result in handling and storage issues. Common densification systems include pellet mills, briquette presses, and cubers—all of which increase the bulk density of the biomass by about four to five times over loose material (Tumuluru et al., 2011). Typically, ground herbaceous biomass has a bulk density of about 80–150 kg/m³ depending upon the particle size and moisture content (Tumuluru et al., 2014b; Yancey et al., 2013), whereas pelleted biomass has a density of about 700–750 kg/m³. The bulk density of the densified products depends on feedstock type, particle size, feedstock moisture content, and process conditions such as preheating temperature, die speed, and steam conditioning (Tumuluru et al., 2016; Tumuluru, 2016). According to Tumuluru et al. (2010), the pellets produced from woody biomass are commonly used for bioenergy applications. The advantage of pellets is that they have a consistent size and shape that is easily handled using the existing grain handling systems. Figure 13 compares the density of bales, ground biomass, and pellets—showing that pellets have four to five times higher bulk density than bales do. In addition to bulk density, another important attribute that influences transportation efficiency is durability.

According to Tumuluru et al. (2011), durability is the ability of the pellets to withstand frictional and impact forces during storage and transportation. Typically, the desired durability of pellets or briquettes for long distance transportation should be > 97.5% to avoid fines generation during storage and transportation. According to the Pellet Fuel Institute and European Committee for Standardization, pellet durability values are in the range of 90 to 97.5%. The durability of densified products depends

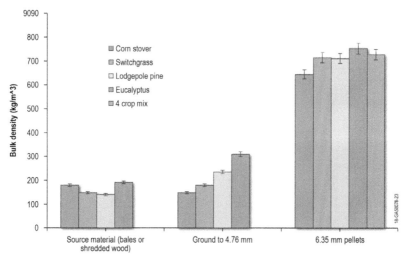

Fig. 13. Bulk density of raw, ground, and pelleted woody and herbaceous material (Yancey et al., 2013).

on feedstock parameters, such as feedstock moisture content and particle size, and process parameters like compression pressure, preheating, and die speed (Tumuluru et al., 2011). Figure 14 indicates the durability values of the woody, herbaceous, and formulated pellets. The study from Yancey et al. (2013) indicated that hardwood pellets have lower durability values as compared to softwood pellets. The same study indicated that when hardwood is blended with lodgepole pine, corn stover, and switchgrass, the durability values are increased to > 97.5% (see Fig. 14).

Another commonly used method for biomass densification is briquetting (Tumuluru et al., 2011). The main difference between pellets and briquettes is that briquettes are bigger than pellets. For the briquetting process, larger particle size material can be used as compared to pelleting. The binding of the biomass during briquetting is mainly due to the mechanical interlocking of the particles (Tumuluru et al., 2015). Figure 15 indicates the bulk density of the briquettes produced using different herbaceous and woody biomass feedstocks, while Fig. 16 indicates the durability of these briquettes. It is clear from the figures that bulk density and durability of both woody and herbaceous biomass briquettes are lower than that of pellets. Many studies conducted by researchers indicated that pellets and briquettes are both suitable for thermochemical and biochemical conversion applications as compared to raw biomass (Ray et al., 2013; Rijal et al., 2012; Yang et al., 2014). In addition to better performance, Tumuluru et al. (2011) indicated that densification of biomass is critical to improving handling and conveying efficiencies throughout the supply system and biorefinery infeed.

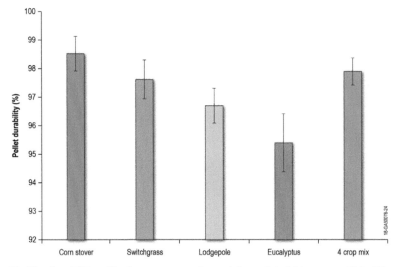

Fig. 14. The durability of herbaceous, woody, and formulated biomass pellets (Yancey et al., 2013).

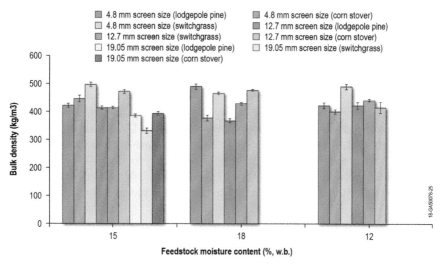

Fig. 15. Bulk density of woody and herbaceous biomass briquettes (Tumuluru et al., 2016).

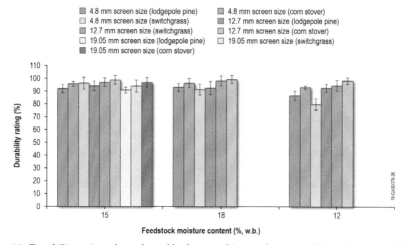

Fig. 16. Durability rating of woody and herbaceous biomass briquettes (Tumuluru et al., 2016).

Currently, biorefineries are experiencing challenges in storage, feeding, handling, and transportation of ground biomass. In addition, variable moisture (e.g., high and low moisture) and less-convenient forms of biomass aggravate these limitations. One common problem associated with raw biomass is the bridging of particles that leads to uneven discharge from the silos and the jamming of the conveyors, which result in an inconsistent feeding to the reactors. These issues are greatly dependent on feedstock moisture content and particle size. Compared to raw ground biomass, pellets have uniform size and shape characteristics with a low-moisture content. Pellets can also be stored more efficiently (e.g., storage footprint reduces by two to five times as compared to baled herbaceous biomass or wood chips). Low moisture in the pellets helps to reduce spontaneous combustion issues that are typically seen in bales and wood chip storage. The major challenge in producing densified biomass from high moisture woody and herbaceous biomass is the cost of drying. Drying takes about 65–70% of the total energy, while Stage-1 grinding takes about 17% (Tumuluru, 2016). Due to this high preprocessing cost, biorefineries are not ready to use high moisture biomass for biofuels production (Searcy et al., 2015). Reducing preprocessing energy consumption, especially grinding and drying, will make the pelleting of biomass a viable option for biorefineries.

2.4 Advances in Biomass Preprocessing

Research and testing are continually identifying improvements in processing methods and equipments to reduce preprocessing costs and improve the quality attributes. This will help to ensure a more economical process model for biomass conversion to fuel and other materials. Idaho National Laboratory, Idaho Falls, Idaho, U.S.A. has developed two recent technologies—fractional milling and high-moisture densification—which have reduced preprocessing costs significantly (see Fig. 17).

Fractional milling uses screening capabilities to take advantage of the size reduction that occurs at the Stage-1 grinder. Studies at Idaho National Laboratory have indicated that about 40–55% of the biomass after the Stage-1 grinding process already meets the size required by the conversion process depending upon the specific requirements for that process. The current preprocessing method sends all of the material to the Stage-2 grinder to ensure that there are no oversized particles. However, this also results in over processing of the material and excess energy being expended. By using a screen to remove the biomass that already meets the size requirements of the process between the Stage-1 and -2 grinders, a more uniform (relative to particle size) product is produced and less energy is consumed.

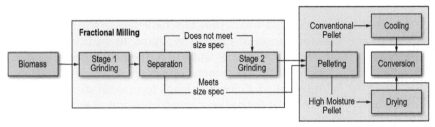

Fig. 17. Material flow in advanced preprocessing, including fractional milling, high-moisture densification, and low-temperature drying.

Biomass moisture removal is one of the main factors contributing to higher preprocessing costs. The typical moisture content for baled corn stover is roughly 20%. However, depending on the year, this can range anywhere from 10–30% moisture content. Under conventional pelleting or cubing processes, a moisture content between 10–12% is suggested as being optimal (Tumuluru et al., 2011). To achieve that moisture content requires active drying, which consumes significant amounts of energy to dry the biomass prior to densification. High-moisture densification has shown that biomass densification can be achieved at a moisture content of more than 30%. Also, studies have shown that the process of fractional milling and high moisture densification will inherently remove significant amounts of moisture (about 15% w.b.) (Tumuluru et al., 2017; Tumuluru, 2014; 2015; 2016). When starting with bales at 30% w.b. moisture, the moisture following Stage-1 and Stage-2 grinding is often about 20% w.b. Additional drying, which occurs during pelleting, can drop the moisture content to between 13 and 15%. To have stable densified biomass, the moisture content should be less than 10%. Once the biomass has been pelletized, it is much easier to reduce this content from 15–10% than it would have been to dry the ground biomass from 30% down to 10% after the Stage-1 grinder using a thermal drying process (e.g., rotary dryer). Another major advantage of high moisture pelleting is that the pellets can be dried using low-temperature drying technologies such as a grain or belt dryer. These dryers operate at temperatures in the range of 60–80°C and are less capital- and energy-intensive as compared to rotary dryers. Preliminary testing has indicated that energy costs are reduced by half by using fractional milling and low temperature drying as compared to using the conventional pelleting method currently followed by the industry.

2.5 Fractional Milling Studies on Corn Stover

Data from many historical runs made on the Process Development Unit (PDU) located at INL indicates that as much as 40% of the processed material from the Stage-1 grinder already met the particle size requirements needed for pelleting using a 6.25 mm screen size grind. To confirm this, a test run using corn stover at three moisture levels (e.g., 6, 20, and 30% w.b.) and using eight milling screen combinations at each of the three moisture contents tested was conducted. These tests confirmed that as much as 60% of the material processed using a 76.2-mm screen in the Stage-1 grinder would be reduced in size enough for feeding a bioconversion process (less than 6 mm) and that 30–35% would be reduced in size enough (less than 2 mm) to be densified. Figure 18 shows the particle size distribution for a conventional Stage-1 and Stage-2 grinding process. It is quite clear from the figure that more than 30% of the material processed by Stage-2 grinding is converted into fines. Fines typically below 600 microns generated during grinding creates feeding problems. Fractional milling helps to generate fewer fines during this process (< 8%) as compared to conventional, which is > 30%. Figure 19 shows the particle size distribution for Stage-1 grinding of high moisture biomass (30% w.b.). It is clear from the figure that more than 50% of the material meets the specification of 6.35 mm screen, which is commonly used for pelleting. Bypassing this material around the Stage-2 grinder will reduce the grinding energy by 50% in Stage-2 and helps to avoid redundant preprocessing, which in turn generates more fines.

In fractional milling, as less material is conveyed to the Stage-2 mill, grinding energy will be reduced proportionally to the amount of material that bypasses the grinder. Figure 20 shows the grinding energy requirements for four different screens and three different moisture contents to grind

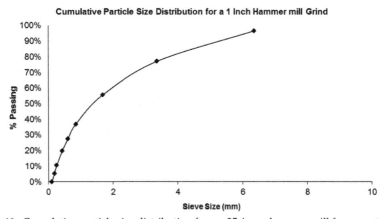

Fig. 18. Cumulative particle size distribution from a 25.4-mm hammer mill for corn stover.

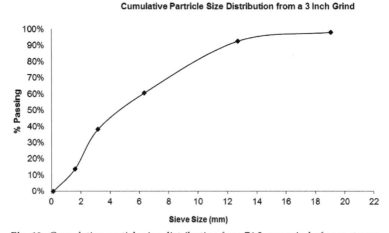

Fig. 19. Cumulative particle size distribution for a 76.2 mm grind of corn stover.

corn stover to a 6.35-mm grind size. In each case, about a 20% reduction in grinding energy was observed for the entire process when fractional milling was used to bypass the Stage-2 grinder. In Fig. 20, the grinding energy of the system without fractional milling is shown with a solid line, while the dashed line shows the grinding system energy using fractional milling. Biomass moisture significantly influences grinding energy. At a low moisture content (e.g., between 5–7%), the grinding energy is about eight times lower than when the moisture content is at 30%.

Studies on corn stover were conducted using a low energy debaler to break the corn stover bale apart. This was followed by grinding the corn stover in the Stage-1 grinder fitted with two different screens—a 152.4-mm screen and a 76.2-mm screen. The results showed that even after debaling about 15% of the dry corn stover meets the particle size requirements of 6.35 mm screen. Material that meets the particle size specification can be bypassed to reduce the grinding energy and avoid fines. Similar trends have been observed with the other screen as well (see Fig. 21).

As stated earlier, fines tend to result in plugging issues during many conversion processes. Ash content is also a concern for most conversion processes, decreasing the value of the feedstock (due to reduced conversion efficiency and equipment wear). Ash content of the ground biomass is typically highest in the fines, where dirt tends to accumulate. Therefore, removal of these fines would create a more beneficial product from a particle

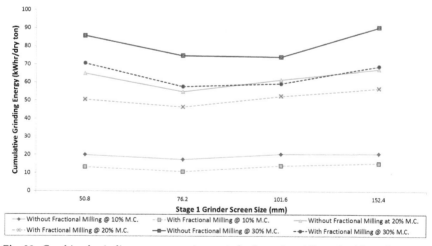

Fig. 20. Combined grinding energy requirements for Stage-1 and Stage-2 without fractional milling.[2]

---

[2] The energy for operating the screening technology was not included in the calculation, but is considered insignificant when compared to grinding, milling, or drying energies.

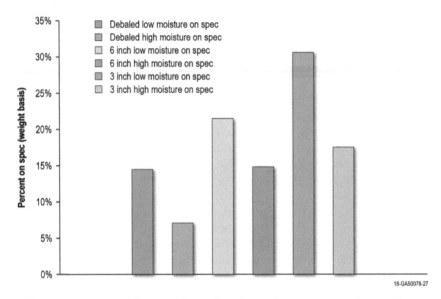

Fig. 21. Percent material on specification based on moisture content and screen size.

size standpoint, which would result in a reduction in the total ash content. As such, it is important to minimize the amount of biomass that is ground into fines so that as the fines are removed, only minimal amounts of the biomass are removed as well, while higher concentrations of ash are being removed. Studies on the concentration of ash based on the particle size of ground corn stover indicated that the concentration of ash in the fines, less than 600 microns, was more than 70%, while the concentration of ash in the fraction greater than 3 mm was as low as 2.5%. Figure 22 shows the ash content of three size fractions tested following the Stage-1 grinder. In this case, the ash content of the corn stover less than 500 microns is greater than 60%. By removing the fraction less than 600 microns, it not only improves the conversion process by removing the fines, which can plug in the system, but a significant reduction in the ash can also be observed. Similar trends were observed for woody biomass in terms of energy savings and minimizing the fines.

2.6 High-Moisture Pelleting Process

In the conventional pelleting method, biomass is dried to about 10% before pelleting. In high-moisture biomass pelleting, the biomass is pelleted at > 18% moisture content and the high moisture pellets produced are further dried in low-temperature drying technologies such as a grain or belt dryer.

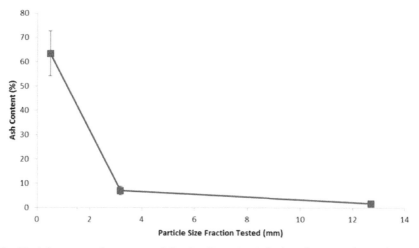

Fig. 22. Ash content of corn stover following Stage-1 grinder based on particle size fraction.

Pelleting biomass at high moistures and then drying the high-moisture pellets using low capital and low-temperature drying methods, such as with a grain or belt dryer, helps to reduce pelleting costs by about 50% (Lamer et al., 2015). High-moisture pelleting eliminates the energy-intensive rotary drying step at the front end with a grain dryer at the back end of the pellet mill (see Fig. 23). Pelleting corn stover, ammonia fiber explosion pretreated corn stover, and municipal solid waste at high-moisture content > 20% w.b. in a flat die and ring die pellet mill indicated that good quality pellets, in terms of density (> 560 kg/m³) and durability > 95%, can be produced (Tumuluru, 2014; 2015; Bonner et al., 2014). In the high-moisture pelleting process, the steam conditioning of biomass is replaced with a short preheating step. Also, Tumuluru (2014; 2015; 2016) indicated that some of the moisture in the biomass is lost due to frictional heat developed in the die during compression and extrusion. When pelleting biomass with a moisture content greater than 20% prior to pellet milling, there is about 5–10% moisture content loss in the biomass during the pelleting process. Further drying of the high-moisture pellets using a grain or belt dryer, which operates at < 90°C (Hellevang, 2013), can significantly influence pelleting costs. Techno-economic analysis of this process compared with conventional methods indicated that there is about a 40% reduction in pelleting costs for herbaceous biomass (Lamers et al., 2015). Tumuluru (2014; 2015; 2016) and Tumuluru et al. (2016) have demonstrated this high moisture pelleting process at moisture content > 28% w.b. in a laboratory scale flat die pellet mill. A recent study by Tumuluru et al. (2017) involving scale-up of high

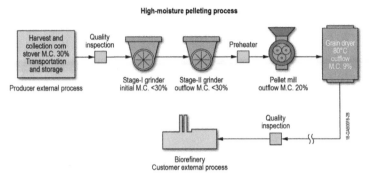

Fig. 23. Schematic of a high-moisture pelleting process (adapted from Lamers et al., 2015).

moisture pelleting in a ring die pellet mill with a throughput of 1 ton/hr has also indicated that this process is scalable. Furthermore, trends have matched with the data obtained with the lab-scale pellet mill.

2.6.1 *Effect of Process Variables on High Moistures Pelleting Process*

Studies conducted by Tumuluru (2014; 2015; 2016) and Tumuluru et al. (2016) indicate that process variables such as moisture content, preheating temperature, die rotational speed, die dimensions, and the addition of binders all impact the quality of the pellets produced. These studies also indicate that moisture loss in the biomass during pelleting depends on the initial moisture content (Tumuluru, 2014; 2015; 2016). Studies on corn stover indicated that a lower feedstock moisture of 28% w.b. and higher preheating temperatures of 110°C employed for four to five minutes further reduced the pellet moisture content to about 14–16% w.b. (Tumuluru, 2014). In addition, a higher feedstock moisture content of 38% w.b. with higher preheating temperatures of 110°C resulted in 28–30% w.b. moisture in the pellets after cooling. Studies conducted by Tumuluru (2016) on ground lodgepole pine at a higher moisture content in the range of 33–39% w.b. corroborated this moisture loss observation. Tumuluru (2016) has argued that the main reason for moisture loss during pelleting of high moisture biomass is due to moisture flash-off and has validated this phenomenon by measuring the expansion ratio of the pellets. His results indicate that high moisture in the biomass during pelleting result in a high expansion ratio and a higher moisture loss (Tumuluru, 2016).

Other quality attributes, which are affected by the high-moisture pelleting process, are density and durability. High-moisture pelleting studies conducted by Tumuluru (2014; 2015; 2016), Hoover et al. (2014), and Tumuluru et al. (2016) at a feedstock moisture content of > 26% indicated that high-feedstock moisture content reduces pellet density and durability. Feedstock moisture content is inversely correlated with density and durability, as shown in Figs. 24, 25, and 26. These studies indicate that at a high-moisture content of 38% w.b., bulk density values were about 350 kg/m³, whereas with a lower feedstock moisture content of 26–28% w.b., bulk density values are in the range of 550 to 600 kg/m³. In the case of specific energy consumption of a high-moisture pelleting process, a higher feedstock moisture content and a lower die speed resulted in a higher specific energy consumption; however, starting with a lower moisture content and a higher die speed resulted in a lower specific energy consumption. Studies by Tumuluru (2016) indicate that 39% w.b. feedstock moisture content and lower die speed of 40 Hz result in higher specific energy consumption values of > 180 kWhr/ton. Lowering the ground lodgepole pine moisture content to 33%, increasing the die speed to 60 Hz, and preheating the temperature to 90°C lowered the specific energy consumption values to < 110 kWhr/ton.

Binders, like lignosulphonate and starch, increase the adhesion between biomass particles, reduce frictional resistance in the die, and increase the overall throughput of the mill. Studies by Tumuluru et al. (2016) on the effect of starch-based binders on high-moisture pelleting of corn stover indicate that the specific energy consumption of this process can be decreased, while the durability of the pellets can be increased with the addition of the starch-based binders. Further studies conducted by Tumuluru et al. (2016) indicate that the durability of corn stover pellets made at the high-moisture content of 33% (green durability) is about 87.2%, whereas an addition of a cornstarch binder at 2 and 4% increased the durability values to 93.2 and

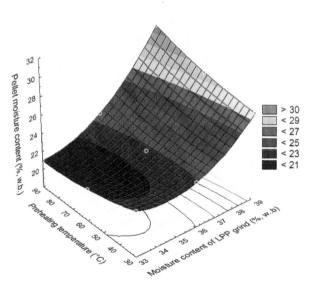

Fig. 24. Impact of preheating temperature and moisture content on lodgepole pine pellet moisture content (Tumuluru, 2016).

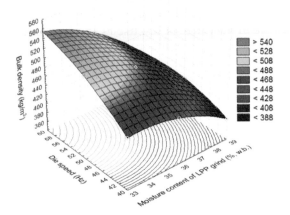

Fig. 25. Effect of feedstock moisture content of lodgepole pine grind and die speed on the pellet bulk density (Tumuluru, 2016).

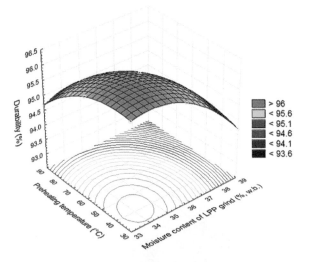

Fig. 26. Effect of preheating temperature and moisture content of lodgepole pine grind on pellet durability (Tumuluru, 2016).

96.1%, respectively. The authors reasoned that it is critical to add binders to high-moisture feedstocks to improve pellet durability, which further enhances storage and handling characteristics. Another major impact of using a binder on the pelleting process is a reduction in pelleting energy. With no binder, the specific energy required for 33, 36, and 39% feedstock moisture content was between 118–126 kWh/ton. Adding a 2% binder reduced the specific energy consumption to about 75–94 kWh/ton. Increasing the binder percentage further to 4% reduced the specific energy consumption to about 68–75 kWh/ton for all feedstock moisture contents that were tested, as shown in Fig. 27.

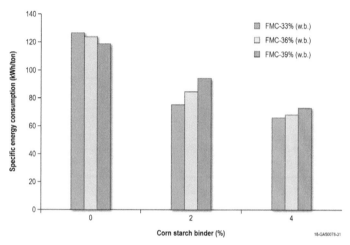

Fig. 27. Effect of binder on the specific energy consumption (Tumuluru et al., 2016).[3]

2.6.2 *Comparison of High-Moisture Pellet Properties with Conventional and International Standards*

Table 2 provides wood pelleting standards from the Pellet Fuel Institute (PFI) and the European Committee for Standardization (CEN). Pellets produced using either the lower moisture or conventional pelleting process meet the highest durability and bulk density standards (> 97.5% and > 700 kg/m³) established by PFI and CEN. In the case of the pellets produced using the high-moisture pelleting process, the pellets meet the different durability standards set by CEN and PFI. According to CEN, bulk density values should be stated when the pellets are traded in volume basis, whereas PFI has established the values needed to meet the different grades. For a "utility grade", pellets bulk density values should be in the range of 576 to 736 kg/m³. Studies by Tumuluru (2016) and Tumuluru et al. (2016, 2017) indicate that high moisture pelleting helps to meet the bulk density requirements for utility grade pellets. The high-moisture pelleting process developed at Idaho National Laboratory has helped to produce pellets with various densities and durabilities suitable for different transportation scenarios. In

---

[3] Feedstock moisture content (FMC).

Table 2. Comparison of pellet properties made from conventional and high moisture pelleting process with CEN and PFI standards (Tumuluru, 2016).

Physical property	Normative			Informative	
	Moisture content (%, w.b.)	Durability (%)	Diameter (mm)	Bulk density (kg/m³)	Unit density (kg/m³)
CEN standard	M10 ≤ 10%	DU97.5 ≥ 97.5 DU95.0 ≥ 95.0 DU90.0 ≥ 90.0	D06 ≤ 6 mm ± 0.5 mm D08 ≤ 8 mm ± 0.5 mm	Recommended to be stated if traded by volume basis	
Pellet Fuel Institute (utility standard)	≤ 10	> 95	≤ 10	576–736	
Conventional wood pellets (Yancey et al., 2013; Tumuluru et al., 2010)	< 8	≥ 97.5	6	700–750	1150–1200
High moisture lodgepole pine wood pellets (Tumuluru et al., 2016)	< 8	≥ 90.0 ≥ 95.0	8	368–562	887–1092

general, for short-distance transportation of pellets using a truck, the higher density obtained using conventional pelleting process may not be needed as trucks are more weight-limited and cannot fill the truck. Therefore, a low density pellet produced using high moisture pelleting process can help to fill the truck. The high-moisture pelleting process that eliminates the rotary drying step can reduce pelleting costs and make the pelleting of woody and herbaceous biomass a viable option for biorefineries.

2.7 Techno-economic Analysis of the Conventional and Advanced Biomass Preprocessing

2.7.1 Conventional Preprocessing System

Research conducted by Lamers et al. (2015) regarding the techno-economic analysis of conventional and high-moisture pelleting indicates that there is about a 35–40% reduction in the cost of high-moisture pelleting vs. the conventional process. In the conventional process, high-moisture biomass is size-reduced to less than a 50-mm particle size, which is further dried to a 10–12% moisture content using a rotary dryer. The dried biomass is then passed through a second-stage grinding process to reduce the average mean particle size to less than 5 mm (typically to 2 mm) using a 6.35-mm screen fitted to the hammer mill. This size-reduced material is then pelleted. The role of the second-stage grinder is to reduce the particle size further to meet particle size distribution requirements for pelleting. In the conventional pelleting process, drying is a major energy consumption unit operation, accounting for about 70% of the total pelleting energy (Tumuluru, 2016). The techno-economic analysis indicated that lower capital and operating costs resulted in a lower cost for high-moisture preprocessing (U.S. $30.80 Mg-1), while convential preprocessing showed higher total costs (U.S. $51.30 Mg-1) (Lamers et al., 2015). High-cost pellet production is heavily dependent on electricity prices for grinders and dryers. The studies conducted by Lamers et al. (2015) also indicate that for the conventional pelleting process, fuel, labor, interest, and depreciation are high due to the rotary dryer.

3. Advanced Preprocessing System

3.1 Fractional Milling

Research at INL has demonstrated that even when the bale moisture content prior to processing is 30%, the process of grinding, separating, and milling will reduce the moisture content to as low as 16–19% when processing through a 6.35-mm screen prior to densification (see Fig. 28). Densification can further reduce the moisture content to 12–14%. Then, a grain dryer can easily reduce the moisture in the pellets further to appropriate levels (< 10%). Combining fractional milling with high-moisture pelleting reduces total preprocessing costs by about 50% as compared to using the conventional method (see Table 3) (Kenney et al., 2013).

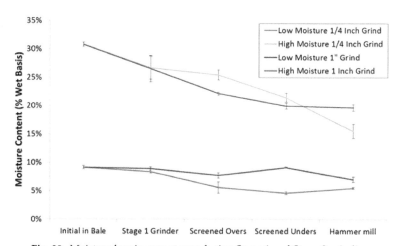

Fig. 28. Moisture loss in corn stover during Stage-1 and Stage-2 grinding.

This analysis indicates that a cost savings of 35% could be achieved by moving from conventional preprocessing techniques to high-moisture preprocessing configurations. The major reason for cost reduction is due to the transition from a rotary dryer in conventional preprocessing to a cross-flow pellet dryer in the advanced process. The other reasons are increasing

Table 3. Conventional preprocessing systems cost (Kenney et al. 2013).

	2013 State of Technology (2011 $/dry T)
Grinder 1	16.80
Grinder 2	11.60
Drying	15.20
Densification	7.70
Totals	51.3

Table 4. Grinding cost demonstrated in 2016 using Stage-1 and Stage-2 grinders and separator.

Cost (in 2014$/dry ton)				
Grinding conditions (bale moisture and screen size)	10% M.C. 76.2 mm screen in Stage-1 and 25.4 mm screen in Stage-2	30% M.C. 76.2 mm screen in Stage-1 and 25.4 mm screen in Stage-2	10% M.C. 76.2 mm screen in Stage-1 and 6.35 mm screen in Stage-2	30% M.C. 76.2 mm screen in Stage-1 and 6.35 mm screen in Stage-2
Stage 1 Grinder Costs	$2.55	$11.67	$2.55	$11.67
Separator	$0.37	$0.37	$0.37	$0.37
Stage 2 Grinder Costs	$1.06	$3.09	$2.33	$13.42
Total	$3.98	$15.13	$5.25	$25.45

the machine throughput, reducing the number of equipment operations necessary to process material, and consequently, lowering capital costs. A techno-economic analysis conducted to demonstrate the benefit achieved using high-moisture densification and fractional milling for low- and high-moisture corn stover is shown in Table 4. Figure 29 indicates the cost of preprocessing biomass at different moisture content and screen size.

3.1.1 High-Moisture Pelleting and Low-Temperature Drying

In the high-moisture pelleting process (HMPP) developed at INL, biomass at about 30% w.b. moisture content is preheated and pelletized. The partially dried pellets are further dried to a moisture of < 9% w.b. using a grain

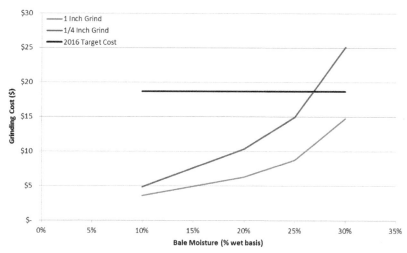

Fig. 29. Corn stover grinding cost using a Stage-1 and Stage-2 grinder for different moistures and screen sizes.

or belt dryer. In this process, the rotary dryer, which is typically used in conventional processes at the front end, is replaced with a grain dryer at the back end. Table 5 provides the results of the conventional pelleting method techno-economic analysis (TEA) using a commercial-scale pellet mill (5 ton/hr) and HMPP using a pilot scale system (1 ton/hr). The TEA takes the pelleting of the material into account at about 20% w.b. moisture content, and then dries the pellets in a grain dryer. The upper limit of 20% w.b. for high-moisture pelleting was selected based on the grinding studies. Additional studies on corn stover bales at 30% w.b. moisture content

Table 5. Techno-economic analysis of conventional and high moisture pelleting process.

Case studies	Densifier ($/dry ton)	Dryer ($/dry ton)	Total ($/dry ton)
Conventional (Kenney et al., 2013)	7.70	15.20 (from an initial moisture of 30%, w.b. to a final moisture of 12%, w.b. using a rotary dryer)	22.50
HMPP with grain dryer	8.31	1.27	9.58
In the conventional case, the dryer is a rotary dryer and the high moisture pelleting case uses a cross-flow grain dryer, both fueled by natural gas at $7.55/MMBTU. Additionally, the type of dryer has changed from the tower grain dryer, with a capital cost of $270,000, in the 2017 design case to a more efficient and less capital intensive cross-flow grain dryer, which has a capital cost of $35,009 (Lamers et al., 2015). The change in the dryer from rotary dryer (conventional case) to crossflow grain dryer has helped to reduce the preprocessing cost.			

lose ≈ 10% w.b. moisture content when passing through the Stage-1 and Stage-2 grinders and a separator (Yancey and Tumuluru, 2015). For the TEA, the total pelleting energy of about 73 kWhr/ton at about 20% (w.b.) moisture content was considered based on the studies conducted on high moisture pelleting of corn stover in the pilot scale ring die pellet mill (1 ton/hr) (Tumuluru et al., 2017). The corn stover pellets produced at 20% w.b. moisture content resulted in pellets with about 14% moisture content. A grain dryer was used in the TEA to reduce the pellet moisture content from 14–9% w.b. The pellets at 14% w.b. moisture content had a green durability (durability immediately after pelleting) value of ≈ 98% and cured durability (durability after drying) of about 98.5% (Tumuluru et al., 2017). The pellets produced at a moisture content of 20% w.b. had bulk density values of about 610 kg/m³ (Tumuluru et al., 2016; Tumuluru et al., 2017). These high green durability values of the pellets suggest that the pellets can be transported even without drying. Once the pellets are received at the bio-refineries, belt or grain dryers can be used to dry the pellet using the waste heat available at the biorefineries. The high-moisture pelleting process makes drying of biomass optional; it can be used when the pellets need to be stored for a long period and transported over a long distance. Two case studies were considered for the TEA: (a) conventional pelleting process, and (b) high-moisture (20% moisture content) pelleting with a grain dryer. The TEA illustrates that the total cost to produce pellets using conventional pelleting is $22.50/dry ton (with just drying and pelleting cost) (see Table 5). The high-moisture pelleting process reduced the cost of pelleting to $9.38/dry ton (see Table 5), reducing the cost by $12.92/dry ton—greater than a 40% reduction when compared to conventional. Table 6 indicates the cost to preprocess a bale with a 30% w.b. moisture content

Table 6. Cost to produce pellets using fractional milling, high-moisture pelleting, and low-temperature drying.

Wet feedstock (30%, w.b.)	Cost ($/ton)	Energy (kWhr/ton)	Moisture loss (%, w.b.)	Mill throughput (dry ton/hr)
Stage-1 (3-inch grind)	10.78	15.9	30	3.0
Stage-2 (1/4-inch grind)	8.52	35.3 (54.3 w/ 35% bypass)	25	2.2
Separator screen size (1/4-inch)	0.21		1	5
Pelleting	8.31	73.0	19	4.8
Drying (grain dryer)	1.27	50	14–9	5
Total	29.09			

using fractional milling with a three-inch screen in Stage-1 and 1/4-inch screen in Stage-2 followed by high-moisture pelleting and low-temperature drying. It is clear from Table 6 that the cost to produce pellets using corn stover bales at a 30% w.b. moisture content using fractional milling, and high moisture pelleting followed by low temperature drying, the cost of producing the pellets can be reduced by as much as 50% compared to the conventional pellet production process (Lamers et al., 2015).

4. Conclusion

From the present research, it was concluded that preprocessing plays a major role in transforming biomass into a feedstock with desired particle size distribution and density specifications. The major challenge in conventional preprocessing is the cost of the process concerning the moisture content. The grinding energy and drying cost increases exponentially with moisture content in the biomass. New grinding technologies and efficient moisture management technologies reduce preprocessing costs significantly. Compared to hammer and collision mills, rotary shear grinders performed more efficiently for high-moisture corn stover. Fractional milling of biomass helps to separate the biomass that has already met the specifications after Stage-1 grinding and only process the biomass through the Stage-2 grinder that has not met the specifications. High-moisture pelleting and subsequent drying using low-temperature drying methods like grain dryer results in high-density (500–600 kg/m³) pellets and makes biomass a commodity-type product. A techno-economic analysis indicated that advanced preprocessing (e.g., fractional milling, high-moisture pelleting, and low-temperature drying) could reduce the pellet production cost by about 50% as compared to the current pellet production process followed by the industry.

Acknowledgments

The authors would like to acknowledge the U.S. Department of Energy (DOE) Bioenergy Technology Office for supporting this work. This work was supported by the DOE, Office of Energy Efficiency and Renewable Energy, under DOE Idaho Operations Office Contract DE-AC07-05ID14517. Accordingly, the U.S. government retains and the publisher, by accepting the article for publication, acknowledges that the U.S. government retains a non-exclusive, paid-up, irrevocable, worldwide license to publish or reproduce the published form of this manuscript or allow others to do so, for U.S. government purposes.

Author Disclosure Statement

No competing financial interests exist. This information was prepared as an account of work sponsored by an agency of the U.S. government. Neither the U.S. government nor any agency thereof, nor any of their employees, makes any warranty, expressed or implied, or assumes any legal liability or responsibility for the accuracy, completeness, or usefulness of any information, apparatus, product, or process disclosed, or represent that its use would not infringe privately owned rights. References herein to any specific commercial product, process, or service by trade name, trademark, manufacturer, or otherwise, do not necessarily constitute or imply its endorsement, recommendation, or favoring by the U.S. government or any agency thereof. The views and opinions of the authors expressed herein do not necessarily state or reflect those of the U.S. government or any agency thereof.

References

American Society of Agricultural and Biological Engineers (ASABE). 2007. ASABE Standards 2007: Standards Engineering Practices Data. ASABE Press. St. Joseph, Michigan. ISBN 18-92769-57-2.

Bell, T. A. 2005. Challenges in the scale-up of particulate processes—An industrial perspective. Powder Technol. 150(2): 60–71.

Bonner, I. J., K. G. Cafferty, D. J. Muth, Jr., M. D. Tomer, D. E. James, S. A. Porter and D. L. Karlen. 2014. Opportunities for energy crop production based on subfield scale distribution of profitability. Energies 7(10): 6509–6526.

Dibble, C. J., T. A. Shatova, J. L. Jorgenson and J. J. Stickel. 2011. Particle morphology characterization and manipulation in biomass slurries and the effect on rheological properties and enzymatic conversion. Biotechnol. Progr. 27(6): 1751–1759.

Granström, K. 2005. Emissions of volatile organic compounds from wood. Ph.D. Dissertation, Karlstad University, Karlstad, Sweden. ISBN 91-85335-46-0.

Hellevang, K. J. 2013. Grain drying. North Dakota State University extension service. North Dakota State University. Fargo, North Dakota.

Hoover, A. N., J. S. Tumuluru, F. Teymouri, J. Moore and G. Gresham. 2014. Effect of pelleting process variables on physical properties and sugar yields of ammonia fiber expansion pretreated corn stover. Bioresource Technol. 164: 128–135.

Jahirul, M. I., M. G. Rasul, A. A. Chowdhury and N. Ashwath. 2012. Biofuels production through biomass pyrolysis—A technological review. Energies 5(12): 4952–5001.

Johansson, A. and A. Rasmuson. 1998. The release of monoterpenes during convective drying of wood chips. Dry. Technol. 16(7): 1395–1428.

Kenney, K. L., K. G. Cafferty, J. J. Jacobson, I. J. Bonner, G. L. Gresham, J. R. Hess, L. P. Ovard, W. A. Smith, D. N. Thompson, V. S. Thompson, J. S. Tumuluru and N. Yancey. 2013. "Feedstock Supply System Design and Economics for Conversion of Lignocellulosic Biomass to Hydrocarbon Fuels. Conversion Pathway: Biological Conversion of Sugars to Hydrocarbons: The 2017 Design Case." Idaho National Laboratory Report. INL/EXT-13-30342.

Lamers, P., M. S. Roni, J. S. Tumuluru, J. J. Jacobson, K. G. Cafferty, J. K. Hansen, K. Kenney, F. Teymouri and B. Bals. 2015. Techno-economic analysis of decentralized biomass processing depots. Bioresource Technol. 194: 205–213.

Lee, D., V. N. Owens, A. Boe and P. Jeranyama. 2007. Composition of Herbaceous Biomass Feedstocks. North Central Sun Grant Center. South Dakota State University Press, Brookings, South Dakota.

Pirraglia, A., R. Gonzalez and D. Saloni. 2010. Techno-economical analysis of wood pellets production from U.S. manufacturers. BioResources 5(4): 2374–2390.

Ray, A. E., A. N. Hoover, N. Nagle, X. Chen and G. L. Gresham. 2013. Effect of pelleting on the recalcitrance and bioconversion of dilute-acid pretreated corn stover under low- and high-solids conditions. Biofuels. 4(3): 271–284.

Rijal, B., C. Igathinathane, B. Karki, M. Yu and S. W. Pryor. 2012. Combined effect of pelleting and pretreatment on enzymatic hydrolysis of switchgrass. Bioresource Technol. 116: 36–41.

Sakkampang, C. and T. Wongwuttanasatian. 2014. Study of ratio of energy consumption and gained energy during briquetting process for glycerin-biomass briquette fuel. Fuel. 115: 186–189.

Searcy, E. and P. Lamers, J. Hansen, J. Jacobson and E. Webb. 2015. Advanced Feedstock Supply System Validation Workshop: Golden, Colorado. Idaho National Laboratory Report. INL/EXT-10-18930.

Tumuluru, J. S., S. Sokhansanj, C. J. Lim, T. Bi, A. Lau, S. Melin, T. Sowlati and E. Oveisi. 2010. Quality of wood pellets produced in British Columbia for export. Appl. Eng. Agric. 26(6): 1013–1020.

Tumuluru, J. S., C. T. Wright, J. R. Hess and K. L. Kenney. 2011. A review of biomass densification systems to develop uniform feedstock commodities for bioenergy application. Biofuel. Bioprod. Bior. 5(6): 683–707.

Tumuluru, J. S. 2014. Effect of process variables on the density and durability of the pellets made from high moisture corn stover. Biosyst. Eng. 119: 44–57.

Tumuluru, J. S., K. G. Cafferty and K. L. Kenney. 2014a. Techno-economic analysis of conventional, high moisture pelletization and briquetting process. American Society of Agricultural and Biological Engineer Annual Meeting, Paper No. 141911360, Montreal, Quebec, Canada. July 13–16, 2014. doi: 10.13031/aim.20141911360.

Tumuluru, J. S., L. G. Tabil, Y. Song, K. L. Iroba and V. Meda. 2014b. Grinding energy and physical properties of chopped and hammer-milled barley, wheat, oat, and canola straws. Biomass Bioenerg. 60: 58–67.

Tumuluru, J. S. 2015. High moisture corn stover pelleting in a flat die pellet mill fitted with a 6 mm die: physical properties and specific energy consumption. Energy Sci. Eng. 3: 327–341.

Tumuluru, J. S., L. G. Tabil, Y. Song, K. L. Iroba and V. Meda. 2015. Impact of process conditions on the density and durability of wheat, oat, canola, and barley straw briquettes. Bioenerg. Res. 8(1): 388–401.

Tumuluru, J. S. 2016. Specific energy consumption and quality of wood pellets produced using high-moisture lodgepole pine grind in a flat die pellet mill. Chem. Eng. Res. Des. 110: 82–97.

Tumuluru, J. S., C. C. Conner and A. N. Hoover. 2016. Method to produce high durable pellets at lower energy consumption using high moisture corn stover and a corn starch binder in a flat die pellet mill. J. Vis. Exp. In Press.

Tumuluru, J. S., N. Yancey, R. McCulloch, C. Fox, C. C. Conner, D. Hartley, M. Dee and M. Plummer. 2017. Biomass Engineering: Size reduction, drying and densification of high moisture biomass. Feedstock supply and logistics platform, DOE, Project Peer Review, US. Department of energy, Bioenergy Technologies Office, March, Denver, Colorado, 7th, 2017.

van Walsum, G. P., S. G. Allen, M. J. Spencer, M. S. Laser, M. J. Antal and L. R. Lynd. 1996. Conversion of lignocellulosics pretreated with liquid hot water to ethanol. Appl. Biochem. Biotech. 57: 157–170.

Yancey, N. A., J. S. Tumuluru and C. Wright. 2013. Grinding and densification studies on raw and formulated woody and herbaceous biomass feedstocks. J. Biobased Mater. Bio. 7(5): 549–558.

Yancey, N. A. and J. S. Tumuluru. 2015. DOE Quarterly Milestone Completion Report. Idaho National Laboratory.

Yang, Z., M. Sarkar, A. Kumar, J. S. Tumuluru and R. L. Huhnke. 2014. Effects of torrefaction and densification on switchgrass pyrolysis products. Bioresource Technol. 174: 266–273.

CHAPTER 3

Effects of Mechanical Preprocessing Technologies on Gasification Performance and Economic Value of Syngas

Amit Khanchi,[1],* Bhavna Sharma,[2]
Ashokkumar Sharma,[3] Ajay Kumar,[4]
Jaya Shankar Tumuluru[5] and Stuart Birrell[6]

1. Introduction

Currently, biochemical and thermochemical conversions are the two major pathways for production of biofuels, biochemicals, and biopower

[1] Department of Agricultural and Biosystems Engineering, 2333 Elings Hall, 605 Bissell Road, Iowa State University, Ames, Iowa.
[2] Department of Agronomy, 1203 Agronomy Hall, Iowa State University, Ames, Iowa.
  Email: sharmabhavnak@gmail.com
[3] Department of Mechanical Engineering Technology, ETAS 227 D, University of Arkansas, Little Rock, Arkansas.
  Email: amsharma@ualr.edu
[4] Department of Biosystems and Agricultural Engineering, 228 Agricultural Hall, Oklahoma State University, Stillwater, Oklahoma.
  Email: Ajay.kumar@okstate.edu
[5] Biofuels Department, 750 University Blvd, Energy Systems Laboratory,, Idaho National Laboratory, Idaho, ID, 83415.
[6] Department of Agricultural and Biosystems Engineering, 2323 Elings Hall, Iowa State University, Ames, Iowa.
  Email: sbirrell@iastate.edu
* Corresponding author: amit@iastate.edu

from lignocellulosic biomass (Tumuluru et al., 2012). The biochemical process involves the conversion of biomass to alcohols and methane by fermentation and anaerobic digestion, respectively (Kumar et al., 2009b). In the thermochemical process, heat is used as the primary source for converting biomass into biofuels or bioenergy. Gasification, pyrolysis, hydrothermal liquefaction, torrefaction, and combustion are the various biomass thermochemical technologies commonly used in development stages (Yan et al., 2012). Among these thermochemical processes, gasification of lignocellulosic biomass has received an increased attention worldwide as a promising renewable energy technology for production of different liquid fuels, chemicals, power and value-added products (Fig. 1). Biomass gasification is a sustainable alternative to displace a significant part of the fossil fuel demand in terms of fuels, chemicals, power, and heat (Lickrastina et al., 2011). Gasification was discovered in early 1600 and commercialized later in 1800 for industrial and residential heating, and street lightning (NETL, 2013). With the development and use of more easily available and economical energy sources such as natural gas, coal, and petroleum, the focus was shifted from gasification. However, in the past few decades, gasification technology has gained increased attention due to shortage of and increase in price of natural gas, oil, and petroleum products. Currently, gasification technology can be utilized for production of electrical energy, synthetic natural gas, liquid fuels, and chemical products from coal, biomass, waste materials and other carbon-containing materials under strict environmental constraints (NETL, 2013).

Gasification involves the conversion of carbonaceous feedstock into a gaseous fuel called syngas or producer gas by partial oxidation at higher temperatures (Hernandez et al., 2010). Gasification consists of a series of sub-processes (Fig. 2). In the first step, moisture evaporates from biomass resulting in drying. As the temperature of the biomass particles increases further beyond 120°C, devolatization (pyrolysis) takes place with the

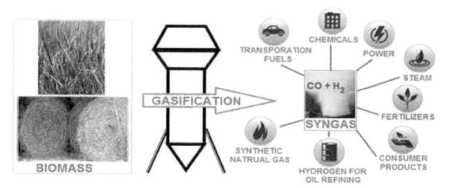

Fig. 1. Gasification and syngas utilization.

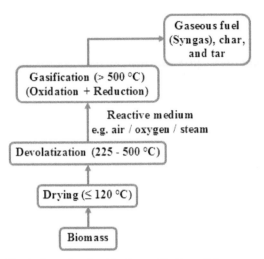

Fig. 2. Steps involved in the gasification of biomass.

disintegration of lignin, hemicellulose, and cellulose into volatile molecules such as hydrocarbons, H_2, CO, CO_2, and H_2O (Tinaut et al., 2008). Finally, the remaining solid portion called char and volatile gases react with the reactive medium, also called oxidizing or gasification agent (air, oxygen, and steam), and undergo series of gasification reactions, i.e., oxidation and reduction reactions. The limited supply of reactive medium prevents the complete conversion of biomass carbon and hydrogen into combustion products (e.g., CO_2 and H_2O) and instead results in the production of combustible gases such as CO, H_2, and CH_4 (Martinez et al., 2012). The inorganic portion of the biomass which does not volatilize in the gasification process is left as ash. The mixture of ash and unconverted biomass carbon is known as char. Tar is the unconverted fraction of biomass devolatilized products condensable at room temperature.

Gasification has benefits of recovering energy from low-grade materials such as biomass wastes and low-value coal which would otherwise cause negative environmental effects (Hernandez et al., 2010). It is also cleaner and more effective than combustion as it provides better electric generation performances (30–32% using gas engines, compared to 22% with a conventional Rankine cycle) and lower NO_x and SO_x emissions (Hernandez et al., 2010). Despite many advantages, the gasification needs to overcome several technical issues before its successful commercialization. The major problems are minimization or removal of undesired syngas trace contaminants (tars and alkali compounds) that occur in the gasification process, feedstock flexibility (the ability of the gasifier to utilize a plethora of different crops, organic wastes, or residues), syngas composition optimization, heat loss, gasifier scale-up, and lack of syngas utilization as a transportation fuel (Kumar et al., 2009b). The effect of biomass properties and the gasifier operating conditions on syngas quality and gasifier performance are not completely understood (Hernandez et al., 2010). The issues related to production, logistics, and preprocessing of biomass are a limiting factor in the successful commercialization of the gasification process (Hernandez et al., 2010). Additionally, there are several drawbacks specifically, associated with biomass feedstock which eventually limit the utilization of diverse biomass materials for the gasification process. Biomass materials are heterogeneous in nature and have low bulk density ranges from 60–80 kg/m³ for grasses and agricultural straws, and 200–400 kg/m³ for wood chips (Tumuluru et al., 2010). Moreover, different biomass feedstocks exhibit dissimilar physiochemical properties (e.g., particle size and shape, chemical composition, and heating value), which consequently result in an irregular material flow within the reactor. This leads to inconsistent operation of the gasifier and hence, the syngas quality. Therefore, a biomass feedstock with consistent quality in terms of size, shape, density, moisture content, and chemical composition is highly desired to achieve a reliable and trouble-free gasifier operation. Furthermore, the difference in the biomass type (such as grassy and woody materials) and physiochemical properties poses difficulty in mechanization of continuous feeding and further in controlling the burning rate (Luo et al., 2011). The difference in the densities of biomass and coal also causes feeding problems of their mixture in the existing feed systems of the coal-based boiler and power plants. Problems associated with co-firing of lignocellulosic biomass with coal also reduce the efficiency of thermochemical conversion processes (Tumuluru et al., 2010).

Due to the issues associated with the raw biomass feedstocks for gasification utilization, preprocessing techniques are required to transform the miscellaneous lignocellulosic biomass materials into more homogeneous and energy dense feedstock for efficient, trouble-free and reliable operation of gasification system. Several preprocessing technologies such as size reduction, pelletization and briquetting, and torrefaction are available, or underdeveloped, to facilitate gasification of a wide range of biomass feedstocks as well as municipal solid wastes. Size reduction of biomass feedstock is necessary for the production of ethanol from lignocellulosic biomass via both biochemical and thermochemical conversion processes (Cadoche and Lopez, 1989). Compaction techniques such as pelletization and briquetting increase the density of biomass by a factor of ten, decreasing transportation and storage costs, and also reduce feeding problems in the downstream production processes (Tumuluru et al., 2010). Torrefaction is another important preprocessing step that improves the physical properties and chemical composition of lignocellulosic biomass. It produces a high-quality biomass feedstock that is suitable for thermochemical conversion processes such as gasification, pyrolysis, and combustion. Torrefied biomass has several desirable properties for, e.g., low moisture content, higher energy density, improved grindability, and lower O/C ratio than the original biomass. These preprocessing techniques improve the characteristics of lignocellulosic biomass materials and transform them into more uniform and suitable feedstocks for the gasification process.

During the last few decades, numerous research efforts have evaluated different biomass preprocessing techniques. Many authors have studied the size reduction of biomass and investigated several parameters such as energy consumption, machinery type, bulk density and compaction behavior, influence of moisture content, and particle size distribution (Bitra et al., 2009a; Bitra et al., 2009b; Chevanan et al., 2010; Igathinathane et al., 2009; Zewei et al., 2010). Kratky and Jirout (2011) reviewed the biomass size reduction machines for enhancing bio-syngas production. The authors reviewed equipment design parameters and their energy requirement in relation to initial and final particle sizes, bulk density, and moisture content of biomass. Tumuluru et al. (2011) performed an extensive review of biomass densification for bioenergy applications and studied: (1) different particle bonding mechanisms during densification; (2) various densification systems (such as pellet mill, briquette press, tablet press, roller press, screw extruder, and agglomerator) and their specific energy consumption; (3) effect of densification variables (such as process type, biomass feedstock, and biomass composition) on quality of the densified products; and (4) effect of biomass pretreatment methods on densification process (Tumuluru et al., 2011a; Tumuluru et al., 2011b). However, the available review studies mainly focus on the biomass preprocessing techniques and lack the important information on the various effects of these preprocessing parameters on the gasification process and gas quality and yield. To our best knowledge, no literature is available which details and evaluates the influence of these preprocessing techniques on the overall performance of gasification process. The goal of this chapter is to highlight the pivotal roles of biomass preprocessing techniques for improving the performance efficiency and economy of biomass gasification process. The specific objective was to perform a detailed analysis of the effects of preprocessing techniques such as size reduction and pelletization on the biomass gasification process and the economic value of the syngas produced from biomass. Unlike previous studies, this chapter provides key insight about how one or combination of preprocessing techniques can noticeably enhance the overall performance of the biomass gasification process.

2. Gasifier Performance Parameters

Before discussing preprocessing techniques, it is important to understand the parameters that affect the quality of syngas and gasifier performance. Biomass, oxidizing agents such as air and location of air injection, together with reactor design, strongly influence the quality of syngas during a gasification process (Martinez et al., 2012; Sharma et al., 2014). The main output parameters that evaluate the performance of the gasification process are gasification efficiency and yield, and composition and calorific value of syngas (Martinez et al., 2012). Gasification efficiency as determined by the lower heating value of syngas is defined either as the hot gas or the cold gas efficiency. Cold gas efficiency accounts only for the chemical energy of the syngas and does not include the sensible heat energy of the hot syngas while the hot gas efficiency accounts both chemical and sensible energies of the syngas (Sharma et al., 2011). A typical value of cold gas efficiency in a downdraft gasification system varies between 50–80% (Martinez et al., 2012). Carbon conversion efficiency is the other important gasification performance parameter which is defined as the percentage of carbon in the biomass successfully converted into carbonaceous gaseous products mainly including CO, CO_2, CH_4, C_2H_2, C_2H_4, and C_2H_6. Yield is also used to measure the specific production of syngas per mass of fuel supplied to the system and is expressed as Nm^3kg^{-1}. During gasification, the yield is influenced by biomass ash content and equivalence ratio (ER), which is defined as the ratio of the actual air supplied to stoichiometric air required per kg of fuel, and the residence time of reacting gases in reduction zone of the gasifier. Equations to determine gasification of cold gas, hot gas, and carbon conversion efficiencies and gas yield can be found elsewhere (Martinez et al., 2012; Sharma et al., 2011).

During gasification, the composition of syngas depends on several parameters such as physical and chemical properties of the biomass feedstock and gasifier operating temperature and pressure. The concentration of combustible gaseous components of the syngas such as CO, H_2, and CH_4 are also affected by ER and chemical kinetics of various gasification reactions, e.g., biomass decomposition reaction Eq. (1), carbon gasification reactions Eqs. (2)–(4), water gas-shift reaction Eq. (5), methane reforming reaction Eq. (6), volatile oxidation reactions Eqs. (7)–(9), and char combustion reactions Eqs. (10)–(11) (Kumar et al., 2014). Their concentrations typically increase with ER and reach the peak levels at the optimal ER condition. However, beyond optimal ER, the reaction shifts towards combustion, thus CO and H_2 concentrations generally decrease while CO_2 concentration increases (Kumar et al., 2009b). The calorific value of the syngas is further affected by the type of oxidizing agent used during gasification. When oxygen, steam, or combination of both oxidizing agents is used, the syngas CO and H_2 concentrations increase thus, resulting in a higher calorific value. However, when the air is used as an oxidizing agent, a low calorific value syngas, highly diluted with N_2, is obtained. Martinez et al. (2012) reported a calorific value of 6 MJ/Nm^3 when the air was used as an oxidizing agent compared to 18 MJ/Nm^3 when O_2 or steam was used as an oxidizing agent (Martinez et al., 2012). The details on gasification performance parameters can be found elsewhere (Kumar et al., 2009a; Kumar et al., 2009b; Kumar, 2014; Sharma et al., 2014; Sharma et al., 2011).

$$CH_xO_yN_zS_s \rightarrow a\ CO + b\ CO_2 + c\ CH_4 + d\ H_2 + e\ NH_3 + f\ H_2S + g\ H_2O + h\ tar + i\ char \qquad \text{Eq. (1)}$$

$$C + CO_2 \leftrightarrow 2CO \qquad \text{Eq. (2)}$$

$$C + H_2O \leftrightarrow CO + H_2 \qquad \text{Eq. (3)}$$

$$C + 2H_2 \leftrightarrow CH_4 \qquad \text{Eq. (4)}$$

$$CO + H_2O \rightarrow CO_2 + H_2 \qquad \text{Eq. (5)}$$

$$CH_4 + H_2O \leftrightarrow CO_2 + 3H_2 \qquad \text{Eq. (6)}$$

$$CO + \tfrac{1}{2}\,O_2 \rightarrow CO_2 \qquad \text{Eq. (7)}$$

$$H_2 + \tfrac{1}{2}\,O_2 \rightarrow H_2O \qquad \text{Eq. (8)}$$

$$CH_4 + 2O_2 \rightarrow CO_2 + 2H_2O \qquad \text{Eq. (9)}$$

$$C + \tfrac{1}{2}O_2 \rightarrow CO \qquad \text{Eq. (10)}$$

$$C + O_2 \rightarrow CO_2 \qquad \text{Eq. (11)}$$

3. Size Reduction

3.1 Biomass size reduction process and influential parameters

Size reduction processes reduce particle size, form uniform shape, improves desired material properties, e.g., bulk density, flow-ability and porosity, and increases surface area (Bitra et al., 2011). Grinding biomass to a mean particle size of 1 mm considerably increases the biomass surface area and hence, the number of reaction sites required for chemical reactions (Bitra et al., 2011). A particle size of 1–2 mm is also recommended for effective hydrolysis in the biochemical conversion process (Kratky and Jirout, 2011). However, size reduction is an energy-intensive process compared to the other processing steps in biomass supply chain. For instance, about one-third of the total power requirement for the conversion of biomass to ethanol is consumed during size reduction (USDOE, 1993). Size reduction of lignocellulosic biomass can be achieved by a combination of chipping, grinding, and milling processes. Generally, the lignocellulosic material is reduced to a size of 10–30 mm after chipping and 0.2–2 mm after grinding or milling (Kratky and Jirout, 2011). The economics of size reduction largely depends on the energy consumed for reduction of the biomass material to a specific particle size. Further, the energy requirements for size reduction depends on the type of milling machine selected, initial and final particle sizes, biomass characteristics (such as moisture content, mechanical strength, and composition), and quantity of biomass to be processed (Kratky and Jirout, 2011). Usually, the energy consumption increases with moisture content. Igathinathane et al. (2008) evaluated the effect of moisture content (51% and 9%, wet basis), knife grid spacing (25.4, 50.8, 101.6 mm), and packed bed depths (50.8, 101.6, 152.4 mm) on cutting strength of switchgrass by a linear knife grid cutting device. They found that high moisture content switchgrass had a higher shear stress of 0.68 MPa as compared to low moisture content switchgrass (0.41 MPa). Their results showed that the high moisture content switchgrass required higher cutting energy (4.5 MJ/ dry kg) as compared to 3.64 MJ/dry kg in case of low moisture content switchgrass. Reduced knife grid spacing and increased packed bed depths further showed an increase in the cutting energy. However, no significant difference in ultimate stress and cutting energy was observed when similar tests were performed on corn stalks with high (78.8%, wet basis) and low (11.3%, wet basis) moisture contents (Igathinathane et al., 2009).

3.2 Currently used equipments for biomass size reduction

Commonly used biomass size reduction equipments are ball mills, vibratory mills, hammer mills, knife mills, two roll mills, colloid mills, attrition mills, and extruders (Kratky and Jirout, 2011). Extruders are not typical size reduction machines, however, their capability for continuous operation, use of high shear forces, rapid heat transfer, and effective mixing have attracted some interest for biomass disintegration (Kratky and Jirout, 2011). The selection of proper grinding or milling machine is mostly influenced by the moisture content of biomass. Kratky and Jirout (2011) reported that colloid mills and extruders are only suitable for biomass having a moisture content in excess of 15–20% (wet basis) whereas, two roll, attrition, hammer, or knife mills are suitable for biomass having a moisture content less than 10–15% (wet basis). Size reduction machines such as ball or vibratory ball mills are universal types of disintegrators and can be used for either dry or wet biomass materials. On the one hand, ball mills and vibratory mills operations have been found to be time-consuming and energy intensive (Kratky and Jirout, 2011), therefore their commercial applications for biofuel processing are limited. On the other hand, hammer mill (Fig. 3) and knife mill (Fig. 4) are the most commonly used machines for the size reduction of lignocellulosic biomass as far as energy consumption is concerned (Cadoche and Lopez, 1989; Kratky and Jirout, 2011). Hammer mills have wide applications for biomass size reduction as they provide high size reduction ratio, ease of adjustment in the particle size range, and good cubic shapes of particles (Kratky and Jirout, 2011). Similarly, knife mills are also the most prevalent size reduction equipment used for loose and low bulk density lignocellulosic materials such as grasses, straws, and forage

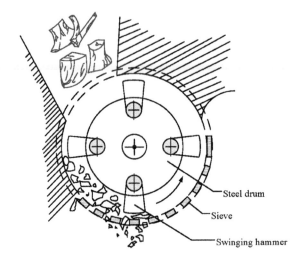

Fig. 3. A hammer mill (Zhang et al., 2012).

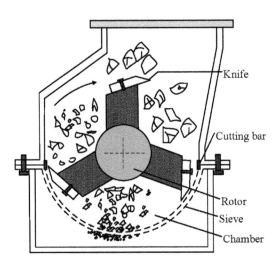

Fig. 4. A knife mill (Zhang et al., 2012).

crop leftovers. Many researchers have studied the energy consumption of knife mills, hammer mills, and other mills discussed earlier. In all cases, the knife mill showed lower energy consumption, thus outperformed the other machine types (Kratky and Jirout, 2011). Cadoche and Lopez (1989) found an increase in the energy consumption (kWh/tonne) from 21 to 29 to 42 for reducing the straw to the particle sizes of 3.2, 2.54, and 1.6 mm, respectively, with a hammer mill. When the same straw was comminuted with knife mill, the energy consumption of 7.5 and 6.4 kWh/tonne was recorded for achieving the straw particle sizes of 1.6 and 2.54 mm, respectively. This shows that hammer mills require more energy than knife mills to reduce a material to the desired final size. The increase in energy consumption by hammer mill is even more prominent in the case of hardwood. A hammer mill consumed considerably higher energy of 120, 115, 95 kWh/tonne as compared to 80, 50, 25 kWh/tonne of energy consumption by the knife mill in reducing the hardwood size to the final sizes of 2.54, 3.20, 6.35 mm, respectively (Cadoche and Lopez, 1989). Since knife mills consume the lowest energy, they appear to be the right solution for size reduction of biomass feedstocks with low moisture contents.

3.3 Effect of size reduction on biomass gasification process

Physiochemical properties of biomass materials (e.g., particle size, shape, moisture content, proximate, and ultimate analyses) considerably influence the gasification process and so does the product quality such as syngas composition, tar, and particulate matters (Sharma et al., 2011). Indeed, biomass particle size plays a major role in the overall gasification process. Compared to the larger particle, a smaller particle offers higher surface area/volume ratio resulting in an effective process in terms of rapid particle heating, improved heat and mass transfers, better diffusion of reactants and products during successive gasification reactions, and enhanced solid-gas reactions. Further, compared to a large particle, the small particle releases higher volatiles leading to the production of more porous char with better reactivity and thus gasification reactions completes up to a greater extent. Furthermore, the gasification time and the size of the reactor are also influenced by the biomass particle size (Hernandez et al., 2010).

3.3.1 Effect of biomass particle size on gasification efficiencies and gas yield

Effects of particle size on gasification performance for different biomass feedstocks are presented in Table 1. The residence time needed to achieve complete cracking of the heaviest and condensable volatile fraction decreases with decreasing particle size. Further, the decrease in the particle size also increases the bulk density of the gasifier feedstock resulting in the higher mass throughput thus, increasing the conversion capacity of the gasifier reactor. In other words, a smaller particle size could result in a reduction in the reactor size. Table 1 shows that decrease in the particle size leads to higher gas yield. Wang and Kinoshita (1993) computed that the residence time essential for 90% carbon conversion efficiency increases linearly with the particle size (Wang and Kinoshita, 1993). Hernandez et al. (2010) reported that reduction in the fuel (dealcoholized marc of grapes)

Table 1. Effect of particle size on gasification performance evaluation parameters.

Fuel used/Gasifier type	Particle size (mm)	Efficiency (%)		Gas yield (kg/kg fuel)	Gas LHV (kJ/Nm³)	Gas composition (vol %/vol %)						Char and tars (wt. % of fuel)	Source
		Carbon conversion	Cold gas			H_2	CO	CH_4	CO_2	C_2H_6	C_2H_4		
Almond shell / Fluidized-bed	1.09	-	-	0.70	-	0.25	0.16	0.06	0.16	-	-	35	(Rapagna and Latif, 1997)
	0.75	-	-	0.80	-	0.33	0.15	0.07	0.2	-	-	22	
	0.53	-	-	1.00	-	0.43	0.2	0.09	0.25	-	-	10	
	0.29	-	-	1.20	-	0.52	0.28	0.12	0.23	-	-	0	
Cynara cardunculus L./Cylindrical Stainless steel reactor	1.6–2	97.5	-	0.86	-	38.3*	11.8*	1.6*	10.4*	0.19*	0.42*	-	(Encinar et al., 2002)[‡]
	1–1.6	96.9	-	0.89	-	38.1	12.3	1.9	10.2	0.21	0.41	-	
	0.63–1	97.2	-	0.88	-	37.8	11.9	1.8	10.4	0.21	0.43	-	
	0.4–0.63	97.3	-	0.87	-	37.3	12.1	1.7	9.9	0.2	0.42	-	
Pine saw dust / Fluidized-bed	0.6–0.9	77.62	-	1.53[a]	6,976	33	37.5	6	21	-	2.5	-	(Lv et al., 2004)[‡]
	0.45–0.6	84.4	-	1.93	7,937	33	37.5	6.5	19	-	2.8	-	
	0.3–0.45	90.6	-	2.37	8,708	32.5	40	7	17	-	3	-	
	0.2–0.3	95.1	-	2.57	8,737	30	41	7.5	16.5	-	4	-	
Pine saw dust / Fixed bed	0.18–0.25	-	-	2.15	-	40	17	5	39	-	-	-	(Feng et al., 2011)[†]
	0.15–0.17	-	-	2.10	-	47	18	4	31	-	-	-	
	0.13–0.15	-	-	2.00	-	49	20	2	29	-	-	-	
	< 0.13	-	-	1.95	-	50	20	1	28	-	-	-	
White Oak / Fluidized bed	25	-	-	0.60 ± 0.3[£]	-	0.011 ± 0.001	0.33 ± 0.02	0.055 ± 0.002	0.21 ± 0.01	-	-	0.17 ± 0.06	(Gaston et al., 2011)
	18	-	-	0.58 ± 0.1	-	0.009 ± 0.001	0.31 ± 0.0	0.056 ± 0.0	0.21 ± 0.01	-	-	0.10 ± 0.04	
	13	-	-	0.68 ± 0.7	-	0.01 ± 0.001	0.37 ± 0.05	0.052 ± 0.005	0.23 ± 0.05	-	-	0.12 ± 0.01	
	6	-	-	0.65 ± 0.5	-	0.011 ± 0.002	0.35 ± 0.01	0.065 ± 0.003	0.22 ± 0.02	-	-	0.10 ± 0.01	

Feedstock / Reactor	Particle size												Reference
Municipal solid waste/Fixed-bed	10 to 20	-	1.28	-	26.5	23.3	4.6	42.8	0.6	2.2	22		(Luo et al., 2010b)[†]
	5 to 10	-	1.3	-	30.6	25.7	5.6	34.5	1	2.6	18		
	< 5	-	1.38	-	32.8	31.1	8.7	22.5	1.2	3.7	12		
Peach tree prunnings/Down draft fixed bed	60–80	-	3.4 ± 0.1[α]	-	12.7 ± 0.4	21.5 ± 0.1	0.4 ± 0.14	10.5 ± 0.1	0.0	0.02	25 ± 5[β]		(Yin et al., 2012)
	40–60	-	3.0 ± 0.2	4.1 ± 0.2	12.5 ± 0.7	24 ± 0.5	0.3 ± 0.1	8.4 ± 0.2	0.0	0.00	15 ± 5		
	20–40	-	2.75 ± 0.2	3.9 ± 0.3	14 ± 1	20 ± 0.5	0.62 ± 0.1	12 ± 1	0.02	0.08	70 ± 5		
	10–20	-	2.6 ± 0.22	4.5 ± 0.24	15.2 ± 1.8	21 ± 0.5	0.94 ± 0.16	11.5 ± 0.2	0.03	0.18	75 ± 5		
	< 10	-	2.25 ± 0.2	4.2 ± 0.2	16 ± 1.5	16 ± 1	1.74 ± 0.14	14 ± 0.8	0.08	0.25	550 ± 30		
Dealcoholized marc of grapes/Entrained flow	8	57.5	10	1.9	1,000[λ]	3	5	1	18.5	-	-	-	(Hernandez et al., 2010)[†]
	4	65	24	2.3	2,250	6	10	2.2	17.5	-	-	-	
	2	74	24.8	2.2	2,300	7	11	2.4	17	-	-	-	
	1	80	26	2.2	2,500	8	12	2.8	16.5	-	-	-	
	0.5	91.4	30	2.1	2,900	9	14	3	16	-	-	-	
Empty fruitbunch Biochar/Fluidized bed	2.0–1.0	-	-	-	38,720[£]	25.3	3.5	2.0	-	-	-	-	(Mohd Salleh et al., 2015)[†]
	1.0–0.5	-	-	-	40,590	26.4	3.7	2.0	-	-	-	-	
	0.5–0.25	-	-	-	44,350	28.9	4.0	1.8	-	-	-	-	
	0.25–0.2	-	-	-	47,840	31.3	4.1	1.6	-	-	-	-	
	< 0.2	-	-	-	52,870	34.8	4.3	1.2	-	-	-	-	

*Gas composition is in mol/kg of fuel feed, £Gas yield and composition in g/g biomass, αGas yield in Nm³/kg, λLHV values in KJ/Kg, χHHV values in kj/kg. βTar and dust content (mg/Nm³), †deviation of values not provided in the literature.

particle size results in better fuel conversion and cold gas efficiency. In their study, the authors found the maximum fuel conversion to be 91.4% for the smallest particle size (0.5 mm) compared to 57.5% for the maximum particle size (8 mm). No significant effect of particle size on the gas yield was reported but a significant increase in the cold gas efficiency was observed with decreasing particle size. The reported cold gas efficiency for a particle size of 8 mm was 10% which increased to 30% for the smaller particle size of 0.5 mm (Hernandez et al., 2010). Tinaut et al. (2008) studied the influence of biomass particle size, air flow velocity, and biomass type on the process propagation velocity and its efficiency. Maximum process efficiency was obtained when smaller particle size and lower air velocity were selected (Tinaut et al., 2008). Luo et al. (2009) evaluated the effect of particle size and bed temperature on the gasification performance in a lab scale fixed bed reactor. The biomass was segregated into five different fractions and the bed temperature was varied from 600 to 900°C. They observed that both dry gas yield and carbon conversion efficiency increased with an increase in the temperature. Further, at every temperature interval, the smallest particle size showed the highest dry gas yield and carbon conversion efficiency. With decreasing the particle size from 0.6–1.2 mm to 0.075 mm at 900°C, an increase in both dry gas yield from 1.3 to 1.7 m^3/kg and carbon conversion efficiency from 76 to 99.87% was observed. The influence of particle size on gas yield and carbon conversion efficiency, however, decreased at higher temperatures. With an increase in the reactor temperature, the impact of particle size on gasification efficiency considerably decreased. The authors further reported that the effect of particle size on dry gas yield and carbon conversion efficiency decreased with an increase in gasification temperature. However, at the same temperature, the gas yield and carbon conversion efficiency for the smallest and largest particles were considerably different. This was clearly evident from the dry gas yield and carbon conversion efficiency data obtained for the smallest and largest particle sizes. For the studied particle sizes, the gas yield at 600°C ranged from about 0.21 m^3/kg (largest size) to 0.76 m^3/kg (smallest size) while a narrow range from nearly 1.38 m^3/kg (largest size) to 1.66 m^3/kg (smallest size) was observed at 900°C. Similarly, the carbon conversion efficiency at 600°C ranged from 30% (largest size) to 58% (smallest size) and from about 82% (largest size) to 99.87% (smallest size) at 900°C (Luo et al., 2009). Yan et al. (2010) studied steam gasification of biomass and stated that solid residence time had a larger effect on the gasifier performance as compared to the particle sizes. Hence, selection of proper biomass particle size and longer residence time were helpful in improving dry gas yield and carbon conversion efficiency. The smaller particle size required less time for reaction completion than the larger particles. The authors also reported that influence of particle size on gasification was minimized at elevated temperatures (Yan et al., 2010).

3.3.2 Effect of particle size on gas composition

Smaller particle size typically results in a higher quality gas composition. The difference in particle sizes significantly affects the syngas composition. As shown in Table 1, the concentration of combustible gases in the syngas increases with the decrease in the fuel particle size (Table 1). Hernandez et al. (2010) observed that with a reduction in particle size from 8 to 0.5 mm, the concentration of combustible gases such as CO, H_2, and CH_4 increased, whereas the concentration of CO_2 decreased. Decreasing the biomass particle size from 8 to 0.5 mm, the authors observed major improvement in the gas heating value with remarkable increase in the syngas CO and H_2 concentrations by 180% and 200%, respectively (Table 1) (Hernandez et al., 2010). A similar study to evaluate the effect of particle size and bed temperature on gas composition was performed by Luo et al. (2010). The authors found that the decrease in particle size from 0.6–1.22 mm to 0.075 mm, the H_2 yield markedly increased by 275% and by 46% at gasification temperatures of 600°C and 900°C, respectively (Table 1) (Luo et al., 2009). Similar trends on gas yield and gas composition were obtained when the study was repeated on municipal solid waste in a fixed bed reactor (Luo et al., 2010). Tinaut et al. (2008) reported that smaller particle size and low air flow velocity resulted in higher fuel/air ratio and, thus, produced higher calorific value gas (Tinaut et al., 2008).

3.3.3 Effect of particle size on char and tar contents of the gas

During gasification, particle size and bed temperature both have a significant effect on the char and tar contents of the gas. Luo et al. (2009) reported that at biomass particle size below 0.075 mm combined with a gasification temperature of 600°C, a negligible amount of char and tar was observed. However, at the same temperature for the large particles (0.6–1.2 mm), a considerably higher char and tar content of 23% was recovered. When the temperature was increased to 900°C, the combined load of char and tar content decreased to 12% for the biomass particles of sizes between 0.6–1.2 mm. These results confirmed that the particle size and gasification temperature both had a significant effect on the tar and char yields during gasification (Luo et al., 2009). Rapagna and Latif (1997) also reported that char and heavy tar production was negligible for the smallest particle size tested in their study. The authors noticed that the char and tar content for the larger particles decreased with an increase in the temperature; however, the large particles still had a noteworthy production of char and tar (35%, Table 1) at a reactor temperature of 800°C.

The primary reason for such high char and tar generation with large biomass particles was that the larger particles have limited heat and mass transfers due to the low surface area available for the reaction causing a drop in the rate of reaction, thus leading to an ineffective biomass conversion during gasification (Rapagna and Latif, 1997). This can be understood from Eq. (12) which clearly shows that the rate of reaction is directly related to the surface area of the particles. In Eq. (12), S_a is the particle surface area and r is the reaction rate of char gasification (Mani et al., 2011). Overall, these findings confirmed that the char and tar contents can be minimized by employing a biomass preprocessing technique (i.e., size reduction step) to obtain sufficiently small biomass particles for subsequent gasification to generate high-quality gas.

$$r = 0.6147 \exp\left(-\frac{18753}{T}\right) S_a (1-X)^{0.9} \qquad \text{Eq. (12)}$$

4. Pelletization

Plant based lignocellulosic biomass has a low bulk density of about 30 kg/m³ (Mani et al., 2006), which increases to typically 100–200 kg/m³ upon bailing in the form of round or square bale (Kaliyan et al., 2009). The moisture content of the lignocellulosic crop also varies from 10% to 70% (wet basis) based on the maturity stage at harvest (Khanchi and Birrell, 2017). Due to low bulk density and varying moisture content, the lignocellulosic biomass is difficult to handle and very expensive to transport, store, and directly utilize for gasification and similar processes. A possible solution to minimize or avoid these problems is the densification of loose and low bulk density biomass into pellets or briquettes. Residues obtained from the sawmill, forest and agricultural crops, as well as whole energy crops such as switchgrass, can be compacted into pellets or briquettes for efficient and economical utilization in downstream processes such as gasification for producing biofuels and biochemicals. Further, waste material can also be densified to the order of 7–10 times to that of normal bales by applying 400–800 MPa pressure (Demirbas and Sahin-Demirbas, 2009). Pellets are cylindrical in shape with 6–8 mm diameter and 10–12 mm long (Mani et al., 2006). During pelletization, the specific density of biomass increases to about 1,000 kg/m³ (Mani et al., 2006). Pelletized biomass is economical to handle, store, and transport. It is uniform in shape and size and also free-flowing, thus offers better flow control during gasification, combustion, and pyrolysis processes (Kaliyan et al., 2009). Densified biomass generates less dust in the production process and has low fire risks as compared to loose and low-bulk density biomass.

Major unit operations in pelletization are biomass drying, size reduction, and densification. The pressure needed for densification highly depends on the moisture content of biomass. The optimal moisture content of biomass at the time of densification varies between 10–25% (wet basis) (Uslu et al., 2008). If the moisture content is outside the optimum range, the pressure needed for densification increases significantly. In the drying process, biomass is dried to about 10% moisture content (wet basis) using a rotary drum dryer, steam dryer, or hot air dryer. Drying increases energy density and, thus reduces biomass transportation and handling costs. Dried biomass is also stable during storage due to less susceptibility to mold development and insect attack (Uslu et al., 2008). After drying, biomass is fed to the hammer mills for size reduction in the range of 3.2–6.4 mm, which is then passed through the press mills to form pellets. The feed material is first fed into the dies (cylindrical cavities) wherein the rotation of the dye and the simultaneous roller pressure compresses the feed material to pass through the die to form pellets. Once the pellets are formed, the length can be adjusted by changing the speed of the knives. There are two main types of pelletizers, namely flat die and ring die pelletizers (Kaliyan et al., 2009). An example of ring die pelletizer is shown in Fig. 5. The true and bulk densities of pellets vary from 1,000–1,200 kg/m^3 and 550–700 kg/m^3, respectively (Mani et al., 2006). The density and durability of pellets are

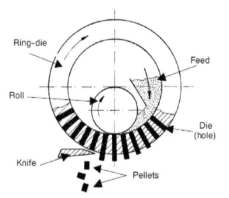

Fig. 5. A ring die pelletizer machine (Kaliyan et al., 2009).

affected by process temperature, applied pressure, and physiochemical properties of biomass. The cellulose present in the biomass is stable below 250°C but the lignin which acts as a glue and binds the cellulose portion begins to soften at 100°C (Uslu et al., 2008). During pelletization, biomass temperature can approach 150°C, which softens lignin and allows molding of biomass into dye shape. Treating sized biomass with steam above 100°C also helps to release natural binders present in biomass.

4.1 Effect of pelletization on biomass gasification

Compared to size reduction, densification of biomass in the form of pellets or briquettes increases the fuel particle size; however, this process is vital for the loose and low bulk density biomass such as bagasse and leafy plant residues which are difficult to gasify in the fixed-bed or fluidized-bed gasification systems (Erlich et al., 2006). One major limitation of the gasifier is the inflexibility to diverse biomass feedstocks (Erlich and Fransson, 2011). The use of pelletized biomass provides flexibility of raw material, as well as automation of the gasifier feed system. Due to less bridging of pellets in the bed, the requirements of working with open ports on the gasifier reactor is reduced, which further improves the health and safety of the operators (Erlich and Fransson, 2011).

4.1.1 Effect of pelletized biomass on gasification performance

Pelletized biomass is a more reliable fuel in fluidized-bed gasifier systems as it provides uniform feeding rates. Further, the increase in the particle size delays or evades the release of sticky volatile matters from the pelletized particles in the feeding line and, thus prevents melting or agglomeration problems in the hot parts of the feeding line (Ruoppolo et al., 2012). Size of the biomass pellet also has a significant impact on the rate of gasification. In general, large pellets result in slow gasification rates compared to small pellets made from same material. Erlich and Fransson (2011) studied downdraft gasification of empty fruit bunch (EFB) and reported that EFB pellets of size 8 mm took a longer time to gasify than 6 mm EFB pellets (Erlich and Fransson, 2011). Erlich et al. (2006) performed gasification of wood and bagasse pellets in a thermogravimetric equipment and found that gasification rate was affected by pellet size and material. The char from wood chips gasified faster than char from wood pellets, and further the 6 mm char pellets have faster gasification rate than 8 mm char pellets. Similarly, for Brazilian bagasse, researchers reported that the 6 mm char pellets have significantly shorter conversion time than 12 mm char pellets. They also observed that char from wood converts more quickly than char from bagasse during steam gasification (Erlich et al., 2006). These findings confirm that faster gasification rates can be achieved by selecting smaller sized pellet made of a specific material such as wood chips.

Erlich et al. (2006) observed that pelletized biomass is more stable during pyrolysis and gasification. They found that among pellets of same size and material, very little variance occurred along the length and diameter of pellets after pyrolysis while a considerable variation was observed along the direction perpendicular to the fiber length in untreated shredded bagasse. These findings suggest that during pyrolysis in a packed bed reactor, pelleted biomass has a more homogeneous physical behavior as compared to untreated biomass. Additionally, the size of the pellet also displayed a larger impact on shrinkage behavior of the pelleted biomass than the original raw biomass. Erlich et al. (2006) also reported that the volume reduction of char obtained from larger sized pellets (8–12 mm) was higher than that observed with char from smaller sized pellets (6 mm) (Erlich et al., 2006). In a downdraft gasification study using wood and wheat straw pellets, Lickrastina et al. (2011) reported that the biomass consumption in the gasifier was influenced by the biomass composition, external heat energy directly supplied to the biomass using propane fired burner system, and air supply rates. The thermal decomposition of pellets and production of volatiles were dependent on the specific material of the pellets. In case of wood pellets, a maximum volatile production rate with peak values of CO and H_2 in produced gas was observed during pyrolysis/gasification stage. However, for wheat straw pellets, the maximum production rate of CO and H_2 was observed during the final stage of char gasification (Lickrastina et al., 2011). With the use of additional heat energy and sub-stoichiometric air supply, faster decomposition rate of pelletized biomass was obtained. An air to fuel ratio of 1.3 to 1.6 resulted in a gas with higher CO and H_2 contents which increased calorific value of the gas. Erlich and Fransson (2011) reported that cold gas efficiency was also dependent on the material and size of the pellets. Wood pellets gave the highest efficiency compared to bagasse and EFB pellets (Table 2). EFB pellets of 6 mm size had a cold gas efficiency of 38% compared to 56% for 8 mm size. For same load series on the blower, more air per kg of fuel was drawn into the gasifier for 8 mm EFB pellets compared to 6 mm EFB pellets. This higher volume flow of gas per kg of fuel spent for 8 mm EFB pellets resulted in higher cold gas efficiency compared to that obtained using 6 mm pellets. Since larger pellets take longer to gasify than smaller pellets composed of the same material, EFB pellets of 8 mm size also caused less fuel consumption per hour than 6 mm sized EFB pellets. For a certain gasifier and blower capacity, they also concluded that size of the pellet was more important than composition for mass consumption rate (Erlich and Fransson, 2011).

4.1.2 *Effect of pelletized biomass on heating value and gas composition*

Pellet material type, composition, and size (diameter and length) effect the composition and heating value of gas during gasification. Erlich and Fransson (2011) reported that wood as a material produced the richest gas with a lower heating value (LHV) of 5.5 MJ/m^3 compared to 5.0 and 4.3 MJ/m^3 produced by bagasse and EFB pellets, respectively. Moreover, the gas produced by EFB pellets of 8 mm diameter had slightly less LHV of 4.1 MJ/m^3 compared to 4.3 MJ/m^3 produced by EFB pellets of 6 mm diameter. An effect of pellet material type on gas composition was also observed in the same study. Compared to wood, the gas generated using bagasse and EFB pellets contained higher concentrations of incombustible gases such as N_2 and CO_2 which decreased LHV of the gas. The authors

Table 2. Effect of pelletization on gasification performance evaluation parameters.

Gasifier type	Pellet material	Gasification performance evaluation parameters											Source
		Cold gas efficiency, %	Gas production		Gas LHV (MJ/Nm³)	Gas composition (v/v %)						Tar, g/Nm³	
			Nm³/h	Nm³/kg		H_2	CO	CH_4	CO_2	C_2H_4	N_2		
Downdraft	Wood sawdust	67.7	94	2.2	5.7	17.6	21.6	2.3	12	0.4	46	-	(Simone et al., 2012)[†]
		67.7	105	2.2	5.5	16.3	21.3	2.3	12.4	0.4	47.2	-	
	50% mix of Sunflower meal and wood sawdust pellets	70	84	2.4	5.6	15.8	19.7	2.3	11.6	0.8	49.5	-	
		70	101	2.4	5.7	17.6	20.6	2.5	12.8	0.5	45.9	-	
	Bagasse, 6 mm	-	-	4.8-6.1	5 ± 0.1	9.9 ± 0.6	23.3 ± 1.2	2.8 ± 0.3	11.4 ± 0.9	-	52.6 ± 0.9	-	(Erlich and Fransson, 2011)[a]
	Wood, 6 mm	-	-	5.2-6.8	5.4 ± 0.3	11.9 ± 1.1	25.7 ± 1.7	2.6 ± 0.2	9.9 ± 1.0	-	50.4 ± 1.7	-	
	Empty fruit bunch, 6 mm	-	-	5.0-6.2	4.3 ± 0.2	13.5 ± 0.8	17 ± 0.9	1.9 ± 0.4	14.5 ± 1.2	-	53.3 ± 2.2	-	
	Empty fruit bunch, 8 mm	-	-	5.0-5.6	4.1 ± 0.2	12.9 ± 0.3	17.4 ± 1.5	1.5 ± 0.2	13.7 ± 0.6	-	55 ± 1.0	-	
	Rice husk (no pellet)	60	-	-	4.54	13.6	14.9	2.3	12.9	-	-	-	(Yoon et al., 2012)[†]
	Rice husk pellet	70	-	-	5.50	18.6	20.2	1.5	8.1	-	-	-	
	DDGS	9.5	-	0.9	1.9	4.9	7.8	1.1	11.2	-	-	-	(Kallis et al., 2013)[†]
	Miscanthus Type 1	30	-	1.4	3.7	11.3	14.3	1.9	12.1	-	-	-	
	Miscanthus Type 2	18.8	-	1.1	2.6	6.1	11.4	1.5	9.9	-	-	-	

Fluidized bed													
Alfalfa, ER 0.23	21.1 ± 1.6	-	1.0 ± 0.7	3.3 ± 0.5	2.9 ± 0.3	14.3 ± 1.1	1.7 ± 0.1	13.6 ± 0.7	-	-	N.D	(Sarker et al., 2015a)
ER 0.30	31.7 ± 1.8	-	1.3 ± 0.4	3.8 ± 0.2	3.5 ± 0.8	15.0 ± 1.7	2.0 ± 0.3	14.6 ± 1.3	-	-	53.1	
ER 0.35	42.5 ± 2.1	-	1.7 ± 0.3	3.7 ± 0.5	3.8 ± 0.7	14.4 ± 1.4	2.6 ± 0.2	15.0 ± 0.9	-	-	34.8	
Wheat straw,												
ER 0.20	25.9 ± 1.3	-	1.1 ± 0.2	4.0 ± 0.4	4.1 ± 0.5	15.2 ± 1.0	2.2 ± 0.3	13.5 ± 1.1	-	-	40.4	
ER 0.25	25.8 ± 1.8	-	1.4 ± 0.2	3.3 ± 0.5	3.6 ± 0.8	10.7 ± 1.1	2.5 ± 0.2	14.5 ± 0.8	-	-	95.7	
ER 0.30	31.2 ± 1.9	-	1.5 ± 0.2	4.1 ± 0.2	3.7 ± 0.3	13.4 ± 0.8	3.0 ± 0.1	16.2 ± 0.4	-	-	58.7	
ER 0.35	33.6 ± 1.2	-	1.7 ± 0.2	3.4 ± 0.3	2.9 ± 0.5	10.9 ± 1.3	2.3 ± 0.2	16.9 ± 0.6	-	-	71.1	
Alfalfa, ER 0.25	34.3 ± 1.2	-	1.4 ± 0.0	4.2 ± 0.2	13.3 ± 0.5	8.4 ± 0.3	2.7 ± 0.1	20.2 ± 0.5	0.9 ± 0.1	52.8 ± 0.9	1.2	(Sarker et al., 2015b)
ER 0.30	38.5 ± 1.4	-	1.5 ± 0.0	4.2 ± 0.2	12.8 ± 0.4	9.1 ± 0.3	2.7 ± 0.2	19.8 ± 0.3	0.9 ± 0.1	52.5 ± 0.9	1.1	
Wood sawdust												(Ruoppolo et al., 2012)[†]
Sand bed	46	-	-	5*	12	18	5	12.5	1.35	-	50	
Catalyst bed	54	-	-	4*	14	17	3	12.5	0.06	-	30	
Biomass and plastic												
Sand bed)	94	-	-	7.5*	16	15	12	10	-	-	60	
Catalyst bed	79	-	-	6.5*	30	14	3	16	0.12	-	28	
Olive husk												
Sand bed	60	-	-	6*	15	18	6	12	-	-	42	
Catalyst bed	61	-	-	5*	17.5	19	7	13	0.69	-	19	
Wood, on bed feeding	-	30.68	1.15	14.23	32	34	11	13	3	-	11.18[£]	(Kern et al., 2013)[†]
In bed feeding	-	27.52	1.04	13.62	40	30	10	14	2	-	1.94	

*LHV in MJ/kg, ªvalues with respective mean square distribution, [£]tar content in g/kg fuel, [†]deviation of values not provided in the literature.

suggested that the gas composition was directly related to the reactivity of the biomass. A biomass with high reactivity results in higher temperature and thus produced a richer gas compared to the low reactive biomass (Erlich and Fransson, 2011). Yoon et al. (2012) observed that much more combustible gases such as H_2 and CO and lower CO_2 were produced when rice husk is converted to pelletized form. From rice husk gasification, a composition of 13.6% H_2, 14.9% CO, 12.9% CO_2, and 2.3% CH_4 was obtained; and from rice husk pellets, a composition of 18.6% H_2, 20.2% CO, 8.1% CO_2, and 1.5% CH_4 was recorded (Yoon et al., 2012). Other studies showing the influence of biomass pelletization on gas heating value and composition are presented in Table 2.

Sarker et al. (2015a) studied an influence of ER on gas composition, heating value, and char and tar yield from alfalfa and wheat straw pellets. In case of alfalfa pellets, an improvement in LHV (4.1 MJ/Nm³), specific gas yield (1.7 Nm³/kg), CGE (42%) and CCE (72%) was observed when ER was increased from 0.2 to 0.35. However, for wheat straw pellets, the improvement was not linear with increase in ER. The optimum performance was observed at ER of 0.3 where a maximum CGE of 37% and LHV of 4 MJ/Nm³ were obtained. Tar yield of 34.8 g/Nm³ was also least at the max ER of 0.35 tested for alfalfa pellets, whereas tar yield of 58.7 g/Nm³ at optimum ER of 0.3 from wheat straw was still greater than tar yield of 40.4 g/Nm³ obtained at ER of 0.2.

4.1.3 Pressure drop during gasification of pelletized biomass

Erlich and Fransson (2011) stated that pressure drop was found to be lower (desirable) for larger pellets and thus more air was drawn in the reactor compared to smaller sized pellets (Erlich and Fransson, 2011). Simone et al. (2012) reported that gasification of pelletized biomass resulted in high and unstable pressure drops which further reduced the gasifier productivity and stability. The reason for the pressure drop was related to the dust formation, as biomass pellets were likely to break up under mechanical and thermal stresses. The formation of dust provided resistance in the flow of biomass pellet in the gasifier. High-pressure drops reduce the gas production and enforce the use of higher discharge rates which results in poor biomass conversion (Simone et al., 2012). Also, fine residues were generated during the gasification of pelleted biomass which reduced the efficiency of wet ash removal systems. Overall, a good gas composition (17.2% H_2, 46% N_2, 2.5% CH_4, 21.2% CO, 12.6% CO_2, and 0.4% C_2H_4), specific gas production (2.2–2.4 Nm³ Kg⁻¹) and cold gas efficiency (67.7–70%) were achieved despite a few performance issues with the pelletized biomass. Good gas composition and global performance indicators suggested that pelletized biomass can be used as a complimentary feedstock for improving the energy content per volume of biomass (Simone et al., 2012). These findings suggest that pelletized biomass can create pressure drop problems in conventional down draft gasifiers if they disintegrate during the gasification process.

4.1.4 Effect of pelletization on co-gasification of biomass

In co-gasification of biomass with coal, the yield and composition of gas are influenced by the composition of biomass material blended with coal. Alzate et al. (2009) reported that co-gasification process behaves considerably different when a pellet composed of pulverized wood and granular coal is employed in a single matrix compared to using pellet blend of 100% wood and granular coal separately (Alzate et al., 2009). Pellets (8X4 mm) were made from pulverized wood and granular coal with varying granular coal proportion, i.e., 0, 5, 10, 20, and 30%.

When the results were compared, similar trends were observed in the evolution of H_2 and CO_2, but different trends were observed for CH_4 and CO. In a single 30% coal and wood matrix the amount of CO produced was 7.4% vol/vol compared to 4.9% vol/vol when a mixture containing 100% wood pellets and coal was used. These findings suggest that the different ways in which the biomass and coal mixture was mixed together, in both cases, considerably affected the production of CO during co-gasification. In the pellets made from wood and coal in a single matrix, both materials are more intimately held together compared to a mixture of 100% wood and pulverized coal mixed in the same proportion. Since the wood occupies a larger volumetric proportion in the wood coal matrix, the reaction of closely packed hidden coal particles somehow get delayed and affected the formation reaction of the CO. In addition, Alzate et al. (2009) reported that pelletized material had low material drag percentage which aided in adequate co-gasification of pelletized wood residues. Due to low material drag, the residency time of biomass in the fluidized bed reactor was increased which led to an improved performance by an increase in the combustible gas per mass unit of the precursor material (Alzate et al., 2009).

4.2 Influence of size reduction and pelletization on economic value of syngas

Syngas from gasification can directly be used for power generation or indirectly converted into transportation liquid fuels using suitable catalysts (e.g., diesel, gasoline, and ethanol) or specialty chemicals such as alcohols, aldehydes, and isobutane (Swanson et al., 2010). Hence, the economic value of syngas substantially relates to the economic value of syngas-derived end-product such as liquid fuel, chemical, electricity or heat, and its respective processing cost. However, in the present review, the economic value of syngas was evaluated based on its energy content. The amount of syngas energy produced per unit biomass fed to the gasification system (unit: MJ/kg biomass) was calculated using the gas yield (Nm^3/kg of biomass) and gas energy content (MJ/Nm^3) data reported in the literature for different biomass feedstocks. Since the cost of syngas is highly dependent on several biomass preprocessing and gasification factors and further no such reliable syngas cost is available in the literature; the cost of natural gas based on industrial applications was used to estimate the cost of syngas energy in $/kJ (EIA, 2015). The cost of syngas produced from each treatment in $ per tonne of biomass is given in Tables 3 and 4 which was calculated from energy equivalent of the current industrial natural gas price of $3.53/1,000 ft^3 (EIA, 2015). The economic value variation of syngas is also reported and was calculated from 5% quantile ($3.59/1000 ft^3), 50% quantile ($5.48/1,000 ft^3), and 95% quantile ($10.04/1,000 ft^3) cost variation of industrial natural gas prices from 2001 to 2015 (EIA, 2015). Further, most gasification work in the peer-reviewed literature was reported based on the gram, lab and pilot scales; no scale-up factor was incorporated in estimating the syngas cost from $/kg to $/tonne. From literature, the costs of processing operations such as pelletization on identical or comparable biomass are provided in Tables 3 and 4. The variation observed in the cost of these preprocessing operations such as wood pellets cost varied from $38–$113 per tonne, which is due to difference in scale of operations considered during their study. Some studies were more detailed and included the several cost components, e.g., biomass, delivery, packaging, land and infrastructure costs while calculating the cost of pellet which in fact increased the cost of the preprocessing step. Additionally, biomass feed rate is also provided in Tables 3–4 to give an estimate on the effect of feed rate (size of gasifier unit) on the overall economic value of syngas produced under different treatments.

When only size reduction pretreatment step was considered for biomass gasification the economic value of syngas based on the current natural gas prices varied from $18.9 to 111.5 per tonne of biomass as shown in Table 3. Remarkably, when biochar was used as the gasifier feedstock, the economic value of syngas increased up to $172 per tonne (Table 3). When pelletized biomass was used for gasification, the economic value of syngas varied from $10.4 to 53.3 per tonne of pelletized biomass under the current natural gas prices (Table 4), whereas the cost of pelletization of biomass varied from $38 to 156 per tonne of biomass based on the available literature (Table 4).

In the present study, the economic value of syngas was estimated using the energy generation per unit biomass (kJ/kg biomass) which was further analyzed as the substitute for natural gas for economic comparison of various biomass preprocessing technologies. Therefore, under current natural gas price scenario, it can be found from Tables 3 and 4 that syngas production from biomass is only economical if the biomass feedstock is directly gasified without any preprocessing or with moderate size reduction

Table 3. Economic analysis of size reduction process.

Author/Biomass/Gasifier type/ Feed rate	Particle size (mm)	HHV (MJ/ Kg)	Economic value ofsyngas	
			$/tonne biomass[t]	$/tonne biomass[α] 5%, 50%, 95% quantile
(Sarkar et al., 2014)/Switchgrass[1]/ Fixed bed/2 g	< 25 mm	11.7	38.1	64.3,61.0,67.4
(Rapagna and Latif, 1997)/Almond shells/ Fluidized bed/1 g/min	Temp. (°C)/ particle size			
	800/1.09	13.5	43.8	44.5,68.0,124.7
	800/0.75	19.6	63.7	64.7,98.8,181.0
	800/0.53	17.7	57.7	58.6,89.5,164.0
	800/0.29	18.2	59.4	60.3,92.1,168.8
	750/1.09	10.9	35.5	36.1,55.1,101.0
	750/0.75	16.2	52.7	53.6,81.8,150.0
	750/0.53	14.4	46.8	47.6,72.7,133.2
	750/0.29	15.3	49.7	50.5,77.1,141.3
	700/1.09	10.8	35.2	35.8,54.6,100.1
	700/0.75	14.0	45.7	46.4,70.9,129.9
	700/0.53	12.6	41.0	41.7,63.6,116.7
	700/0.29	13.8	44.8	45.5,69.4,127.3
	650/1.09	7.9	25.6	26.0,39.7,72.7
	650/0.75	9.0	29.4	29.8,45.5,83.5
	650/0.53	11.5	37.3	37.9,57.9,106.0
	650/0.29	15.4	50.0	50.8,77.6,142.1
	600/1.09	5.8	18.9	19.2,29.3,53.7
	600/0.75	7.8	25.2	25.6,39.1,71.7
	600/0.53	8.7	28.2	28.6,43.7,80.1
(Encinar et al., 2002)/ Cynara cardunculus L./ Cylindrical reactor/3.3 g/hr	1.6–2	15.7	51.1	51.9,79.3,145.3
	1–1.6	16.1	52.2	53.1,81.0,148.5
	0.63–1	15.8	51.3	52.1,79.6,145.8
	0.4–0.63	15.6	50.7	51.6,78.7,144.3
(Lv et al., 2004)/Pine saw dust/ FB/0.5 kg/h	0.75	18.1	59.0	60.0,91.6,167.9
	0.53	23.6	76.7	78.0,119.0,218.1
	0.38	30.2	98.3	99.9,152.5,279.5
	0.25	34.2	111.5	113.3,173.0,316.9

Table 3 contd. ...

...*Table 3 contd.*

Author/Biomass/Gasifier type/ Feed rate	Particle size (mm)	HHV (MJ/ Kg)	Economic value of syngas	
			$/tonne biomass[i]	$/tonne biomass[a] 5%, 50%, 95% quantile
(Luo et al., 2010a)	10 to 20	11.9	38.7	39.3,60.0,109.9
Municipal waste	5 to 10	14.0	45.5	46.2,70.6,129.3
gasified at 900°C/ Fixed bed/(5 g/min)	> 5	18.7	60.8	61.7,94.3,172.8
(Yin et al., 2012)/Peach tree prunings/Downdraft fixed bed/ 25 kg/h	60–80	13.0	42.4	43.0,65.7,120.4
	40–60	13.1	42.7	43.4,66.3,121.5
	20–40	11.7	38.0	38.6,59.0,108.1
	20–10	12.2	39.7	40.3,61.5,112.8
	> 10	10.3	33.4	33.9,51.8,95.0
(Feng et al., 2011)/Pine saw dust/ Fixed bed/Bench scale	0.18–0.25	18.2	59.3	60.3,92.1,168.7
	0.15–0.17	19.0	61.9	62.9,96.0,175.9
	0.13–0.15	17.6	57.2	58.1,88.8,162.7
	< 0.13	16.7	54.2	55.1,84.1,154.1
(Gaston et al., 2011)/White Oak/ FB/8.3 g	25	8.0	25.9	26.3,40.2,73.7
	18	7.5	24.5	24.9,38.0,69.7
	13	8.1	26.3	26.6,40.7,74.6
	6	8.7	28.4	28.8,44.0,80.7
(Mohd Salleh et al., 2015)/EFB biochar/FB	2.0–1.0	38.7	126.0	128.1,195.6,358.4
	1.0–0.5	40.6	132.1	134.3,205.0,375.7
	0.5–0.25	44.4	144.3	146.7,224.0,410.5
	0.25–0.2	47.8	155.7	158.3,241.7,442.8
	< 0.2	52.9	172.1	174.9,267.1,489.4

[i]Assuming cost of industrial natural gas is $3.53/1000 ft[3] (EIA, 2015), [a]Economic value variation of syngas based on variation in industrial natural gas cost from $3.59 (lower 5% quantile), $5.48 (50% quantile), $10.04 (upper 95% quantile) per 1000 ft[3] from 2001 to 2015 (EIA, 2015). EFB = Empty fruit bunch, FB = Fluidized bed. [i]Cost of size reduction ($/tonne biomass) for switchgrass: 5 (grinding cost), 55 (grinding, storage and transportation cost) (Shastri et al., 2014).

as reported in a few studies (Table 3). On the other hand, the syngas production from pelletized biomass is not economically viable. Even though the use of pelletized biomass reduces transportation cost and increases the throughput capacity of gasifier; conversely, the overall process cost is higher than the economic value of syngas alone that is produced from the combined process (i.e., pelletization + gasification). Syngas production from biomass

Table 4. Cost comparison of pelletization process with economic value of syngas produced during different treatments.

Author/Gasifier type/ Biomass pellet size (mm)/ Feed rate	Treatment	HHV	Economic value of syngas		Cost of pelletization
		MJ/kg	$/tonnebiomass‡	$/tonne biomass[a] 5%, 50%, 95% quantile	$/tonne biomass
Simone et al. (2012)/Downdraft 6x10–30/54.2 kg/h	Wood saw dust (WSP)	12.4	40.4	41,62.7,114.8	
10x30–60/44.7 kg/h	Sunflower meal (SMP)	13.1	39.1	43.2,65.9,120.8	
	50% mix of WSP and SMP	13.6	42.5	44.9,68.5,125.6	
Φ(Ruoppolo et al., 2012)/ FB/3 to 5.2 kg/h	ER=0.3, S/F=0, Catalyst	4.0	13.0	13.2,20.2,37.0	44[1,£]
Wood pellet (6x20)	ER=0.3, S/F=0.65, Catalyst	4.9	16.0	16.2,24.7,45.3	113[2,£]
	ER=0.3, S/F=0.91, Catalyst	4.2	13.7	13.9,21.2,38.8	
	ER=0.3, S/F=0, Quartzite	5.0	16.3	16.5,25.2,46.2	
	ER=0.3, S/F=0.65, Quartzite	4.0	13.0	13.2,20.2,37.0	
	ER=0.3, S/F=0.91, Quartzite	4.1	13.3	13.5,20.7,37.9	
	ER=0.17, S/F=0.65, Catalyst	6.0	19.5	19.8,30.3,55.5	
	ER=0.17, S/F=0, Quartzite	7.0	22.8	23.1,35.3,64.7	
	ER=0.17, S/F=0.65, Quartzite	6.0	19.5	19.8,30.3,55.5	
Olive husk (6x20)	ER=0.17, S/F=0, Catalyst	5.8	18.9	19.1,29.3,53.6	
	ER=0.09, S/F=0.44, Catalyst	8.2	26.7	27.1,41.4,75.9	
	ER=0.17, S/F=0, Quartzite	6.1	19.9	20.1,30.8,56.4	
	ER=0.09, S/F=0.44, Quartzite	6.0	19.5	19.8,30.3,55.5	

Table 4 contd. ....

...*Table 4 contd.*

Author/Gasifier type/ Biomass pellet size (mm)/ Feed rate	Treatment	HHV	Economic value of syngas		Cost of pelletization
		MJ/kg	$/tonne biomass[‡]	$/tonne biomass[a] 5%, 50%, 95% quantile	$/tonne biomass
Biomass/Plastic pellet (6x20)	ER=0.23, S/F=0, Catalyst	6.5	21.2	21.5,32.8,60.1	
	ER=0.19, S/F=0, Catalyst	4.0	13.0	13.2,20.2,37.0	
	ER=0.12, S/F=0.6, Catalyst	7.9	25.7	26.1,39.9,73.1	
	ER=0.23, S/F=0, Quartzite	7.5	24.4	24.8,37.8,69.4	
	ER=0.19, S/F=0, Quartzite	6.0	19.5	19.8,30.3,55.5	
(Erlich and Fransson 2011)/ Downdraft/EFB/(2 to 2.7 kg/h)	Pellet size (mm): 8 X11	10.2	33.2	33.7,51.5,94.4	
EFB (2.4 to 3.3 kg/h)	Pellet size (mm): 6 X 12	8.9	28.7	29.5,45.0,82.6	
Bagasse (2.7 to 3.5 kg/h)	Pellet size (mm): 6 X 12	8.8	28.7	29.1,44.4,81.4	
Wood (2.7 to 3.2 kg/h)	Pellet size (mm): 6 X 13	11.2	36.4	37.0,56.5,103.6	38–51[3,£]
(Kern et al., 2013)/FB	In bed feeding of pellets	14.2	46.1	46.8,71.5,131.1	44[4,£], 98[5,£]
Wood pellets/18.6 kg/h	Pellets fed from top of bed	16.4	53.3	54.1,82.6,151.4	
(Sarker et al., 2015a)/FB	ER 0.23	3.2	10.4	10.5,16.1,29.5	156[6]
Alfalfa/0.2 kg/h	ER 0.30	4.7	15.2	15.4,23.5,43.0	
6x25	ER 0.35	7.1	23.0	23.4,35.7,65.5	
Wheat straw (8x20)/0.3 kg/h	ER 0.20	4.2	13.7	13.9,21.3,39.0	130
0.24 kg/h	ER 0.25	4.7	15.2	15.4,23.6,43.2	
0.18 kg/h	ER 0.30	6.1	19.9	20.1,30.8,56.4	
0.18 kg/h	ER 0.35	5.6	18.3	18.5,28.3,51.9	

(Sarker et al., 2015b)/ FB	ER 0.25	5.9	19.1	19.4,29.6,54.3	156[6]
Alfalfa/ 4.7 kg/h	ER 0.30	6.4	20.8	21.1,32.2,59.0	
(Yoon et al., 2012)/	ER 0.22	4.2	13.8	13.9,21.3,39.1	
Downdraft/	ER 0.25	4.8	15.5	15.7,24.0,44.1	
Rice husk/	ER 0.28	5.2	16.8	17.0,26.0,47.6	
40–60 Kg/h	ER 0.3	5.9	19.3	19.6,30.0,54.9	
	ER 0.32	5.9	19.1	19.3,29.5,54.2	
	ER 0.36	5.3	17.2	17.4,26.6,48.8	
	ER 0.43	5.0	16.3	16.6,25.3,46.4	

[+]Assuming cost of industrial natural gas is \$3.5/1000ft[3] (EIA 2015), [a]Economic value variation of syngas based on variation in industrial natural gas cost from \$3.59 (lower 5% quantile), \$5.48 (50% quantile), \$10.04 (upper 95% quantile) per 1000ft[3] from 2001 to 2015 (EIA, 2015). EFB = Empty fruit bunch, FB = Fluidized bed, [ɸ]Heating values are presented as LHV instead of HHV. [ɛ]Converted from Euro to US dollar. References used in cost of pelletization column: [1]Uslu et al. (2008), [2]Nilsson et al. (2011), [3]Uasuf and Becker (2011), [4]Uslu et al. (2008), [5]Thek and Obernberger (2004), [6]Sultana and Kumar (2012), [7]Sultana et al. (2010).

can be viable in future as fossil fuel prices are expected to increase with the continuous depletion of the limited fossil resources. It should be noted that above evaluation of syngas economic value is based on the current natural gas price (which are currently lower than 50% quantile prices of the last 15 yr); although, syngas production using preprocessed biomass can be economically viable for producing specialty high-value end products and transportation liquid fuels by further downstream synthesis of biomass-derived syngas under proper catalyst, but it needs further analysis.

5. Conclusions

Diverse lignocellulosic biomass materials are the potential energy feedstocks for production of biofuels and value-added bioproducts, and further to mitigate the emission of greenhouse gas such as CO_2. However, lignocellulosic feedstocks are heterogeneous in nature and bear different physical and chemical properties, which result in their inefficient and uneconomical utilization in the downstream biomass conversion processes such as gasification and pyrolysis. In the present study, several biomass preprocessing techniques and their effects on the gasification process were discussed and summarized. Size reduction is an essential preprocessing step to achieve desired biomass particle size as suitable for the gasification process. Size reduction considerably improved the gasification performance in terms of cold gas efficiency and gas composition, and further lowered the amount of gas impurities such as tar and char. Compared to larger particle size, the gasification of smaller biomass particle showed a better gas heating value, gas H_2/CO ratio, cold gas efficiency, and biomass conversion. Pelletization is the other important preprocessing step that yield a superior quality biomass feedstock with high energy density and of uniform shape and size, which eventually aids in transportation and handling of lignocellulosic biomass feedstocks for gasification utilization. Pelletization step increases uniformity and energy density of heterogeneous lignocellulosic feedstocks and crop-residues. In gasification, smaller sized biomass pellet improved the gas yield, biomass conversion efficiency, and gas heating value. From comparison with the current natural gas price ($3.5/1,000 ft³), it was observed that pelletization of biomass before gasification is not economically viable for producing syngas as a direct fuel (based on its energy content). But reducing the pelleting production cost can make pelleting an economically viable option. Lamers et al. (2015) have suggested that high moisture pelleting can help reduce the pellet production cost by about 40% compared to the current pelleting method followed by the industry. Therefore, advances in mechanical preprocessing can have a significant impact on the quality of syngas as well as on the economic feasibility. Syngas production using preprocessed biomass can be economically viable for producing specialty high-value end-products and transportation liquid fuels by further downstream synthesis of biomass-derived syngas under proper catalyst, but it needs further analysis.

References

Alzate, C. A., F. Chejne, C. F. Valdés, A. Berrio, J. D. L. Cruz and C. A. Londoño. 2009. CO-gasification of pelletized wood residues. Fuel 88: 437–445.

Bitra, V. S. P., A. R. Womac, C. Igathinathane, P. I. Miu, Y. C. T. Yang, D. R. Smith, N. Chevanan, and S. Sokhansanj. 2009a. Direct measures of mechanical energy for knife mill size reduction of switchgrass, wheat straw, and corn stover. Bioresource Technol. 100: 6578–6585.

Bitra, V. S. P., A. R. Womac, Y. C. T. Yang, P. I. Miu, C. Igathinathane, N. Chevanan and S. Sokhansanj. 2011. Characterization of wheat straw particle size distributions as affected by knife mill operating factors. Biomass Bioenerg. 35: 3674–3686.

Bitra, V. S. P., A. R. Womac, N. Chevanan, P. I. Miu, C. Igathinathane, S. Sokhansanj and D. R. Smith. 2009b. Direct mechanical energy measures of hammer mill comminution of switchgrass, wheat straw, and corn stover and analysis of their particle size distributions. Powder Technol. 193: 32–45.

Cadoche, L. and G. D. Lopez. 1989. Assessment of size-reduction as a preliminary step in the production of ethanol from lignocellulosic wastes. Biol. Waste. 30: 153–157.

Chevanan, N., A. R. Womac, V. S. P. Bitra, C. Igathinathane, Y. T. Yang, P. I. Miu and S. Sokhansanj. 2010. Bulk density and compaction behavior of knife mill chopped switchgrass, wheat straw, and corn stover. Bioresource Technol. 101: 207–214.

Demirbas, K. and A. Sahin-Demirbas. 2009. Compacting of biomass for energy densification. Energ. Source. Part A. 31: 1063–1068.

EIA. 2015. United States Natural Gas Industrial Price (Dollars per Thousand Cubic Feet) Available at: https://www.eia.gov/dnav/ng/hist/n3035us3m.htm.

Encinar, J. M., J. F. González and J. González. 2002. Steam gasification of Cynara cardunculus L.: influence of variables. Fuel Process. Technol. 75: 27–43.

Erlich, C., E. Bjornbom, D. Bolado, M. Giner and T. H. Fransson. 2006. Pyrolysis and gasification of pellets from sugar cane bagasse and wood. Fuel. 85: 1535–1540.

Erlich, C. and T. H. Fransson. 2011. Downdraft gasification of pellets made of wood, palm-oil residues respective bagasse: Experimental study. Appl. Energ. 88: 899–908.

Feng, Y., B. Xiao, K. Goerner, G. Cheng and J. Wang. 2011. Influence of particle size and temperature on gasification performance in externally heated gasifier. Smart Grid Renew. Energ. 2: 158–164.

Gaston, K. R., M. W. Jarvis, P. Pepiot, K. M. Smith, W. J. Frederick Jr. and M. R. Nimlos. 2011. Biomass pyrolysis and gasification of varying particle sizes in a fluidized-bed reactor. Energ. Fuel. 25: 3747–3757.

Hernandez, J. J., G. Aranda-Almansa and A. Bula. 2010. Gasification of biomass wastes in an entrained flow gasifier: Effect of the particle size and the residence time. Fuel Process. Technol. 91: 681–692.

Igathinathane, C., A. R. Womac, S. Sokhansanj and S. Narayan. 2008. Knife grid size reduction to pre-process packed beds of high- and low-moisture switchgrass. Bioresource Technol. 99: 2254–2264.

Igathinathane, C., A. R. Womac, S. Sokhansanj and S. Narayan. 2009. Size reduction of high- and low-moisture corn stalks by linear knife grid system. Biomass Bioenerg. 33: 547–557.

Kaliyan, N., R. V. Morey, M. D. White and A. Doering. 2009. Roll press briquetting and pelleting of corn stover and switchgrass. T. ASABE. 52: 543–555.

Kallis, K. X., G. A. Pellegrini Susini and J. E. Oakey. 2013. A comparison between miscanthus and bioethanol waste pellets and their performance in a downdraft gasifier. Appl. Energ. 101: 333–340.

Kern, S., C. Pfeifer and H. Hofbauer. 2013. Gasification of wood in a dual fluidized bed gasifier: Influence of fuel feeding on process performance. Chem. Eng. Sci. 90: 284–298.

Khanchi, A. and S. Birrell. 2017. Drying models to estimate moisture change in switchgrass and corn stover based on weather conditions and swath density. Agr. Forest. Meteorol. 237-238: 1–8.

Kratky, L. and T. Jirout. 2011. Biomass size reduction machines for enhancing biogas production. Chem. Eng. Technol. 34: 391–399.

Kumar, A., K. Eskridge, D. D. Jones and M. A. Hanna. 2009a. Steam–air fluidized bed gasification of distillers grains: Effects of steam to biomass ratio, equivalence ratio and gasification temperature. Bioresource Technol. 100: 2062–2068.

Kumar, A., D. D. Jones and M. A. Hanna. 2009b. Thermochemical biomass gasification: A review of the current status of the technology. Energies 2: 556–581.

Kumar, A., A. M. Sharma and P. Bhandari. 2014. Biomass gasification and syngas utilization. *In*: L. Wang (ed.). Sustainable Bioenergy Production. CRC Press, Florida, USA.

Lamers, P., M. S. Roni, J. S. Tumuluru, J. J. Jacobson, K. G. Cafferty, J. Hansen, K. K. Kenney, F. Teymouri and B. Bals. 2015. Techno-economic analysis of decentralized biomass processing depots. Bioresource Technol. 194: 205–213.

Lickrastina, A., I. Barmina, V. Suzdalenko and M. Zake. 2011. Gasification of pelletized renewable fuel for clean energy production. Fuel. 90: 3352–3358.

Luo, S. Y., B. Xiao, X. J. Guo, Z. Q. Hu, S. M. Liu and M. Y. He. 2009. Hydrogen-rich gas from catalytic steam gasification of biomass in a fixed bed reactor: Influence of particle size on gasification performance. Int. J. Hydrogen Energ. 34: 1260–1264.

Luo, S. Y., B. Xiao, Z. Q. Hu, S. M. Liu, Y. W. Guan and L. Cai. 2010. Influence of particle size on pyrolysis and gasification performance of municipal solid waste in a fixed bed reactor. Bioresource Technol. 101: 6517–6520.

Luo, S. Y., C. Liu, B. Xiao and L. Xiao. 2011. A novel biomass pulverization technology. Renew. Energ. 36: 578–582.

Lv, P. M., Z. H. Xiong, J. Chang, C. Z. Wu, Y. Chen and J. X. Zhu. 2004. An experimental study on biomass air-steam gasification in a fluidized bed. Bioresource Technol. 95: 95–101.

Mani, S., S. Sokhansanj, X. Bi and A. Turhollow. 2006. Economics of producing fuel pellets from biomass. Appl. Eng. Agric. 22: 421–426.

Mani, T., N. Mahinpey and P. Murugan. 2011. Reaction kinetics and mass transfer studies of biomass char gasification with CO_2. Chem. Eng. Sci. 66: 36–41.

Martinez, J. D., K. Mahkamov, R. V. Andrade and E. E. S. Lora. 2012. Syngas production in downdraft biomass gasifiers and its application using internal combustion engines. Renew. Energ. 38: 1–9.

Mohd Salleh, M. A., H. K. Nsamba, H. M. Yusuf, A. Idris and W. A. W. A. K. Ghani. 2015. Effect of Equivalence ratio and particle size on EFB char gasification. Energ. Source. Part A. 37: 1647–1662.

NETL. 2013. National Energy Technology Laboratory. Introduction to gasification-History of gasification, Vol. 2013. Available at: http://www.netl.doe.gov/research/coal/energy-systems/gasification/gasifipedia/history-gasification.

Nilsson, D., S. Bernesson and P.-A. Hansson. 2011. Pellet production from agricultural raw materials—A systems study. Biomass Bioenerg. 35: 679–689.

Rapagna, S. and A. Latif. 1997. Steam gasification of almond shells in a fluidised bed reactor: The influence of temperature and particle size on product yield and distribution. Biomass Bioenerg. 12: 281–288.

Ruoppolo, G., P. Ammendola, R. Chirone and F. Miccio. 2012. H2-rich syngas production by fluidized bed gasification of biomass and plastic fuel. Waste Manage. 32: 724–732.

Sarkar, M., A. Kumar, J.S. Tumuluru, K.N. Patil and D.D. Bellmer. 2014. Gasification performance of switchgrass pretreated with torrefaction and densification. Appl. Energ. 127: 194–201.

Sarker, S., J. Arauzo and H. K. Nielsen. 2015a. Semi-continuous feeding and gasification of alfalfa and wheat straw pellets in a lab-scale fluidized bed reactor. Energ. Convers. Manage. 99: 50–61.

Sarker, S., F. Bimbela, J. L. Sánchez and H. K. Nielsen. 2015b. Characterization and pilot scale fluidized bed gasification of herbaceous biomass: A case study on alfalfa pellets. Energ. Convers. Manage. 91: 451–458.

Sharma, A. M., A. Kumar, K. N. Patil and R. L. Huhnke. 2011. Performance evaluation of a lab-scale fluidized bed gasifier using switchgrass as feedstock. T. ASABE. 54: 2259–2266.

Sharma, A. M., A. Kumar and R. L. Huhnke. 2014. Effect of steam injection location on syngas obtained from an air–steam gasifier. Fuel. 116: 388–394.

Shastri, Y. N., Z. Miao, L. F. Rodríguez, T. E. Grift, A. C. Hansen and K. C. Ting. 2014. Determining optimal size reduction and densification for biomass feedstock using the biofeed optimization model. Biofuels, Bioprod. Biorefin. 8: 423–437.

Simone, M., F. Barontini, C. Nicolella and L. Tognotti. 2012. Gasification of pelletized biomass in a pilot scale downdraft gasifier. Bioresource Technol. 116: 403–412.

Sultana, A., A. Kumar and D. Harfield. 2010. Development of agri-pellet production cost and optimum size. Bioresource Technol. 101: 5609–5621.

Sultana, A. and A. Kumar. 2012. Ranking of biomass pellets by integration of economic, environmental and technical factors. Biomass Bioenerg. 39: 344–355.

Swanson, R. M., J. A. Satrio, R. C. Brown, A. Platon and D. D. Hsu. 2010. Techno-economic analysis of biofuels production based on gasification National Renewable Energy Laboratory, Available at: http://www.nrel.gov/docs/fy11osti/46587.pdf.

Thek, G. and I. Obernberger. 2004. Wood pellet production costs under Austrian and in comparison to Swedish framework conditions. Biomass Bioenerg. 27: 671–693.

Tinaut, F. V., A. Melgar, J. F. Perez and A. Horrillo. 2008. Effect of biomass particle size and air superficial velocity on the gasification process in a downdraft fixed bed gasifier. An experimental and modelling study. Fuel Process. Technol. 89: 1076–1089.

Tumuluru, J. S., C. T. Wright, K. L. Kenney and R. J. Hess. 2010. A Technical Review on Biomass Processing: Densification, Preprocessing, Modeling and Optimization. ASABE Annual International Meeting, Pittsburgh, Pennsylvania. Paper Number: 1009401.

Tumuluru, J. S., C. T. Wright, J. R. Hess and K. L. Kenney. 2011. A review of biomass densification systems to develop uniform feedstock commodities for bioenergy application. Biofuel Bioprod. Bior. 5: 683–707.

Tumuluru, J. S., S. Sokhansanj, J. R. Hess, C. T. Wright and R. D. Boardman. 2011a. A review on biomass torrefaction process and product properties for energy applications. Ind. Biotechnol. 7: 384–401.

Tumuluru, J. S., C. T. Wright, J. R. Hess and K. L. Kenney. 2011b. A review of biomass densification systems to develop uniform feedstock commodities for bioenergy application. Biofuels, Bioprod. Biorefin. 5: 683–707.

Tumuluru, J. S., J. R. Hess, R. D. Boardman, C. T. Wright and T. L. Westover. 2012. Formulation, pretreatment, and densification options to improve biomass specifications for co-firing high percentages with Coal. Ind. Biotechnol. 8: 113–132.

Uasuf, A. and G. Becker. 2011. Wood pellets production costs and energy consumption under different framework conditions in Northeast Argentina. Biomass Bioenerg. 35: 1357–1366.

USDOE. 1993. Assessment of costs and benefits of flexible and alternative fuel use in the U.S. transportation sector. Technical report eleven: Evaluation of a potential wood to ethanol process, (Ed.) United States Department of Energy. Energy. Washington, pp. DOE/EP-0004.

Uslu, A., A. P. C. Faaij and P. C. A. Bergman. 2008. Pre-treatment technologies, and their effect on international bioenergy supply chain logistics. Techno-economic evaluation of torrefaction, fast pyrolysis and pelletisation. Energy 33: 1206–1223.

Wang, Y. and C. M. Kinoshita. 1993. Kinetic-model of biomass gasification. Sol. Energy 51: 19–25.

Yan, F., L. G. Zhang, Z. Q. Hu, G. Cheng, C. C. Jiang, Y. L. Zhang, T. Xu, P. W. He, S. Y. Luo and B. Xiao. 2010. Hydrogen-rich gas production by steam gasification of char derived from cyanobacterial blooms (CDCB) in a fixed-bed reactor: Influence of particle size and residence time on gas yield and syngas composition. Int. J. Hydrogen Energ. 35: 10212–10217.

Yan, W., S. Islam, C. J. Coronella and V. R. Vásquez. 2012. Pyrolysis kinetics of raw/ hydrothermally carbonized lignocellulosic biomass. Environ. Prog. Sustain. Energy 31: 200–204.

Yin, R., R. Liu, J. Wu, X. Wu, C. Sun and C. Wu. 2012. Influence of particle size on performance of a pilot-scale fixed-bed gasification system. Bioresource Technol. 119: 15–21.

Yoon, S. J., Y. I. Son, Y. K. Kim and J. G. Lee. 2012. Gasification and power generation characteristics of rice husk and rice husk pellet using a downdraft fixed-bed gasifier. Renew. Energ. 42: 163–167.

Zewei, M., E. G. Tony, C. H. Alan and K. C. Ting. 2010. Specific energy consumption of biomass particle production and particle physical property. ASABE annual International Conference, Pittsburgh, Pennsylvania. Paper No: 1008497.

Zhang, M., X. X. Song, P. F. Zhang, Z. J. Pei, T. W. Deines and D. H. Wang. 2012. Size reduction of cellulosic biomass in biofuel manufacturing: a study on confounding effects of particle size and biomass crystallinity. J. Manuf. Sci. E-T ASME. 134: 11009.

CHAPTER 4

Mechanical Fractionation of Biomass Feedstocks for Enhanced Pretreatment and Conversion

Jeffrey A. Lacey

1. Introduction

Mechanical fractionation of plant material has been thoroughly developed and is used widely in the food industry. Early on in human history, man was actively separating grains from the non-edible plant material using simple yet effective methods such as wind-assisted winnowing to remove the chaff from the grain and sieving to isolate the smaller grain fraction from the larger leaves and stems. These separations were possible due to the natural differences in the size, shape, density, and surface area of the different parts of the plant. Since these ancient times, separation technologies have become more advanced and effective at isolating the high-value food products.

While mechanical fractionation is widely used to isolate the edible part of the plant, the non-edible or less valuable portions of the plant are typically collected together regardless of their specific chemical or physical properties. The plant material that we collectively call "biomass" is the shell of a once-living organism that at one time performed complex physiological functions using a variety of different tissues and organs. These different

Staff Scientist, Idaho National Laboratory.
  Email: jeffrey.lacey@inl.gov

functions each required unique physical and chemical characteristics of the involved tissues and organs. These differences include physical features such as rigidity, density, or porosity; and chemical features such as diverse lignin or ash content. These differences can make the individual tissues or organs better suited for specific conversion processes, and can also provide the physical features necessary to effectively separate the tissues or organs into physically and chemically unique feedstock fractions using mechanical fractionation (Table 1).

Most biomass utilization schemes treat all of the biomass in a similar way, grinding the material and processing it as if it were a homogeneous feedstock (Fig. 1a). While this approach can be effective for size reduction, some benefits are lost when the tissue-specific differences within the plant are ignored. If applied to the biomass prior to conversion, mechanical fractionation has the potential to take advantage of these natural differences in tissues and plant parts found in most biomass (Fig. 1b and 1c). Rather than treating the biomass as a homogeneous feedstock, the material could be viewed as a collection of ingredients, pairing those specific physical and chemical compositions to a specific purpose. The challenge then becomes developing low-cost methods to separate the unique fractions to isolate the higher value or pathway-specific components.

During a casual inspection of corn stover, one can identify several distinct plant parts including cobs, leaves, stalks, and husks. The visible differences are obvious and the differences in the physical traits are easily detected upon handling the material. These visible and physical differences often correlate to chemical differences as well. In a more detailed analysis, Li et al. (2012) described physical and chemical differences found in these anatomical fractions of corn stover. The corn stalk rind had good fiber characteristics making it a strong candidate for paper feedstock. The stalk rind was high in lignin and cellulose but low in hemicellulose. They also found that the stalk rind was high in potassium and chlorine, while the leaf and stalk pith were higher in silica. Lacey et al. (2016) found highly diverse ash concentrations throughout the anatomical fractions of corn stover with leaves containing nearly seven times as much ash as the cobs (Table 2a). These differences also were seen in the different size fractions of the same biomass (Table 2b). The differences in composition can even be detected at the tissue and cellular levels. Sun et al. (2011) identified tissue-specific chemical differences in lignin and cellulose abundance in corn stover sclerenchyma, tracheids, epidermal cells, bundle sheath cells, and parenchyma cells.

These differences are not isolated from the visibly obvious differences that can be found in corn stover. Similar differences in physical and chemical characteristics have been identified in many other biomass feedstocks. Hu et al. (2011b) performed a detailed chemical analysis of the anatomical fractions

Table 1. Compositions of anatomical fractions found in the literature. (Garlock et al., 2009; Hu et al., 2011a; Jin et al., 2013; Motte et al., 2014; Normark et al., 2014; Papatheofanous et al., 1998; Shinners et al., 2007; Sun et al., 2014; Zeng et al., 2012; Zhang et al., 2014a.)

Feedstock	Fraction	Glucan	Xylan	Arabinan	Galactan	Mannan	Lignin	Acid-insoluble lignin	Acid Soluble lignin	Ash	Other	Source
Corn Stover	Early Leaves	27.5	17.8					13.2		7.3	34.2	Garlock et al., 2009
	Early Stem	35.1	19					14.9		3.4	27.6	
	Late Leaves	35.3	21.8					13.6		6	23.3	
	Late Stem	37.8	23.6					16.9		2.4	19.3	
	Late Husk	39	26.5					11.6		2.1	20.8	
	Late Cob	27.5	32.3					25		1.1	13.3	
Corn Stover	Rind	38.1	20.5				19.8					Sun et al., 2014
	Husk	29.2	25.4				15.8					
	Leaf	31.6	21.2				18.8					
	Pith	37.5	20.5				14.3					
Corn Stover	Leaf	34.3	25.1	4.5	*		19	16.6	2.4	5.9		Zeng et al., 2012
	Rind	41.6	21.9	2	*		27.4	25.4	2	3.1		
	Pith	42.9	22.9	3.3	*		18.7	16.4	2.3	4.4		
Corn Stover	Cob	31.9	28.3	2.7	1.4	1	12.1					Shinners et al., 2007
	Husk	33.2	23.7	3.7	2	0.6	11.4					
	Leaf	31.2	18.8	3.4	1.8	0.6	9.7					
Wheat Straw	Whole Straw	32.1	29.2**				16.4			4.8	17.5	Papatheofanous et al., 1998
	Internodes	34.7	29.4**				17.9			4.3	13.7	
Wheat Straw	Internodes	32	17	1			10					Motte et al., 2014
	Leaves	25	17	2			6					

Table 1 contd. ...

...Table 1 contd.

Feedstock	Fraction	Glucan	Xylan	Arabinan	Galactan	Mannan	Lignin	Acid-insoluble lignin	Acid Soluble lignin	Ash	Other	Source
Wheat Straw	Nodes	26	17	3			8					Jin et al., 2013
	Chaff	29	18	3			7					
	Rachis	28	21	2			7					
Wheat Straw	Whole Straw	37.9	19.9	2.9			21.5	19	2.5	8		
	Stem	40.4	20.1	2.5			23	20.8	2.2	7.6		
	Leaf	34.9	19.8	3			18.1	15.4	2.7	11.3		
Wheat Straw	Leaf	32.4	18.5	3.4	1.3	1.1		23.2		7.8		Zhang et al., 2014
	Stem	41	17.7	2.1	0.5	0.4		28.1		2.5		
Switchgrass	Leaves	30.7	15.2	2.9	1.5			19.5	3.5			Hu et al., 2011
	Internodes	42.6	20.7	1.7	0.7			20.2	1.8			
	Nodes	40.5	26.8	2.8	1			24.7	2.1			
	Seedhead	36.6	19.8	2.9	1.5			23.3	4.1			
Scots Pine	Juvenile Heartwood	42.7	6.6	2	3.1	11.8	29.5	27.9	1.6	0.2		Normark et al., 2014
	Mature Heartwood	42.2	5.3	1.8	3	12.1	29.1	27.7	1.4	0.2		
	Juvenile Sapwood	39.7	6.2	1.9	2.8	10.5	27.4	25.8	1.7	0.3		
	Mature Sapwood	41.8	5.5	1.7	2.3	14	28.4	26.9	1.5	0.2		
	Bark	41.8	5.4	2.4	2.8	11.7	30	28.2	1.9	0.9		
	Top Parts	41.4	6.6	2	3.4	11.4	29.7	28.1	1.6	0.3		
	Knotwood	38.2	6.6	2.2	4.1	11.9	31.5	30	1.6	0.3		

*Included in Xylan total.
**Total Hemicellulose.

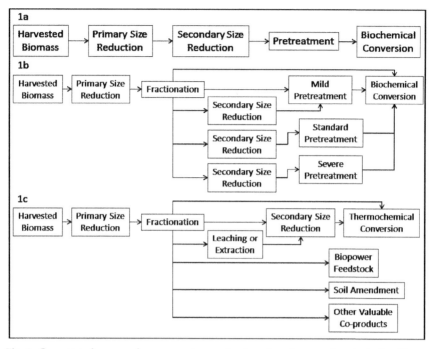

Fig. 1. Current and proposed preprocessing, pretreatment, and conversion strategies. Figure 1a shows the current approach to biomass processing and biochemical conversion, where biomass is treated as a homogeneous feedstock and subjected to a single set of preprocessing, pretreatment, and conversion conditions. Figure 1b shows the proposed "fractional conversion" approach to biomass processing and biochemical conversion, where fractionation is used to isolate physical and chemical differences in the biomass and specific preprocessing, pretreatment, and conversion strategies are developed to maximize efficiency and conversion yields from each individual fraction. Figure 1c shows the proposed "fractional conversion" approach to biomass preprocessing and thermochemical conversion. In this example, some fractions, based on their physical or chemical characteristics, have been identified as valuable co-products and are sold to other markets.

Table 2a. Differences in ash content found in the hand separated anatomical fractions of corn stover. Adapted from Lacey et al., 2016.

Anatomical Fraction	% of Plant Mass*	Ash content
Leaves	20%	10.4 ± 0.5%
Sheath	10%	6.9 ± 0.2%
Nodes	10%	4.0 ± 0.3%
Husk	10%	3.7 ± 0.2%
Internode	30%	3.5 ± 0.2%
Cob	20%	1.5 ± 0.2%
Composite Total	100%	4.9 ± 0.1%

*Estimated from values found in the literature.

Table 2b. Differences in ash content found in the size fractions of ground corn stover. Baled corn stover was grounded in a hammer mill fitted with a 6″ screen. Adapted from Lacey et al., 2016.

Size fraction	% of Total mass	Ash content
9.5 mm	31.0 ± 4.2%	3.5 ± 0.4%
6.0 mm	18.8 ± 3.7%	4.6 ± 0.9%
2.0 mm	27.5 ± 1.4%	5.1 ± 0.5%
0.6 mm	14.4 ± 6.0%	5.9 ± 0.2%
0.425 mm	3.5 ± 0.9%	4.8 ± 0.7%
0.150 mm	3.9 ± 0.9%	7.5 ± 2.5%
< 0.150 mm (pan)	0.8 ± 0.2%	22.5 ± 10.8%

of switchgrass. They found differences between leaves and internodes in ash content and elemental ash content, solvent extractives, and carbohydrate content. Additionally, differences were observed in lignin concentrations and the molecular characteristics of the lignin found in different plant parts. Hu et al. (2010) found that switchgrass stems were low in ash and high in lignin, while leaves were high in ash and hemicellulose and low in cellulose. Recognizing the potential for enhanced biochemical conversions, they suggested anatomical fraction specific pretreatment chemistries as a way to obtain maximum yields from the different fractions. Park et al. (2003) observed that most of the silica in rice husks was present in the outer epidermal cells. Howard (1973) found that ash was spread differentially throughout slash pine tree parts. Needles were found to have the highest ash content, while stemwood ash content was very low. Bark was highest in lignin content and lowest in cellulose content, while the stemwood was highest in cellulose content. Lacey et al. (2015) found similar patterns of ash distribution in pine forest residues, with pine needles containing over five times as much ash as the clean stemwood. The low ash content of stemwood is the reason that clean stemwood is the feedstock of choice for many thermochemical conversion platforms.

While the differences in physical and chemical properties of anatomical and tissue fractions of biomass feedstocks may be obvious or detectable, these differences must be worth exploiting before economic justifications can be made to spend the time and energy to develop the fractionation methods and perform the necessary separations. A discussion follows in subsequent sections that will describe the targets and benefits of isolating these sections to support the development or implementation of mechanical separations, and a discussion regarding the developed and developing mechanical fractionation technologies that could be used in an industrial setting. This includes implementation in a feedstock preprocessing depot

(Eranki et al., 2011; Lamers et al., 2015) or at the biorefinery, to fractionate biomass into higher quality or higher value fractions.

2. The Case for Mechanical Fractionation

There are many potential benefits to mechanical fractionation of biomass feedstocks. As with any additional biomass feedstock processing step, the added cost of the processing step must be outweighed by improvements gained in downstream process costs. Those improvements could come in the way of improved material handling characteristics, improved grinding characteristics, or improved performance in pretreatment and conversion processes. The following sections will discuss the benefits that mechanical fractionation if developed and implemented could have on biomass pretreatment, enzymatic hydrolysis, and biochemical and thermochemical conversion processes.

2.1 Ash removal

Ash from biomass feedstocks comes from two primary sources: physiological ash and introduced ash. Physiological ash is a natural byproduct of plant biology. Many ash minerals are required to sustain life throughout the plant's life including macronutrients (calcium (Ca), magnesium (Mg), potassium (K), sulfur (S), nitrogen (N), phosphorous (P)), micronutrients (zinc (Zn), iron (Fe), manganese (Mn), copper (Cu), chlorine (Cl), boron (B), molybdenum (Mo), nickel (Ni)) (Kochian, 2000; Taiz and Zeiger, 2002; Tao et al., 2012), and non-essential beneficial elements (sodium (Na), silicon (Si)) (Currie and Perry, 2007; Epstein, 1999; Ma and Yamaji, 2006; Ma and Yamaji, 2008; Richmond and Sussman, 2003; Subbarao et al., 2003; Taiz and Zeiger, 2002). Physiological ash is typically found inside biomass cells and may be unevenly distributed within the different tissues and anatomical features of the plant. Introduced ash is not a natural part of the biomass feedstock but enters the feedstock supply chain during the harvest and collection process and storage. Introduced ash typically consists of soil and rocks and is enriched in Si, aluminum (Al), Fe, Na, and Ti (Lacey et al., 2015; Vassilev et al., 2010; Vassilev et al., 2012). The amount of introduced ash is highly dependent upon harvest and collection methods. Stover that is collected using a multi-pass method is well known to contain higher ash concentrations compared to stover collected in a single pass manner (Shinners et al., 2012). Introduced ash is primarily found on the surface of the biomass pieces and may or may not be adhered to the surface of the particles.

Ash is an undesirable component of biomass in every conversion process. Conversion processes benefit from biomass feedstocks with low ash content for several reasons. At best, ash represents unconvertable material, and at worst, ash destroys equipment and degrades the conversion products. Ash, especially in herbaceous feedstocks, can comprise over 10% of the biomass. If not removed, this unconvertable material must be collected, purchased, transported, processed, fed into the reactor, and ultimately disposed of when the conversion processes are completed since nothing is gained from its inclusion in the conversion process. Additional feedstock must also be purchased to replace this fraction of the unconvertable material, increasing process costs further. In biochemical conversions, certain elements of ash can have a buffering effect on the reaction conditions, requiring more acid or base to be added to the reaction mixture to reach the ideal pretreatment or conversion conditions (Weiss et al., 2010). The addition of more acid or base also increases the amount of waste ash that must be discarded.

In thermochemical bioenergy reactions, ash still represents an unconvertable fraction of the biomass, however specific elements of ash can be very harmful in conversion processes. Elevated bulk ash concentrations during pyrolysis reactions lead to decreased process yields (Das et al., 2004; Fahmi et al., 2008). During pyrolysis, trace levels of K, Na, Ca, and Mg can catalyze the thermal degradation of biomass feedstocks to undesirable light gasses, decreasing the yield of the valuable primary conversion products (Azeez et al., 2010; Carpenter et al., 2014; Evans and Milne, 1987). Potassium has been found to depolymerize and fragment products of pyrolysis reactions (Fuentes et al., 2008). Biomass combustion reactions are significantly impacted by the presence of ash. Increased ash concentrations in combustion reactions lead to decreased energy conversion and destruction of the equipment. Slagging and fouling of the combustion reactors regularly occur due to the presence of the most common ash elements including Si, K, Na, S, Cl, P, Ca, Mg, and Fe (Baxter, 1993; Jenkins et al., 1998). Combinations of these elements at high temperatures enhance low temperature melting or softening of these elements and lead to their deposit on the surfaces of the reactor, impeding heat transfer. Additionally, Cl can be incorporated into the deposits where it corrodes the surface of the reactor's components (Nielsen et al., 2000; Tillman et al., 2009).

Ash is not a desirable component in biochemical or thermochemical reactions and mechanical fractionation represents an opportunity to remove ash prior to pretreatment and conversion to improve reaction conditions, preserve conversion products, enhance conversion yields, and maximize economic gains from the conversion of biomass to bioenergy.

2.2 Improved pretreatment, enzymatic hydrolysis, and conversion

Mechanical separations also present an opportunity to generate fractions of biomass with specific chemical and physical traits that could be advantageous in pretreatment, hydrolysis, or conversion processes. Some anatomical and tissue fractions of biomass feedstocks can be ideally suited for specific reactions or conditions within those reactions, and understanding these benefits can improve biomass performance in pretreatment, enzymatic hydrolysis, and conversion reactions.

2.2.1 Pretreatment and enzymatic hydrolysis improvements

Typically, the first step in the biochemical conversion of raw biomass to energy products involves a pretreatment step to break up the lignocellulose matrix of the biomass and release some of the sugars followed by an enzymatic hydrolysis reaction to release the remaining sugars. Anatomical fractions of corn stover have been found to perform differently in pretreatment and conversion reactions and some of these differences are very significant. Crofcheck and Montross (2004) found dramatic differences in glucose released from non-pretreated anatomical fractions of corn stover during enzymatic hydrolysis with cellulase. Cobs, leaves, and husks produced over 300% more glucose than stalks during hydrolysis. Leaves averaged 91% conversion efficiency even without a pretreatment step, compared to 63% and 33% for cobs and stalks, respectively. This would indicate that leaves may require little to no pretreatment during the conversion process, and cobs could benefit from a reduced severity pretreatment. Because of the possible usage of the cost saving tissue-specific pretreatment conditions, they indicated that fractionation of stover could increase the value of the feedstock for glucose production. In another study, these same authors suggested that the collection of only those parts of the stover with the highest glucose potential could produce a higher quality feedstock and lead to an improved glucose production in the pretreatment systems (Crofcheck and Montross, 2004). This time, using a mild sodium hydroxide pretreatment followed by enzymatic hydrolysis, cobs released significantly more glucose than any other part of the plant. They suggested that to collect the highest quality stover while maintaining soil health, all of the cobs and 74% of the leaves and husks should be collected during harvest. Duguid et al. (2009) found that cobs, husks, and leaves responded very well to room temperature acid and alkali pretreatment followed by enzymatic hydrolysis, with husks releasing over 90% of the available glucan compared to 45% of similarly treated stalks. They suggested that using fractionation within a biorefinery could minimize the costs of ethanol production through the customization of pretreatment and enzymatic hydrolysis reactions for specific anatomical fractions. Garlock et al. (2009) suggest using a selective harvest strategy to increase glucose production efficiency finding that husks and leaves yielded the most glucose in an ammonia fiber expansion (AFEX) pretreatment followed by enzymatic hydrolysis. Using aqueous ammonia pretreatment followed by enzymatic hydrolysis with cellulose, Sun et al. (2014) found that the stover rind released the lowest amount of sugars due to a high content of crystalline cellulose in the rind. They showed that by adding xylanase to their hydrolysis mixture, sugar yields from stover rind could be improved. Zeng et al. (2011) found that hot water treatment followed by enzymatic hydrolysis of corn stover anatomical fractions showed different conversion efficiencies, with stalk rind releasing only about 50% of its glucose compared to 90% for stalk pith. They also identified chemical differences between fractions, with rind and pith being highest in cellulose and leaves being highest in hemicellulose. The rind was also highest in lignin and lowest in ash. They suggested anatomical fractionation as a method to reduce enzyme cost and improve glucose yield in pretreatment and hydrolysis.

Pine and pine forest residues represent a plentiful biomass source that could be fractionated for enhanced pretreatment and enzymatic hydrolysis reactions. Normark et al. (2014) found that different fractions of pine had some chemical differences and behaved differently in pretreatment reactions. Knotwood was found to be high in extractives, while the bark performed well in hydrolysis reactions without pretreatment.

Although the visible differences of anatomical fractions in herbaceous feedstocks such as wheat stover and switchgrass are not as obvious, there are still physical and chemical differences between their anatomical and tissue fractions that can affect pretreatment reactions. Zhang et al. (2014a) found that under low-severity pretreatment conditions, increase in the leaf stem ratio led to an improvement in enzymatic hydrolysis of wheat straw. Increased enzyme dosage was most effective if the leaf/stem ratio was greater than 50%. They suggested employing low-severity pretreatment conditions when pretreating wheat straw feedstocks with high leaf stem ratios. Zhang et al. (2014b) found that leaves of wheat stover were far more degradable than the stems, likely due to higher glucan accessibility in the leaves. Glucan in the stems became more accessible after hydrothermal pretreatment, while glucan accessibility in the leaves was unaffected by the same treatment. This finding indicates that leaves may need little to no pretreatment prior to enzymatic hydrolysis. Jin et al. (2013) found that leaves pretreated with sodium carbonate released about 16% more sugar than the stem under similar conditions.

2.2.2 Conversion improvements

Utilization of mechanical fractionation in conversion reactions has shown to improve product yields and to lower conversion costs. Tests involving biochemical conversion reactions showed increased performance when using specific anatomical fractions, primarily the leaves, husks, and cobs. Hansey et al. (2010) found strong evidence that the leaves of corn stover performed far better as a feedstock due to their high digestibility and conversion efficiency. They suggested a selective harvest scheme collecting only the stover leaves that would improve conversion yields, but acknowledged that the amount of leaves available might not be adequate to feed conversion reactions economically. Duguid et al. (2007) also suggested a selective harvest strategy for wheat stover, again collecting only the leaves. They suggested that this strategy would increase conversion efficiencies while also protecting the soil by leaving some biomass on the ground. Additional studies on selective harvest would be needed to determine whether the value gained from these enhanced conversions exceeded the value lost due to the collection of less biomass per acre.

In anaerobic digestion reactions, corn husks proved to be the best material for methane and biogas production, utilizing about 70% of the total dry matter, compared to 53% dry matter utilization in the stalks (Menardo et al., 2015). Motte et al. (2014) reported a strong correlation between wheat straw anatomical fractions and their performance in anaerobic digestion. They found that wheat straw with higher contents of leaves, rachis, and chaffs could increase methane yields. They suggested either using a wheat variety with a high leaf/stem ratio or some fractionation step before conversion to increase the proportion of leaves in the feedstock.

Anatomical and different tissue fractions can perform differently in thermochemical reactions as well. Lizotte et al. (2015) suggested that a partial cob and husk harvest strategy would yield less biomass per acre. However, the quality of the fuel derived from these parts of the stover would be considerably higher due to the low ash content of these anatomical fractions. Medic et al. (2012) found that the anatomical fractions of corn stover performed differently in torrefaction reactions, leading to differing dry matter loss and final energy values. They suggested that these differences were related to actual physical differences and fiber composition in the plant tissues. Wang and Dibdiakova (2014) found that the melting point of ash elements in Norway spruce branches and twigs were 100–200°C lower than the ash elements from stem wood and bark. Utilization of these parts of the trees in combustion reactions would present specific challenges. Lengowski et al. (2014) found that pine needles were a good candidate resource for cooking briquettes and food additives.

3. Mechanical Fractionation Methods and Applications

Based on the present information, there is a clear role of mechanical fractionation in improving the performance of biomass feedstocks in bioenergy conversion systems. There are many beneficial mechanical fractionations for biomass quality improvement that are currently under development; however, few have been adopted by industry.

3.1 Size reduction and screening

The primary mechanical fractionation technique used currently is size fractionation. Different anatomical fractions and plant tissues have diverse physical properties. These differences become apparent when plant tissue is placed in a grinder or shredder to reduce the particle size as different tissues break apart differently. Physical properties impact deconstruction properties and, therefore different tissues break down differently. Lacey et al. (2016) investigated particle size distributions produced by anatomical fractions of corn stover when the fractions were knife milled. They found that cobs were most likely to remain in larger pieces, while more of the husks were found in the smaller size fractions. Ash was also found to increase as particle size decreased. Particles smaller than 0.150 mm contained an average of 22% ash, while the largest particle size fraction (> 9.5 mm) contained only 3.5% ash on average. They suggested a combination of size fractionation and anatomical fractionation as an effective method to reduce the ash content of the feedstock. Chundawat et al. (2007) found that the particle size separations of ground corn stover could affect AFEX pretreatment and enzyme digestibility. Larger size fractions were found to be enriched in cobs and stalks and did not perform well in hydrolysis reactions. The smaller size fractions were rich in leaves and husks, and pretreatment methods were very effective for these fractions. They suggested using fractionation methods to separate feedstocks into two different product streams: easily hydrolyzable (leaves and husks) and recalcitrant (stalks and cobs) fractions; and using more or less severe pretreatment conditions customized to each fraction. Tamaki and Mazza (2010) found that the particle size was directly correlated to some chemical characteristics. As particle size increased, cellulose and hemicellulose increased in concentration and protein and extractives decreased in concentrations. Miranda et al. (2012) reported that the milling process of the barks of pine trees was not uniform due to different tissue, anatomical, physical, and chemical properties of the bark. The resulting size fractions were chemically unique, with the fine particle-containing higher concentrations of ash and extractives and the larger particles containing more cellulose. Bridgeman et al. (2007) reported that the process of size reduction did not yield particle size fractions with uniform physical or chemical characteristics. Cellulose, hemicellulose, and lignin tended to have a higher concentration of larger particle sizes, while ash was most concentrated in the smallest particle size fractions. Silva et al. (2011) found chemical differences between anatomical fractions of wheat straw with leaves, nodes, and chaff having more ash and proteins than the stems. Stems were highest in cellulose and lignin. Upon grinding and sieving, they found that small particle fractions were typically higher in ash and protein concentrations, while larger size fractions were higher in cellulose. They suggested that sieving could give rise to different product streams with different applications and that anatomical fractionation could compose another step in biomass fractionation that would improve the effectiveness of conversions. Zhang et al. (2012) used size fractionation to separate the bark from the wood in Douglas fir resulting in an effective separation of lignin from cellulose. Bark, lignin, and ash were enriched in the smaller particle sizes. They found that this separation enhanced the enzymatic digestibility of the larger particle sizes that were enriched in stemwood.

The type of size reduction and screening equipment also plays a significant role in how biomass deconstructs. As an example, corn husks, when running through a knife mill, tend to be more evenly distributed among size fractions (Lacey et al. 2016). However, when running through a hammer mill, corn husks tend to remain in the mill for longer periods of time. As a result, the fibers of the husks remain long and intact and the tissues between the fibers become pulverized and end up in the smallest size fractions (Lacey, personal communication). Papatheofanous et al. (1998) used a disk mill and size selection to fractionate wheat straw into anatomical fractions. They found that the chips, or larger particles, were primarily comprised of the internodes, while the meal or fines contained most of the nodes and leaves. The internode fraction was enriched in cellulose and lignin and contained less ash than the fines and unfractionated material. A trommel screen was used by Dukes et al. (2013) to remove fine particles from forest residues. As a result, the energy density was increased by selectively removing higher ash material that was concentrated in the fine particles. Using this process, the energy density was increased and ash content was decreased. Cutshall et al. (2012) investigated size selection to remove fines that were high in ash as a method to improve quality in woody biomass. Their methods were effective at reducing ash content; however economic models showed that the use of the screen was not always economically advantageous.

3.2 Density separations

Separation methods based on particle density, morphology, or surface area are not well-developed for biomass feedstocks although, as noted earlier, these techniques are well-developed for the food industry and have been used since ancient times during harvest. The natural variation found in tissue and anatomical fractions affords the opportunity to separate biomass feedstocks based on their differing densities and particle characteristics. Some work has been done with density separations on biomass feedstocks to improve the quality of the feedstocks and create unique fractions of feedstocks that may have different preferred end uses. Bootsma and Shanks (2005) used air classification to separate the pith from the fiber of finely ground corn stover. The separation was effective; however, their experiments found that the separated biomass performed no better in their hydrolysis reactions than the whole stover that was tested. Lacey et al. (2015) used air classification to remove ash from pine forest residues. They found that the lightest fractions of their feedstock contained the highest concentrations of ash. By strategically removing the highest ash fractions from the bulk of the biomass, they were able to remove 40% of the ash while only removing about 7% of the total biomass. Air classification had concentrated the ash into a very small fraction. They reported that air classification could be performed on biomass feedstocks for as little as $2.23 per ton.

Using a combination of mechanical fractionation techniques, Thompson et al. (2016) reported that air classification and size fractionation were used to mitigate the high ash concentrations found in herbaceous feedstocks including corn stover, switchgrass, and residential lawn clippings. Formulation techniques were then used to create mixtures of the fractions from the various feedstocks to produce feedstock blends that met specific quality and cost specifications. They also reported that the high ash fraction that was collected would make a good biopower feedstock due to its specific ash elemental makeup.

4. The Future of Mechanical Fractionation

The differences in biomass feedstock fractions are identifiable. Some of these differences do affect conversion efficiencies and methods exist to begin using mechanical fractionation to meet cost, quality, and conversion specifications.

It is critical that the improvements to the feedstock are justified by added value. Thus far, the implementation of some advanced techniques has been limited due to excessive process costs (i.e., energy requirements, disposal costs) or expensive equipment (i.e., pressure vessels, acid or base resistant reactors, advanced electronics). However, given the right conditions, feedstock or high-value co-product, some of the advanced techniques could be employed to improve process economics for biomass feedstocks.

Optical sorting is one such technology that is used widely in the mining, food, and pharmaceutical industries; however, it has not been tested with biomass feedstocks. Optical sorting is an automated process that uses the optical properties of particles to determine whether to "accept" or "reject" that individual particle. An image is taken of each particle moving down a conveyor, the image is analyzed by a computer, and depending upon the parameters set by the user, the particle is either allowed to proceed to the "accept" stream or the particle is hit with a jet of air and diverted into the "reject" stream. These sorters can be set to select particles based on size, shape, color, UV or IR characteristics, hyperspectral characteristics, or X-ray characteristics. Examples of some of the more simple optical sorting capabilities could include separation of bark from stemwood based on color or isolation of pine needles based on shape. Advanced separations could include the detection of specific chemical signatures such as carbohydrate content or lignin content, using UV, IR, or hyperspectral imaging. The capabilities of this technology in the bioenergy industry are yet to be fully realized.

5. Conclusions

While industries such as the food or mining have been developing mechanical fractionation technologies for many years, the biomass industry has not widely adopted mechanical fractionation as a method to improve biomass quality for enhanced conversion reactions. Much of the bioenergy industry is still in the developmental stages and most processes have the luxury of only working with the most pristine feedstocks because there is little to no competition for some of these. As the number of operational thermochemical and biochemical biorefineries begin to increase and as the supply of these "ideal" feedstock streams become scarcer, there will be a need to begin to adopt mechanical fractionation techniques into the normal feedstock handling process, either in a feedstock preprocessing depot setting or at the biorefinery itself. Lower quality feedstocks will be needed to meet an increasing feedstock demand and the characteristics that make the low feedstock quality will be needed to be mitigated. Mechanical fractionation can provide effective and affordable methods to improve the quality of feedstocks, thus making less expensive feedstocks available for use in the feedstock supply chain.

Incorporating mechanical fractionation into the biomass feedstock supply chain could be added in-line with existing feedstock processing equipment. The challenge to getting mechanical fractionation adopted at the industrial level is demonstrating its effectiveness to industrial partners. This can be done using findings from many of the referenced studies and new studies and incorporating these results into economic models that show how mechanical fractionation can be implemented cost-effectively. The groundwork has begun on this front (Lacey et al., 2015; Thompson et al., 2016) and it is anticipated that additional studies will further justify using mechanical fractionation in biomass feedstock depot settings or industrial biomass conversion facilities.

References

Azeez, A. M., D. Meier, J. Odermatt and T. Willner. 2010. Fast pyrolysis of African and European lignocellulosic biomasses using Py-GC/MS and fluidized bed reactor. Energy & Fuels 24: 2078–2085.

Baxter, L. L. 1993. Ash deposition during biomass and coal combustion—a mechanistic approach. Biomass & Bioenergy 4: 85–102.

Bootsma, J. A. and B. H. Shanks. 2005. Hydrolysis characteristics of tissue fractions resulting from mechanical separation of corn stover. Applied Biochemistry and Biotechnology 125: 27–39.

Bridgeman, T. G., L. I. Darvell, J. M. Jones, P. T. Williams, R. Fahmi, A. V. Bridgwater, T. Barraclough, I. Shield, N. Yates, S. C. Thain and I. S. Donnison. 2007. Influence of particle size on the analytical and chemical properties of two energy crops. Fuel 86: 60–72.

Carpenter, D., T. L. Westover, S. Czernik and W. Jablonski. 2014. Biomass feedstocks for renewable fuel production: a review of the impacts of feedstock and pretreatment on the yield and product distribution of fast pyrolysis bio-oils and vapors. Green Chemistry 16: 384–406.

Chundawat, S. P. S., B. Venkatesh and B. E. Dale. 2007. Effect of particle size based separation of milled corn stover on AFEX pretreatment and enzymatic digestibility. Biotechnology and Bioengineering 96: 219–231.

Crofcheck, C. L. and M. D. Montross. 2004. Effect of stover fraction on glucose production using enzymatic hydrolysis. Transactions of the ASAE 47: 841–844.

Currie, H. A. and C. C. Perry. 2007. Silica in plants: Biological, biochemical and chemical studies. Annals of Botany 100: 1383–1389.

Cutshall, J. B., S. A. Baker and W. D. Greene. 2012. Improving woody biomass feedstock logistics by reducing ash and moisture content. *In* 35th Council on Forest Engineering Annual Meeting, New Bern, NC.

Das, P., A. Ganesh and P. Wangikar. 2004. Influence of pretreatment for deashing of sugarcane bagasse on pyrolysis products. Biomass & Bioenergy 27: 445–457.

Duguid, K. B., M. D. Montross, C. W. Radtke, C. L. Crofcheck, S. A. Shearer and R. L. Hoskinson. 2007. Screening for sugar and ethanol processing characteristics from anatomical fractions of wheat stover. Biomass & Bioenergy 31: 585–592.

Duguid, K. B., M. D. Montross, C. W. Radtke, C. L. Crofcheck, L. M. Wendt and S. A. Shearer. 2009. Effect of anatomical fractionation on the enzymatic hydrolysis of acid and alkaline pretreated corn stover. Bioresource Technology 100: 5189–5195.

Dukes, C. C., S. A. Baker and W. D. Greene. 2013. In-wood grinding and screening of forest residues for biomass feedstock applications. Biomass & Bioenergy 54: 18–26.

Epstein, E. 1999. Silicon. Annual Review of Plant Physiology and Plant Molecular Biology 50: 641–664.

Eranki, P. L., B. D. Bals and B. E. Dale. 2011. Advanced regional biomass processing depots: a key to the logistical challenges of the cellulosic biofuel industry. Biofuels Bioproducts & Biorefining-Biofpr. 5: 621–630.

Evans, R. J. and T. A. Milne. 1987. Molecular characterization of the pyrolysis of biomass. 1. Fundamentals. Energy & Fuels 1: 123–137.

Fahmi, R., A. Bridgwater, I. Donnison, N. Yates and J. M. Jones. 2008. The effect of lignin and inorganic species in biomass on pyrolysis oil yields, quality and stability. Fuel 87: 1230–1240.

Fuentes, M. E., D. J. Nowakowski, M. L. Kubacki, J. M. Cove, T. G. Bridgeman and J. M. Jones. 2008. Survey of influence of biomass mineral matter in thermochemical conversion of short rotation willow coppice. Journal of the Energy Institute 81: 234–241.

Garlock, R. J., S. P. S. Chundawat, V. Balan and B. E. Dale. 2009. Optimizing harvest of corn stover fractions based on overall sugar yields following ammonia fiber expansion pretreatment and enzymatic hydrolysis. Biotechnology for Biofuels 2.

Hansey, C. N., A. J. Lorenz and N. de Leon. 2010. Cell Wall Composition and ruminant digestibility of various maize tissues across development. Bioenergy Research 3: 28–37.

Howard, E. T. 1973. Physical and chemical properties of slash pine tree parts. Wood Science 5: 312–317.

Hu, Z., R. Sykes, M. F. Davis, E. C. Brummer and A. J. Ragauskas. 2010. Chemical profiles of switchgrass. Bioresource Technology 101: 3253–3257.

Hu, Z., M. Foston and A. J. Ragauskas. 2011a. Comparative studies on hydrothermal pretreatment and enzymatic saccharification of leaves and internodes of alamo switchgrass. Bioresource Technology 102: 7224–7228.

Hu, Z., M. B. Foston and A. J. Ragauskas. 2011b. Biomass characterization of morphological portions of alamo switchgrass. Journal of Agricultural and Food Chemistry 59: 7765–7772.

Jenkins, B. M., L. L. Baxter, T. R. Miles and T. R. Miles. 1998. Combustion properties of biomass. Fuel Processing Technology 54: 17–46.

Jin, Y. C., T. Huang, W. H. Geng and L. F. Yang. 2013. Comparison of sodium carbonate pretreatment for enzymatic hydrolysis of wheat straw stem and leaf to produce fermentable sugars. Bioresource Technology 137: 294–301.

Kochian, L. V. 2000. Molecular physiology of mineral nutrient acquisition, transport, and utilization. pp. 1204–1249. *In*: W. G. R. J. B. Buchanan (ed.). Biochemistry & Molecular Biology of Plants. American Society of Plant Physiologists, Rockville, MD.

Lacey, J., J. Aston, T. Westover, R. Cherry and D. Thompson. 2015. Removal of introduced inorganic content from chipped forest residues via air classification. Fuel 160: 265–273.

Lacey, J. A., R. M. Emerson, D. N. Thompson and T. L. Westover. 2016. Ash reduction strategies in corn stover facilitated by anatomical and size fractionation. Biomass and Bioenergy 90: 173–180.

Lamers, P., M. S. Roni, J. S. Tumuluru, J. J. Jacobson, K. G. Cafferty, J. K. Hansen, K. Kenney, F. Teymouri and B. Bals. 2015. Techno-economic analysis of decentralized biomass processing depots. Bioresource Technology 194: 205–213.

Lengowski, E. C., S. Nisgoski, W. L. Esteves de Magalhaes, G. Capobianco, K. G. Satyanarayana, and G. I. Bolzon de Muniz. 2014. Characterization of Pinus sp. of needle to assess their possible industrial applications. Journal of Biobased Materials and Bioenergy 8: 192–201.

Li, Z. Y., H. M. Zhai, Y. Zhang and L. Yu. 2012. Cell morphology and chemical characteristics of corn stover fractions. Industrial Crops and Products 37: 130–136.

Lizotte, P.-L., P. Savoie and A. De Champlain. 2015. Ash content and calorific energy of corn stover components in Eastern Canada. Energies 8: 4827–4838.

Ma, J. F. and N. Yamaji. 2006. Silicon uptake and accumulation in higher plants. Trends in Plant Science 11: 392–397.

Ma, J. F. and N. Yamaji. 2008. Functions and transport of silicon in plants. Cellular and Molecular Life Sciences 65: 3049–3057.

Medic, D., M. Darr, A. Shah and S. Rahn. 2012. The effects of particle size, different corn stover components, and gas residence time on torrefaction of corn stover. Energies 5: 1199–1214.

Menardo, S., G. Airoldi, V. Cacciatore and P. Balsari. 2015. Potential biogas and methane yield of maize stover fractions and evaluation of some possible stover harvest chains. Biosystems Engineering 129: 352–359.

Miranda, I., J. Gominho, I. Mirra and H. Pereira. 2012. Chemical characterization of barks from Picea abies and Pinus sylvestris after fractioning into different particle sizes. Industrial Crops and Products 36: 395–400.

Montross, M. D. and C. L. Crofcheck. 2004. Effect of stover fraction and storage method on glucose production during enzymatic hydrolysis. Bioresource Technology 92: 269–274.

Motte, J. C., R. Escudie, N. Beaufils, J. P. Steyer, N. Bernet, J. P. Delgenes and C. Dumas. 2014. Morphological structures of wheat straw strongly impacts its anaerobic digestion. Industrial Crops and Products 52: 695–701.

Nielsen, H. P., F. J. Frandsen, K. Dam-Johansen and L. L. Baxter. 2000. The implications of chlorine-associated corrosion on the operation of biomass-fired boilers. Progress in Energy and Combustion Science 26: 283–298.

Normark, M., S. Winestrand, T. A. Lestander and L. J. Jonsson. 2014. Analysis, pretreatment and enzymatic saccharification of different fractions of Scots pine. Bmc Biotechnology 14.

Papatheofanous, M. G., E. Billa, D. P. Koullas, B. Monties and E. G. Koukios. 1998. Optimizing multisteps mechanical-chemical fractionation of wheat straw components. Industrial Crops and Products 7: 249–256.

Park, B. D., S. G. Wi, K. H. Lee, A. P. Singh, T. H. Yoon and Y. S. Kim. 2003. Characterization of anatomical features and silica distribution in rice husk using microscopic and micro-analytical techniques. Biomass & Bioenergy 25: 319–327.

Richmond, K. E. and M. Sussman. 2003. Got silicon? The non-essential beneficial plant nutrient. Current Opinion in Plant Biology 6: 268–272.

Shinners, K. J., G. S. Adsit, B. N. Binversie, M. F. Digman, R. E. Muck and P. J. Weimer. 2007. Single-pass, split-stream harvest of corn grain and stover. Transactions of the Asabe 50: 355–363.

Shinners, K. J., R. G. Bennett and D. S. Hoffman. 2012. Single- and two-pass corn grain and stover harvesting. Transactions of the ASABE 55: 341–350.

Silva, G. G. D., S. Guilbert and X. Rouau. 2011. Successive centrifugal grinding and sieving of wheat straw. Powder Technology 208: 266–270.

Subbarao, G. V., O. Ito, W. L. Berry and R. M. Wheeler. 2003. Sodium—A functional plant nutrient. Critical Reviews in Plant Sciences 22: 391–416.

Sun, L., B. A. Simmons and S. Singh. 2011. Understanding tissue specific compositions of bioenergy feedstocks through hyperspectral raman imaging. Biotechnology and Bioengineering 108: 286–295.

Sun, Z. P., X. Y. Ge, D. L. Xin and J. H. Zhang. 2014. Hydrolyzabilities of different corn stover fractions after aqueous ammonia pretreatment. Applied Biochemistry and Biotechnology 172: 1506–1516.

Taiz, L. and E. Zeiger. 2002. Plant Physiology. Sinauer Associates, Inc., Sunderland, MA.

Tamaki, Y. and G. Mazza. 2010. Measurement of structural carbohydrates, lignins, and microcomponents of straw and shives: Effects of extractives, particle size and crop species. Industrial Crops and Products 31: 534–541.

Tao, G. C., P. Geladi, T. A. Lestander and S. J. Xiong. 2012. Biomass properties in association with plant species and assortments. II: A synthesis based on literature data for ash elements. Renewable & Sustainable Energy Reviews 16: 3507–3522.

Thompson, V. S., J. A. Lacey, D. Hartley, M. A. Jindra, J. E. Aston and D. N. Thompson. 2016. Application of air classification and formulation to manage feedstock cost, quality and availability for bioenergy. Fuel 180: 497–505.

Tillman, D. A., D. Duong and B. Miller. 2009. Chlorine in solid fuels fired in pulverized fuel boilers—sources, forms, reactions, and consequences: a literature review. Energy & Fuels 23: 3379–3391.

Vassilev, S. V., D. Baxter, L. K. Andersen and C. G. Vassileva. 2010. An overview of the chemical composition of biomass. Fuel 89: 913–933.

Vassilev, S. V., D. Baxter, L. K. Andersen, C. G. Vassileva and T. J. Morgan. 2012. An overview of the organic and inorganic phase composition of biomass. Fuel 94: 1–33.

Wang, L. and J. Dibdiakova. 2014. Characterization of ashes from different wood parts of Norway spruce tree. Chemical Engineering Transactions 37: 37–42.

Weiss, N. D., J. D. Farmer and D. J. Schell. 2010. Impact of corn stover composition on hemicellulose conversion during dilute acid pretreatment and enzymatic cellulose digestibility of the pretreated solids. Bioresource Technology 101: 674–678.

Zeng, M. J., E. Ximenes, M. R. Ladisch, N. S. Mosier, W. Vermerris, C. P. Huang and D. M. Sherman. 2012. Tissue-specific biomass recalcitrance in corn stover pretreated with liquid hot-water: Enzymatic hydrolysis (part 1). Biotechnology and Bioengineering 109: 390–397.

Zeng, Y. J., J. Conner and P. Ozias-Akins. 2011. Identification of ovule transcripts from the Apospory-Specific Genomic Region (ASGR)-carrier chromosome. Bmc Genomics 12: 15.

Zhang, C., J. Y. Zhu, R. Gleisner and J. Sessions. 2012. Fractionation of forest residues of douglas-fir for fermentable sugar production by SPORL pretreatment. Bioenergy Research 5: 978–988.

Zhang, H., J. U. Fangel, W. G. T. Willats, M. J. Selig, J. Lindedam, H. Jorgensen and C. Felby. 2014a. Assessment of leaf/stem ratio in wheat straw feedstock and impact on enzymatic conversion. Global Change Biology Bioenergy 6: 90–96.

Zhang, H., L. G. Thygesen, K. Mortensen, Z. Kadar, J. Lindedam, H. Jorgensen and C. Felby. 2014b. Structure and enzymatic accessibility of leaf and stem from wheat straw before and after hydrothermal pretreatment. Biotechnology for Biofuels 7: 11.

CHAPTER 5

Biomass Gasification and Effect of Physical Properties on Products

Sushil Adhikari,* Hyungseok Nam and Avanti Kulkarni

1. Introduction

Gasification is a thermochemical process of converting a carbonaceous fuel into a gaseous product, called producer gas, in the operating temperature ranging from 600 to 1200°C in a controlled amount of air, oxygen and/or steam. Major reactions during gasification include dehydration, devolatilization, and a series of gas, liquid and solid phase formation. Producer gas primarily consists of carbon dioxide, carbon monoxide, methane, hydrogen and a trace amount of contaminants (tar, hydrogen sulfide, carbonyl sulfide, ammonia, hydrogen cyanide, and other N and S derived species) (Abdoulmoumine et al., 2014; Basu, 2010; Carpenter et al., 2010). Although strictly speaking, synthesis gas (or syngas) is a mixture of only carbon monoxide and hydrogen, producer gas and syngas are interchangeably used to refer gaseous products from gasifiers. Herein, we will refer syngas for the gaseous products from biomass gasification. Produced syngas can be used to run gas turbines and internal combustion engines for electric power generation. In addition, syngas can be converted into liquid fuels and chemicals such as methanol, ethanol, or higher hydrocarbons through the Fischer-Tropsch process and others.

Biosystems Engineering Department, Auburn University, Auburn, Alabama.
* Corresponding author: sushil.adhikari@auburn.edu

2. Biomass Properties

Biomass feedstocks such as agriculture residues, energy crops, woody biomass, and municipal solid wastes (MSWs) are used for gasification. The choice of biomass is region specific. However, as an initial step for gasification, biomass properties need to be analyzed. The common biomass characterization methods include proximate analysis of ash, volatile combustible matter (VCM), and fixed carbon (FC) contents, ultimate (or elemental) analysis of C, H, N, and S contents, and compositional analysis to determine the amount of lignin, sugars, protein, and extractives. These analyses are required to understand the reactivity of biomass and determine the gasification operating conditions. In addition, biomass physical properties such as particle size distribution, bulk density, and particle density need to be understood for designing material handling system and selecting appropriate gasifier with which the size of biomass hopper, feeding pipe, and motor power can be determined.

Biomass feeds differ in their proximate composition, elemental composition, sugar composition, and lignin content as shown in Table 1. The proximate analysis of biomass is typically measured by ASTM E872 (VCM), and ASTM E1755 (ash) standards. Briefly, VCM is measured by placing in an electric furnace at 950°C for 7 min with bone-dry biomass sample in a platinum crucible with a cap. The percent of VCM is important content during thermal conversion process as it produces a large amount of combustible gases. FC mainly results from lignin in the biomass, which is calculated by subtracting the percent weight of VCM and ash from 100% (the initial bone-dry sample weight). Understanding the inorganic composition of ash is helpful to predict ash agglomeration, fouling, and slagging behavior during gasification process, which have been major problems resulting in an unscheduled shutdown or defluidization with a fluidizing bed reactor (Maglinao Jr. and Capareda, 2010; Liu et al., 2009). If the feedstock is composed of a large amount of alkali minerals and a small amount of silicate, a higher chance of ash fusion takes place in a gasifying reactor under a certain range of gasification temperature. There are two routes of bed agglomeration: (1) melt-induced and (2) coating-induced

Table 1. Elemental and proximate analyses, and higher heating value (HHV) of selected biomass samples on dry basis (Kulkarni et al., 2016; Nam and Capareda, 2015).

Biomass	Ultimate analysis, wt%				Proximate analysis, wt.%			
	C	H	N	O	VCM	FC	Ash	HHV, MJ/kg
Pine	51.13	7.15	0.44	40.96	84.94	14.53	0.53	20.18
Cotton stalk	43.10	5.94	0.90	49.90	76.80	16.80	6.40	16.90
Torrefied pine	58.61	6.27	0.35	34.46	75.74	23.10	1.17	23.60
Torrefied cotton stalk	60.42	4.73	1.72	16.69	48.08	35.85	16.07	23.10

agglomerations (Visser, 2004). A melt-induced agglomeration is formed with molten ash that binds two or more fluidizing medium particles, whereas the coating-induced agglomeration is caused by the bed particles coated with ash that binds together by the AAEM (alkali and alkaline earth metals) of ash that bind together in a high temperature gasification condition. The moisture content of gasification feedstock is also critical. The recommended moisture content of the biomass is lower than 10–15% for an efficient gasification. Biomass with 15% MC (moisture contents) was able to be gasified (Cross et al. 2018). And biomass less than 10% MC is a better condition for gasification. The ultimate analysis provides the elemental composition of the biomass. This information is necessary to determine an empirical chemical formula of biomass, which is required for determining the amount of air or oxidizing medium for gasification. Carbon, hydrogen, and oxygen are typically predominant elements in biomass, which can also be represented by H/C and O/C ratios. High H/C and low O/C ratios correspond to a high HHV (higher heating value) of biomass according to Boie's formula (Eq. 1) (Boie, 1953).

HHV (MJ/kg) = 0.3515 C + 1.1617 H + 0.06276 N + 0.1046 S – 0.1109 O (1)

The chemical formula for the biomass can be defined as $C_aH_bO_cN_dS_e$, where a, b, c, d, and e represent the moles of carbon, hydrogen, oxygen, nitrogen, and sulfur in the given samples, respectively. Agricultural biomass wastes have a large amount of oxygen (30–50% on dry basis) as compared to coal or peat.

Molecular Formula Calculation of Biomass

When the weight percent of the primary elements is known, the respective molar concentration is calculated by dividing the weight percent of the element with its of molecular weight.

Example 1

Calculate the molecular weight of biomass with following elemental composition on dry ash free basis: C = 46.3%, H = 6.88%, N = 1.98%, S = 0.78% and O = 44.06%.

Solution: Number of moles of C 'a' = (46.3 g/12 g/mol = 3.85 mol, where molecular weight of carbon is 12. Similarly, b, c, d and e are calculated as 6.88, 0.14, 0.02 and 2.75. Thus, molecular formula of the biomass sample is: $C_{3.85}H_{6.88}O_{2.75}N_{0.14}S_{0.02}$ (or $CH_{1.79}O_{0.71}N_{0.04}S_{0.005}$)

The preferred physical properties such as particle size and density of biomass are different depending on the type of gasifiers used. For example, downdraft and updraft gasifiers can usually handle biomass such as woodchip and medium/high density compared to raw lignocellulosic biomass (low density). It is difficult to feed low-density biomass such as ground corn stover or wheat straw in downdraft and updraft gasifiers which easily lead to bridging during in-feed. To overcome this issue, pelletization is often required in size from 20 mm to 80 mm. Typically, the size reduction and pelletization are performed together with low-density biomass such as agricultural residues and herbaceous biomass (see Sections 4 and 7 for further explanation). The particle size and densification of the feedstock is especially critical to a fluidized bed gasifier for their proper fluidization with the gasification bed materials, which will be further explained in Section 4.

The syngas composition of biomass gasification varies depending on a number of factors such as biomass type, temperature, particle size, gasifier type, and an oxidizing agent. Table 2 shows the gasification syngas

Table 2. Performance of four types of biomass samples on syngas composition under similar conditions (Carpenter et al., 2010).

	Corn stover	Vermont wood	Wheat straw	Switchgrass
CO, Vol %	24.7	23.5	27.5	33.2
CO_2, Vol %	23.7	24.0	22.0	19.4
CH_4, Vol %	15.3	15.5	16.3	17
H_2, Vol %	26.9	28.6	25.4	23.5
C_n, Vol %	1.07	28.6	25.4	23.5
Gas yield (kg/kg of feed)	0.54	0.74	0.54	0.62

compositions from agriculture residues, woody biomass wastes, and energy crops under similar operating conditions, which indicate that the different properties of biomass result in the different syngas compositions.

3. Gasifying Agent

Gasifying agents such as oxygen, air, and steam play a very important role in syngas production. Since air is widely and freely available, it is mostly used for many gasification studies. The amount of oxidizing agent needed for gasification is defined by an equivalence ratio (ER), where ER is defined as a ratio of supplied air to stoichiometric air required for burning unit kg of biomass as shown in Eq. 2 (Basu, 2010). Also, cold gas efficiency (Eq. 3) and carbon conversion efficiency (Eq. 4) are important parameters to evaluate the efficacy of the gasification process.

$$ER = \frac{(\frac{Air}{Fuel})_{actual}}{(\frac{Air}{Fuel})_{stoichiometric}} \tag{2}$$

Cold gas efficiency (CGE), η_{cold} (%)

$$\eta_{cold} = \frac{M_{syngas} \times LHV_{syngas}}{M_{biomass} \times LHV_{biomass}} \qquad (3)$$

M_j = mass flow rate (kg/min) where j *is* syngas or biomass

Carbon conversion efficiency (CCE), η_{carbon} (%)

$$\eta_{carbon} = \frac{12}{V_m} \times f_{syngas} \times \frac{Y_{CO} + Y_{CO_2} + Y_{CH_4} + 2 \times (Y_{C_2H_2} + Y_{2CH_4})}{C\% \times M_{biomass}} \qquad (4)$$

Y_i = %mole (v/v) of syngas of each gas component where i is each combustible gas
$C\%$ = the mass percentage of carbon in the ultimate analysis of biomass fuel in dry basis
$M_{biomass}$ = mass flow rate (kg/min)
f_{syngas} = syngas volumetric flow rate (Nm³/min)

Superheated steam is frequently used as a gasifying medium to produce a high amount of H_2 gas. The quantity of steam used for gasification process can be expressed as steam to carbon ratio (S/C) or steam to biomass ratio (S/B). An increase in the S/B ratio increases total gas yield and H_2, while reducing CO and CH_4 (Gil et al., 1999). However, an excess amount of steam escapes without any reaction, therefore it is critical to supply the right amount of steam during gasification. Table 3 shows the comparative experiments of air, steam-O_2, and steam gasification using a bubbling fluidized bed gasifier. The steam-O_2 gasification produces the

Table 3. Comparison among performance of air, steam, and steam-oxygen as gasifying media (Gil et al., 1999).

	Air	Steam-O_2	Steam
ER	0.18–0.45	0	0.24–0.51
S/B	0.08–0.66	0.53–1.10	0.48–1.11
Temperature, °C	780–830	785–830	750–780
Gas composition, vol %			
CO	9.9–22.4	17–32	42.5–52.0
CO_2	9.0–19.4	13–17	14.4–36.3
CH_4	2.2–6.2	7–12	6.0–7.5
H_2	5.0–16.3	38–56	13.8–31.7
C_2H_n	0.2–3.3	2.1–2.3	2.5–3.6
Steam	11–34	52–60	38–61
N_2	41.6–61.6	0	0
LHV MJ/Nm³	3.7–8.4	12.2–13.8	10.3–13.5

highest amount of combustible gases (CO, CH_4, and H_2) resulting in the highest LHV (lower heating value) followed by steam and air gasification. In contrast, air gasification produces the lowest concentration of CO, H_2, and CH_4, which accordingly leads to the lowest LHV that is mainly due to the syngas dilution with N_2 gas. In addition, the lowest amount of C_2 gases indicates that the air gasification produces the cleanest (least amount of tar) syngas (Abdoulmoumine et al., 2014; Kulkarni et al., 2016). Typically, a higher amount of tar is generated from steam gasification. However, the presence of O_2 during steam gasification (steam-O_2 gasification) helps cracking heavy hydrocarbon gases (tar), which accordingly leads to lower C_2 gas as compared to that from steam gasification. A typical range of air gasification syngas LHV is 4 to 6 MJ/N•m³ whereas oxygen and/or steam gasification produce a higher syngas LHV (10 to 14 MJ/N•m³).

CO_2 is also used as a gasifying agent. One of the advantages of CO_2 gasification is its consumption during the process rather than being released to the atmosphere (CO_2 is a major greenhouse gas). However, CO_2 gasification process is endothermic and a continuous external heat supply is required to maintain the reactor's temperature. With CO_2 as a gasifying medium, more CO is produced compared to air and steam gasification. The typical temperature range for CO_2 gasification is 700–1000°C, similar to others. Many studies of steam and CO_2 gasification (Butterman and Castaldi, 2009; Everson et al., 2006; Ye et al., 1998; Zhang, 2006) have been carried out with low-grade biomass fuel sources.

Among many gasification reactions, the following four major reactions are typically used to explain the syngas formation at varied gasification conditions. The most important reactions are methanation (Eq. 5) and water-gas shift (Eq. 6) during the gasification process. The other reactions are Boudouard and water-gas reactions (Eq. 7 and Eq. 8) (Rodriguez-Alejandr et al., 2016).

Methanation reaction $C + 2H_2 \leftrightarrow CH_4$ –75 MJ/kmol   (5)

Water-gas shift reaction $CO + H_2O \leftrightarrow CO_2 + H_2$ –41 MJ/kmol   (6)

Boudouard reaction $C + CO_2 \leftrightarrow 2CO$ +172 MJ/kmol   (7)

Water-gas reaction $C + H_2O \leftrightarrow CO + H_2$ +131 MJ/kmol   (8)

4. Types of Gasifiers

There are several types of gasifiers but the commonly used ones are downdraft, updraft, fluidized bed, and entrained flow. The advantages and disadvantages of major gasifiers are listed in Table 4. The major advantage of downdraft and updraft gasifiers is that they are easy to operate and cheap to construct, whereas in a fluidized bed gasifier better fuel mixing

Table 4. Advantages and disadvantages of different types of gasifiers (Bridgwater, 1995).

Type of gasifier	Advantages	Disadvantages
Downdraft gasifier	Simple and reliable Easy construction Clean gas (low tar) High exit temperature	Design specific for feedstock Low moisture fuel needed High ash can cause slugging & plugging Scale up potential low
Updraft gasifier	Simple and robust Low exit temperature High thermal efficiency Can accommodate moderate ash content Good scale-up potential	High tar content Gas clean up needed to downstream
Bubbling fluidized bed gasifier	Can tolerate wide range of feedstock and particle size Good temperature control and gas-solid mixing In-bed catalyst use possible	Moderate amount of tar in gas High particulates in the product gas
Circulating fluidized bed gasifier	Good temperature control and gas-solid mixing Good carbon conversion Good scale-up potential	Operation can be difficult than fixed-bed
Entrained flow gasifier	Good gas quality Good scale up opportunity	Feed preparation costly Low feedstock options Ash content can cause slagging

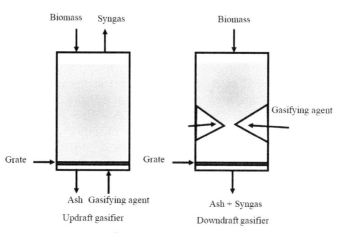

Fig. 1. Fixed bed gasifiers.

results in a high gas yield and efficiency. The entrained flow gasifier has an advantage on the high LHV syngas production.

Updraft/Downdraft gasifier: The updraft and downdraft gasifiers are classified depending on the location of the syngas exhaust as shown in Fig. 1.

In an updraft gasifier, the syngas exits from the top of the gasifier. The biomass drops downward in the reactor allowing the different reaction zones from the top of the reactor where biomass feeds. Thus, the updraft gasifier has distinct temperature zones such as biomass dehydration (100 to 150°C) at the top, devolatilization (300 to 500°C) at the middle, and the gaseous and char reactions (700 to 1000°C) at the bottom. As the produced syngas flows toward the exhaust pipe from the upper reactor, the hot syngas gets in contact with biomass particles that are at low temperature, which results in the condensation of heavy hydrocarbons. This increases the amount of tar inside the reactor as well as in the syngas. However, the cooled syngas is an advantage of an updraft gasifier (Basu, 2010).

A downdraft gasifier has a syngas exhaust outlet at the bottom of the gasifier and produces hot syngas as the syngas passes through a region of high temperature at the bottom of the reactor. This syngas passage over high-temperature region results in the further thermal cracking of heavy hydrocarbons (or tars) into lighter and cleaner gases. Thus, the main advantage of a downdraft gasifier as compared to an updraft gasifier is that it produces cleaner syngas. Both updraft and downdraft gasifiers are simple in operation and require fewer pretreatment processes for the feedstock as compared to bubbling fluidized bed gasifiers.

Particle size is an important physical property of biomass during gasification. The particle size not only impacts the formation of syngas product during gasification but it also dictates the selection of the gasifier. The updraft and downdraft gasifiers with a smaller biomass particle lead to more active gasification reactions, which produce more H_2 and CO gas (Luo et al., 2009). The effect of particle size on the reactions is related to the water-gas shift reaction, carbon gasification reactions, the secondary cracking reactions (tar compounds reacting with steam to yield CO and H_2), and the Boudouard reaction leading to an increase in CO, CH_4, and H_2 concentrations. Along with the changes in syngas compositions, the syngas yield increases and char yield decreases with a decrease in particle size according to the thermodynamic equilibrium model, which was also validated experimentally in a bench-scale fixed bed reactor (Demirbas, 2004). This is also supported by Luo et al. (2009) in which low char yields were obtained with a smaller particle size in a lab-scale fixed bed gasifier. Depending on the conditions, the temperature can be more influential than particle size on the gasification process, especially with a fixed bed gasifier. Rapagna and Latif (1997) noted that the effect of particle size was minimized on syngas composition and yield at a higher temperature above 900°C as indicated in Fig. 2. When it comes to carbon conversion efficiency, a higher carbon conversion efficiency of around 60% was obtained with the smallest particle size comparable to 35% with the largest particle size at 700°C as shown in Fig. 2. This is because smaller particles had a higher surface area

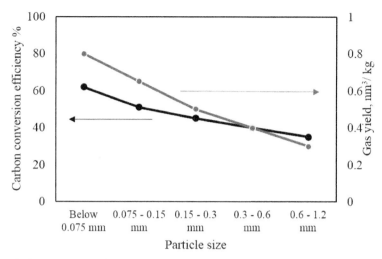

Fig. 2. Carbon conversion efficiency and gas yield as a function of particle size for pine sawdust at 700°C in a lab-scale fixed bed gasifier as extrapolated (Luo et al., 2009).

available for active reactions. Similar to carbon conversion efficiency, an increase in energy content and cold gas conversion efficiency were obtained with a decrease in particle sizes due to evenly distributed heat and mass transfer across the finer particles.

Fluidized bed: Fluidized bed is typically composed of the feeder, main bed, freeboard, and cyclone. The major advantage of the fluidized bed over downdraft and updraft gasifiers is the temperature uniformity along the reactor. One important physical component for a fluidized bed reactor is the selection of bed materials for the fluidization with a gasifying agent (air, steam, or carbon dioxide). The physical properties of bed material (such as material size, density, sphericity, and voidage) are keys to determining the operating conditions of fluidization. Based on the properties of bed material, the minimum fluidization velocity is determined. It is suggested that the actual fluidization velocity should be 2–4 times higher than the minimum fluidization velocity. An excellent mixing and temperature uniformity due to the fluidization of bed materials can be easily achieved and leads to high carbon conversion efficiency. The freeboard connected to the main fluidization bed allows the produced syngas to reduce the flow speed before escaping from the gasifier. The fluidized bed gasifier is more tolerant with high ash contenting biomass because it does not let the ash melt inside the gasifier like an updraft or downdraft gasifier due to the continuous gasifying medium mix. One of the major disadvantages of fluidized bed gasifiers is complex design and difficult operation (Basu, 2010). There are two representative types of fluidized bed gasifiers: circulating fluidized bed and bubbling fluidized bed gasifier (shown in Fig. 3). The fluidization velocity for bubbling fluidized

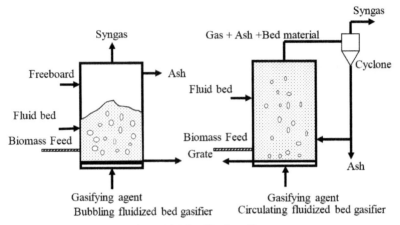

Fig. 3. Fluidized bed gasifier.

bed gasifier is about 0.5–1.0 m/s, while the circulating fluidized bed gasifiers operate under the fluidization velocity of 3.5–5.5 m/s. The circulating fluidization bed regime is longer than bubbling fluidized bed gasifier, and the longer main fluidization bed without a freeboard leads to a higher elutriation of solids out of the main bed. The solids and syngas products then enter a cyclone as shown in Fig. 3, where the solid bed material and char particles are trapped and recirculated into the main fluidization reactor zone. The fluidizing bed medium density should be heavier than the fuel density for proper operation (Geldart, 1973; Nam et al., 2016).

Syngas composition as a function of fluidizing bed particle size has been reported by several authors (Luo et al., 2009; Hernández et al., 2010). When a particle size is smaller, the external surface area/volume ratio becomes larger. Thus, the produced char becomes more porous due to the release of its high volatile combustible matter of biomass wastes. It was also reported by Babu and Chaurasia (2004) that the devolatilization of a smaller particle size takes place faster than that of larger particles so that more active thermochemical reactions of biomass fuel and char particles occur during biomass gasification process. Also, the smaller particles exhibit even more uniform heat and mass transfer compared to larger particles, which leads to a uniform temperature distribution throughout. Thus, a close to complete reaction takes place over all the fine biomass particles resulting in syngas' quality improvement. Lv et al. (2004) examined the effect of particle size on syngas yields and carbon conversion efficiencies. Larger particles' fuel decreased the carbon conversion efficiency and syngas yield in a fluidized bed gasifier. For example, the gas yield decreased from 2.57 to 1.53 Nm³/kg and the carbon conversion efficiency from 95 to 77% when the particle size increased from 0.2–0.3 mm to 0.6–0.9 mm. Rapagna and Latif (1997) also reported a similar trend of results from a bench-scale fluidized bed gasifier with ground almond shells. On the one hand, an increased temperature in a fluidized bed reactor similar to a fixed-bed gasifier reduced the effect of particle size on the carbon conversion efficiency and syngas yield. On the other hand, the effect of bigger particle size after pelletization led to different gasification performance (Bronson et al., 2016). The use of pellets for the gasification helped to produce higher H_2 to CO ratio, while smaller particles led to a lower ratio.

Entrained flow gasifier: In an entrained flow gasifier (as shown in Fig. 4), the pulverized fuel particles are suspended from the top of the reactor with a stream of oxidant. Entrained flow gasifiers are widely used for coal gasification but not much for biomass gasification. The biomass particles used in an entrained flow gasifier need to be very small in size since the residence time for reactions is very short (Basu, 2010).

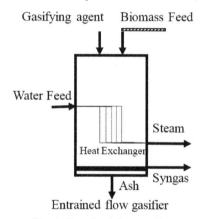

Fig. 4. Entrained flow gasifier.

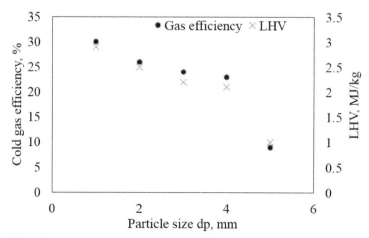

Fig. 5. Effect of particle size on LHV and gas efficiency at 1050°C in an entrained air flow gasifier with pine sawdust.

The entrained gasifier showed better syngas yields and compositions with a reduction in the size of the feed, similar to other gasifiers discussed above. Hernandez et al. (2010) reported that a decreasing mean fuel particle diameter from 8.0 to 0.5 mm increased the fuel conversion efficiency from 60 to 90%. The cold gas efficiency was also increased with a decrease in the particle size of the biomass as shown in Fig. 5 (Hernandez et al., 2010).

5. Effect of Temperature and ER on Gasification

Temperature is an important parameter during gasification which influences gasification chemical reactions including Eq. 5–8. Figure 6 shows the effect of bed temperature (700–850°C) on the produced syngas and its LHV from

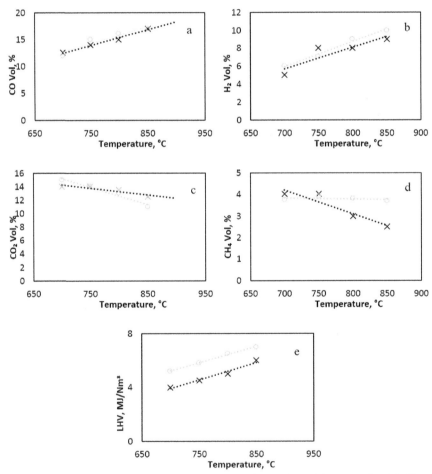

Fig. 6. Effect of temperature on gas composition (a, b, c & d) and heating value (e). (Data obtained from Narvaez et al. (1996) for ×, and Xue et al. (2014) for o.

a fluidized bed gasifier (Narvaez et al., 1996; Xue et al., 2014). The CO and CO_2 gases are inversely related at higher temperature due to multiple reactions including water-gas shift reaction (Eq. 6), Boudouard reaction (Eq. 7), and water-gas reaction (Eq. 8). Xue et al. (2014) reported that an increase in temperature helped increase carbon conversion efficiency as well as the cold gas conversion efficiency. Nam et al. (2016) statistically showed that the effect of temperature on the variation of syngas composition was higher as compared to the effect of ER. It is also reported that the temperature also largely influenced the total gas yield. An increase in the temperature helped degrade the feedstock and reduce the total tar content, which eventually increases the gas yield. Figure 7 shows the effect

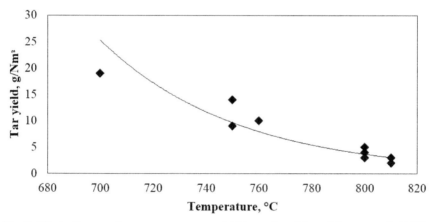

Fig. 7. Effect of increase in temperature on tar yield (reproduced from (Narvaez et al., 1996)).

of temperature on tar content as adapted from Narvaez et al. (1996). At a higher gasification temperature, primary (oxygenates, phenolics) tar and secondary (alkyl, phenolics, heterocyclic) tar are decomposed and converted into the tertiary (polynucleic, aromatic, hydrocarbons) tar products. The amount of benzene (one of tertiary tar) increased at a higher temperature while others such as toluene, naphthalene, biphenylene, fluoranthene and pyrene decreased (Abdoulmoumine et al., 2014).

ER is another important parameter during gasification process. Typical ER for gasification is between 0.2 and 0.4. A higher ER (> 0.4) for gasification produces more combustion products (CO_2 and H_2O), while a lower ER produces more combustible gases such as H_2, CO, and CH_4 as shown in Fig. 8. Accordingly, the syngas LHV decreases at a higher ER (Gil et al., 1999) and total syngas yield increases at a higher ER due to an elevated CO_2 concentration as shown in Fig. 9 (Narvaez et al., 1996). A study on the torrefied miscanthus gasification (Xue et al., 2014) showed that an increase in ER significantly increased CO_2 gas along with a slight increase in CO, which led to a decrease in LHV and an increase in carbon conversion efficiency

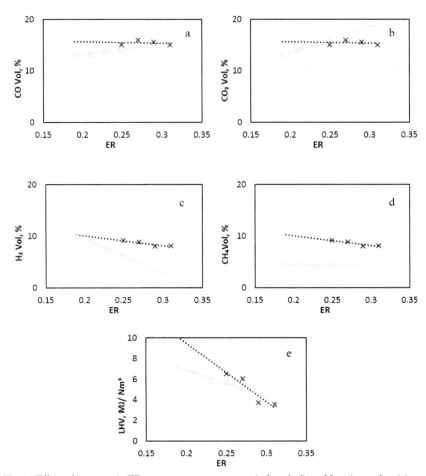

Fig. 8. Effect of increase in ER on syngas components (a, b, c & d) and heating value (e) at a constant temperature (800°C). (Data obtained from Narvaez et al., 1996 for × and Xue et al., 2014 for o.)

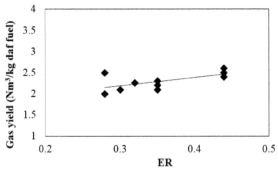

Fig. 9. Effect of ER on gas yield (reproduced from (Narvaez et al., 1996)). (daf=dry ash free)

(73.5 to 82.5%). Our previous study (Abdoulmoumine et al., 2014) showed that the effect of ER on tar content was more significant than the effect of temperature during the pinewood gasification. A higher ER caused an active decomposition of tar compounds due to combustion reactions. Narvaez et al. (1996) also reported a decrease in tar yield with an increase in ER.

6. Contaminants

Tar: Tar is an undesirable and unavoidable product in the gasification process. It condenses in the low-temperature zone due to its property of a thick black viscous fluid. In a gasifier, it clogs the gas pipe which leads to system disruption. The syngas components that have molecular weight heavier than benzene are identified as tar (Knoef and Ahrenfeldt, 2005). Tar can be classified as primary, secondary, and tertiary, which are grouped based on its molecular structures. Once the air is fully reacted with feedstock, the absence of oxygen during the gasification process promotes primary tars (phenols, acids, sugars, and ketones) from the breakage of hemicelluloses, cellulose, and lignin. Secondary tars are formed mainly by rearrangement of primary tar molecules at a temperature above 500°C. Tertiary tars, formed at even higher temperatures, are methyl derivatives of aromatics, which can be transformed into polyaromatic hydrocarbons (PAH). Milne et al. (1998) stated that the tertiary hydrocarbons are formed at the expenses of primary hydrocarbon as well. At temperatures between 750 and 800°C, primary tars are completely consumed, whereas tertiary and secondary tars still remain. The complete removal of tar via thermal cracking does not take place below 1100°C (Milne et al., 1998). Applications of syngas in an internal combustion engine, gas turbine, and hydrogen production impose tight limits for tar as shown in Table 5. It should be noted that the direct combustion has no limitation on the tar content while the internal combustion engines are more tolerant than gas turbines (Milne et al., 1998).

Tar concentration can be reduced by two routes. One is an *in situ* tar reduction (during gasification), and the other is tar reduction after gasification. The *in situ* tar reduction methods include the selection of

Table 5. Acceptable limit of biomass tar (Basu, 2010).

End use application	Permissible tar content (g/ Nm³)
Direct combustion	No limit specified
Syngas production	0.1
Expansion in gas turbine	0.05–5
Internal combustion engine	50–100
Pipeline transport	50–500 for compressor
Fuel cells	< 1.0

proper gasification temperature, bed material, and catalysts. The operating parameters such as temperature, ER, and residence time also play an important role in tar elimination (Kinoshita et al., 1994; Devi et al., 2003). It was observed that a number of tar species were reduced with an increase in temperature (from 700 to 900°C). The ways to reduce tar after gasification process include spray tower, wet scrubber, wet cyclones, electrostatic precipitators (ESP) and catalytic tar reforming.

Nitrogen contaminants: The NH_3 and HCN concentrations in syngas are proportional to fuel nitrogen content of biomass (Zhou et al., 2000). In addition, these concentrations are affected by an oxidizing agent (steam/ O_2 or air), temperature, and ER (Zhou et al., 2000; Aljbour and Kawamoto, 2013). In general, an increase in temperature and ER, decrease the concentration of NH_3 and HCN (Abdoulmoumine et al., 2014; Zhou et al., 2000; Leppälahti, 1993). However, there have been instances where NH_3 and HCN did not follow this behavior (Aljbour and Kawamoto, 2013) as NH_3 concentrations increased from 353 to 404 ppmv and HCN increased from 2 to 6 ppmv over the ER ranges from 0 to 0.3. This was caused by the conversion of fuel-nitrogen into volatile nitrogen gas over the destruction of the fuel at a higher temperature and ER conditions. The ranges of NH_3 and HCN concentrations of switchgrass gasification increased from 5800 to 10000 ppm and from 400 to 2500 ppm, respectively, when gasification operating conditions changed from ER = 0.21 at 700°C to ER = 0.32 at 880°C (Broer et al., 2015). In contrast, some other (Abdoulmoumine et al., 2014; Aljbour and Kawamoto, 2013; Van der Drift et al., 2001) studies reported lower concentrations from 300 to 1800 ppmv for NH_3 and less than 70 ppmv for HCN. The primary incentive for ammonia removal is the reduction in the NO_x emission from combustion.

Sulfur contaminants: Sulfur contaminants show less problematic during gasification compared to combustion process (Basu, 2010) because easily convertible sulfur gas (H_2S and COS) are produced during gasification process. The gases H_2S and COS can be converted into elemental sulfur or H_2SO_4 that can be sold in the market. The oxygen purged gasification process converted 93% of sulfur into the H_2S and the remaining sulfur was carbonyl sulfide (COS) (Higman et al., 2003). However, the combustion system produces less market potential ash-mixed $CaSO_4$ and/or SO_2. A fluidized bed combustion study at around 700 and 800°C reported that 400–500 ppm of the SO_2 emission was produced with coal-biomass (up to 23 wt.%) feedstock (Xie et al., 2007). The sulfur contaminants need to be removed for downstream application since they can cause catalyst poisoning and equipment corrosion.

Hydrogen Halides: The HCl and HF are produced from the reactions between halides and hydrogen both present in the biomass. These halides also react with the alkali metals such as Na and K in the biomass ash to form their respective salts. Pinewood biomass gasification produced much lower concentrations of HCl (13.6 ppm at 790°C and 8.9 ppm at 934°C (Abdoulmoumine et al., 2014)) as compared to the HCl concentration (600 ppm) from coal gasification (Duong et al., 2009). This is due to the low concentration of halides present in biomass feedstock. However, even a few ppm volume of HCl could cause corrosion on filters, heat exchangers, and turbine blades during downstream applications due to the high reactivity of HCl.

7 The Importance of Feedstock Densification

The utilization of biomass wastes can be a great way to resolve waste management issues and produce an additional profit. However, there are some challenges such as transportation costs, low energy density, high oxygen and moisture content, and storage issues as well as irregular physical properties. Thus, fuel pretreatment such as pelletization and torrefaction can be considered for the gasification process.

7.1 Prevention of a bridge over a hopper opening

Many fuel bridging issues occur in a fuel hopper regardless of the type of reactor connected to the auger/hopper conveyor system. Pretreatments are important for the gasification process not just to improve the syngas quality but also to prevent fuel bridges (arch and rathole) as indicated in Fig. 10. Arching takes place in the outlet of a hopper when the fuel particles

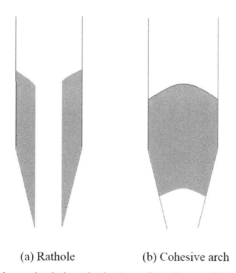

(a) Rathole   (b) Cohesive arch

Fig. 10. Physical shape of rathole and cohesive arch in a hopper (Woodcock and Mason, 1987).

are mechanically locked due to a large fuel size compared to the size of the opening (interlocking arch) or when the fuel particles bond together (cohesive arch). Similarly, a rathole bridge happens when a vertical free-flowing channel develops in sufficiently bonded fuels around the channel.

The bridging tendency over fuel entrance openings can be induced by the physical properties of the fuel or the mechanical issues of the conveyor system. The physical properties of the fuel include particle size, size distribution, shape, moisture content, bulk density, the depth of the fuel bed, and the angle of friction. A study of different physical pretreatments of municipal solid waste (MSW) was performed by Maglinao and Capareda (2010) to solve bridging and clogging issues. The MSW was processed into fluff, shredded, or made into pellets as a fluidized bed gasification fuel. The fluff MSW clogged the 10 cm diameter auger conveyor, while the shredded MSW sample partially relieved the clogging in the auger. However, the shredded MSW produced a bridge or a rathole in the hopper even with the extra help of biomass agitators.

Other agricultural and woody type biomass wastes still require physical pretreatments for smooth feeding, especially in a pilot or commercial scale gasification system. Figure 11 shows the relationship between the particle length and the tendency of bridging in a hopper (Mattsson, 1997).

A larger number of opening lengths indicates a higher potential for fuel bridging. The slot opening lengths for grass and straw biomass at different particle sizes place close between 600 and 800 mm, while the woodchips show a much lower bridge tendency as it locates near 100 mm. This can be

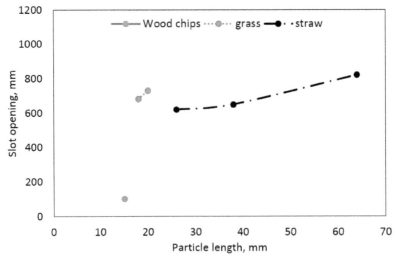

Fig. 11. Bridge tendency of different feedstocks at different particle length of fuel (Mattsson, 1997).

Table 6. The fuel characteristics of bulk density and friction angle.

Feedstock	Treatment	Bulk density (kg/m³)	Friction angle (degree)	Bridging tendency	Reference
Grass/straw	Chopped	≤ 50	N/A	High	(Mattsson, 1997)
Woodchip	Chopped	≤ 200	N/A	Low	(Mattsson, 1997)
MSW	Fluff	33	33	High	(Maglinao Jr., 2013)
MSW	Shredded	62	25	High	(Maglinao Jr., 2013)
MSW	Pellet	421	18	Low	(Maglinao Jr., 2013)
Dairy Manure	Ground	267	43	Low	unpublished

N/A: not available.

explained by the difference in bulk density of straw and wood types, which vary by more than four times as indicated in Table 6. Even though bulk density cannot be a representative indicator to judge the bridging potential, heavy bulk density wastes such as woodchips, MSW pellets, and dairy manure show a low tendency for bridging. The physical pretreatments of grinding and pelletization prevent flow obstruction of fuel into the reactor as well as improve the quality of the syngas.

7.2. The effect of fuel densification on the gasification process

Densification is a preferred way to improve gasification process especially regarding feeding biomass in the reactor because of the uniform shape, controlled particle size, and very low moisture. The two main densification pretreatments for relieving storage and transportation issues are pelletization and torrefaction.

Pelletization is widely used not just to relieve fuel obstructions in a fuel hopper but also to help in stable gasification reactions. The pelletized MSW with a bulk density of 421 kg/m³ was the only treatment that helped with uniform feeding and smooth temperature control during fluidized bed gasification (Maglinao Jr., 2013). A gasification comparison study with rice husk and its pellet was conducted in a downdraft gasifier (Yoon et al., 2012). The syngas composition with rice husk pellets showed to be much more stable compared to rice husk which showed a large variation. The uniform shape and density of the pellets helped smooth feeding and gasification reactions. Less than twice the air was needed for pellet gasification and produced a better quality syngas with a higher H_2 and CO which resulted in a higher syngas heating value. Also, the cold gas efficiency was 60% for rice husk gasification and 70% for pelleted rice husk gasification.

In an approach that is different from pelletization, torrefaction treatment can be done thermally with the temperatures ranging from 200 to 300°C. After the torrefaction process, the properties of biomass waste are enhanced for better grindability and hydrophobicity by losing their oxygen and hydrogen, functional groups. Berrueco et al. (Berrueco et al., 2014) reported on the effect of torrefaction severities on the produced syngas from a fluidized bed gasifier. The syngas from torrefied woody biomass (Norwegian spruce and Norwegian forest residues) showed an increase in gas and char yields and a decrease in tar yield compared to raw

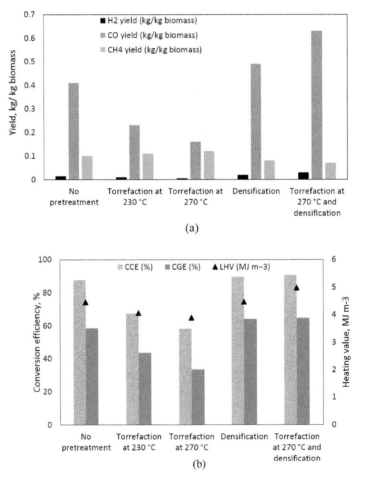

Fig. 12. Effect of pelletization and torrefaction on (a) syngas yields and (b) carbon conversion efficiency (CCE), cold gas efficiency (CGE) and lower heating value (LHV) of syngas.

biomass. When it comes to the syngas composition, more hydrogen gas was produced with torrefied materials. However, the effect of torrefaction on CO and CO_2 yields was limited or slightly increased. The changes of syngas compositions and gas product yields can be explained by the char and lignin gasification instead of devolatilization during the torrefied biomass gasification. Furthermore, a study by Sarkar et al. (2014) was done on the effects of a combination of pelletization and torrefaction of switchgrass on the product yields and product composition as indicated in Fig. 12a. The pelletized and torrefied syngas composition showed the highest H_2 and CO yields, followed by the syngas from the densified raw biomass gasification. Gasification with torrefied switchgrass produced the highest amount of CH_4 compared to other pretreated biomass gasification. When it comes to efficiencies, the pretreatments of pelletization and the combination of pelletization and torrefaction for gasification showed the highest carbon and cold gas conversion efficiencies as depicted in Fig. 12b. Thus, densification of biomass after pretreatment is beneficial in increasing efficiencies.

8. Gasification Syngas Applications

Coal gasification syngas has been used for the production of liquid transportation fuels, which helped to diversify fuel supply capability, energy security, and sustainability. Coal or biomass is initially gasified to produce syngas that can be converted into liquid fuel or alcohols, which is also called indirect liquefaction or simply gas to liquid (GTL) process. Because of the syngas cleaning process, very small amount of sulfur and mercury impurities remain in the synthesized fuels and burn cleaner than conventional gasoline and diesel fuels (National Energy Technology Laboratory, 2017). One of the most widely used syngas for liquid synthesis technologies is the Fischer-Tropsch (FT) process. Carbon monoxide (CO) and hydrogen (H_2) in syngas are synthesized into hydrocarbons or alcohols over a choice of catalysts at different operating conditions such as reactor temperature, pressure, and feed composition. Typical FT products include paraffin, olefins, aromatics, and others which can be refined into gasoline and diesel ranged fuels. A recent FT synthesis was performed with biomass-derived syngas using aniron carbide catalyst at various reaction conditions. The conversion rate of CO and H_2 was over 85%, which produced C_{5+} liquid hydrocarbons (over 65% selectivity) at the condition of 310°C, 1000 psig, and 3000 h^{-1} (Lu et al., 2017). The major C_{5+} chemical was determined to be an olefin group. A different composited catalyst (100Fe/4Cu/4K/6Zn) for biomass syngas FT synthesis was used by another group which obtained a high C_{5+} selectivity (Mai et al., 2015). Another synthesis technology is the ExxonMobil methanol to gasoline (MTG) process which utilized methanol produced from syngas to produce gasoline fuel. Li et al. (2015) fabricated a nanocrystalline Fe, Al/ZSM-5 catalyst for the MTG process. The gasification syngas is also fermented to produce biofuels (mainly ethanol) over various microbial catalysts such as *C. ljungdahlii, C. autoethanogenum, A. woodii, C. carboxidivorans* and *P. productus* (Munasinghe and Khanal, 2010).

9. Industrial Scale Gasification Plants

As gasification technology has been developed over the past few decades, there are more than 272 operating gasification plants with 686 gasifiers in the world and 33 of these plants are in the United States (Gasification and Syngas Technologies Council, 2017). The Indian River Bioenergy Center (Vero Beach, Florida) uses a hybrid thermochemical-fermentation procedure to process 250,000 tons of yard and wood waste per year. The gasification syngas is initially fermented and distilled to produce a final product of cellulosic ethanol (8 million gallons per year). An LCA (life cycle assessment) indicates that the E-85 fuel from the produced ethanol emits 65% fewer greenhouse gases than petroleum-based fuels (DOE, 2013). Edwardsport IGCC (Edwardsport, Indiana) was completed in 2007 and produces 618 MW of energy from coal. Clean syngas is obtained with an activated carbon bed for mercury absorption, two heat recovery steam generators for catalytic nitrogen oxide reduction, and a multiple cell cooling tower. The InEnTec Columbia Ridge Facility (Arlington, Oregon) utilizes municipal solid wastes (MSW) to produce renewable hydrogen gas with a 2 MW gasifier. The University of South Carolina gasifier (1.38 MW) and ORNL biomass gasifiers (60 MMBtu/hr biomass) were completed in 2007 and 2012, respectively, and were decommissioned due to frequent failures of plant equipment (Nexterra, 2018).

10. Conclusions

Biomass gasification, as one of the most practical technologies, is gaining attention because of its ability to deal with an increasing amount of annual solid wastes through its massive handling capacity and high energy conversion efficiency. Based on the targeting products or feedstock, a proper gasifier and gasifying agent have to be selected. This chapter described the basic concept of gasification as well as various gasification performances at different operating conditions such as equivalence ratio and temperature. In addition, the effects of the physical properties of biomass on syngas compositions were discussed as well as the importance of physical pretreatments of biomass feedstock. Syngas contaminants during the gasification process were introduced. Also, a short discussion regarding the recent technologies of gasification syngas utilization and industrial gasification plants was included.

References

Abdoulmoumine, N., A. Kulkarni and S. Adhikari. 2014. Effects of temperature and equivalence ratio on pine syngas primary gases and contaminants in a bench-scale fluidized bed gasifier. Ind. Eng. Chem. Res. 53: 5767–5777.

Aljbour, S. H. and K. Kawamoto. 2013. Bench-scale gasification of cedar wood–Part II: effect of operational conditions on contaminant release. Chemosphere 90: 1501–1507.

Babu, B. and A. Chaurasia. 2004. Heat transfer and kinetics in the pyrolysis of shrinking biomass particle. Chemical Engineering Science 59: 1999–2012.

Basu, P. 2010. Biomass Gasification and Pyrolysis: Practical Design and Theory, Academic press.

Berrueco, C., J. Recari, B. M. Güell and G. del Alamo. 2014. Pressurized gasification of torrefied woody biomass in a lab scale fluidized bed. Energy 70: 68–78.

Boie, W. 1953. Fuel technology calculations. Energietechnik 3: 309–316.

Bridgwater, A. 1995. The technical and economic feasibility of biomass gasification for power generation. Fuel 74: 631–653.

Broer, K. M., P. J. Woolcock, P. A. Johnston and R. C. Brown. 2015. Steam/oxygen gasification system for the production of clean syngas from switchgrass. Fuel. 140: 282–292.

Bronson, B., P. Gogolek, P. Mehrani and F. Preto. 2016. Experimental investigation of the effect of physical pre-treatment on air-blown fluidized bed biomass gasification. Biomass Bioenergy 88: 77–88.

Butterman, H. C. and M. J. Castaldi. 2009. CO_2 as a carbon neutral fuel source via enhanced biomass gasification. Environ. Sci. Technol. 43: 9030–9037.

Carpenter, D. L., R. L. Bain, R. E. Davis, A. Dutta, C. J. Feik, K. R. Gaston, W. Jablonski, S. D. Phillips and M. R. Nimlos. 2010. Pilot-scale gasification of corn stover, switchgrass, wheat straw, and wood: 1. Parametric study and comparison with literature. Ind. Eng. Chem. Res. 49: 1859–1871.

Cross, Phillip et al. 2018. Bubbling fluidized bed gasification of short rotation Eucalyptus: Effect of harvesting age and bark. Biomass and Bioenergy 110: 98–104.

Demirbas, A. 2004. Effects of temperature and particle size on bio-char yield from pyrolysis of agricultural residues. J. Anal. Appl. Pyrolysis 72: 243–248.

Devi, L., K. J. Ptasinski and F. J. J. G. Janssen. 2003. A review of the primary measures for tar elimination in biomass gasification processes. Biomass Bioenergy 24: 125–140.

DOE. 2013. INEOS-New planet: Indian River Bioenergy Center (https://www.energy.gov/eere/bioenergy/ineos-new-planet-indian-river-bioenergy-center), Office of Energy Efficiency & Renewable Energy.

Duong, D. N., D. A. Tillman, F. W. NA and N. Clinton. 2009. Chlorine issues with biomass cofiring in pulverized coal boilers: sources, reactions, and consequences—a literature review 31.

Everson, R. C., H. W. Neomagus, H. Kasaini and D. Njapha. 2006. Reaction kinetics of pulverized coal-chars derived from inertinite-rich coal discards: gasification with carbon dioxide and steam. Fuel. 85: 1076–1082.

Gasification & Syngas Technologies Council. 2017. The Gasification Industry (http://www.gasification-syngas.org/resources/the-gasification-industry/).

Geldart, D. 1973. Types of gas fluidization. Powder Technol. 7: 285–292.

Gil, J., J. Corella, M. P. Aznar and M. A. Caballero. 1999. Biomass gasification in atmospheric and bubbling fluidized bed: effect of the type of gasifying agent on the product distribution. Biomass Bioenergy 17: 389–403.

Hernández, J. J., G. Aranda-Almansa and A. Bula. 2010. Gasification of biomass wastes in an entrained flow gasifier: effect of the particle size and the residence time. Fuel Process Technol. 91: 681–692.

Higman, C. and M. van der Burgt. 2003. Gasification, Gulf Professional Publsihing. United Kingdom.

Kinoshita, C., Y. Wang and J. Zhou. 1994. Tar formation under different biomass gasification conditions. J. Anal. Appl. Pyrolysis 29: 169–181.

Knoef, H. and J. Ahrenfeldt. 2005. Handbook Biomass Gasification, BTG biomass technology group. The Netherlands.

Kulkarni, A., R. Baker, N. Abdoulmomine, S. Adhikari and S. Bhavnani. 2016. Experimental study of torrefied pine as a gasification fuel using a bubbling fluidized bed gasifier. Renewable Energy 93: 460–468.

Leppälahti, J. 1993. Formation and behaviour of nitrogen compounds in an IGCC process. Bioresour. Technol. 46: 65–70.

Li, J., P. Miao, Z. Li, T. He, D. Han, J. Wu, Z. Wang and J. Wu. 2015. Hydrothermal synthesis of nanocrystalline H[Fe, Al]ZSM-5 zeolites for conversion of methanol to gasoline. Energy Conversion and Management 93: 259–266.

Liu, H., Y. Feng, S. Wu and D. Liu. 2009. The role of ash particles in the bed agglomeration during the fluidized bed combustion of rice straw. Bioresour. Technol. 100: 6505–6513.

Lu, Y., Q. Yan, J. Han, B. Cao, J. Street and F. Yu. 2017. Fischer–Tropsch synthesis of olefin-rich liquid hydrocarbons from biomass-derived syngas over carbon-encapsulated iron carbide/iron nanoparticles catalyst. Fuel 193: 369–384.

Luo, S., B. Xiao, X. Guo, Z. Hu, S. Liu and M. He. 2009. Hydrogen-rich gas from catalytic steam gasification of biomass in a fixed bed reactor: influence of particle size on gasification performance. Int. J. Hydrogen Energy 34: 1260–1264.

Lv, P., Z. Xiong, J. Chang, C. Wu, Y. Chen and J. Zhu. 2004. An experimental study on biomass air–steam gasification in a fluidized bed. Bioresour. Technol. 95: 95–101.

Maglinao Jr., A. L. and S. Capareda. 2010. Predicting fouling and slagging behavior of dairy manure (DM) and cotton gin trash (CGT) during thermal conversion. Transactions of the ASABE 53: 903–909.

Maglinao Jr., A. L. 2013. Development of a segregated municipal solid waste gasification system for electrical power generation. Texas A&M University. College Station.

Mai, K., T. Elder, L. H. Groom and J. J. Spivey. 2015. Fe-based Fischer Tropsch synthesis of biomass-derived syngas: effect of synthesis method. Catalysis Communications 65: 76–80.

Mattsson, J. E. 1997. Biomass quality for power production tendency to bridge over openings for chopped Phalaris and straw of Triticum mixed in different proportions with wood chips. Biomass Bioenergy 12: 199–210.

Milne, T. A., N. Abatzoglou and R. J. Evans. 1998. Biomass Gasifier" Tars": Their Nature, Formation, and Conversion. National Renewable Energy Laboratory Golden, CO.

Munasinghe, P. C. and S. K. Khanal. 2010. Biomass-derived syngas fermentation into biofuels: Opportunities and challenges. Bioresource Technology 101: 5013–5022.

Nam, H. and S. Capareda. 2015. Experimental investigation of torrefaction of two agricultural wastes of different composition using RSM (response surface methodology). Energy 91: 507–516.

Nam, H., A. L. Maglinao Jr., S. C. Capareda and D. A. Rodriguez-Alejandro. 2016. Enriched-air fluidized bed gasification using bench and pilot scale reactors of dairy manure with sand bedding based on response surface methods. Energy 95: 187–199.

Narvaez, I., A. Orio, M. P. Aznar and J. Corella. 1996. Biomass gasification with air in an atmospheric bubbling fluidized bed. Effect of six operational variables on the quality of the produced raw gas. Ind. Eng. Chem. Res. 35: 2110–2120.

National Energy Technology Laboratory, Gasoline & Diesel. 2017. https://www.netl.doe.gov/research/coal/energy-systems/gasification/gasifipedia/fuels.

Nexterra, Projects (http://www.nexterra.ca/files/projects.php), 2018.

Rapagna, S. and A. Latif. 1997. Steam gasification of almond shells in a fluidised bed reactor: the influence of temperature and particle size on product yield and distribution. Biomass Bioenergy 12: 281–288.

Rodriguez-Alejandr, D. A., H. Nam, A. L. Maglinao Jr., S. C. Capareda and A. F. Aguilera-Alvarado. 2016. Development of a modified equilibrium model for biomass pilot-scale fluidized bed gasifier performance predictions. Energy 115: 1092–1108.

Sarkar, M., A. Kumar, J. S. Tumuluru, K. N. Patil and D. D. Bellmer. 2014. Gasification performance of switchgrass pretreated with torrefaction and densification. Appl. Energy 127: 194–201.

Van der Drift, A., J. Van Doorn and J. Vermeulen. 2001. Ten residual biomass fuels for circulating fluidized-bed gasification. Biomass Bioenergy 20: 45–56.

Visser, H. J. M. 2004. The Influence of Fuel Composition on Agglomeration Behaviour in Fluidised-Bed Combustion, Energy research Centre of the Netherlands ECN Delft.

Woodcock, C. and J. Mason. 1987. Bulk Solids Handling—An Introduction to the Practice and Technology: Leonard Hill.

Xie, J., X. Yang, L. Zhang, T. Ding, W. Song and W. Lin. 2007. Emissions of SO_2, NO and N_2O in a circulating fluidized bed combustor during co-firing coal and biomass. Journal of Environmental Sciences 19: 109–116.

Xue, G., M. Kwapinska, A. Horvat, W. Kwapinski, L. Rabou, S. Dooley, K. Czajka and J. Leahy. 2014. Gasification of torrefied Miscanthus× giganteus in an air-blown bubbling fluidized bed gasifier. Bioresour. Technol. 159: 397–403.

Ye, D., J. Agnew and D. Zhang. 1998. Gasification of a South Australian low-rank coal with carbon dioxide and steam: kinetics and reactivity studies. Fuel. 77: 1209–1219.

Yoon, S. J., Y. Son, Y. Kim and J. Lee. 2012. Gasification and power generation characteristics of rice husk and rice husk pellet using a downdraft fixed-bed gasifier. Renewable Energy 42: 163–167.

Zhang, L., J. Huang, Y. Fang and Y. Wang. 2006. Gasification reactivity and kinetics of typical Chinese anthracite chars with steam and CO_2. Energy Fuels 20: 1201–1210.

Zhou, J., S. M. Masutani, D. M. Ishimura, S. Q. Turn and C. M. Kinoshita. 2000. Release of fuel-bound nitrogen during biomass gasification. Ind. Eng. Chem. Res. 39: 626–634.

CHAPTER 6

The Impacts of Biomass Pretreatment Methods on Bio-oil Production

Yang Yue and *Sudhagar Mani**

1. Introduction

Biomass is the abundantly available renewable and sustainable resource to produce fuels and value-added chemicals via thermo-chemical and biochemical approaches with net zero greenhouse gas emission. The US Department of Energy (DOE) and the Department of Agriculture (USDA) have estimated that more than one billion tons of lignocellulose biomass is sustainably available annually in the United States, which if converted efficiently, could replace 43% domestic petroleum consumption (Bond et al., 2014). Among that, 47 million dry tons of forest biomass can be potentially supplied, in the form of logging thinnings, logging residues, and pulpwood with the delivered price of approximately $40 per dry ton (Carpenter et al., 2014). The agricultural residues are another crucial fraction of lignocellulose biomass resource with large diversity, including but not limited to energy crops and sugar crops/bagasse with various yields and delivered costs. Corn stover is a representative agricultural residue with the largest availability of 196 million tons annually in the US and has a similarly delivered price as forest biomass (Graham et al., 2007). However, the low bulk density,

College of Engineering, University of Georgia, Athens, GA 30602.
* Corresponding author: smani@engr.uga.edu

low energy density, high moisture and ash compositional variations, and structural heterogeneity of lignocellulosic biomass pose major challenges for economic supply and efficient conversion into competitive biofuels.

Among various conversion pathways, pyrolysis is a promising thermo-chemical technology to convert a wide range of biomass into bio-oil, similar to petroleum crude to produce drop-in biofuels and biochemicals. Biomass pyrolysis is a process of rapidly decomposing lignocellulosic biomass under atmospheric pressure at around 500°C in the absence of oxygen. Three main products are consequently obtained when the volatile vapors are condensed: (i) a solid residue, called biochar containing fixed carbon and ash; (ii) non-condensable gases, such as CO and CO_2, mainly formed through decarboxylation reaction of structural carbohydrates and lignin; (iii) a liquid product, called bio-oil containing number of water-soluble sugars, organic acids and hydrocarbon mixtures (Carpenter et al., 2014). Figure 1 shows the typical products of pyrolysis derived from pine woodchips.

Pyrolysis can be categorized as slow pyrolysis or fast/flash pyrolysis, depending on the rate of heating. Slow pyrolysis aims to produce solid biofuel, biochar, with very low heating rates ranging from 0.01 K/s to up to 2 K/s; lower temperature is applied to maximize the yield of solid product. For fast pyrolysis, the goal product is a liquid fuel, bio-oil, from as much biomass feedstock as possible; fast pyrolysis requires a much higher heating rate of up to 1000 K/s (Bartek et al., 2012) with higher pyrolysis temperature to promote the evaporation of volatile organic fractions from

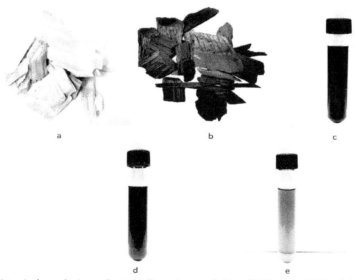

Fig. 1. A typical pyrolysis products (a) Raw pinewood chip, (b) Bio-char, (c) Raw bio-oil, (d) Oil phase, (e) Aqueous phase.

biomass. Fast pyrolysis is the most common type of technology developed at pre-commercial scale to produce bio-oil for further downstream conversion or to co-process with petroleum crude-oil. Flash pyrolysis is derived from fast pyrolysis with a very rapid and higher heating rate of up to 10^4 K/s and a shorter holding time of less than 0.5 s. Similar to fast pyrolysis, flash pyrolysis also aims to produce bio-oil and minimize the formation of char and gaseous products. The bio-oil yields of fast pyrolysis and flash pyrolysis are comparable. The detailed reaction's conditions of slow, fast and flash pyrolysis are listed in Table 1.

Fast pyrolysis has been comprehensively studied in the last decades for its thermal decomposition of renewable biomass into liquid fuel. It displays promising potential for petroleum alternative production with net carbon emission into the atmosphere. Compared to other thermal conversion technologies, such as hydrothermal liquefaction, fast pyrolysis has higher bio-oil yield and retains most of the energy from biomass feedstock (Liu et al., 2014). The pyrolysis yield from various biomass types is summarized in Table 2. The yield of bio-oil could reach up to 74% (wt) from lignocellulosic biomass (Lede et al., 2007). However, the physicochemical properties of bio-oil are undesirable for engine combustion or to be used as miscible fuel with hydrocarbon because of the low vapor pressure, low heating value, low stability, high acidity, and viscosity (Liu et al., 2014). Although the above challenges could be ameliorated with *in situ* or *ex situ* catalytic upgrading process, economic feasibility is still a hurdle to bio-oil commercialization. Diversification of low-cost forest and agriculture biomass and consistent quality of feedstock and its derived bio-oil characteristics are critical for efficient and economical upgrading bio-oil into biofuels/chemicals (Carpenter et al., 2014).

The heterogeneity of raw biomass is reflected in the variability in chemical compositions and physical properties (Westover et al., 2013). Although water is formed during pyrolysis, the higher moisture content (25–60 wt.%) of biomass results in higher water content in bio-oil, which has shown negative impacts on engine combustion such as reducing the heating

Table 1. The reaction conditions for slow and fast pyrolysis technologies.

Reaction conditions	Pyrolysis		
	Slow pyrolysis	**Fast pyrolysis**	**Flash pyrolysis**
Temperature [°C]	~400	400–600	400–600
Residence time [s]	> 5 s for volatiles Minutes to hours for solids	< 2 s for liquid fuels < 1 s for specialty chemicals	< 0.5 s for liquid fuels
Heating rate [°C/s]	0.01–2 K/s	10–1000 K/s or above	~10^4 K/s
Quenching rate [°C/s]	Minutes to hours to room temperature	100–1000 K/s	100–1000 K/s

Table 2. The yield of pyrolysis products from various lignocellulosic biomass.

Biomass feedstock	Pyrolysis reactor	Temperature [°C]	Bio-oil yield [%]	Char yield [%]	Gas yield [%]	Reference
Woody biomass						
Pine	Auger	450	60.1	19.1	20.8	Kenney et al., 2013
Oak	Vortex	525	55.3	12.2	12.4	Raveendran et al., 1995
Polar	Fluidized bed	504	66.2	11.8	11.0	Chang et al., 2013
Spruce	Fluidized bed	500	66.5	12.2	11.0	Chang et al., 2013
Eucalyptus wood	Fluidized bed	500	59.5	18.3	22.2	Wan Isahaka et al., 2012
Agricultural biomass						
Switchgrass	Vortex	525	34.3	19.9	21.7	Raveendran et al., 1995
Corn stover	Fluidized bed	500	52.7	15.9	15.1	Fahmi et al., 2007
Corn cob	Fluidized bed	470	56.2	14.7	27.2	Schell and Harwood, 1994
Rice husk	Conical spouted bed	450	70.0	5.0	25.0	Alvarez et al., 2014

value of fuel, and dropping the flame temperature, and causing extended ignition delay (Lede et al., 2007). Ash content is another important chemical composition ranging from 0.1 to 6.4% (mean 1.9%) for woody biomass and 1.0 to 26.2% (mean 7.0%) for forest residues and herbaceous biomass (Tao et al., 2012). The composition of ash, specifically the alkaline content, results in the rapid catalytic polymerization reaction and the formation of fuels coagulation, which increases bio-oil viscosity and lowers the quality of bio-oil (Das et al., 2004). Moreover, alkaline ash also damages the pyrolysis reactor by fouling reactor surfaces and heat transfer tubes, which finally diminishes the overall thermal efficiency.

Energy density (8–14 MJ/kg), bulk density (60–100 kg/m^3), and particle size distributions (depending on the type of grinding unit) are other factors to consider for pyrolysis. These physical characteristics do not alter the pyrolysis reaction pathways or intermediates but significantly impact reaction severity, thereby changing the bio-oil yield and distribution. Moreover, low bulk density causes an uneven mixture of feedstock and heating agent (sand) in fluid bed reactor. Also, large particle size inhibits heat transfer; whereas, small particle size causes diffusion of the reaction zone. Some physical and chemical characteristics of biomass are not compatible with effective pyrolysis process. Therefore, preprocessing/pretreatment prior to pyrolysis is necessary to reduce the heterogeneity of the untreated biomass, alter the physical characterization and chemical composition, remove the undesirable components from feedstock and modify the reaction route of biomass thermal decomposition.

Feedstock pretreatment is an economic approach to produce bio-oil with high quality and yield, while reducing the downstream upgrading cost (Hassan et al., 2009). Currently, three pretreatment methods can be employed prior to pyrolysis: thermal pretreatment, physical pretreatment, and chemical pretreatment. Thermal pretreatment involves drying and torrefaction processes that mainly aim to reduce the formation of oxygenated compounds during pyrolysis and increase the energy density of bio-oil. Physical pretreatment, such as grinding and densification, focusses on improving the reaction process' efficiency. Chemical pretreatment, such as water washing, acid and base leaching, and steam explosion, typically concentrates on optimizing bio-oil physicochemical property and product distributions through removing inorganic elements and altering compositions of lignocellulose feedstock.

2. Biomass Compositions

2.1 Biomass feedstock type

Various types of biomass are available for conversion into energy. Due to the presence of significantly high moisture content (up to 90% after harvesting) for aquatic biomass, pyrolysis process is not a preferred form of conversion pathway to produce bio-oil. Hydrothermal liquefaction of aquatic biomass under critical water pressure is typically used to produce algae-based bio-crude oil. Lignocellulosic biomass, containing woody and herbaceous biomass, is composed of three major components: cellulose, hemicellulose, and lignin. The former two structural carbohydrates are also termed as holocellulose. Cellulose constitutes more than 40% (wt) of plant biomass with glucose units linked through β-1,4-glycosidic bonds. Crystalline structure cellulose is more tolerant to thermal decomposition and acid hydrolysis than its amorphous form. Hemicellulose is a heterologous polymer, consisting of various five-carbon sugar monomers and uronic acid; it accounts for 20–40% (wt) of plant biomass. Lignin is an amorphous aromatic heteropolymer with p-hydroxyphenylpropanoid units, accounting for 20–30% of the plant biomass. In lignocellulosic biomass, holocellulose and lignin twist together and form a tenacious matrix structure. During pyrolysis, the hemicellulose is first decomposed followed by amorphous cellulose and crystalline around 300°C. Lignin decomposition occurs at wide temperature ranges although some lignin fractions are thermal resistant and tolerant to approximately 400°C. The structural carbohydrates and lignin contents in woody and herbaceous biomasses differ greatly and are summarized in Table 3. Besides structural carbohydrates and lignin, the remaining materials are extractives and inorganic ash. The extractives only account for a small fraction of dry weight including oils, gums, waxes, pectin, and proteins (NREL Determination of Extractives in Biomass, 2008). The ash is composed of mineral matter and mainly composed of silica and alkaline salts.

The properties of biomass feedstocks include elemental composition, moisture, volatile matter, fixed carbon, ash content, energy density and biochemical compositions. The characteristics of some biomass are listed in Table 3.

3. Thermal Pretreatment

Thermal pretreatment of biomass feedstock is employed at elevated temperatures to remove moisture and some thermal susceptible components prior to pyrolysis with or without the presence of an oxygen-limited atmosphere. Depending on the temperature ranges, thermal pretreatment methods can range from simple drying to severe treatments. Simple drying is a nondestructive thermal process, which evaporates free and bound water from biomass. The typical severe pretreatment for pyrolysis is torrefaction, during which devolatilization, depolymerization, and carbonization reactions occur mainly on hemicellulose, and thereby change the compositional structure of raw biomass feedstock. Although multiple decompositions and cross-reactions occur among biomass components during torrefaction, this pretreatment is considered to be a thermal approach, as no additional chemical agent is used. Both drying and torrefaction pretreatments have demonstrated that biomass characteristics such as toughness, energy density, and hydrophilicity can be altered to mitigate biomass heterogeneity. As a result, multiple feedstock types can be thermally pretreated and blended to generate pyrolysis ready feedstock for efficient bio-oil production.

3.1 Drying

The drying process involves two temperature regions: 50–150°C for nonreactive drying and 150–200°C for reactive drying. During nonreactive drying, the change in physical properties includes particle size shrinkage and porosity reduction due to water evaporation. During reactive drying, biomass thermally degrades. For example, lipophilic extractives and volatile compounds begin to evolve rapidly around 200°C (Westover et al., 2013). When the biomass temperature increases to 120–150°C, a fraction of lignin

Table 3. The compositions of some lignocellulosic biomass.

Analysis		Woody biomass		Agricultural biomass				
		Pine Wood (Naik et al., 2010)	Poplar wood (U.S. DOE 2006; Slopiecka et al., 2012)	Corn stover (Tillman et al., 2009; Vassilev et al., 2012)	Rice Straw (Lee et al., 2005)	Switch-grass (Tillman et al., 2009; Vassilev et al., 2012)	Wheat straw (Bridgemana et al., 2008)	Sugarcane bagasse (Das et al., 2004)
Ultimate analysis [% wt]	Carbon	49.0	45.5	49.7	39.2	50.7	47.3	56.3
	Hydrogen	6.4	6.3	5.9	4.8	6.3	6.8	7.8
	Nitrogen	0.1	1.0	1.0	1.6	0.8	0.8	0.9
	Sulfur	0.0	ND	0.1	0.7	0.2	ND	ND
	Oxygen	44.4	47.2	42.6	53.7	41.0	37.7	27.5
Proximate analysis [% wt]	Moisture	5.8	9.6	8.0	6.6	9.8	4.1	ND
	Volatile Matter	82.4	75.5	69.7	63.3	69.1	76.4	84.8
	Fixed Carbon	10.3	11.2	15.4	17.9	12.9	17.3	13.3
	Ash	1.5	3.7	6.9	11.1	9.8	6.3	1.9
Composition analysis [% wt]	Hemicellulose	34.0	20.4	23.5	32.5	35.3	30.8	23.3
	Cellulose	39.0	49.9	36.3	54.6	44.8	41.3	31.0
	Lignin	12.0	18.1	17.5	12.8	11.9	7.7	21.8

ND: not determined; NA: not applicable.

becomes soft and acts as a binder, which results in a glossy appearance on biomass particle surface. The bulk density of biomass tends to increase. At this stage, the changed physical structure could be recovered with rewetting. When the temperature reaches approximately 200°C, physically bound water begins to release from structural carbohydrates. A small fraction of lipophilic extractives and volatiles evaporate with the breakage of hydrogen and carbon bonds. Partial hemicellulose depolymerization and structural deformity occur at this stage, and the physical and chemical changes could not be revised by rewetting. Mass loss is also more obvious post-drying (Westover et al., 2013). The physical and chemical properties of lignocellulosic biomass feedstock after nonreactive drying, reactive drying, and destructive drying (typically torrefaction) are respectively described in Table 4.

The influence of drying pretreatment on pyrolysis is mainly reflected in decreased moisture content of bio-oil. Untreated biomass contains around 25 to 60% of water (Pang and Mujumdar, 2010); without drying pretreatment, the majority of water from raw biomass is condensed after quenching and harvested as a fraction of the water phase in bio-oil (the other fraction of the water phase comes from dehydration reactions). The moisture content of fast pyrolysis bio-oil reaches from 15 to 30%, much higher than that of crude oil (0.1%). The high moisture content of bio-oil reduces the heating

Table 4. The physical and chemical changes during thermal pretreatments of biomass (Carpenter et al., 2014; Westover et al., 2013).

	Non-reactive drying	Reactive drying	Destructive drying (torrefaction)
Temperature	50–150°C	150–200°C	240–320°C
Property change	Physical change, glossy appearance on surface	Initial thermal degradation on lipophilic extractives and volatile compounds	Extensive thermal degradation with oxygen removal
Tissue	Cell structure is initially destructed and shrunk with porosity reduction	Structural deformity	Fibrous tissue is destructed into brittleness and water affinity reduction
Moisture	Most moisture evaporates	physically bound water releases	Dehydration from decomposed hemicellulose
Hemicellulose	Simply drying	Depolymerization	Deacetylation and devolatilization
Cellulose	Simply drying	Simply drying	Partial depolymerization and carbonization
Lignin	glass transition	Partial depolymerization	Partial demethoxylation cleavage on aryl-ether linkages, and carbonization

value (16–19 MJ/kg as compared with crude oil 44 MJ/kg), and the flame temperature increases ignition delay and lowers combustion rate (Liu et al., 2014). Drying improves bio-oil quality by reducing its water content. The impact of drying pretreatment on the bio-oil property from pine chips is described in Table 5. The decrease in the moisture content of wood chips from 7.4% to approximate 3% reduced the water content in the bio-oil from 14% to 10% (Westover et al., 2013). However, it is not clear that the drying process has any influence on the organic fraction of bio-oil.

Table 5. The impact of drying pretreatment on feedstock and pyrolysis bio-oil moisture content (Westover et al., 2013).

Feedstock pretreatment	Feedstock Pine chip	Fast pyrolysis bio-oil*							
		Oil fraction				Aqueous fraction			
	Moisture [wt%]	yield [wt%]	Water content [%]	Carbon [wt%]	Total acid number	Yield [wt%]	Water content [%]	Carbon [wt%]	Total acid number
Untreated	7.4	87	14	53	80	13	76	13	47
120°C drying	4.1	93	9.8	57	78	7	74	11	37
180°C drying	3.2	91	11	57	72	9	77	11	35

*Bio-oil was obtained at 480°C, continuous feed bubbling bed reactor.

3.2 Torrefaction

Torrefaction, low temperature (200–300°C) pyrolysis process is carried out under the anoxic condition that produces torrefied solid biomass and condensable liquids and non-condensable gases. During torrefaction, the fibrous structure and tenacity of biomass are destroyed (van der Stelt et al., 2011), oxygen and water are removed from lignocellulose through carboxylation and dehydration reactions; thus nearly doubling the energy density of the torrefied solid biomass, which could be further used as the feedstock for gasification and co-firing.

Torrefaction can be severed alone as a mild slow pyrolysis process, which occurs at temperatures between 200–320°C in the absence of oxygen under ambient pressure for solid torrefied biomass production with the byproducts of condensable liquids and non-condensable gases. However, different from slow pyrolysis occurring around 500°C, the majority of cellulose and lignin are tolerant and remain in the solid torrefied biomass. This means that torrefaction can also work as a thermal pretreatment process with the purpose to alter the composition of biomass feedstock prior to fast pyrolysis.

During torrefaction, most of the hemicellulose in the raw biomass is decomposed, and the structures of cellulose and lignin are modified. It was also observed that fraction of lignin is degraded during torrefaction

(Zhang et al., 2012; Haiping et al., 2007). Removal of hydroxyl groups during torrefaction reduces the oxygen content, thus increasing the energy density. With the devolatilization and carbonization of structural carbohydrates, carboxylation and dehydration reactions occur to remove oxygen from biopolymers or their degraded intermediates in the formation of CO, CO_2, and H_2O. The torrefied biomass yield depends on the torrefaction operation condition and biomass type. In general, approximately 70% of initial biomass remains as a solid product with up to 90% of the initial energy. The liquid product contains volatile compounds derived from holocellulose and partial lignin. The dominant components in the liquid product are water, carboxylic acids, ketones, aldehydes alcohol, and phenol derivate. The color of the biomass turns dark or even black, depending the torrefaction severity, and the biomass fibers become brittle and hydrophobic. Torrefaction also reduces the energy consumption for grinding and benefits storage. The compositional characterizations of some torrefied biomass feedstocks are listed in Table 6.

Several studies have investigated the effect of torrefaction on pyrolysis bio-oil conversion. Pyrolysis of torrefied biomass considerably altered the bio-oil yield and composition. Higher torrefaction severity resulted in less solid byproducts and high-quality bio-oil with less oxygenated compounds after fast pyrolysis (Neupane et al., 2015; Zheng et al., 2013; Meng et al., 2012).

The bio-oil yield and compositional distribution from several torrefied lignocellulosic biomasses are listed in Table 7. Neupane et al. (2015) employed pinewood as feedstock to estimate the effect of torrefaction severity on bio-oil composition at 550°C. It was found that the phenolic yield in pyrolysis gas, chromatography-mass spectrometry (pyro-GC/MS) increased to 2.7 folds with torrified pinewood at 275°C for 15 min under non-catalytic condition. The microporous material of HZSM-5 has been most extensively studied for catalytic conversion of hydrocarbons from lignocellulosic biomass due to its strong acidity and shape selectivity. It was reported under the catalytic condition of H-ZSM5 that the torrefied biomass at 250°C for 15 minutes produced the maximum hydrocarbon yield of 35.3%, which was 67% higher than the non-torrefied pinewood. It was also illustrated under moderate torrefaction temperature that the degradation of hemicellulose and de-methoxylation of lignin were the dominant reactions and cellulose did not start to decompose. Relatively higher cellulose content in the torrefied biomass maximized hydrocarbon content in the bio-oil (Neupane et al., 2015).

Zheng et al. (2013) investigated pyrolysis with torrefied corncob at 470°C in the fluidized bed reactor and found that the higher heating value of bio-oils increased with an increase in torrefaction severity. The higher heat value of bio-oil from raw corncobs was 14.9 MJ/kg, and this value was

Table 6. The change in chemical composition of biomass after torrefaction.

Biomass	Torrefaction condition		Composition			Reference
	Temperature [°C]	Time [minutes]	Hemicellulose	Cellulose	Lignin	
Pinewood	-	-	21.4	46.4	27.3	Zheng et al., 2012
	240	40	17.5	50	19.5	
	260	40	13.4	51.4	33.5	
	280	40	8.1	49.0	40.5	
	300	40	5.5	41.0	49.0	
	320	40	1.6	37.0	60.5	
Pinewood	-	-	24.1	39.1	30.7	Neupane et al., 2015
	225	15	25.4	41.2	29.8	
	225	30	17.0	39.7	23.5	
	225	45	5.4	27.9	38.5	
	250	15	18.9	40.3	27.5	
	250	30	4.9	32.9	37.4	
	250	45	1.6	19.6	44.7	
	275	15	9.1	34.8	40.9	
	275	30	ND	15.4	46.5	
	275	45	ND	1.9	46.6	
Pinewood	-	-	15.2	48.6	26.2	Phanphanich and Mani, 2011
	225	30	12.9	41.2	38.4	
	250	30	6.9	41.9	45.7	
	275	30	1.0	39.5	53.3	
	300	30	0.6	12.8	80.0	
Log residue	-	-	13.3	37.5	26.2	Phanphanich and Mani, 2011
	225	30	14.8	41.0	33.2	
	250	30	5.9	38.6	42.5	
	275	30	5.2	34.1	52.8	
	300	30	1.0	6.1	85.1	

increased to 16.5 MJ/kg from torrefied corncob at 275°C for 20 min and 17.2 MJ/kg at 300°C for 20 minutes. This was because, with severe torrefaction pretreatment, most structural carbohydrates were decomposed; the components in torrefied corncobs were more concentrated in pyrolytic lignin with less water content. The O/C ratio, therefore, dropped from 1.13 of raw corncob to 0.60. The oxygen in raw corncobs was mainly removed in the form of water, acetic acid, methanol, and CO_2 after torrefaction. This

Table 7. The bio-oil yield and its compositions after torrefaction pretreatment.

Biomass	Torrefaction condition	Pyrolysis condition	Bio-oil yield wt.%	Bio-oil HHVs MJ/Kg	Bio-oil composition	Reference
Pinewood	-	Pyro-GC/MS 550°C 90 s	-	-	0.1% aromatics 3.1% furans 1.0% phenolics 2.3% guaiacol 0.2% ketones	Neupane et al., 2015
	225~275°C 15~45 min	Pyro-GC/MS 550°C 90 s	-	-	0.1~0.3% aromatics 0.1~2.9% furans 1.4~3.5% phenolics 0.7~3.5% guaiacol 0.0~0.5% ketones	
Hardwood	-	Bubbling fluidized bed 500°C	71.1	21.5	9.3% acetic acid 9.5% acetol 0.2% furfural 0.2% phenol 9.2% levogulcosan	Boateng and Mullen, 2013
	230°C 30 min	Bubbling fluidized bed 500°C	52.6	22.7	7.7% acetic acid 5.8% acetol 0.2% furfural 0.2% phenol 11.0% levogulcosan	
Loblolly pine	-	500°C 1.5 s	67.2	21.0	12.8% acetyloxyacetaldehyde 7.3% acetol 6.6% acetic acid 1.9% furfural 1.6% phenol	Meng et al., 2012

Table 7 contd. ....

...Table 7 contd.

Biomass	Torrefaction condition	Pyrolysis condition	Bio-oil yield wt.%	Bio-oil HHVs MJ/Kg	Bio-oil composition	Reference
Loblolly pine	270~330°C 2.5 min	500°C 1.5 s	33.5~54.5	21.2~28.7	5.3~8.0% acetyloxyacetaldehyde 5.5~6.2% acetol 5.1~5.7% acetic acid 1.3~1.9% furfural 1.6% phenol	Meng et al., 2012
Switchgrass	-	Bubbling fluidized bed 500°C	57.1	19.4	6.9% acetic acid 12.4% acetol 0.1% furfural 0.2% phenol 4.8% levogulcosan	Boateng and Mullen, 2013
	250°C 90 min	Bubbling fluidized bed 500°C	50.6	22.8	4.6% acetic acid 5.8% acetol 0.1% furfural 0.1% phenol 5.6% levogulcosan	
Corncob	-	Fuidized bed reactor 470°C	57.2	14.9	7.2% acids 8.2% ketones 1.0% furans 2.5% phenols	Zheng et al., 2013
	250~300°C 10~60 min	Fuidized bed reactor 470°C	39.4~55.2	15.1~17.2	3.7~6.9% acids 4.3~7.7% ketones 0.4~1.0% furans 3.1~5.3% phenols	

reduced the formation of oxygenated compounds during the pyrolysis process and increased the H/C-ratio in the bio-oil. Also, the acetic acid, hydroxyacetone, and furfural compositions in the bio-oil decreased, but with increased phenol derivatives. However, the bio-oil yield decreased from 57.2% for raw corncob to 22.2% and 21.1% with torrefied corncob at 275°C for 20 minutes and 300°C for 20 minutes, respectively. Although torrefaction is considered as an effective pretreatment approach, the lower bio-oil yield, to some extent, offsets its advantage on bio-oil quality improvement (Zheng et al., 2013).

Meng et al. (2012) drew a similar conclusion with torrefied loblolly pine chips used as a pyrolysis feedstock. They torrefied pine chips at 270, 300, and 330°C separately for 2.5 minutes and carried out the pyrolysis at 500°C. The bio-oil yield decreased from 67.2 wt.% (non-torrefied pine chips) to 33.5 wt.% (torrefied pine chips at 330°C) with the enhancement of torrefaction severity. However, the higher heating value of the above bio-oils increased from 21.0 MJ/kg for raw pine chips to 28.7 MJ/kg for torrefied pine chips at 330°C. All identified compounds in the bio-oil from raw pine chips existed in the bio-oil from torrefied pine chips. As two significant precursors of hydrocarbons, the contents of phenolics and anhydrosugars were enriched in bio-oil after torrefaction pretreatment; this benefited the production of hydrocarbons after following hydrotreating process. A remarkable decline of 48%~81% in light oxygenates yield was observed, such as acids, ketones, and aldehydes. Besides the reason mentioned by Zheng et al. (2013), the authors provided a more detailed explanation that torrefaction pretreatment modified the compositional structure of biomass feedstock and altered their pyrolysis behaviors (Meng et al., 2012).

Torrefaction is usually combined with other pretreatment processes, such as grinding and densification to improve the biomass feedstock properties for the operational and logistical cost. The overall advantage of torrefaction is listed in Table 8. Torrefaction increases the energy density by about 40–60%, which can benefit transport and storage. The brittleness of biomass tissue and lignin glass transition remarkably reduces energy consumption to about 60% during grinding and pelleting (Phanphanich and Mani, 2011). The destruction of lignocellulose increases the pellet bulk density between 1.2 and 1.7 folds and the biomass reaches an energy density of 17.7 GJ/M^3 (Bergman et al., 2005).

However, a challenge of torrefaction is that more than 30% of initial mass biomass is lost, which limits the application. Finding an efficient approach to effectively convert the condensable liquid and non-condensable gas products from torrefaction into value-added chemicals or fuels will assist the acceleration of torrefaction for its commercialization.

Table 8. The change in biomass properties and benefits of torrefaction pretreatment method.

Properties	Mechanism	Improvement
Physical		
Brittleness	Fiber destruction and coherence reduction with hemicellulose degradation	50~90% energy saved for size reduction (Phanphanich and Mani, 2011)
Moisture content	Free and physical bound water was removed with temperature	Lower moisture in pyrolysis bio-oil, increasing around 20~35% higher heating value of pretreated biomass (Meng et al., 2012; Phanphanich and Mani, 2011)
Energy density	Oxygen was removed with decarboxylation and dehydration in the form of hemicellulose decomposed intermediates	40~70% lower O/C ratio (Zheng et al., 2013; Boateng and Mullen, 2013), which also increases higher heating values of pretreated biomass
Reduced moisture affinity	Breakage of the hydroxyl group on the polysaccharide chains reduced polarity	Around 60% of moisture removed (Phanphanich and Mani, 2011)
Chemical		
Holocellulose and lignin	Deacetylation and decomposition on hemicellulose with limited devolatization and carbonization on partial lignin (demethoxylation) and cellulose	35~50% less carboxylic acids (Boateng and Mullen, 2013) and 30% less furans (Meng et al., 2012) in pyrolysis bio-oil with 2.5 times increased phenolic compounds. Lower acidity of bio-oil
Ash content	Mineral materials were tolerant to torrefaction, its content increases with hemicellulose decomposition	Slight promotion of catalytic polymerization on pyrolytic compounds. Higher viscosity in bio-oil

4. Physical Pretreatment

Particle size and bulk density of biomass are two interrelated properties affecting the pyrolysis' behaviors. The particle size of biomass is highly controlled by the type of grinder used. The higher bulk density of biomass feedstock benefits the transport, storage, and pyrolysis feeding operations. Bulk density could be increased with densification from 40–200 kg/m^3 to the final of 600–1400 kg/m^3 (Chen et al., 2015) in the form of pellets and briquettes. The densified biomass is commonly used for biochar production using a slow pyrolysis process.

4.1 Grinding

Particle size strongly impacts heat transfer and mass transfer, which separately affect the feedstock inner heating rate and decomposed volatiles diffusion in pyrolysis. Most lab-scale pyrolysis studies use the fine particle size with a narrow range to satisfy the rapid heating rate. The desirable average particle diameter and size deviation could be obtained with grinding and screening operations. However, particle size reduction requires considerable energy consumption. The use of fine particle size at the industrial scale pyrolysis plant pose major safety issues as well. Appropriate feedstock particle size is required to meet pyrolysis specifications and to reduce pretreatment operation cost. This promptly demands a comprehensive understanding of the effect of biomass particle size on bio-oil yield and chemical compositional distribution, as well as inter- and intra-particle secondary reactions of diffused bio-oil precursors during biomass decomposition.

Grinding is a common preprocessing method used to reduce biomass feedstock particle size such that it can be a pyrolysis feedstock. Essentially, biomass grinding is an energy-intensive process with low-value products. Biomass type (hardwood, softwood, or herb), moisture content, and in-feed and out-feed size are the important factors affecting the specific energy consumption. It is reported the energy requirement for hardwood biomass has an order of magnitude higher than straw (Cadoche and López, 1989) mainly because of the compositional structure and fiber content. Drying is favored before grinding to reduce energy consumption, increase production rate, avoid overheat, and screen blinding (Schell and Harwood, 1994). Some data on specific energy consumption and resulting particle sizes are listed in Table 9. The effect of grinding pretreatment on pyrolysis is presented through the impact of feedstock particle size on either bio-oil conversion (yield and compositional distribution) of fast pyrolysis or biochar conversion of slow pyrolysis.

During fast pyrolysis, biomass feedstock with fine powder is rapidly heated and is expected to reach the desired temperature instantly. This means that the thermal decomposition of biomass occurs simultaneously on the surface and the center of the feedstock. However, for large particle size (> 0.3 mm), the actual heating rate decreases rapidly with feedstock diameter at the level of two orders of magnitude (Blasi, 2002). This leads the pyrolysis decomposition to occur on the surface first to form char coat before the inside bio-oil precursor diffusion. This non-uniform pyrolysis behavior greatly decreases the yield of bio-oil due to various heating rate distribution within the feedstock. More specifically, lignin-derived oligomers are thereby not completely fragmented into phenol derivate intermediates and more aptly to repolymerize as residues during diffusion. Therefore,

Table 9. The grinding energy consumption and particle size for various lignocellulosic biomass (Vidal Jr. et al., 2011).

Biomass	Mill	Size [mm]	Specific energy [kWh/t]	Reference
Woody				
Poplar	Hammer	< 1	85	Esteban and Carrasco, 2006
Pine	Hammer	< 1	118	Esteban and Carrasco, 2006
Pine bark	Hammer	< 1	20	Esteban and Carrasco, 2006
Herbaceous				
Straw	Hammer	1.6	37	Bitraa et al., 2009
	Hammer	3.2	28–35	Bitraa et al., 2009
	Hammer	1.6	42	Cadoche and López, 1989
	knife	1.5	7.5	Cadoche and López, 1989
Corn stover	Hammer	1.6	15	Mani et al., 2004
	Knife	3.2	20	Cadoche and López, 1989
Switchgrass	Hammer	1.6	52	Mani et al., 2004

in most cases, the decline of final bio-oil yield is the result of the reduction of lignin decomposition vapors. Additionally, large particle size feedstock increases mass transfer resistance. The generated volatiles from the plant cell wall tends to diffuse along the fiber structure and are captured at the end in the large particle. It is the length of the large particle size that resists the diffusion of the bio-oil precursor into the reactor; and the diameter of the large particle size impedes the uniform pyrolysis temperature within biomass (Shen et al., 2009; Brackmann et al., 2003).

Shen et al. (2009) investigated the bio-oil yield and distribution with 8 different ranges of particle size (0.18–5.6 mm) from mallee wood biomass at 500°C in a bench-scale fluidized-bed reactor. The bio-oil yield increased to 12–14% when the average particle size decreased from 1.5 to 0.3 mm. With the increased bio-oil yield, char and gas yield decreased separately by 5% and 8–10%. However, the yield of bio-oil, char, and gas indicated stability with the larger particle size increasing from 1.5 to 5.2 mm. The relationship between bio-oil yield and particle size is indicated in Fig. 2 (Shen et al., 2009). Besides particle size, the authors also provide another possible explanation that the destructed cellular structure by intensive grinding pretreatment to small particle size (0.18–0.6 mm) resulted in intensification of secondary reactions in the closed cells (Shen et al., 2009), led to decreased bio-oil yield. In addition,

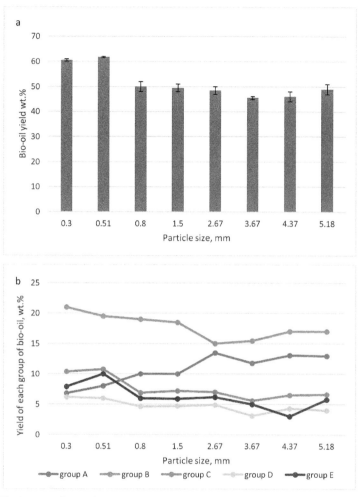

Fig. 2. The impact of particle size on bio-oil yield (a) and composition (b) (Shen et al., 2009). Group A: volatile organic compounds, mainly hydroxyacetaldehyde, formic acid, and methanol; group B: water and organic compounds with boiling points close to water, such as acetic acid, acetol, and propionic acid; group C: phenols and furans; group D: levoglucosan and polyaromatics; group E: water-soluble oligosugars.

the bio-oil composition and its distribution were also influenced by particle size. Based on Shen's results, the yield of small molecular volatile compounds, such as hydroxyacetaldehyde, formic acid, and methanol was decreased half with size reduction from 5.2 to 0.3 mm; whereas, the yield of less volatile compounds such as acetic acid, acetol, and propionic acid increased slightly. The yields of phenols, furans, anhydrosugars and oligomer sugars showed a similar sharp-increase trend with the particle size reduction from 0.8 to 0.5

mm; but slightly varied when particle sizes between 0.8 and 5.2 mm, as seen in Fig. 2 (Shen et al., 2009). During slow pyrolysis, the yield of biochar decreases with particle size reduction. A smaller particle size reflects a larger specific surface area of char for intra-particle interaction and thereby formation of more vapors and diffusion from residues into the reactor. Haykiri-Acma investigated the impact of particle size on biochar formation with hazelnut shell and found both biochar yield and apparent activation energy increased with the particle size from 0.15 to 1.4 mm (Haykiri-Acma, 2006). Some major results of Haykiri-Acma's research are shown in Fig. 3, which had a lot of similarity with the Demirbas' study, that applied agricultural residues such

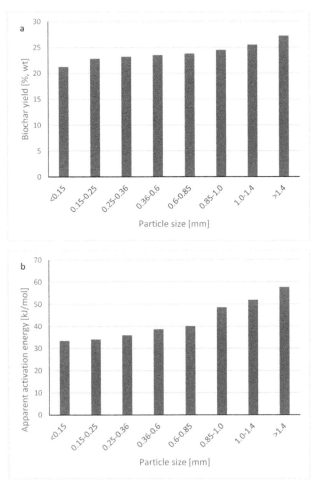

Fig. 3. The impact of particle size distribution on biochar yield and apparent activation energy (Haykiri-Acma, 2006). (a), the impact of particle size on biochar yield; (b), the impact of particle size on apparent energy activity.

as olive husk, corncob, and tea waste as feedstock with the particle sizes ranging from 0.5 to 2.2 mm (Demirbas, 2004).

4.2 Densification

Densification is a common preprocessing method used to increase the bulk density of biomass. During this process, the durability is enhanced, moisture content is reduced, and fiber is compacted into uniformly sized solid that leads to effective transport, storage, and handling. Moreover, the volumetric energy density of untreated biomass dramatically increases after densification. Pelleting, as a representative densification treatment is thereby commercialized for power generation and residential heating from woody biomass in Europe. The pellet quality is determined by multiple factors, such as fiber sources, particle size, moisture content, and operating parameters (Bissen, 2009). During pelleting, a small fraction of lignin is softened, and the glass transition of lignin acts as a binder to consolidate the pellet shape. The notable effect of densification pretreatment on pyrolysis is achieved through the density enhancement of biomass feedstock. Biomass feedstock with low bulk density shows stronger surface cohesive force and tends to suppress the flowability of feedstock during feeding into the reactor. However, densification only benefits the operating process but indicates no direct impact on pyrolysis bio-oil quality and yield. The quality of biochar generated from pelleted biomass is influenced by char yield and mechanical stability of pellets (Erlich et al., 2006; Vassilev et al., 2012).

5. Chemical Pretreatment

5.1 Effect of ash on pyrolysis

Ash content of biomass ranges from 0.8 to 26%, depending on the plant species, soil condition, and fertilization. Agricultural biomass, such as straw and grass, has more ash content than woody biomass. For woody feedstocks, de-barking is favored before pyrolysis, as the bark is high in ash and mineral materials, which catalyzes the polymerization reaction of pyrolytic compounds in the bio-oil and increases the viscosity. The ash contents of various biomass types are listed in Table 10. It is widely demonstrated that when the total ash content is higher than 1% wt the bio-oil yield is reduced, while char yield increases.

The main metal elements of ash composition include silicon, potassium, and calcium, followed by aluminum, magnesium, and sodium in the form of silicates, oxy-hydroxides, sulfates, phosphates, carbonates, chlorides, and nitrates (Vassilev et al., 2012). The impacts of several aforementioned metal elements on pyrolysis behavior have been investigated. Silicon is relatively inert, and its content has negligible influence on pyrolysis thermal

Table 10. The ash content and composition of lignocellulosic biomass.

Biomass	Ash content*	Elemental composition (% of ash)									Reference
		Si	P	S	Na	Mg	Al	K	Ca	Fe	
Wood	0.8~2.5	22.2	3.5	2.8	2.9	6.1	5.1	10.8	43.0	3.4	Carpenter et al., 2014
Corn cob	2.8	45.3	2.1	0.1	0.7	7.8	0.0	43.0	0.8	0.1	Raveendran et al., 1995
Corn stalk	6.8	37.5	6.0	1.6	18.1	16.6	5.4	0.1	13.1	1.5	Raveendran et al., 1995
Miscanthus	2~8	66.3	3.9	0.8	0.4	4.7	2.2	9.6	10.2	1.4	Carpenter et al., 2014
Sugarcane bagasse	1.83	49.8	0.8	2.3	0.7	23.9	0.2	9.6	4.8	0.2	Das et al., 2004
Rice straw	19.8	88.2	0.4	0.1	2.6	3.2	0.0	2.7	2.4	0.1	Raveendran et al., 1995
Rice husk	23.5	93.6	0.1	0.1	0.1	0.7	0.0	3.8	0.8	0.2	Raveendran et al., 1995
Switchgrass	5.49	66.3	3.9	0.8	0.4	4.7	2.2	9.6	10.2	1.4	Raveendran et al., 1995
Wheat straw	11.2	45.9	0.2	0.8	8.1	4.5	2.5	29.9	7.9	0.1	Raveendran et al., 1995

*on the basis of dried biomass.

decomposition of structural carbohydrates and lignin. After pyrolysis, the residual silicon is accumulated and constitutes the dominant ash component in the char. The alkali and alkaline earth metallic cations indicate a negative effect on the bio-oil yield but promote the yield of char by altering the pyrolysis mechanism as a catalyst. It includes potassium and sodium from alkali metals and calcium and magnesium from alkaline earth metals. However, magnesium and iron have negligible impact on the bio-oil yield.

Potassium has different catalytic effects on the pyrolysis mechanism of structural carbohydrates and lignin. For hemicellulose and cellulose, potassium promotes the decomposition of depolymerized anhydrosugars intermediates (levoglucosan) into small molecular water-soluble products, such as acetic acid, formic acid, hydroxyacetaldehyde, and acetol (Fuentes et al., 2008). For lignin, potassium significantly promotes the pyrolytic decomposition of lignin into phenol derivate as a catalyst. Subsequently, alkali and alkaline metallic cations, especially potassium and calcium, in the entrained char induce repolymerization reaction among those monomeric phenol derivate in the presence of small molecular compounds such as acetic acid, formaldehyde, and alcohols into oligomeric compounds. These oligomers increase bio-oil viscosity and decrease bio-oil yield with minimal change in the bio-oil chemical distribution. Therefore, de-ashing, especially the removal of the alkali and alkaline cations (potassium and calcium) is favored prior to pyrolysis.

5.2 Washing and leaching

The presence of alkaline and metallic ions in biomass significantly changes the mechanism of pyrolysis decomposition resulting in different physiochemical properties of bio-oil. Typically, a diverse variety of highly oxygenated carbonyls and hydroxyl compounds are formed. These oxygenated compounds include but are not limited to hydroxyacetaldehyde, acetic acid, and acetol (Scott et al., 2001). The presence of cations catalytically breaks lignin and holocellulose units, promotes the formation of undesirable compounds, and finally impairs the bio-oil quality.

The majority of soluble ashes and inorganic mineral materials could be removed through washing and leaching process with water, diluted acid, or a mild base before fast pyrolysis. Water is recommended for leach alkali sulfates, carbonates, and chlorides. Diluted HCl is more suitable for carbonates and sulfates of alkaline earth metal, while a mild base, such as ammonia is favored to remove Mg, Ca, K, and Na (Dai et al., 2008). Compared to dilute acid and base leaching, the washing condition with hot water is moderate but is able to effectively remove the majority of ions by minimizing the operation cost. Therefore, it is always preferred to be conducted prior to pyrolysis. Scott et al. (1994), applied deionized poplar powder as feedstock and found the yield of levoglucosan to increase dramatically from 3.0% to 17.1% in bio-oil; while the yields of hydroxyacetaldehyde and acetic acid decreased, respectively from 10.0 and 5.4% to 2.0 and 1.3% (Piskorz et al., 1994). Das et al. (2004), investigated the water leaching effect on bio-oil yield at 500°C with the feedstock of sugarcane bagasse. Water leaching for 1 to 2 hr removed 40% of indigenous ash (mainly potassium and calcium) and 49.2% extractives as well as remaining soluble sugars. The yield of oil fraction in bio-oil from de-ashed sugarcane bagasse increased from 19.5% to 30%, compared to that of untreated sugarcane bagasse. The higher heating value and moisture did not influence water leaching (Das et al., 2004). Fahmi et al. (2007), removed 71–79% of alkali metals (mainly K) with water leaching from grass feedstock and found that the yield of levoglucosan in bio-oil remarkably increased at the expense of hydroxyacetaldehyde. Pyrolytic lignin compounds were also enriched on phenol derivate without the catalytic polymerization from alkali metals. Those lignin decomposed volatiles were dominant in phenol and 2-methyloxy-4-vinyl phenol (Fahmi et al., 2007).

The use of a dilute acid and mild base solution provide a stronger capability to remove metallic and alkaline cations. The cellular structure could be destroyed and physiologically bound metals could be washed off from plant tissue after acid and base treatment. However, a crucial impact of acid and base pretreatment on biomass should be explained: the dilute acid and mild base solution cause hydrolysis of hemicellulose (favored by acid leaching) and partial lignin (favored by base leaching) before pyrolysis. While acid or base washing led massloss, the washed biomass showed an increase in bio-oil yield. Hassan et al. (2009), investigated the effect of acid and base leaching pretreatments on bio-oil conversion with six solutions (dilute phosphoric acid, dilute sulfuric acid, sodium hydroxide, calcium hydroxide, ammonium hydroxide, and hydrogen peroxide). The pinewood powder with the size from 2 to 6 mm was washed with above solutions at 80–100°C for 60 minutes, then washed with distilled water till neutralization before pyrolysis at 450°C. The characterization of bio-oil was affected by agents. Final pH (2.81~3.76) and acid values (51.1~95.6) of bio-oils fluctuated as compared to that of untreated pinewood bio-oil. It was also reported that the viscosities of acid pretreated bio-oil increased to 53% and 65% for the phosphoric and sulfuric acid treatments, respectively. The hydrolysis and removal of hemicellulose with acid and base pretreatment prior to pyrolysis led to a remarkable decline in furfural yield. The yield of anhydrosugars such as levoglucosan also dropped from 5.06% to around 2% after acid leaching (Hassan et al., 2009). Zhou et al. (2015), investigated the effect of calcium hydroxide on lignin pyrolysis at 400–600°C. Without calcium hydroxide pretreatment, the lignin tended to agglomerate during pyrolysis due to the presence of phenolic hydroxyl, carboxylic acids, and aldehyde groups. Calcium hydroxide pretreatment inhibited lignin agglomeration behavior and converted the above functional groups into hydroxylcalcium phenoxides, phenolic alcohols, and phenolic carboxylate salts. The mechanism of calcium hydroxide pretreatment was indicated in Fig. 4. It was also found that the lignin pyrolysis oil was enriched with 38% of phenol derivate compounds (Zhou et al., 2015).

In general, both acid and base pretreatment methods can be applied to biomass to produce selective high-value chemicals from pyrolysis. These chemicals include but are not limited to furfural, 2,3-Dihydrobenzofuran (DHB), 4-Vinyl guaiacol (VGO), 1-Hydroxy-2-butanone (HBO), and levoglucosenone (LGO). Wang et al. (2015), studied the optimal pretreatment condition for above chemicals with leached corncob through pyrolysis at 500°C. Rather than neutralizing treated feedstocks with water washing, residual acid and base were retained on purpose to promote structural carbohydrates and lignin pyrolysis process and water leaching (neutral in Table 11) was used as a control. It was found that the highest bio-oil yield of 52.5% was obtained at neutral pretreatment compared to that of acid and

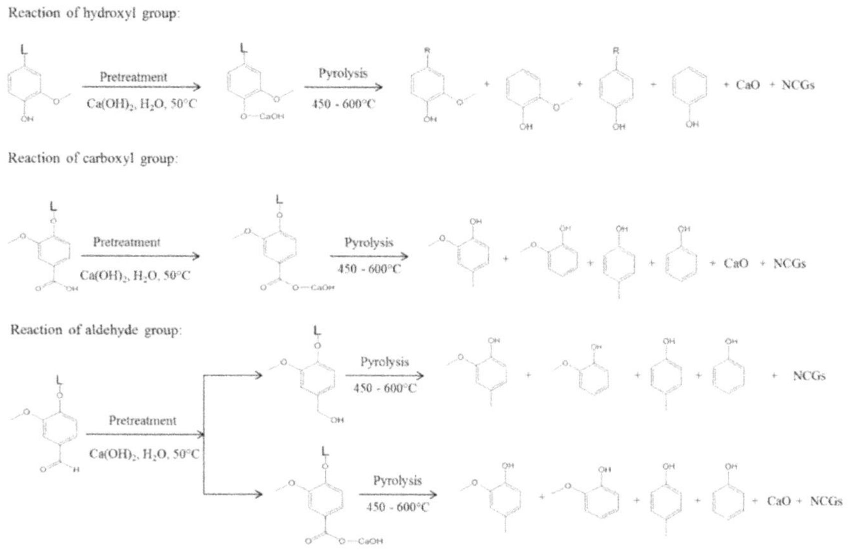

Fig. 4. The impact of calcium hydroxide on lignin pyrolysis. Cited from (Zhou et al., 2015), reproduced by permission of The Royal Society of Chemistry.

Table 11. The yield of chemicals in bio-oil from pretreated corncob with acid or base pretreatments (Wang et al., 2015).

Chemicals in the bio-oil	Pyrolysis chemicals yield (%)								
	H_2SO_4 solution [H⁺] [mol/L]				Neutral	NaOH solution [OH⁻] [mol/L]			
	0.2	0.02	0.1%	0.01%		0.008%	0.08%	0.8%	8%
Acetic acid	3.9	8	9.5	10.1	10.6	11	11.8	12.2	13.3
Furfural	29.9	16.9	5.4	4	3.7	3.1	3.5	4.3	
Levoglucosenone		3.5							
1,2-cyclopentanedione			2.8	2.5					
2,3-dihydrobenzofuran			2.7	2.4					
4-vinyl guaiacol			2.8						
Benzene					4.1	4	5.2	6.2	9.9
Methanol						3.8	4	3.9	5.5
1-hydroxy-2-butanone						2.6	2.9	3.1	3.4

alkaline pretreated biomass due to the cracking reaction of pyrolysis vapors. The bio-oil composition also showed significant variability depending on the solution (Table 11). It was observed that LGO yield was about 3.5% only at moderate acid pretreatment condition (H_2SO_4:corncob = 0.01:1). A mild acid (H_2SO_4:corncob = 0.001:1) promoted the formation of CDO,

DHB, and VGO from lignin and lignin branch with a higher yield of 2.8%, 2.6%, and 2.8% respectively. Benzene, methanol, and HBO were produced only with base treatment and their yields gradually increased. Acetic acid was produced with all pretreatments, its yield showed a similar trend as benzene, methanol, and HBO, reaching 13.3% under the condition of NaOH:corncob = 0.08:1. In contrast, the yield of furfural decreased with OH⁻ and disappeared when NaOH:corncob = 0.08:1. It was deduced from the above results that acid and base pretreatment prior to pyrolysis provide more probability of contact with lignocellulose efficiently. Acid and base chemicals worked as phase transfer catalysts in pyrolysis and affected the liquid product distribution (Wang et al., 2015).

5.3 Steam explosion

Steam explosion is a thermo-chemical treatment method used to permeate lignocellulosic biomass tissues with saturated steam at moderate temperature (180~240°C) and high pressure (7000 kPa). It was first tested for bioethanol and wood pellet production followed by pyrolysis. During steam explosion process, lignin is broken into smaller polymer units of 400 to 8000; hemicellulose is more susceptible than lignin and is decomposed into oligomeric saccharides, which are predominantly water soluble and washed off (Carpenter et al., 2014). Severe than water leaching, physiologically bound metallic and alkaline cations can be substantially removed with high pressurized steam permeation. Therefore, the steam explosion has recently attracted attention in pyrolysis pretreatment. Since the hemicellulose is removed with steam, the mass loss of feedstock, as well as the change in composition, affects pyrolysis bio-oil yield and its compositional distribution.

Biswas et al. (2011) investigated the steam explosion effect on pyrolysis behavior when using woodchips as a feedstock. The wood chip was pretreated with steam at the temperature of 205 to 228°C for 6 to 12 minutes before pyrolysis with thermogravimetric analysis (TGA). It was found that the treated feedstock was more reactive during pyrolysis due to the decomposition of hemicellulose, amorphous cellulose, and a fraction of lignin (Biswas et al., 2011). Wang et al. (2011) compared the effects of three pretreatments (dilute acid, dilute alkali, and steam explosion) on pyrolysis bio-oil conversion with pine wood at 450°C. Pine wood powder was pretreated at 100°C for 1 hour with 0.5% and 1% H_2SO_4 or 0.5% NaOH for acid and alkali pretreatment. The pressure of 1.3 MPa was applied in the steam explosion pretreatment from 173 to 193°C for 10 minutes. It was found that the dilute acid pretreated pine wood had the highest bio-oil yield of 63%, while the dilute alkali and steam explosion pretreated feedstock had lower bio-oil yields of 49% and 44%, respectively than that of untreated biomass (54%). The absence of hemicellulose was considered to be the determining reason of decreased bio-oil yield during steam explosion pretreatment. The higher yields of levoglucosan and other anhydrosugars in bio-oil were observed from acid and steam explosion pretreated biomass due to the removal of indigenous metallic and alkaline cations (Wang et al., 2011).

6. Mechanical Preprocessing

Mechanical preprocessing aims to produce biomass feedstock with proper physical properties prior to fast pyrolysis. It is commonly employed in pilot plant scale or industrial scale. Similar to physical pretreatment, mechanical preprocessing concentrates on improving particle size and bulk density but has no impact on the composition of biomass. However, with mechanical preprocessing, typically air classification, ununiformed fractions in raw biomass such as clays, stones, and dust could be migrated from the majority of biomass feedstock. Therefore, mechanical preprocessing indicates indirect impacts on pyrolysis behavior and bio-oil composition through the removal of exogenous ash from biomass. Currently, as an alternative unit operation, mechanical preprocessing is more attractive as an economic and effective de-ash pretreatment than physical improvement. The impacts of ash on pyrolysis performance and bio-oil composition was described in chemical pretreatment. In this section, the process and economic analysis of air classification were concentrated as a typical example of mechanical preprocessing prior to pyrolysis.

6.1 Air classification

Air classification is a mechanical approach to roughly separate particles with air flow. It has accomplished the separation of particles below 100 to 300 microns which sieving cannot effectively complete (Voshell, 2015). After air classification, the particles with different size and density are separated into the fine product and coarse product, respectively, collected from the top/side and bottom outlets of air classifier. Particle size, shape, and density are the main physical properties, impacting the separation effect. However, for uniformed materials, individual particles have similar density that avoids the influence of density on separation. When the density of the particle is approximating the same, size, and shape are the main factors influencing the classification result. It was reported that the possibility of flaky particle to be collected as the fine product is more than spherical particle because of different drag coefficient (Johansson and Evertssonn, 2014).

The factor of cut size d_{50} is employed to characterize the separation result. Particles with sizes less than d_{50} constitute fine fraction, while over d_{50} compose coarse particles. During the air classification process, the particle behavior in separation zone can be modeled and calculated. Four basic types of separation zones have been investigated as gravitational-counterflow, gravitational-crossflow, centrifugal-counterflow, and centrifugal-crossflow (Shapiro and Galperin, 2005). The structural schematics and particle force analysis are indicated in Fig. 5.

Currently, air classification is also employed as a potentially effective and economical mechanical treatment approach to mitigate ash from biomass prior to thermochemical conversions. Ash is deleterious for corrosion, fouling and slagging; it also impedes heat transfer and advocates catalyst poison (Lacey et al., 2015). Air classification accomplishes ash reduction and enhances the biomass feedstock quality through mitigating most exogenous ash with the least amount of fine fractions biomass from the bulk. The fine fractions concentrate multiple elemental ash components from soil contamination such as sodium, alumina, silica, iron; and could be migrated at the expense of slight biomass loss. However, the intrinsic ash with alkali and alkaline elements of calcium, potassium, and magnesium are distributed evenly throughout all fraction, and could not be effectively removed with air classification.

Prior to the pretreatment of air classification, dryness and size reduction are both required for green biomass. The desirable moisture content was reported to be approximately 10 wt.% with the size less than 1 inch

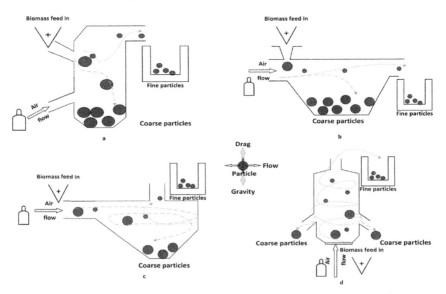

Fig. 5. Separation zones: (a) gravitational-counterflow zone, (b) gravitational-crossflow zone, (c) centrifugal-counterflow zone, (d) centrifugal-crossflow zone. Cited from (Shapiron and Galperin, 2005).

(Thompson et al., 2016). Large particles of mineral components such as stones and rocks are also required to sieve away from the bulk biomass before air classification. The dried and ground biomass is fed into the chamber and separated with adjustable airflow velocity into the good product (coarse particle) and fine particles. The isolation of high-quality biomass from low-quality biomass enables the specific utilization of both fractions. With industrial scale, a cyclone or a baghouse is always employed after air classifier to deposit fine particles. The energy is mainly consumed by fan or blower, which provides an upward air stream for separation.

The compositional and physical characterizations of different anatomic fractions vary significantly among a diversity of lignocellulosic biomass, such as forest residues, agricultural residues and energy crops. Anatomic fractionation indicated most exogenous ash distributed non-evenly in discrete fractions. An example of loblolly pine indicated the total ash content and alkali and alkaline metal elements (AAMEs, mainly Ca, Na, K, and Mg) content concentrated in the fine particles fraction. The needle fraction had the highest total ash content of 2.21 wt.% (by biomass) and highest AAMEs contents of 0.631 wt.% with the mass percentage of 1.3 wt.%; while white wood fraction contained lowest ash content of 0.38 wt.% and lowest AAMEs content of 0.155 wt.% with the majority mass percentage of 59.6 wt.% (Lacey et al., 2015). The contents of total ash and AAMEs from bark, twig, branch, and cambium were in the range of needle and white wood. Thereby, a selective ash separation is possible as a compromise strategy to mitigate exogenous ash at the expense of slight biomass loss. This strategy had been achieved by adjusting the airflow rate, which was controlled by the fan controller with the unit of Hz. Lacey et al. (2015), illustrated a linear relationship between fan controller setting (Hz) and air velocity with the air cleaner obtained from Key Technologies Company. When the fan controller setting increased from 10 to 60 Hz, the airflow velocity reached from 400 to 1000 ft³/min. The maximum ash reduction was 65% at the expense of 25% mass loss at 22 Hz. It was optimized to mitigate 49% of total ash and 22% of AAMEs with the mass loss of 9 wt.% at the controller setting of 12 Hz (Lacey et al., 2015). Higher airflow velocity led to the organic yield loss excessed ash reduction. The impacts of controller setting on ash removal and organic yield loss are indicated in Fig. 6.

Air classification was economically feasible with the cost of $1.24–$7.42 per metric ton forest residue biomass with the plant capacity of 180,000 dry tons annually (Hu et al., 2017). The cost assumptions for air classification was indicated in Table 13. Among the total cost, process cost was constant at 0.83 $/dry ton. This cost was the sum of capital and operating costs. The capital investment contributed 0.21 $/dry ton to the final cost, mainly from the equipment cost of the classifier and hopper and conveyer system.

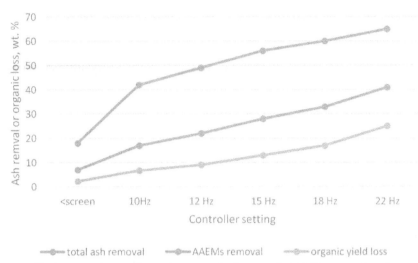

Fig. 6. The impacts of controller setting on total ash removal, AAMEs removal, and organic yield loss (Hu et al., 2017).

The operating cost was 0.62 $/dry ton, three times higher than the capital investment cost. The other cost of air classification was organic yield loss cost, ranging from 0.41 to 6.59 $/dry ton with the biomass loss of 2.0 to 25 wt.% at controllers setting of 10 to 22 Hz. Under the optimized condition of 22 Hz, the cost of organic yield loss was 1.90 $/dry ton with nine wt.% biomass loss; the total cost (sum of process cost and organic yield loss cost) was 2.73 $/dry ton. The relationship of total cost and controller setting is shown in Fig. 7 (Hu et al., 2017).

Air classification as a dry separation approach is recognized as adaptable for ash reduction from biomass. It has been demonstrated effectively to mitigate major exogenous inorganic dust and fractional intrinsic ash. The isolated fractions have enabled the possibility for specific processes and distinct utilization. However, because of the separation mechanism, air classification is usually associated with other mechanical or chemical preprocessing such as a sieve, water washing, and wet chemical leaching. The limitations of air classification on ash reduction were mainly caused by the non-uniformed properties of biomass and separation process design:

1) Air classification is more suitable for the reduction of exogenous ash without breaking different-sized fractions. However, it is less effective to remove intrinsic ash adhering to the compositional structure of carbohydrates and lignin.

2) Air classification works effectively only over a small range of particle densities with similar sizes. Sieving pretreatment is preferred to assist with removing larger and heavier exogenous minerals such as stones and rocks. Pretreatments of dry and size reduction are also commonly

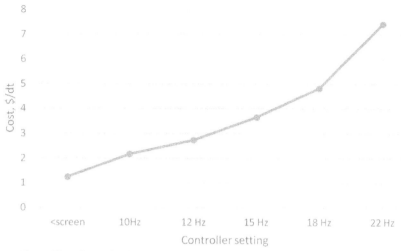

Fig. 7. The relationship between total cost and controller setting (Hu et al., 2017).

required prior to air classification to ameliorate the compositional and physical properties of raw biomass.

3) Low throughput of air classifier is another issue of the application. The demanded large volumes of air require larger fans and powerful motors. A cyclone or baghouse is also suggested for industrial-scale separation to avoid the emission of fine particles into the atmosphere.

7. Summary

A multitude of parameters (such as physical and chemical properties of biomass, pretreatment methods, pyrolysis conditions, reactor configurations) affect the pyrolysis behavior of biomass that leads to variation in bio-oil yield and its compositions. Pretreatment, as one of the commonly employed approaches to improve the physical and compositional properties of biomass is summarized in this chapter. To minimize the negative effects of undesirable components in biomass and non-uniformed feedstock, thermal, chemical, and physical pretreatments, as well as mechanical preprocessing were applied with respective mechanisms and effects. Chemical pretreatment mainly aims to migrate ashes and modify the structural components of biomass. Demineralization of high ash content feedstock could be achieved by washing and leaching with water, acid, or alkali solution. Demineralization pretreatment method also increases anhydrosugars content that leads to decline in oligomeric phenols in bio-oil. Washing with rigorous acid or alkali may hydrolyze hemicellulose and a small fraction of lignin. This will remove some highly oxygenated carbonyls or hydroxyl compounds at the expense of a loss in bio-oil yield. Thermal pretreatment (torrefaction) and steam explosion also alter the structure and components of lignocellulosic biomass; resulting in the reduction of hemicellulose content in feedstock and decreasing the formation of carboxylic acids. The crucial different effect of torrefaction and steam explosion pretreatment is the residual moisture content in pretreated biomass. Torrefaction, almost completely removes moisture from biomass. Physical pretreatment method (grinding & densification) and mechanical preprocessing do not substantially change the biomass composition. However, grinding of biomass reduces the particle size and results in the enlargement of surface area, which benefits pyrolysis behavior with efficient heat transfer. Besides physical improvement, mechanical preprocessing such as air classification also works as a promising approach for exogenous ash removal with slight loss of biomass. Table 12 summarizes the impacts of various pretreatment methods on the pyrolysis technology.

In practice, several of the preprocessing/pretreatment methods are always employed prior to pyrolysis to improve feedstock quality. The choice of appropriate combination of pretreatment methods depend on the entire biomass and biofuel supply chains, economic benefits, and

Table 12. The major effects of pretreatment methods on the bio-oil yield.

Pretreatment		Major effects	Outcomes	Ref.
Thermal	Drying	Remove moisture	Moisture content in bio-oil reduces 10~14%, thus increases the energy density of bio-oil	Westover et al., 2013
	Torrefaction	Removes hemicellulose	In bio-oil, phenols and hydrocarbons contents increase by more than 60% with half O/C ratio, higher heating value increases 1.4 folds; 30% initial biomass feedstock loses	Zheng et al., 2013; Meng et al., 2012
Physical	Grinding	Size reduction	Bio-oil yield increases 12–14%, char and gas yield decrease by 5% and 10%, no significant difference on bio-oil component	Shen et al., 2009
Chemical	Washing and leaching	Remove alkaline and metallic actions	Bio-oil yield increases 1.5 folds, hydroxyacetaldehyde and acetic acid contents reduce half in bio-oil	Piskorz et al., 1994
	Steam explosion	Breaks lignin into units and removes hemicellulose	Acid value of bio-oil dropped around 30%, viscosity declines 40%, bio-oil yield decreases 20% due to the loss of hemicellulose	Wang et al., 2011

Table 13. Cost assumptions for air classification (Hu et al., 2017).

Assumption	Value	
Plant life	30 years	
Interest rate	8%	
Construction period	1 year	
Benefits and general overhead	90% of total salaries	
Maintenance	3% of fixed capital investment	
Insurance and taxes	0.7% of fixed capital investment	
Cost component	**Unit cost or consumption**	**Cost ($/dt biomass)**
Installed equipment cost	$484,000	
Fixed capital investment	$841,000	
Total capital investment	$1,124,000	0.21
Electricity for blowers	2X 15 HP	0.10
Salaries	$52,700/year	0.18
Other fixed costs	$55,000/yr	0.34
Total		0.83

pyrolysis reactor configurations. Comprehensively, to produce high-quality bio-oil, multiple integrated processes are mandatorily demanded including harvest, pretreatments, pyrolysis, stabilization/upgrading, and fuel synthesis. A pathway of integrated pretreatments is illustrated in Fig. 8. Additionally, the selection of suitable preprocessing and pretreatment methods could also benefit the production of other high-value chemicals

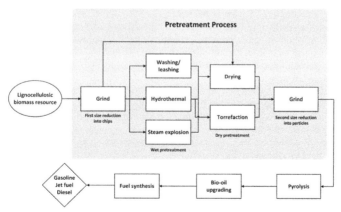

Fig. 8. A pathway to integrate pretreatment methods with high-quality bio-oil production including harvest, pretreatments, pyrolysis, stabilization/upgrading, and fuel synthesis. Cited from (Carpenter et al., 2014).

(such as levoglucosan) along with pyrolytic bio-oil production. Appropriate design of pretreatment process is significant for the rapid commercialization of pyrolysis technology.

Acknowledgment

The authors would like to thank the financial support received from the USDA-NIFA sustainable bioenergy research grant program project number GEOX-2010-03868.

References

Alvarez, J., G. Lopez, M. Amutio, J. Bilbao and M. Olazar. 2014. Bio-oil production from rice husk fast pyrolysis in a conical spouted bed reactor. Fuel 128: 162–169.

Bartek, R., M. Brady and D. Stamires. 2012. Biomass pretreatment for fast pyrolysis to liquid. US Patent 8,236,173.

Bergman, P. C., A. R. Boersma, R. W. Zwart and J. H. Kiel. 2005. Project no. 2020-02-12-14-001, Energy Research Centre of the Netherlands (ECN).

Bissen, D. 2009. Biomass densification document of evaluation (July 13). Minneapolis, MN, USA: Zachry Engineering Corporation.

Biswas, A. K., K. Umeki, W. Yang and W. Blasiak. 2011. Change of pyrolysis characteristics and structure of woody biomass due to steam explosion pretreatment. Fuel Processing Technology 92: 1849–1854.

Bitraa, V. S. P., A. R. Womaca, N. Chevanana, P. I. Miua, C. Igathinathaneb, S. Sokhansanjc and D. R. Smith. 2009. Direct mechanical energy measures of hammer mill comminution of switchgrass, wheat straw, and corn stover and analysis of their particle size distributions. Powder Technology 193: 32–45.

Blasi, C. D. 2002. Modeling intra- and extra-particle processes of wood fast pyrolysis. Aiche Journal 48: 2386–97.

Boateng, A. A. and C. A. Mullen. 2013. Fast pyrolysis of biomass thermally pretreated by torrefaction. Journal of Analytical and Applied Pyrolysis 100: 95–102.

Bond, Q., A. A. Upadhye, H. Olcay, G. A. Tompsett, J. Jae, R. Xing, D. M. Alonso, D. Wang, T. Zhang, R. Kumar, A. Foster, S. M. Sen, C. T. Maravelias, R. Malina, S. R. H. Barrett, R. Lobo, C. E. Wyman, J. A. Dumesic and G. W. Huber. 2014. Production of renewable jet fuel range alkanes and commodity chemicals from integrated catalytic processing of biomass. Energy and Environmental Science 7: 1500–1523.

Brackmann, C., M. Alden, P. E. Bengtsson, K. O. Davidsson and J. B. Pettersson. 2003. Optical and mass spectrometric study of the pyrolysis gas of wood particles. Applied Spectroscopy 57: 216–222.

Bridgemana, T. G., J. M. Jonesa, I. Shieldb and P. T. Williams. 2008. Torrefaction of reed canary grass, wheat straw and willow to enhance solid fuel qualities and combustion properties. Fuel 87: 844–856.

Cadoche, L. and G. D. López. 1989. Assessment of size reduction as a preliminary step in the production of ethanol from lignocellulosic wastes. Biological Wastes 30: 153–157.

Carpenter, D., T. L. Westover, S. Czernik and W. Jablonski. 2014. Biomass feedstocks for renewable fuel production: a review of the impacts of feedstock and pretreatment on the yield and product distribution of fast pyrolysis bio-oils and vapors. Green Chemistry 16: 384–406.

Chang, S., Z. Zhao, A. Zheng, X. Li, X. Wang, Z. Huang, F. He and H. Li. 2013. Effect of hydrothermal pretreatment on properties of bio-oil produced from fast pyrolysis of eucalyptus wood in a fluidized bed reactor. Bioresource Technology 138: 321–328.

Chen, W., J. Peng and X. Bi. 2015. A state-of-the-art review of biomass torrefaction, densification and applications. Renewable and Sustainable Energy Reviews 44: 847–866.

Dai, J., S. Sokhansanj, J. Grace, X. Bi, C. Lim and S. Melin. 2008. Overview and some issues related to co-firing biomass and coal. The Canadian Journal of Chemical Engineering 86: 367–386.

Das, P., A. Ganesh and P. Wangikar. 2004. Influence of pretreatment for deashing of sugarcane bagasse on pyrolysis products. Biomass and Bioenergy 27: 445–457.

Demirbas, A. 2004. Effects of temperature and particle size on bio-char yield from pyrolysis of agricultural residues. Journal of Analytical and Applied Pyrolysis 72: 243–248.

Erlich, C., E. Björnbom, D. Bolado, M. Giner and T. H. Fransson. 2006. Pyrolysis and gasification of pellets from sugar cane bagasse and wood. Fuel 85: 1535–1540.

Esteban, L. S. and J. E. Carrasco. 2006. Evaluation of different strategies for pulverization of forest biomasses. Powder Technology 166: 139–151.

Fahmi, R., A. V. Bridgwater, L. I. Darvell, J. M. Jones, N. Yates, S. Thain and I. S. Donnison. 2007. The effect of alkali metals on combustion and pyrolysis of Lolium and Festuca grasses, switchgrass and willow. Fuel 86: 1560–1569.

Fuentes, M. E., D. J. Nowakowski, M. L. Kubacki, J. M. Cove, T. G. Bridgeman and J. M. Jones. 2008. Survey of influence of biomass mineral matter in thermochemical conversion of short rotation willow coppice. Journal of the Energy Institute 81: 234–241.

Graham, R. L., R. Nelson, J. Sheehan, R. D. Perlack and L. L. Wright. 2007. Current and potential U.S. Corn Stover Supplies. Agronomy Journal 99: 1–11.

Haiping Yang, Rong Yan, Hanping Chen, Dong Ho Lee and Chuguang Zheng. 2007. Characteristics of hemicellulose, cellulose and lignin pyrolysis. Fuel 86: 1781–1788.

Hassan, E. B., P. Steele and L. Ingram. 2009. Characterization of fast pyrolysis bio-oils produced from pretreated pine wood. Applied Biochemistry and Biotechnology 154: 182–192.

Haykiri-Acma, H. 2006. The role of particle size in the non-isothermal pyrolysis of hazelnut shell. Journal of Analytical and Applied Pyrolysis 75: 211–216.

Hu, H., T. L. Westover, R. Cherry, J. E. Aston, J. A. Lacey and D. N. Thompson. 2017. Process simulation and cost analysis for removing inorganics from wood chips using combined mechanical and chemical preprocessing. Bioenerg. Res. 10: 237–247.

Johansson, R. and M. Evertsson. 2014. CFD simulations of a centrifugal air classifier used in the aggregate industry. Published in Mineral Engineering (Journal).

Kenney, K. L., W. A. Smith, G. L. Gresham and T. L. Westover. 2013. Understanding biomass feedstock variability. Biofuels 4: 111–127.

Lacey, J. A., J. E. Aston, T. L. Westover, R. S. Cherry and D. N. Thompson. 2015. Removal of introduced inorganic content from chipped forest residues via air classification. Fuel 160: 265–273.

Lede, J., F. Broust, F. T. Ndiaye and M. Ferrer. 2007. Properties of bio-oils produced by biomass fast pyrolysis in a cyclone reactor. Fuel. 86: 1800–1810.

Lee, K. H., B. S. Kang, Y. P. Park and J. S. Kim. 2005. Influence of reaction temperature, pretreatment, and a char removal system on the production of bio-oil from rice straw by fast pyrolysis, using a fluidized bed. Energy and Fuels 19: 2179–2184.

Liu, C., H. Wang, A. M. Karim, J. Sun and Y. Wang. 2014. Catalytic fast pyrolysis of lignocellulosic biomass. Chemical Society Reviews 43: 7594–7623.

Mani, S., L. G. Tabil and S. Sokhansanj. 2004. Grinding performance and physical properties of wheat and barley straws, corn stover and switchgrass. Biomass and Bioenergy 27: 339–352.

Meng, J., J. Park, D. Tilotta and S. Park. 2012. The effect of torrefaction on the chemistry of fast-pyrolysis bio-oil. Bioresource Technology 111: 439–446.

Min Zhang, Fernando L. P. Resende, Alex Moutsoglou and Douglas E. Raynie. 2012. Pyrolysis of lignin extracted from prairie cordgrass, aspen, and Kraft lignin by Py-GC/MS and TGA/FTIR. Journal of Analytical and Applied Pyrolysis 98: 65–71.

Naik, S., V. V. Goud, P. K. Rout, K. Jacobson and A. K. Dalai. 2010. Characterization of Canadian biomass for alternative renewable biofuel. Renewable Energy 35: 1624–1631.

Neupane, S., S. Adhikari, Z. Wang, A. J. Ragauskas and Y. Pu. 2015. Effect of torrefaction on biomass structure and hydrocarbon production from fast pyrolysis. Green Chemistry 17: 2406–2417.

NREL Determination of Extractives in Biomass [Report]: Technical Report/U.S DOE and Office of Energy Efficiency and Renewable Energy. - [s.l.] : NREL/TP-510-42619, 2008. - p. http://www.nrel.gov/biomass/pdfs/42619.pdf.

Pang, S. and A. S. Mujumdar. 2010. Drying of woody biomass for bioenergy: drying technologies and optimization for an integrated bioenergy plant. Drying Technology 28: 690–701.

Phanphanich, M. and S. Mani. 2011. Impact of torrefaction on the grindability and fuel characteristics of forest biomass. Bioresource Technology 102: 1246–1253.

Piskorz, J., D. Radlein, D. S. Scott and A. V. Bridgwater (eds.). 1994. Proceedings of International Symposium on Advances in Thermochemical Biomass Conversion. Blackie Academic. London, p. 1432.

Raveendran, K., A. Ganesh and K. C. Khilar. 1995. Influence of mineral matter on biomass pyrolysis characteristics. Fuel 74: 1812–1822.

Schell, D. J. and C. Harwood. 1994. Milling of lignocellulosic biomass: results of pilot-scale testing. Applied Biochemistry and Biotechnology 45: 159–168.

Scott, D. S., L. Paterson, J. Piskorz and D. Radlein. 2001. Pretreatment of poplar wood for fast pyrolysis: rate of cation removal. Journal of Analytical and Applied Pyrolysis 57: 169–176.

Shapiro, M. and V. Galperin. 2005. Air classification of solid particles: a review. Chemical Engineering and Processing: Process Intensification 44: 279–285.

Shen, J., X. Wang, M. Garcia-Perez, D. Mourant, M. J. Rhodes and C. Li. 2009. Effects of particle size on the fast pyrolysis of oil mallee woody biomass. Fuel 88: 1810–1817.

Slopiecka, K., P. Bartocci and F. Fantozzi. 2012. Thermogravimetric analysis and kinetic study of poplar wood pyrolysis. Applied Energy 97: 491–497.

Tao, G. C., T. A. Lestander, P. Geladi and S. J. Xiong. 2012. Biomass properties in association with plant species and assortments I: A synthesis based on literature data of energy properties. Renewable and Sustainable Energy Reviews 16: 3481–3506.

Thompson, V. S., J. A. Lacey, D. Hartley, M. A. Jindraa, J. E. Aston and D. N. Thompson. 2016. Application of air classification and formulation to manage feedstock cost, quality and availability for bioenergy. Fuel 180: 497–505.

Tillman, D. A., D. Duong, B. G. Miller, L. C. Bradley and R. T. Wincek. 2009. Proc. Power-Gen Internat. Las Vegas, NV.

U.S. DOE Breaking the Biological Barriers to Cellulosic Ethanol: A Joint Research Agenda, DOE/SC-0095 [(www.doeegenomestolife.org/biofuels/)]. - Rockville, MD: U.S. Department of energy Office of Science and Office of Energy Efficiency and Renewable Energy, 2006.

van der Stelt, M. J. C., H. Gerhauser, J. H. A. Kiel and K. J. Ptasinski. 2011. Biomass upgrading by torrefaction for the production of biofuels: A review. Biomass and Bioenergy 9: 3748–3762.

Vassilev, S. V., D. Baxter, L. K. Andersen, C. G. Vassileva and T. J. Morgan. 2012. An overview of the organic and inorganic phase composition of biomass. Fuel 94: 1–33.

Vidal, B. C. Jr., B. S. Dien, K. C. Ting and V. Singh. 2011. Influence of feedstock particle size on lignocellulose conversion—A review. Applied Biochemistry and Biotechnology 164: 1405–1421.

Voshell, S. W. 2015. Reuse and Air Classification of Bioash. Atlto University, School of Chemical Technology. Thesis.

Wan Isahaka, W. N. R., M. W. M. Hishama, M. A. Yarmoa and T. Y. Hinb. 2012. A review on bio-oil production from biomass by using pyrolysis method. Renewable and Sustainable Energy Reviews 16: 5910–5923.

Wang, H., R. Srinivasan, F. Yu, P. Steele, Q. Li and B. Mitchell. 2011. Effect of acid, alkali, and steam explosion pretreatments on characteristics of bio-oil produced from pinewood. Energy Fuels 25: 3758–3764.

Wang, X., S. Leng, J. Bai, H. Zhou, X. Zhong, G. Zhuang and J. Wang. 2015. Role of pretreatment with acid and base on the distribution of the products obtained via lignocellulosic biomass pyrolysis. Royal Society of Chemistry 5: 24984–24989.

Westover, T. L., M. Phanphanich, M. L. Clark, S. R. Rowe, S. E. Egan, A. H. Zacher and D. Santosa. 2013. Impact of thermal pretreatment on the fast pyrolysis conversion of southern pine. Biofuels 4: 45–61.

Zheng, A., Z. Zhao, S. Chang, Z. Huang, F. He and H. Li. 2012. Effect of torrefaction temperature on product distribution from two-staged pyrolysis of biomass. Energy and Fuels 26: 2968–2974.

Zheng, A., Z. Zhao, S. Chang, Z. Huang, X. Wang, F. He and H. Li. 2013. Effect of torrefaction on structure and fast pyrolysis behavior of corncobs. Bioresource Technology 128: 370–377.

Zhou, S., R. C. Brown and X. Bai. 2015. The use of calcium hydroxide pretreatment to overcome agglomeration of technical lignin during fast pyrolysis. Green Chemistry 17: 4748–59.

Thermal Preprocessing

CHAPTER 7

Steam Treatment of Cellulosic Biomass for Pelletization

Shahab Sokhansanj,* Hamid Rezaei,
Pak Sui (Wilson) Lam, Tang Yong, Bahman Ghiasi
and Zahra Tooyserkani

1. Introduction

Steam treatment of biomass is used extensively as a hydrolysis pretreatment for biochemical conversion (Lam et al., 2015; Zimbardi et al., 1999). This treatment exposes biomass to steam at temperatures between 180–240°C (1.03–3.45 MPa) for several minutes, followed by a depressurization to the ambient condition. Auto-hydrolysis and explosive depressurization alter dimensions, size distribution, shape, and chemical composition of produced particles. Steam-induced particle size reduction and changes in chemical composition enhance the performance of hydrolysis and ethanol yield (Boussaid et al., 2000; Mabee et al., 2007). The percentage of fines smaller than 0.074 mm sieve opening (#200 mesh) increases from 5% to 90% by weight with increasing pre-treatment severity of steam treated Douglas Fir wood chips (Boussaid et al., 2000).

The influence of physiochemical alteration of biomass by a steam explosion on pelletizing properties has been evaluated for poplar wood and steam exploded wood, which have an average pellet density of 1226 kg/m³, compared to the untreated wood with an average density of 1086 kg/

Department of Chemical and Biological Engineering, University of British Columbia, Canada V6T 1Z3.
* Corresponding author: shahab.sokhansanj@ubc.ca

m³ (Shaw et al., 2009). The conventional mechanical methods are reported to require roughly 70% more energy to achieve the same size reduction as explosive depressurization (Holtzapple et al., 1989). Adapa et al. (2010) compared the properties of pellets made from the untreated materials and materials treated at a specific steam explosion condition. They found the geometric mean particle size and bulk density of untreated wheat straw were larger than those properties measured for steam-exploded straw.

The objectives of the current review is (1) to present the state of the art in principles of steam treatment, (2) the effects that saturated steam temperature and (3) the effects a steam treatment would have on particle size and moisture content of cellulosic biomass and the density of pellets made from the treated materials.

2. Steam and its Properties

Water has three phases below the critical point: solid (ice), liquid, and vapor (steam). A combination of temperature and pressure defines whether water is in ice, liquid, or steam phase. Figure 1 shows the temperature-enthalpy (T-H) diagram of water in three states of subsaturated water, a mixture of vapor and liquid water (wet steam), and superheated (dry steam) steam. Saturation conditions are along the bubble line and dew line that separate these states of water. For example, the saturated liquid is along the bubble line that separates the sub-saturated liquid water and the wet steam. The saturated vapor is along the dew line that separates the superheated steam and the wet steam.

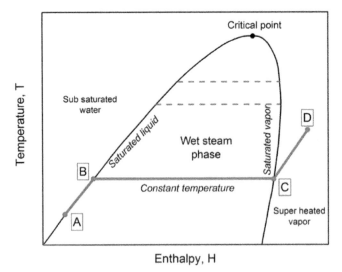

Fig. 1. Temperature-enthalpy diagram for water in three states of liquid, a mixture of vapor and liquid water (wet steam), and dry steam.

Figure 1 shows the relation between the temperature and heat content of water (or enthalpy). In an isobaric process of turning sub-cooled water (state A) to super-heated steam (state D), water takes the heat and follows the red line in Fig. 1. The subsaturated water (state A) absorbs heat and reaches the saturated water state (state B) at a specific temperature (e.g., T_b = 100°C and P_{sat} = 101.3 kPa). Further, heating vaporizes water to change the phase from liquid to gas, starting from saturated water (state B) to saturated vapor (state C). Theoretically, at a constant pressure, water temperature does not change until the complete evaporation of water to steam (phase change B-C) and requires the specific enthalpy of 2257 kJ/kg. Once all water is converted to saturated steam, a further addition of heat increases the steam temperature. At temperatures beyond point C, the steam becomes superheated steam and is known as dry steam.

To produce steam at or beyond the saturation point is that both temperature and pressure of steam should be elevated. This increase in pressure is important especially when the steam is transported. The friction with pipe wall decreases the steam pressure which may even cool the steam to a point to become liquid. The steam should have a pressure higher than its saturation conditions in order to overcome the line pressure drop due to friction. Additionally the steam should remains at a pressure higher than its saturation throughout the pipe especially at the exit in order to prevent condensation. Hence, the required margin depends on the characteristics of the pipes such as size, amount of friction, and steam velocity.

The relationship between saturation pressure and temperature of the water is provided by the following approximate equation (Engineering toolbox, 2014).

$$P_s = \frac{exp\left(77.345 + 0.0057 T_{abs} - 7235/T_{abs}\right)}{T_{abs}^{8.2}} \tag{1}$$

where

T_{abs} = absolute temperature (K)
P_s = saturation vapor pressure (Pa)

Liquid water has a density of about 1000 kg/m³. The density of water decreases when water is converted to the vapor. The following Eq. (2) (Engineering toolbox, 2014) is used to estimate the density of steam as a function of partial vapor pressure and dry bulb temperature.

$$\rho_w = \frac{0.0022 P_w}{T_{abs}} \tag{2}$$

where

P_w = partial pressure water vapor (Pa)

ρ_w = density water vapor or steam (kg/m³)
T_{abs} = absolute dry bulb temperature (K)

Equation (2) yields the density of steam at saturation pressure ($P_s = P_w$) and temperature. Equation (1) is important because it gives the mass of steam required to fill out space. Figure 2 presents the graph of steam pressure and density vs. temperature for saturated steam.

Note that at 100°C the steam pressure is about 101 kPa (the same as atmospheric pressure) and its density is 0.59 kg/m³. The pressure and density of the saturated steam rapidly increase with the increase in temperature. At a temperature of 200°C, the pressure and density of steam are 1530 kPa and 7.1 kg/m³, respectively. As the temperature increases in a boiler, the production of steam increases. The pressure increases if the flow of steam out of the boiler is restricted. Superheated conditions occur

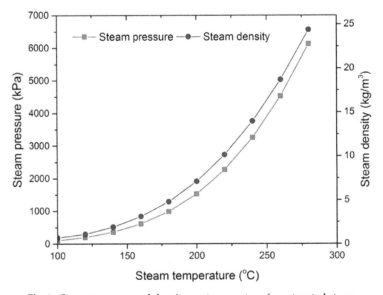

Fig. 2. Steam pressure and density vs. temperature for saturated steam.

when no more water is evaporated and the pressure of the steam inside the boiler increases. The degree of superheat depends on the temperature/ pressure of the above saturation.

3. Pellet Production

Pellets are made by compacting biomass ground particles. The regular pellets in North America are 6.3 mm in diameter with lengths varying from 6 to more than 18 mm. The specific gravity of a single pellet is about 1.2, yielding a bulk density of around 650 kg/m^3. The moisture content of pellets is low about 6–8% (wet mass basis). Pellets are primarily used as a solid fuel for combustion. Recent research shows that pellets can also be used as a feedstock for biochemical and thermochemical conversion to produce (primary/secondary) fuels and chemicals (Kumar et al., 2012; Lam et al., 2013a). U.S. and Canada produced more than 8 million metric tons of pellets in 2016 that are mostly exported to Europe and Pacific Rim regions. However, domestic demand of pellets for residential heat and industrial power production is increasing.

Figure 3 shows a schematic of a typical pellet plant. The raw biomass arrives at the plant in the form of chips. The received material is ground and dried down from a moisture content of 55% (w.b.) to around 10% (w.b.). In a typical pelletization operation, ground biomass is dried in rotary dryers where the biomass is exposed to the hot air stream at 200–600°C.

The drying of pellets consumes 300–3500 MJ/tonne of pellets depending upon the initial moisture content of the biomass (Adapa et al., 2004). Grinding uses 100–180 MJ/tonne of pellets and pelleting uses 100–300 MJ/

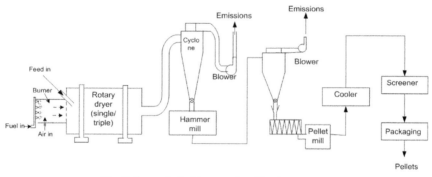

Fig. 3. Layout of a biomass-densification system.

tonne of pellets (Mani et al., 2006). The energy input to produce a tonne of pellet may vary from a minimum of 500 MJ to 4500 MJ or from 2.5% to 25% of the calorific value of pellets (18 GJ/dry tonne). According to the mentioned data, drying operation is the main energy sink in this process. However, pre-processing and transporting the raw biomass to the plant are excluded. Including these two stages make the input/output net energy more unfavorable.

One of the unique advantages of pelletized biomass over a loose biomass is its ease of handling and storage. The International Organization of Standardization (ISO) has developed standards defining the quality of pellets with respect to its handling and use (ISO standard, 2015). Table 1 lists the physical structure and composition of pellets that are the important factors in defining pellet quality for combustion.

Table 1. Quality of produced pellets in North America with respect to their application as a solid biofuel.

Quality factor	Explanation	Values
Moisture content	Moisture content should be low for maximum combustion efficiency and usable heat energy	10%
Calorific value	Less fuel requirement and smaller plant design	19 MJ/kg
Pellet dimensions	Diameter and length of pellets affect feeding and combustion. Smaller diameter pellets are easier to combust. Shorter pellets are easier to be fed to the combustion chamber	4–10 mm
Pellet density	Density of a single pellet determines its burning characteristics or its comminution for combustion	1.2 g/cm^3
Ash melting temperature	Minimizes fouling of the combustion chamber and heat exchange surfaces	850°C
Ash content	Minimizes particle emissions and ash removal	0.5%
Elemental composition	Cl, N, S, K, Mg, Ca, heavy metals. Some of these elements affect the combustion and/or constitute pollutants	mg/kg
Fungi spores	Health risks during materials handling	0

Table 2. Published ISO 17225 grades for wood pellets.

Pellet property	Units	ENplus A1	ENplus A2	ENplus A3
Length > 38 mm	%	≤ 1	≤ 1	≤ 1
Diameter	Mm	3.15–40	3.15–40	3.15–40
Bulk density	kg/m^3	≥ 600	≥ 600	≥ 600
Durability	%	≥ 97.5	≥ 97.5	≥ 96.5
Fines	%	≤ 1	≤ 1	≤ 1
Ash content	%	≤ 0.7	≤ 1.5	≤ 3.0
Moisture content	% (w.b.)	≤ 10	≤ 10	≤ 10
Calorific value	MJ/kg	≥ 16.5	≥ 16.3	≥ 16.0
Chloride	Ppm	≤ 200	≤ 300	≤ 300

Table 2 also gives three published European grades for wood pellets. Pellets directly inherit most of the elemental compositions of the original biomass.

The manufacturing process controls the final moisture content, durability, and density of the pellet. Durability is a mechanical property of wood pellet that represents the ability of the material to not to break into smaller portions. Less durable pellet produces a higher amount of dust when a pellet is exposed to the transportation tensions. The durability of

Table 3. The effect of die thickness on the durability of pellet derived from agricultural and woody biomass (The results present average result of 5 duplicates).

Sample	Die thickness (in)	Raw m.c. (%)	Pellet m.c. (%)	HHV[1] (MJ/kg)	Durability (%)	Pellet density (g/cm³)	Pellet bulk Density[2] (kg/m³)
Douglass fir	0.75	14	7.1	19.2	98.1	1.17	660
Mountain pine	0.75	14	6.8	18.95	98.0	1.15	670
Switchgrass	0.75	14	9.2	18.7	Pellets did not form		
Switchgrass	1.25	17	7.5	18.7	98.0	1.19	630
Wheat straw	0.75	14	10.0	18.2	Pellets did not form		
Wheat straw	1.25	17	8.9	18.2	99.0	1.22	650
Corn Stover	1.25	17	8.35	17.9	99.5	1.15	650

[1]Dry basis.
[2]The woody biomass produced longer pellet than agricultural material using the same die set up.
[3]It was difficult to make pellet from agricultural biomass using thin die in this condition.

a pellet is often used interchangeably with its density though they are not the same. A durable product may not be dense and vice versa.

Due to their lower bulk density, straws require a longer residence time in the pelletization die than the woody biomass. To make dense pellets, the length of the die needs to be increased in order to increase the compression ratio. Our experiments (Table 3) on making pellets using a small pelleting machine (CPM Model CL3) showed that in order to form pellets, the die thickness had to be increased from 0.75 inches to 1.25 inches. Additionally, the material had to be conditioned to higher moisture content to form pellets.

4. Steam Treatment to Produce Durable Pellets

Research and practical experience have led to the consensus that steam conditioning of biomass prior to pressing enhances the durability of pellets. The inclusion of steam would induce many complex thermal and physical mechanisms within the exposed biomass. For super-heated steam, it involves vapor diffusion through the inter-particular voids when the particles are discharged from the pressured steam line. Furthermore, condensation of vapor happens on the surface of the mash to give out sensible and latent heat. The mash undergoes a rewetting process that elevates the moisture content of the mash. This diffused moisture changes the physical and thermal properties of the mash as conditioning proceeds.

The literature on the fibrous feedstock in the particular area of mash conditioning have been briefly reviewed by Sokhansanj and Wood (1991).

A critical review of steam conditioning and its effect on pellet durability is given by Tabil and Sokhansanj (1996). Early elucidation of the factors in conditioning pellet mash and the effect of steam on pellet durability dates back to Bartikoski (1962). Most studies relate to conditioning animal feed that is mainly protein and nonstructural carbohydrates. Information on biomass condition is scarce. Dobie (1959) appraised a hay pelleting operation and found that meal moisture condition was critical to the operation. Dobie (1959) noticed that the pellet throughput decreased at 10% hay moisture content. At moisture contents between 12% and 16%, both the pelleting operation and output were satisfactory. Beyond 18% moisture content, difficulty in fine grinding of the hay hardened the production of durable pellets. Dobie (1959) also found that the density and pelleting capacity were proportionate to the fineness of hay grinds.

The effect of steam-conditioning rate on pelleting variables was studied by Skoch (1981) using a poultry layer-diet containing soyameal, yellow corn, and sorghum. Winowiski (1985) tested the conditioning temperatures for a variety of feed rations and gave tips on how to optimize the feed temperature during the conditioning process prior to pelleting. Hill and Pulkinen (1988) found that 3.5% to 8% meal moisture content prior to steam conditioning did not affect pellet durability, but the power consumption of the pelletizer decreased by one-fold with an increased moisture content possibly due to the lubrication effect of moisture. Increasing the mash temperature from 60°C to 104°C increased the durability and reduced the power consumption. Maier and Gardecki (1993) evaluated industrial conditioning process through a survey of pellet mill problems categorized by steam supply, steam regulation, and conditioner maintenance. They found that only 22.7% of 88 mills evaluated were fully functional. Most pellet mills had one or more problems with steam supply regulation or conditioner maintenance. It was concluded that there was a need for improved educational out-reach to feed main manufacturers in steam supply regulation and conditioning.

The effect of particle size distribution on pellet durability and the inclusion of natural binders during alfalfa conditioning were studied by Tabil (1996), Tabil and Sokhansanj (1996), Tabil et al. (1997) and Adapa et al. (2002, 2004). Based on their work on the binding characteristics of alfalfa, the following recommendations for further study of the steam conditioning of alfalfa grind were made: (1) characterization of the moisture sorption of alfalfa particles, (2) determination of the physical properties, thermal properties, particle size distribution of alfalfa grind, and (3) a detailed study of the steam conditioning process. It is noted from a review of the literature that although steam conditioning of feed particulates has been practiced for decades, it is still more of an art than a science. Most work hitherto done in the steam conditioning of feed materials has been centered around examining the effect of conditioning parameters on pellet quality indices such as durability and color.

Following an extensive research on feed processing for animal nutrition (Lewis et al., 2015) at Kansas State University, Stark and Ferket (2011) summarized their research by developing a pie chart from the literature showing the impact each of the five material properties and processes have on the quality of pellets (nutrition and physical quality). Figure 4 is a redrawn image of their proposed percent of impacts.

Feed formulation has the strongest effect (40%) on pellet quality followed by particle size and conditioning (each 20%). Die specification has 15% and cooling pellets have 5%. Ingredients in a formulation that affect pellet quality consist of protein, fat or oil, fiber, ingredient variability, and abrasiveness. An increase in crude protein content from 15% to 25% enhances the durability from 97.0% to 97.5%. Increasing die temperature from 70°C to 85°C increases durability slightly from 97.0% to 97.3% at crude protein of 15%. Lipids drop durability from 97% to 75% when the fat content increases from 0% to 6% for poultry fat. In conversation with wood pellet producers, the chart generally applies to woody feedstock as well but moisture has the strongest factor. It is not clear whether feed formulations for woody biomass has the same effect as feed formulations for animal feed.

Most of the studies presented above focus on animal feeds instead of lignocellulosic materials. When applying steam conditioning or the related conditioner designed for ground feed particles to lignocellulosic particles, extra binders or severe conditions are needed to obtain a durable product. This is because the raw lignocellulosic biomass lacks an adequate amount of binding components (Lu et al., 2014). Common additives such as starch,

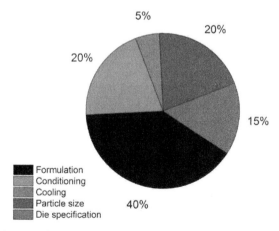

Fig. 4. Effects of various factors on the quality of pellets produced from animal feed (Stark and Ferrket, 2011). (A similar analysis for making wood pellets has not been published.) The graph shows that the conditioning of feed is 20% responsible for making durable feed pellets.

protein, crude glycerol and lignosulphate have been assessed to enhance the inter-particle bonding of lignocellulosic particles (Lu et al., 2014).

The physical changes in structural carbohydrates hemicellulose solubilization, lignin relocation, and size reduction require a treatment temperature over 180°C. Steam treatment affects the degree of these structural changes. Brownell et al. (1986) stated that saturated steam is better for increasing both hemicelluloses hydrolysis and cellulose accessibility than superheated steam. Additionally, the explosion part of the steam treatment improved the interlocking between particles by contributing more size reduction.

5. Steam Explosion and its Applications

The steam explosion, also named as Masonite technology (DeLong, 1981), is usually used to yield pulp from biomass. The process consists of a short-time steam treatment in the temperature range of 180–240°C followed by an explosive decompression of the biomass. Steam explosion uses a high pressure saturated steam ranging from 150 to 500 psi (1.034–3.447 MPa) to heat up biomass rapidly and with or without a rapid decompression (explosion) to rupture the rigid structure of the biomass. Batch and continuous pilot plants for the steam explosion of lignocellulosic material have been commercialized by SunOpta bioprocess Inc. (former company name: StakeTech Ltd.).

Steam explosion aims to increase the accessible sites of cellulose for the enzymes to hydrolyze the material for ethanol production (Sendelius, 2005; Shevchenko et al., 2001; Bura et al., 2002; Bura et al., 2003). In some cases, the addition of acidic gases or dilute acid as catalyst, e.g., SO_2 or H_2SO_4 is useful for enhancing the hydrolysis rate of hemicelluloses (Boussaid et al., 2000; Shevchenko et al., 2001; Bura et al., 2002). Several patents have been granted to this process (Delong, 1983; DeLong, 1990; Foody, 1984).

The quality of the currently produced pellets in North America needs to be improved with respect to durability (i.e., better binding) and better water resistance (i.e., higher hydrophobicity). Suzuki et al. (1998) and Bouajila et al. (2005) showed that the steam pretreatment improved the mechanical properties and hydrophobicity of fiberboards. The improvement was attributed to the self-binding property of plasticized lignin (Bouajila et al., 2005; Startsev and Salin, 2000). Shaw et al. (2009) reported that the pellets made from steam-treated poplar and straw feedstock had an improved tensile strength. The improved durability of pellets was attributed to the particle size reduction after steam treatment and to increased lignin content available for binding.

A steam explosion process can be conducted either in batch or in continuous mode. The batch mode is usually used in the laboratory to pre-treat the biomass for scientific research while industrial processes mostly

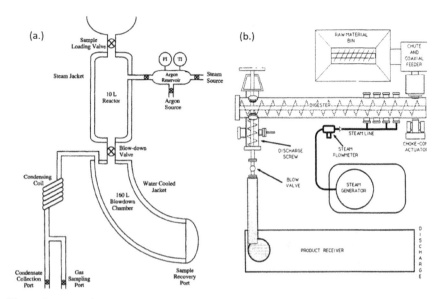

Fig. 5. Steam explosion equipment: (a) batch (Turn et al., 1998) and (b) continuous system Stake II pilot plant facility having a maximum capacity of 4 t/h located in Sherbrook, Quebec, Canada (Lam, 2011).

use continuous mode. The batch system is originally manufactured by the Stake Technologies, Ontario, Canada. The left diagram of Fig. 5 shows a modified batch system. After loading, the valve is closed and steam begins to enter the reactor at a preset pressure. After a preset residence time, the ball valve is opened and the steam-saturated biomass is discharged into a water-cooled discharge chamber. The discharge chamber is vented through a water-cooled condensing coil. The sample of gas emitted is collected into the sampling bags over the duration of the discharge.

The Stake II pilot facility, the right diagram of Fig. 5, is a continuous system of biomass processing (Lam, 2013b). The raw material is conveyed to a storage bin before being fed into the coaxial feeder. The feeder compresses the material forming a dense plug that enters the digester by exerting a pressure against the choke-cone. The back pressure of the choke cone against the plug is adjusted hydraulically. The steam flow transports the material through the digester. The digester residence time is controlled by adjusting the rotational speed of the auger. At the end of the digester, the auger discharges the materials through a Kamyr-blow valve. For each valve opening and closing cycle, the high steam pressure inside the digester generates a small detonation inside the structure of moving materials.

The wet treated materials are disengaged from the steam by a tangential cyclone located just at the entrance of the receiving bin. The suddenly enlarged volume of steam causes the material to undergo a sudden expansion as well. The disintegrated wet material falls onto the floor of the bin and conveyed to the discharge point. The steam is recycled through an aspiration-condensation sampling train for condensation. A vortex-shedding flowmeter is attached to the steam line to measure the steam flow rate and consumption.

6. Steam Treatment Severity

The major operational parameters of the steam explosion are particle size of the biomass feedstock (d_p), applied reaction pressure (P), reaction temperature (T) and residence time (t). A combination of the reaction parameters causes definite changes in the biomass structure and chemical composition. A severity index (R_0) developed by Overend and Chornet (1987), Eq. (3), is widely used for optimizing the steam treatment process. The equation has been developed based on modeling that follows complex reaction systems by assuming each reaction is homogenous and Arrhenius-dependence rate law. The temperature function are linearized by Taylor series (Abatzoglou et al., 1992; Montane et al., 1998).

$$R_0 = \int_0^t \exp\left(\frac{(T-100)}{14.75}\right) dt \qquad (3)$$

where

t = residence time (minute)
T = reaction temperature (°C)

Equation (3) does not include terms to represent moisture content and particle size of the feedstock. These terms also affect the kinetics of physical and chemical changes of the biomass. Brownell et al. (1986) found that high moisture in the wood slows down the surface condensation and heat condition in the wood. Air dried wood of 10% m.c. consumed only half as much steam compared to moist wood. For an initial moisture content of 50% (wet mass basis), the void volume of aspen wood was filled with condensate prior to reaching the final target treatment temperature (Saddler et al., 1983). Further heating was then by the slow conduction mechanism.

The range of R_0 in Eq. (1) depends on the process conditions of end products. The goal of producing more bioethanol is to recover hemicelluloses at lower pretreatment severities. At low severity ($R_0 < 2$), the deconstruction of the biomass begins. The degree of polymerization of the cellulose is greatly decreased for $R_0 > 3$. Severe dehydration and condensation reaction of the hemicellulose occurs for Ro > 4. The degraded byproducts usually inhibit the reaction rates of enzymatic hydrolysis and subsequent fermentation. Therefore, optimization of the steam explosion treatment within the range of R_o of 2 to 4 is the typical objective of preparing the fuel for biochemical conversions.

Zimbardi et al. (1999) compared the operations of the batch and continuous steam explosion reactors. The bach reactor had a capacity of 0.5 kg/cycle and the continuous unit had a throughput of 150 kg/h. The biomass straw was exposed to similar treatment severity in the two reactors. In addition to a wide ranging particle sizes from the two processes, the product collected from the continuous reactor was denser than the biomass from the batch reactor. The dense material was favored for high loading (200 g/L of the substrate) to maximize the recovery of valuable chemicals from the treated biomass.

Zimbardi et al. (1999) developed Eq. (4) relating batch to continuous steam treatments,

$$\log R_{o,Batch} = 1.50 * \left(\log R_{o,Continuous} - 1 \right) \tag{4}$$

where

$R_{o,Batch}$ = severity of the reaction inside the batch reactor
$R_{o,Continuous}$ = severity of the reaction inside the continuous reactor (dimensionless defined in Eq. (3))

The properties of biomass after steam treatment is not yet well researched and understood. Further experiments coupled with rigorous modeling of the steam explosion are required. These efforts must consider all of the physical and thermal properties of the material, e.g., bulk density, particle density, heat capacity, and process variables such as temperature, pressure, residence time.

7. Physical, Chemical, and Morphological Changes of Biomass

Steam explosion pretreatment consists of physical, chemical, and mechanical effects on modifying the biomass structure. The rigid structure of biomass is opened up by steam explosion pretreatment through swelling and is best described in Fig. 6 (Hsu et al., 1988).

Fiber bundles begin to separate at the middle lamella due to the softening of lignin above 135°C in the presence of water (Goring, 1963). The fibers deconstruct to fragments as the treatment severity increases. Scanning Electron Microscope (SEM) and Transmission Electron Microscope (TEM) revealed fiber separation at the middle lamella of hardwood and formation of uniformly distributed lignin beads on the surface of the fibers (Biermann et al., 1987; Angles et al., 2001).

Ramos (2003) reviewed the detailed chemistry involved in the steam treatment of the lignocellulosic materials and concluded that the process is an autohydrolysis process in the absence of an acidic catalyst. The role of

Effect of Pretreatment

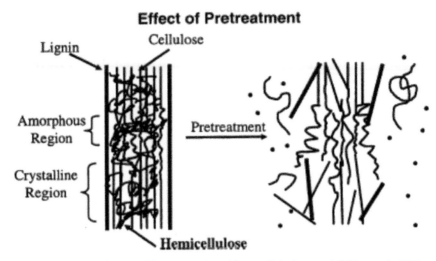

Fig. 6. Schematic of goals of deconstruction of lignocellulosic material (Hsu et al., 1988).

steam and other chemicals released to open up the biomass structure during steaming reaction are discussed in the following sections. Hemicellulose has an order of degree of polymerization (DOP) of about 200–300, lower than cellulose. Its amorphous structure favors the OH groups to be more reactive with steam/dilute acid. Hemicellulose is first hydrolyzed among cellulose and lignin. The released hemicelluloses are soluble in the condensed steam after exploding out of the reactor and can be recovered after flash cooling.

Upon a mild steam explosion, the cellular structure of the biomass is destroyed resulting in a slight increase in crystallinity of the cellulose. Bhuiyan et al. (2000) reported that heat increased the crystallinity of wood cellulose. They also suggested that other components accompanying wood cellulose were involved in the increase of crystallinity.

Lignin imparts mechanical strength to the cell wall. This helps to prevent pests attack and against diseases. The presence of lignin in cell wall inhibits the enzymatic hydrolysis of cellulose. Lignin is responsible for providing stiffness to the cell wall and also serves to bond individual cells together in the middle lamella region. Although lignin is relatively rigid at room temperature, it undergoes glass transition at around 140°C, and the presence of moisture in the cell wall additionally serves as a plasticizer for the lignin network. The presence of the moisture in the cell wall opens up the structure of the lignin. Lignin has a low concentration of hydroxyl groups compared to the polysaccharide components.

During the steam explosion, main lignin reactions involve cleavage of the β-O-4 and β-5 aryl ether linkages of the high molecular weight lignin (acid insoluble lignin). The lignin melts, redistributes, condenses, and forms beads on the surface of the cellulose microfibrils. These beads increase the porosity of the micro-fibers (Donohoe et al., 2008).

Shevchenko et al. (1999, 2001) suggested that the softwood lignin transformations involve condensation reactions via benzyl cation and increase the carbonyl content after the steam explosion. Miranda et al. (1979) proposed the mechanism of depolymerization/repolymerization of the lignin via carbonium ion and their result—the apparent increase in total lignin content of the product which happens due to the hemicellulose degradation product, furfural, and lignin polymerization.

According to Startsev and Salin (2000), fibers of the steam-exploded wood are defiberated and the surfaces are chemically modified after pretreatment. The lignin-cellulose chains form a new chemical bonding to give an improved binding and hydrophobicity. Bouajila et al. (2005) reported that the lignin plasticization contributed to the improved mechanical properties of the pressed fiberboard. A high pressure and temperature of the press caused the cross-linking of lignin-lignin and lignin-polysaccharides. The deform fibers created enough contact surface area to produce an intimate wood-wood contact.

Lam (2011) investigated different steam explosion severity on the Douglas fir ground particles treated at 200–220°C for 5–10 minute in a closed batch reactor followed by a decompression to release the treated particles to the atmospheric conditions. The pellets produced from treated particles were evaluated for their mechanical strength and moisture sorption resistance. It was reported that the steam-treated wood required 12–81% more energy to compact into pellets than the untreated wood (Table 4).

Table 4. Forces and input energy to make steam-exploded wood pellets and density of the produced pellet (Lam et al., 2011).

Treatment (steam temperature/ treatment duration)		Max force (N)	Compression energy[1] (J)	Extrusion energy[2] (J)	Solid density (g/cm³)
Untreated	Avg.	4290	22.3	0.049	1.43
	Std dev.	70	1.3	0.003	0.00
200°C-5 min	Avg.	4190	25.0	0.118	1.42
	Std dev.	145	1.0	0.030	0.01
200°C-10 min	Avg.	4120	31.8	0.117	1.43
	Std dev.	39	1.6	0.014	0.00
220°C-5 min	Avg.	4297	38.8	0.178	1.42
	Std dev.	54	4.4	0.045	0.00
220°C-10 min	Avg.	4079	40.4	0.235	1.42
	Std dev.	27	1.6	0.003	0.00

[1]n = 15; [2]n = 5

This was attributed to the binding ability of mono-sugar and the surface roughness of the fiber to the die surface causing the increase in friction and thereby increasing the compaction energy (Lam et al., 2013b). Pellets made from steam-treated wood had a breaking strength of 1.4–3.3 times the strength of pellets made from untreated wood. Steam-treated pellets had a reduced equilibrium moisture content of 2–4% and a reduced expansion after pelletization. There was a slight increase in the high heating value from 18.94 to 20.09 MJ/kg for the treated samples. Steam-treated pellets exhibited a higher lengthwise rigidity compared to untreated pellets.

8. Superheated Steam Treatment

Steam explosion with saturated steam increases the moisture content of the treated biomass up to two times (Saddler et al., 1983; Brownell et al., 1986; Lam, 2011). Substantial energy is required to remove this excess moisture from biomass for pelletization and other processes that require dry material. The required drying makes the economy of steam explosion questionable (Lam, 2011). The question is whether the natural moisture available inside a wet particle can be heated to a level that would act as steam.

Forest residues are generally wet with an initial moisture content of about 50% (wet basis). When such a moist solid that is saturated with water is heated, the internal water content is converted to steam inside the internal pores. The high pressure facilitates the bound and free water to diffuse out of the wood matrix resulting in a lower moisture content (drying) of the solid material. In the previous studies (Tooyserkani, 2013), the authors integrated the drying in a pressurized reactor with the dual objectives of obtaining both dry and pretreated feedstock and ultimately producing high-quality pellets. Steam was generated from the moisture inside the material and no external steam was applied for the hydrothermal treatment. The thermodynamics of the system and the effect of temperature were investigated. The effect of this treatment on the physical properties of the material such as particle size distribution, bulk density, and moisture adsorption capacity was studied. Elemental analysis was done and the composition was correlated with the calorific heating value of the produced material. The mechanical strength and quality of produced pellets were investigated to assess the effectiveness of this hydrothermal pretreatment prior to pelletization.

The biomass material tested in this research consisted of white softwood species, Douglas fir (*Pseudotsugamenziesii*). The freshly cut stem wood pieces, provided by Malcolm Knapp Research Forest in Maple Ridge, BC, were brought to the lab and debarked manually. The pieces were naturally dried by spreading them on a stack of wire mesh trays in the laboratory environment from about 50% (w.b.) moisture content to about 20% (w.b.). This natural drying was necessary for any further grinding in this research. The wood pieces were ground by a hammer mill with a screen size of

Table 5. Drying effect of the hydrothermal treatment for n = 3, average value (standard deviation).

Pretreatment conditions	Initial moisture content (w.b.) (%)	Final moisture content (w.b.) (%)	Calculated final moisture content (w.b.) (%)	Mass loss (%)
Sat[1]-Treated at 180°C	47.8 (0.0)	41.0 (0.1)	42.7	2.5 (0.5)
Super[2]-Treated at 200°C	49.6 (0.0)	13.5 (0.1)	11.5	7.6 (0.5)
Sat-Treated at 200°C	49.5 (0.1)	40.2 (0.2)	41.7	12.1 (0.8)
Super-Treated at 220°C	49.8 (0.0)	3.4 (0.1)	-[3]	15.9 (0.4)
Sat-Treated at 220°C	49.5 (0.1)	34.6 (0.1)	36.5	18.5 (0.8)
Super-Treated at 250°C	48.3 (0.2)	2.3 (0.0)	-	23.3 (1.0)
Sat-Treated at 250°C	49.5 (0.0)	25.6 (0.1)	21.7	27.0 (0.9)
Super-Treated at 280°C	49.1 (0.1)	2.1 (0.1)	-	24.3 (1.0)
Sat-Treated at 280°C	49.5 (0.1)	4.2 (0.1)	-	30.3 (1.1)

[1]Saturated steam; [2]Superheated steam; [3]It was difficult to calculate final moisture content in this condition.

1.6 mm (Model 10 HMBL, Glen Mills Inc., Clifton, NJ) and the moisture content of the ground sample was adjusted back to about 50% (m.c.) by spraying distilled water on the sample inside a glass jar. The glass jar was sealed completely and kept in a refrigerator at 4°C for about a month while it was shaken each day for mixing and penetration of moisture inside particles before any further treatment. The amount of material conditioned was about 800 g.

Table 5 summarizes the results of the hydrothermal treatment of Douglas fir that has an initial moisture content of 50% (w.b.). No external steam was applied in any of the treatment runs and steam was generated from the moisture inside the particles. Also, by application of cloth mesh bags, an almost complete separation of solid products from the gaseous fraction was achieved.

As expected, the moisture content of material was reduced from the initial value of around 50% (w.b.) to the final moisture content in the range of 2.1% to 41.0% (w.b.) depending on the set treatment temperature. It was expected that in case of superheated steam treatment, the final moisture content was reduced to the bound water content of wood that is lower at higher temperatures (Fyhr and Rasmuson, 1996; Johansson et al., 1997). Table 4 shows this reduction in the final moisture contents at higher temperatures in the case of superheated steam treatments. For superheated steam treatment, the final moisture content can be calculated and well-predicted, by solving the following Eqs. (5) and (6) together. Equations (5) and (6) are the conservative equations, showing mass and energy balance of the system.

$$M_{in}\frac{m_0}{1+m_0}+Q_s-M_{in}\frac{m}{1+m_0}=\rho_d V\frac{dm}{dt} \tag{5}$$

$$c_d\frac{M_{in}}{1+m_0}T_0+c_w m_0\frac{M_{in}}{1+m_0}T_0+c_s(T_s-100)Q_s+Q_s h_{fg}+Q_s c_w 100$$

$$-Q_s c_w T-c_d\frac{M_{in}}{1+m_0}T-c_w m\frac{M_{in}}{1+m_0}T=c_p\rho V\frac{dT}{dt} \tag{6}$$

where

m_0 = initial moisture content of alfalfa grind (0.0701) (w/w, db)
M_{in} = mass flow rate of alfalfa grind at intake (kg/s)
m = moisture content of conditioned alfalfa grind (w/w, db)
t = time(s)
ρ_{mean} = bulk density of alfalfa grind in the conditioner (kg/m³)
ρ_d = mean bulk density for the dry matter of alfalfa grind (kg/m³)
V = effective volume of conditioner for steam conditioning (m³)
c_d = specific heat capacity of dry matter (kJ. Kg⁻¹.K⁻¹)
c_w = specific heat capacity of water (kJ. Kg⁻¹.K⁻¹)
c_s = specific heat capacity of super-heated steam (kJ. Kg⁻¹.K⁻¹)
c_p = mean specific heat capacity of alfalfa grind (kJ. Kg⁻¹.K⁻¹)
T_0 = initial temperature of alfalfa grind (°C)
T_s = temperature of super-heated steam (°C)
T = temperature of conditioned alfalfa grind (°C)
h_{fg} = latent heat of vaporization of water at 100°C and atmospheric pressure (kJ/kg)
Q_s = Flow rate of condensed steam in the chamber (kg/s)

Yang (1998) simplified the Eqs. (3) and (4) using the lumped consideration and the fact that both temperature and moisture were observed to equalize very quickly after the injection of steam (5–10 seconds). The lumped assumption means that no spatial variation of parameters exists inside the biomass particles. All parameters only change by time. Yang developed the following semi-empirical Eqs. (7) to (10) to predict the relationship of real moisture content and temperature versus time.

$$m=m_0+\frac{K[1+\frac{1}{\ln(t)}](1+m_0)}{M_{in}}(1-e^{-At}) \tag{7}$$

$$T = \frac{D}{E} + (T_0 - \frac{D}{E})e^{-Et} \tag{8}$$

$$D = \frac{(c_d + c_w m_0)\dfrac{M_{in}}{1+m_0}T_0 + [c_s(T_s - 100) + h_{fg} + 100c_w]K[1+\dfrac{1}{\ln(t)}]}{c_P \rho V} \tag{9}$$

$$E = \frac{K[1+\dfrac{1}{\ln(t)}]c_w + c_d\dfrac{M_{in}}{1+m_0} + cw\dfrac{M_{in}}{1+m_0}\left\{m_0 + \dfrac{K[1+\dfrac{1}{\ln(t)}](1+m_0)}{M_{in}}(1-e^{-At})\right\}}{c_P \rho V} \tag{10}$$

At a range of steam pressure from 3.4 to 34.5 kPa and a range of flow rate from 2.7 to 10.8 kg/s, the models on Eqs. (7) and (8) was verified by a pilot-scale experiment using means of the inverse method. Regression statistics indicated that these models were capable of predicting the change of temperature and moisture of alfalfa grind during steam conditioning. See Yang (1998) for extensive data on experimental pelletization unit and the validated model.

9. Steam Treatment for Torrefaction

Ghiasi (2015) reports the results of experiments on the steam treatment of biomass to produce durable torrefied pellets. Douglas fir and Mountain Pine wood chips with an average size of 30 to 50 mm and shredded switchgrass, wheat straw, and corn stover with an average length between 50 and 80 mm were used. The initial moisture content of wood chips was 40% (w.b.) and for agricultural samples was 15–20% m.c. (w.b.). The as-received biomass was dried to reach 15% moisture content (w.b.) using the hot air convection oven and ground using a lab scale hammer mill with a 6.35 mm screen to produce a distributed particle size under 2 mm for pellets production.

Steam pretreatment was conducted in a 2 L StakeTech III steam gun (Stake Technologies, Norvall, ON, Canada) in the Forest Products Biotechnology/Bioenergy Laboratory at the University of British Columbia. Briefly, batches of the dry weight of 200 g of biomass were steam treated to collect about 3 kg dry materials of each species to have enough material for pelletization. The recovered solid portion was dried in an oven and pelletized by using a lab scale California Pellet Mill. The pelletizer was equipped with two dies with different thickness (19 mm and 31.5 mm) for pellet production. Similar to the grinder, the pelletizing machine was connected to a computer for recording the power consumption during the process. The flow rate of the material was controlled by adjusting the vibration of the feeding plate thus avoiding exceeding the maximum power consumption of the machine. The ground untreated material had

Table 6. Woody (Douglas fir, DF) and agricultural (switchgrass) pellet properties (raw, steam treated, pelletized, and torrefied).

Conditions	HHV (MJ/kg)	Density (g/cm³)	Durability tumbler (%)
Pellets made from steam treated feedstock and torrefied DF pellet (Torrefaction at 280°C, 15 min)	22.8	1.23	99.2
Torrefied DF pellet made from untreated biomass (Torrefaction at 280°C, 15 min)	23.2	1.14	98.4
Pellet made from steam treated DF but not torrefied	20.3	1.29	99.7
DF pellets made from untreated biomass and not torrefied (Control)	18.8	1.17	98.1
Steam treated and torrefied switch grass pellet (260°C, 15 min)	21.3	1.2	99.0
Torrefied switch grass pellet (260°C, 15 min)	21.8	1.05	98.2
Steam treated switch grass pellet (210°C, 5 min)	19.0	1.3	99.3
Switch grass raw pellet	17.3	1.15	97.8

conditioned in different moisture content varying from 12–18% (w.b.) prior to pelletization.

Pellets made from untreated feedstock and pellets made from steam treated biomass (Steam treated pellets) were torrefied in the oven at 280°C for Douglas fir and at 260°C for switchgrass (Kumar et al., 2012). Table 6 lists the quality of pellets showing that steam explosion improved the properties of pellets derived from Douglas fir and switchgrass, including higher heat value, single density, and durability. The enhanced durability is due to the structure changes led by steam explosion. Steam explosion increases the contact surface area of particles and decreases the materials' contact distance, thus increasing the chance of physical interlocking and chemical bonding between the particles (Lam et al., 2013b). Besides, steam explosion relocated lignin on the surface of particles and produced sugars that could act as binders. These changes also increase the pellet durability.

10. Energy Input for Hydrothermal Treatment

Although steam pretreatment is an energy-intensive process, the benefits gained from enhancing the quality of biomass may offset extra energy input and associated costs. There are published literature on techno-economic and energetic of steam pre-treated woody biomass for enzymatic hydrolysis and fermentation for bioethanol production (Shevchenko et al., 2001; Zimbardi et al., 2002; Mabee et al., 2007). These literatures do not address steam treatments for making pellets.

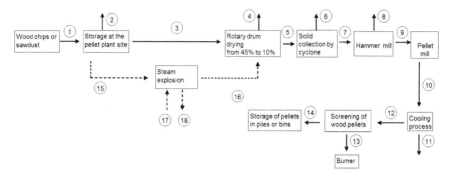

Fig. 7. Biomass preprocessing with and without steam explosion pretreatment (dash line shows steam explosion pretreatment, numbers indicate the streamline to indicate the mass balance).

Lam (2011) conducted an energy balance analysis for a steam treated pelletization process. The mass balance and energy input to produce commercial pellets from steam-treated and untreated feedstock were calculated. Figure 7 shows a typical biomass pelleting plant consisting of seven major unit operations. The major unit operations are receiving and sorting raw biomass, drying, grinding and blending, pelleting, cooling, screening, storing, and shipping. A new steam explosion process unit is added to the system after receiving and sorting and before the drying operation. In this unit, the wood chips are subjected to a steam explosion at conditions that will be specified later in this paper. Treated and untreated biomass are dried, ground, and pelletized in a similar manner.

Production data including power and energy input to each unit operation were taken from a commercial pelleting plant in Princeton, B.C. Canada (Mani, 2005). The pellet plant had an annual production of 45000 tonnes. The plant processed 10.35 t/h of sawdust at 45% moisture content to 6 t/h of pellets at 10% moisture content. The plant operated 24 h a day, 310 days per year that corresponds to an annual utilization of 85%. The calculation of the total heat required to evaporate 1 tonne of water from the biomass includes the use of pellet crumbles and sawdust as a fuel for dryer assuming the combustion efficiency at 80% and a heating value of 18 MJ/kg. Although the raw feedstock for this plant was sawdust and shavings from an adjacent sawmill, we also include debarked wood chips at 45% moisture content in calculations. We envision that a steam explosion plant would also process wood chips as the supply of sawdust dwindles (Cullis et al. 2004). The debarked wood chips are sized to 12 mm by 12 mm by 2 mm (length, width, thickness). In the following, we describe the two systems of conventional and steam explosion process in parallel.

Table 7 lists the four levels of severities calculated by different combinations of steam treatment, time, and temperature using Eq. (3). The enthalpy of steam consists of 3 parts: raising the water temperature

Table 7. Direct energy input for steam explosion at different severity with 0.22 steam to biomass ratio.

Severity	Unit	3.64	3.94	4.23	4.53
Treatment temperature (T)	°C	206	204	229	223
Treatment time (t)	s	302	603	302	603
Steam pressure in boiler (P1)	kPa	1750	1762	2562	2804
Steam pressure in cooker (P2)	kPa	1236	1243	1777	1935
Properties of steam					
Saturated liquid density	kg/m³	857.56	859.47	828.42	836.31
Saturated vapor density	kg/m³	8.8637	8.5869	13.739	12.295
Specific volume—sat. liq.	m³/kg	0.001166	0.001164	0.001207	0.001196
Specific volume—sat. vap.	m³/kg	0.11282	0.11646	0.072785	0.081334
Enthalpy—sat. vap. (h$_g$)	kJ/kg	2795.3	2794.5	2802.8	2801.7
Number of moles (n)		310.47	313.25	425.63	469.23
Mass of steam	g	5.59	5.64	7.66	8.45
Energy input					
Energy to generate saturated steam from boiler (E$_1$)	kJ	15.62	15.76	21.47	23.66
Energy to sustain temperature of cooker (E$_2$)	kJ	0.05	0.10	0.06	0.11
Total energy per unit mass[1] (d.b.)	MJ/kg	0.56	0.57	0.78	0.86

[1]: refers to the mass of biomass to be treated, which is 25 g from the experiment.

to a boiling point at 100°C, evaporation of water at 100°C to steam, and raising the steam temperature from 100°C to 200°C under pressure. The thermal electric energy required to generate 5.6 g saturated steam for steaming at 200°C for 5 minutes was 15.6 J. More steam around 7.7 to 8.5 g was generated at 220°C inside the closed system corresponding to the increased pressure of 1.79 and 1.93 MPa, respectively. The energy required to generate the saturated steam in the boiler depended upon temperature only. However, maintenance heat supplied to the steam chamber increased with both steaming temperature and steaming time. At the same temperature, the energy required to maintain the reaction temperature increased proportionally to the steaming time. The total energy input to steam explosion pretreatments were 0.56, 0.57, 0.78 and 0.86 MJ/kg for the severity of 3.64, 3.94, 4.23 and 4.53, respectively.

Table 8 lists the direct energy input to each unit operation for biomass pelletization process with a steam explosion at different treatment severity with 0.22 steam to biomass ratio. The production rate of pellets using untreated biomass is 6000 kg/h. The production rate of pellets made from

Table 8. Direct Energy Input to the Biomass Pelletization Process with the Laboratory Data of Steam Explosion at Different Treatment Severity and Pelletization with 0.22 Steam to Biomass Ratio.

Severity	Units	Untreated --		200°C, 5 min 3.64		200°C, 10 min 3.94		220°C, 5 min 4.23		220°C, 10 min 4.53	
		Feedstock	Pellets	Feedstock	Pellets	Feedstock	Pellets	Feedstock	Pellets	Feedstock	Pellets
Raw material collection/Transportation	MJ/kg	0.16	0.27	0.16	0.36	0.16	0.36	0.16	0.38	0.16	0.38
Steam explosion	MJ/kg	0	0	0.35	0.80	0.35	0.81	0.45	1.08	0.49	1.20
Drying	MJ/kg	1.68	2.90	1.87	4.30	1.83	4.26	1.73	4.19	1.67	4.12
Size reduction	MJ/kg	0.18	0.31	0.15	0.34	0.15	0.34	0.14	0.34	0.14	0.35
Pelletization	MJ/kg	0.64	1.11	0.73	1.67	0.79	1.83	0.92	2.23	0.94	2.31
Cooling	MJ/kg	0.05	0.08	0.04	0.10	0.04	0.10	0.05	0.11	0.04	0.11
Screener	MJ/kg	0.05	0.08	0.05	0.11	0.05	0.11	0.05	0.11	0.05	0.12
Miscellaneous units	MJ/kg	0.05	0.08	0.04	0.10	0.04	0.10	0.05	0.11	0.04	0.11
Total energy input per unit kg of pellets	MJ/kg	2.80	4.83	3.38	7.78	3.40	7.91	3.53	8.55	3.52	8.70
Production rate of pellets	kg/hr	6000		4487		4444		4271		4185	
Energy content of wood pellet	MJ/kg	19.64		19.73		20.11		20.56		20.73	
Energy ratio		7.00	4.07	5.84	2.54	5.91	2.54	5.82	2.41	5.88	2.38
Energy percentage of steam explosion over pellet heating value	%	0	0	2	4	2	4	2	5	2	6
Energy percentage of preprocessing over the pellet heating value	%	14	25	17	39	17	39	17	42	17	42

steam treated biomass decreased from 4487 to 4185 kg/h. Depending on the severity of steam treatment, the solid yield and volatile loss are the contributors to reduce the steam exploded pellet production rate. The volatiles can be separated from the saturated steam, captured, and recycled to be burnt as a fuel for the whole process. Future work can be considered in the study of the heating value of the volatiles and their combustion performance.

The direct energy input per unit kilogram of the feedstock of raw material collection and transportation in all cases was 0.16 MJ/kg. For steam explosion pretreatment, the direct energy input per unit kilogram of feedstock increased from 0.35 MJ/kg to 0.49 MJ/kg with increasing severity. For drying, the energy required to dry the sawdust per unit kilogram of feedstock from 45% to 10% moisture content (w.b.) for the untreated case was 1.68 MJ/kg. The energy required to dry the steam exploded sawdust with 50% moisture content to 10% moisture content decreased from 1.87 MJ/kg to 1.67 MJ/kg with increasing treatment severity. It is because some materials were lost as volatiles during the steam explosion pretreatment. The energy required to size-reduce the untreated sawdust per unit feedstock was 0.18 MJ/kg. The energy required to size-reduce the steam treated sawdust decreased slightly from 0.15 MJ/kg to 0.14 MJ/kg with increasing the treatment severity. However, the energy required for pelletization in a pellet plant per unit feedstock increased with steam explosion pretreatment severity from 0.73 MJ/kg to 0.94 MJ/kg.

In general, the energetic ratio of the biomass densification process with untreated woodchip based on feedstock was 7.00. The energetic ratios of the biomass densification process with steam explosion pretreatment at 200°C for 5 and 10 minutes were 5.84 and 5.91, respectively. Drying was the most energy consuming unit operation, consuming nearly 60% of the entire energy consumption for the untreated case. With steam explosion pretreatment, the energy consumption of drying was the most energy intensive unit operation between 50%–55% of the entire process. For all steam treatment cases, the second most energy-intensive process is the pelletization which consumes 22%–27% of the entire direct energy input. The steam explosion was the third most energy-intensive process compared to other unit operations. It ranged 10%–14% of the total primary input to the whole process.

The steam explosion unit used in commercial scale should be continuous with insulation against heat losses. Operational characteristics of a continuous unit would be different from the batch reactor we used in this study. The maintenance heat was proportional to the treatment time. This huge energy input can be neglected in a continuous system. Without maintenance heat, the direct energy input of steam explosion per unit feedstock was calculated to be 0.34 MJ/kg for the treatment severity at 200°C for both 5 minutes and 10 minutes. For treatment severity at 220°C, the direct energy input of steam explosion per unit feedstock was between 0.45–0.48 MJ/kg. The direct energy input of steam explosion per unit feedstock for all severities ranged between 11% and 16% of the total direct energy input to the whole process.

11. Economic Analysis

Many previous studies have evaluated the economics of biomass-based energy from the perspective of generic models (Gregg, 1996; Dassanayake and Kumar, 2012). The cost of producing pellets from sawdust has been reported by Mani et al. (2006) who calculated pellets produced from sawdust at a cost of 51 $ t[1] at a plant capacity of 45 kt. A European pellet production scenario has been reported by Thek and Obernberger (2004). They predicted the production cost of sawdust-based pellets in a European setting andreported a production cost of 95.56 $ t[1] of pellets at a plant capacity of 24 kt (Thek and Obernberger, 2004). Urbonowski (2005) used his study to evaluate the capital cost of a regular pellet production plant.

A few research studies evaluate the production costs of steam pretreated pellets or compare the production costs of regular and steam pretreated pellets other than the work reported by Lam (2011) and Tooyserkani (2013). In addition, there is a limited focus on the effects of the economic optimum size of the feedstock on both processes. While life-cycle analyses have been carried out by many researchers (Pa et al., 2012), till date there has been no techno-economic assessment of steam pretreatment processes. There is a need to evaluate the economics of pretreated biomass-based pellets.

Shahrukh et al. (2016) developed a data-intensive techno-economic model to estimate pellet production costs and optimum pellet plant size based on three feedstocks. The agricultural residue, forest residue, and energy crops were considered for two pelletization processes, regular and steam pretreated. The total cost was calculated from biomass harvesting to pellet production. The focus of the study was to apply a specific cost number methodology to feedstocks in western Canada. Costs associated with processing within the plant are capital cost, energy cost, employee cost, and consumable cost. The plant life considered in the model was 30 years based on previous studies (Kumar et al., 2003; Kumar et al., 2008; Kumar, 2009; Sultana and Kumar, 2011). The pellet production cost is the sum of the delivered feedstock cost and the pellet plants' production costs (Shahrukh et al., 2016). In the base case, the pellet plant was assumed to run at 6 t h[-1] with an annual production capacity of 44 kt. This size was based on earlier studies on pellet plants (Sultana and Kumar, 2011). Table 9 lists the itemized costs.

For the base case scenario, the model showed an economic optimum plant size of 190 kt for regular pellets and 250 kt for steam pretreated pellets (Table 9). From the sensitivity analysis, Shahrukh et al. (2016) concluded

Table 9. Economic optimum size and components of the production cost of pellet production from three feedstocks.

Cost items	Regular Pellet	Steam pretreated	Regular pellet	Steam pretreated	Regular pellet	Steam pretreated
Optimum size (kt y^1)	190	250	190	290	190	230
Pellet cost ($ t^1)	96.09	148.30	91.35	144.20	91.50	151.50
Capital recovery	7.53	12.06	8.53	16.02	10.82	16.95
Maintenance cost	2.08	2.90	1.84	2.83	2.03	2.96
Field cost	41.21	46.26	22.20	25.26	20.13	23.14
Transportation cost	26.69	33.85	29.17	39.56	35.73	45.42
Premium	0.00	0.00	5.03	4.33	4.20	4.83
Employee cost	5.72	6.18	5.18	5.32	5.72	6.72
Energy cost	4.26	38.51	11.61	42.35	4.26	42.91
Consumable item cost	8.61	8.55	7.80	8.54	8.61	8.57
Pellet transportation	4.96	4.33	4.88	4.29	5.35	4.80
Total pellet cost ($ t^1)	101.06	152.63	96.23	148.50	96.85	156.31

that pellet production cost was most sensitive to field cost followed by transportation cost. Their model's uncertainty analysis showed that there was greater variation with steam pretreatment than with regular pellet processing because additional energy was required for steam pretreated pellet production which can be minimized by improving drying and steam pretreatment efficiency.

12. Conclusions

In the course of this study, it became evident that the drying rate of the steam treated particles decreased with treatment severity. A similar trend was observed for pellets made from steam treated biomass where more energy is needed to form pellets. Decreased drying rate and increased energy requirement affect the economics of the steam explosion and densification. Future experiments need to be designed to understand the processes that govern the decrease in drying rate and increase in energy to make pellets.

The effect of steam to biomass ratio on the reaction rate during the steaming process is not well-understood. If the concentration of steam is too low (i.e., superheated steam), there may not be enough moisture to condense on the particle surface and facilitate the hydrolysis reactions during steaming. If the steam supply is too high, the operation cost of the process will increase. There is a need to study the optimized steam to biomass ratio for the best pellet quality.

The current study was based on the Douglas Fir in British Columbia. Other wood species, e.g., Pine, Spruce, Poplar, and Willow, would have different physical and chemical properties brought about by steam explosion. Further research towards understanding the effect of steam explosion on forming durable pellets from mixed wood species and other lignocellulosic biomass is essential.

The particles used in the current study were ground particles. The powder generated from wood chips after steam explosion can be made to pellets and evaluated for their properties. Larger opening of the steam treatment reactor needs to be used in order to perform the experiments. More work on modeling increases in pseudo-lignin by reactions between multi-component degraded products from different polysaccharides (arabinan, mannan, xylan) and the solubilized lignin.

The effect of steam explosion at different severities and its optimization on pellet quality was investigated in the current study. However, the optimization of pelletization on steam-exploded wood powder can be carried out. The proposed study can be the effect of several different die temperatures, the different moisture content of the feedstock, and different compressive forces on pellet density.

Compaction behavior of the steam-exploded wood pellets treated at different severities was different. Future work can be done to compare the compression data from the experiment with the predicted values with the model equation for biomass densification.

There is a strong odor from steam-exploded wood powder and wood pellets. The most severely treated wood pellets gave out the strongest smell. It is important to evaluate the storage ability of the steam exploded pellets made at different severities by offgasing and self-heating experiments. Identification of any toxic offgases is important.

There is a volatile loss after steam explosion pretreatment for the wood particles by proximate analysis. It may imply that the combustion behavior of steam exploded wood pellets possess a better performance with less tar formation in the exhausts. This is beneficial to improve the product image for the household heating application. Future work can be done on evaluating the combustion performance and gas emissions of steam exploded wood pellets and compared to the untreated wood pellets made in the same condition.

The surface structure of the steam-exploded wood particles were found to have more pores in this study. High porous particles may increase the gasification efficiency. The effect of the steam-exploded wood particles made at different severities on gasification can be evaluated.

Wood pellets have uniform physical properties. They are a good feedstock supply for biorefineries. In addition, the steam explosion had been studied as an economical pretreatment prior to ethanol fermentation. The evaluation of steam exploded wood pellets made at different severities as a feedstock for ethanol production can be studied.

The amount of steam used for treating biomass was not optimized in this research. The provisions for recovering heat from spent steam was not considered. The numbers on energy use for steam explosion represented a worse scenario. A more accurate mass and energy balance with a sensitivity analysis on energy input to the energy content of pellet are recommended.

Minimum production cost and optimum plant size are determined for pellet plants for forest residue, agricultural residue, and energy crops. The cost varies from 95 to 105 $ t^1 for regular pellets and 146 to 156 $ t^1 for steam pretreated pellets. These costs include harvest and transport costs. The difference in the cost of producing regular and steam pretreated pellets per unit energy is in the range of 2–3 $ GJ^1. The economic optimum plant size (i.e., the size at which pellet production cost is minimum) is found to be 190 kt for regular pellet production and 250 kt for a steam pretreated pellet. Sensitivity and uncertainty analyses were carried out to identify sensitivity parameters and effects of the model error.

References

Adapa, P. K., L. G. Tabil, G. J. Schoenau, B. Crerar and S. Sokhansanj. 2002. Compression characteristics of fractionated alfalfa grinds. Powder Handl. Process 14: 252–259.

Adapa, P. K., G. J. Schoenau, L. G. Tabil Jr. and S. Sokhansanj. 2004. Fractional drying of alfalfa leaves and stems: review and discussion. In Dehydration of Products of Biological Origin. A. S. Mujumdar, ed. Enfield, NH: Science Publishers, Inc.

Adapa P., L. Tabil and G. Schoenau. 2010. Physical and frictional properties of non-treated and steam exploded barley, canola, oat and wheat straw grinds. Powder Technology 201(3): 230–241.

Abatzoglou, N., E. Chornet, K. Belkacemi and R. P. Overend. 1992. Phenomenological kinetics of complex systems: The development of a generalized severity parameter and its application to lignocellulosics fractionation. Chem. Eng. Sci. 47: 1109–1122.

Angles, M. N., F. Ferrando, X. Farriol and J. Salvado. 2001. Suitability of steam exploded residual softwood for the production of binder-less panels. Effect of the pre-treatment severity and lignin addition. Biomass Bioenerg. 21: 211–224.

Bartikoski, R. 1962. The effect of steam on pellet durability. Pages 42–47.

Bhuiyan, M. T. R., N. Hirai and N. Sobue. 2000. Changes of crystallinity in wood cellulose by heat treatment under dried and moist conditions. J. Wood Sci. 46: 431–436.

Bura, R., S. D. Mansfield, J. N. Saddler and J. R. Bothast. 2002. SO_2 catalyzed steam explosion of corn fiber for ethanol production. Appl. Biochem. Biotech. 98–100: 59–72.

Bura, R., R. J. Bothast, S. D. Mansfield and J. N. Saddler. 2003. Optimization of SO_2 catalyzed steam pretreatment of corn fiber for ethanol production. Appl. Biochem. Biotech. 105-108: 319–335.

Boussaid, A., R. Esteghlalian, J. Gregg, K. H. Lee and J. N. Saddler. 2000. Steam pretreatment of douglas fir wood chips. Appl. Biochem. Biotech. 84-86: 693–705.

Bouajila, J., A. Limare, C. Joly and P. Dole. 2005. Lignin plasticization to improve binder-less fiberboard mechanical properties. Polym. Eng. Sci. 45: 809–816.

Brownell, H. H., E. K. C. Yu and J. N. Saddler. 1986. Steam explosion pretreatment of wood: Effect of chip size, acid, moisture content and pressure drop. Biotechnol. Bioeng. 28: 792–801.

Biermann, C. J., D. McGinnis and P. Schultz. 1987. Scanning electron microscopy of mixed hardwoods subjected to various pretreatment process. J. Agri. Food Chem. 35: 713–716.

Cullis, I. F., J. N. Saddler and S. D. Mansfield. 2004. Effect of initial moisture content and chip size on the bioconversion efficiency of softwood lignocellulosics. Biotechnol. Bioeng. 85: 413–421.

Dassanayake, G.D.M.N A. Kumar. 2012 Techno-economic assessment of triticale straw for power generation. Appl. Energy 98: 236e245.

DeLong, E.A. 1981. Method of rendering lignin separable from cellulose and hemicellulose in lignocelluosic material and the product so produced. Canadian Patent No. 1096374.

DeLong, E. A. 1983. A method of rendering lignin separable from cellulose and hemicellulose in land the product so produced. Canadian Patent No. 1141376.

DeLong, E. A., E. P. Delong, G. S. Ritchie and W. A. Rendall. 1990. Method for extracting the chemical components from dissociated lignocellulosic material. US Patent No. 4908098.

Dobie, John, B. 1959. Engineering appraisal of hay pelleting. Agric. Eng. 40.

Donohoe, B. S., S. R. Decker, M. P. Tucker, M. E. Himmel and T. B. Vinzant. 2008.Visualizing lignin coalescence and migration through maize cell walls following thermochemical pretreatment. Biotechnol. Bioeng. 101: 913–925.

Engineering Toolbox, Water vapor and saturation pressure in humid air 2014. Available at: http://www.engineeringtoolbox.com/water-vapor-saturation-pressure-air-d_689.html/ [accessed May 21, 2014].

Foody, P. 1984. Method for increasing the accessibility of cellulose in lignocellulosic materials, particularly hardwoods agricultural residues and the like. US Patent No. 4461648.

Fyhr, Christian and Anders Rasmuson. 1996. Mathematical model of steam drying of wood chips and other hygroscopic porous media. AIChE Journal 42.9: 2491–2502.

Ghiasi, Bahman , C. J. Lim and Shahab Sokhansanj. 2015. Feasibility of pelletization and torrefaction of agricultural and woody biomass. CSBE 15-76. CSBE/SCGAB 2015 Annual Conference Delta Edmonton South Hotel, Edmonton, Alberta 5-8 July 2015. 12 pages.

Gregg, D. J. 1996. The Development of a Techno-economic Model to Assess the Effect of Various Process Options on a Wood-to-ethanol Process [dissertation], University of British Columbia, Vancouver (BC), 1996.

Goring, D. A. I. 1963. Thermal softening of lignin, hemicelllulose and cellulose. Pulp&Paper Magazine Canada 64(12): 517–527.

Hill, B. and D. A. Pulkinen. 1988. A Study of the Factors Affecting Pellet Durability and Pelleting Efficiency in the Production of Dehydrated Alfalfa Pellets. A Special Report. Saskatchewan Dehydrators Association, Tisdale, SK, Canada, pp. 25.

Holtzapple, M. T., A. E. Humphrey and J. D. Taylor. 1989. Energy Requirements for the size reduction of poplar and aspen wood. Biotechnology and Bioengineering 33: 207–210.

Hsu, W. E., W. Schwald, J. Schwald and J. A. Shields. 1988. Chemical and physical changes required for producing dimensionally stable wood-based composites. Wood Sci. and Technol. 22: 281–289.

ISO Standard. Solid Biofuels—Determination of mechanical durability of pellets and briquettes —Part 1: Pellets (17831-1: 2015).

Johansson, Anders, Christian Fyhr and Anders Rasmuson. 1997. High temperature convective drying of wood chips with air and superheated steam. Int. J. Heat Mass Tran. 40.12: 2843–2858.

Kumar, A., J. B. Cameron and P. C. Flynn. 2003. Biomass power cost and optimum plant size in western Canada. Biomass Bioenergy 24: 445–464.

Kumar, A., P. Flynn and S. Sokhansanj. 2008. Biopower generation from mountain pine infested wood in Canada: an economical opportunity for greenhouse gasmitigation, Renew. Energy 33: 1354–1363.

Kumar, A. 2009. A conceptual comparison of bioenergy options for using mountainpine beetle infested wood in Western Canada. Bioresour. Technol. 100: 387–399.

Kumar, L., Z. Tooyserkani, S. Sokhansanj and J. N. Saddler. 2012. Does densification influence the steam pretreatment and enzymatic hydrolysis of softwoods to sugars? Bioresour. Technol. 121: 190–198.

Lam, P. S. 2011. Steam explosion to biomass to produce durable wood pellets. Ph.D. Thesis, University of British Columbia, Vancouver, Canada.

Lam, P. S., Z. Tooyserkani, L. J. Naimi and S. Sokhansanj. 2013a. Pretreatment and pelletization of woody biomass. pp. 93–116. *In*: Z. Fang (ed.). Pretreatment Techniques for Biofuels and Biorefineries. Springer, Berlin Heidelberg.

Lam, P. S., P. Y. Lam, S. Sokhansanj, X. Bi and C. J. Lim. 2013b. Mechanical and compositional characteristics of steam-treated Douglas fir (Pseudotsugamenziesii L.) during pelletization. Biomass Bioenerg. 56: 116–126.

Lam, P. S., P. Y. Lam, S. Sokhansanj, X. T. Bi, C. J. Lim and S. Melin. 2015. Effect of steam explosion pretreatment on size reduction and pellet quality of woody and agricultural biomass. pp. 27–53. *In*: K. Tannous (ed.). Innovative Solutions in Fluid-Particle Systems and Renewable Energy Management. IGI Global, Hershey, PA, U.S.A.

Lewis, L. L., C. R. Stark, A. C. Fahrenholz, J. R. Bergstrom and C. K. Jones. 2015. Evaluation of conditioning time and temperature on gelatinized starch and vitamin retention in a pelleted swine diet. J. Anim. Sci. 93: 615–619.

Lu, D., L. G. Tabil, D. Wang, G. Wang and S. Emami. 2014. Experimental trials to make wheat straw pellets with wood residue and binders. Biomass Bioenerg. 69: 287–296.

Mabee, W. E., D. J. Gregg, C. Arato, A. Berlin, R. Bura, N. Gilkes, O. Mirochnik, W. Pan, E. K. Pye and J. N. Saddler. 2007. Updates on softwood-to-ethanol process development. Applied Biochemistry and Biotechnology. 129(1-3):55-70.

Mani, S. 2005. A systems analysis of biomass densification process. Ph.D. dissertation. Vancouver, Canada: University of British Columbia, Chemical and Biological Engineering.

Mani, S., L. G. Tabil and S. Sokhansanj. 2006. Specific energy requirement for compacting corn stover. Bioresour. Technol. 97: 1420–1426.

Maier, D. E. and J. Gardecki. 1993. Evaluation of pellet conditioning: understanding steam. Feed Management 44.7: 15.

Montane, D., R. P. Overend and E. Chornet. 1998. Kinetic models for non-homogenous complex systems with a time dependent rate constant. Can. J. Chem. Eng. 76: 58–68.

Overend, R. P. and E. Chornet. 1987. Fractionation of lignocellulosics by steam aqueous pretreatments. Philos. T. R. Soc. A. 321: 523–536.

Pa, Ann, Jill Craven, Hsiaotao T. Bi, Staffan Melin and Shahab Sokhansanj. 2012. Environmental Footprints of British Columbia wood pellets from a simplified life cycle analysis. International Journal of Life Cycle Assess 17(2012): 220–231 (DOI) 10.1007/s11367-011-0358-7.

Ramos L. P. 2003. The chemistry involved in the steam treatment of lignocellulosic materials. Quimica Nova. 26: 863–871.

Saddler, J. N., M. Mes-Hartree, E. K. C. Yu and H. H. Brownell. 1983. Biotechnol. and Bioeng., Symposium Series 13: 225.

Sendelius, J. 2005. Master of Science Thesis: Steam pretreatment optimization for sugarcane bagasse in bioethanol production. Department of Chemical Engineering, Lund Institute of Technology, Sweden.

Shaw M. D., C. Karunakaran and L. G. Tabil. 2009. Physicochemical characteristics of densified untreated and steam exploded poplar wood and wheat straw grinds. Biosyst. Eng. 103: 198–207.

Shahrukh, Hassan, Adetoyese Olajire Oyedun, Amit Kumar, Bahman Ghiasi, Linoj Kumar and Shahab Sokhansanj. 2016. Techno-economic assessment of pellets produced from steam pretreated biomass feedstock. Biomass and Bioenergy 87(2016): 131–143.

Shevchenko, S. M., R. P. Beatson and J. N. Saddler. 1999. The nature of lignin from steam explosion/enzymatic hydrolysis of softwood. Appl. Biochem. Biotech. 77-79: 867–876.

Shevchenko, S. M., K. Chang, D. G. Dick, D. J. Gregg and J. N. Saddler. 2001. Structure and properties of lignin in softwoods after SO$_2$ catalyzed steam explosion and enzymatic hydrolysis. Cellul. Chem. and Technol. 35: 487–502.

Skoch, E. R., K. C. Behnke, C. W. Deyoe and S. F. Binder. 1981. The effect of steam-conditioning rate on the pelleting process. Anim. Feed Sci. Tech. 6: 83–90.

Sokhansanj, S. and H. C. Wood. 1991. Engineering aspects of forage processing. Advances in Feed Technology 5(1): 7–23.

Stark Charles and Peter Ferket. 2011. Conditioning, Pelleting, and Cooling. Power point https://www.ncsu.edu/project/feedmill/presentations/WS%202011/PDF2011/Conditioning%20Pelleting%20Cooling%202011.pdf.

Startsev, O. V. and B. N. Salin. 2000. Polycondensation of the components of the lignincarbohydrate complex of steam exploded wood. Doklady Chemical Technology 373-375(1): 30–33.

Sultana, A. and A. Kumar. 2011. Development of energy and emission parameters for densified form of lignocellulosic biomass. Energy 36: 2716–2732.

Suzuki, S., H. Shintani, S. Y. Park, K. Saito, N. Laemasak, M. Okuma and K. Iiyama. 1998. Preparation of binder-less boards from steam exploded pulps of oil palm (ElaeisguneensisJaxq.) Fronds and structural characteristics of lignin and wall polysaccharides in steam exploded pulps to be discussed for self-binding. Holzforschung. 52(4): 417–426.

Tabil, L. G. 1996. Binding and pelleting characteristics of alfalfa. PhD thesis. Department of Agricultural and Bioresource Engineering. University of Saskatchewan. Saskatoon.

Tabil, L. and S. Sokhansanj. 1996. Process conditions affecting the physical quality of alfalfa pellets. Appl. Eng. Agric. 12: 345–350.

Tabil, L. G., S. Sokhansanj and R. T. Tyler. 1997. Performance different binders during alfalfa pelleting. Can. Agr. Eng. 39: 17–23.

Thek, G. and I. Obernberger. 2004. Wood pellet production costs under Austrian and in comparison to Swedish framework conditions. Biomass Bioenerg. 27(6): 671e693.

Tooyserkani, Z. 2013. Hydrothermal pretreatment of softwood biomass and bark for pelletization. Ph.D. Thesis, University of British Columbia, Vancouver, Canada.

Turn, S. Q., C. M. Kinoshita, W. E. Kaar and D. M. Ishiimura. 1998. Measurements of gas phase carbon in steam explosion of biomass. Bioresource Technology. 64(1): 71–75.

Urbanowski, E. 2005. Strategic Analysis of a Pellet Fuel Opportunity in Northwest British Columbia [dissertation], Simon Fraser University, Vancouver (BC), 2005.

Winowiski, T. 1985. Optimizing pelleting temperature. Feed Management 36.7: 28–33.

Yang, Weihua. 1998. Characteristics of ground alfalfa in relation to steam conditioning. Doctoral thesis. University of Saskatchewan Saskatoon, 1998.

Zimbardi, F., D. Viggiano, F. Nanna, M. Demichele, D. Cuna and G. Cardinale. 1999. Steam explosion of straw in batch and continuous systems. Appl. Biochem. Biotech. 77-79: 117.

Zimbardi, F., E. Ricci and G. Braccio. 2002. Technoeconomic Study on steam explosion application in biomass processing. Applied Biochemistry and Biotechnology 98–100(2): 89–99.

CHAPTER 8

Hydrothermal Carbonization (HTC) of Biomass for Energy Applications

S. Kent Hoekman,[1,*] *Amber Leland*[2] *and Larry Felix*[3]

1. Introduction

In recent years, a thermochemical process called hydrothermal carbonization (HTC) has been increasingly investigated and employed as a means of converting raw biomass into a higher-value solid fuel. HTC involves treatment in hot, pressurized, liquid water—typically in the temperature range of 175–275°C. To ensure that water remains in the liquid state, the HTC process is conducted within pressure vessels, usually under autogenous pressure (Pressures of approximately 2–6 MPa correspond to reaction temperatures of 175–275°C). While the HTC process has been known for over a century, interest in it is now rapidly increasing because of its potential to convert a wide variety of low-value feedstocks, such as forest waste and agricultural waste, into renewable solid fuels for generation of heat and electrical power. This increase in interest is illustrated in Fig. 1, which shows the number of technical papers regarding HTC that have been published since 1970. Throughout the literature, HTC is also known by other names, which has led to significant confusion in terminology. Among these other names are hot compressed water (HCW) treatment, hydrothermal pre-treatment, coalification, wet pyrolysis, and wet torrefaction. (By including

[1] Desert Research Institute, Reno, NV 89512, USA.
[2] Coordinating Research Council, Alpharetta, GA 30022, USA.
[3] Gas Technology Institute, Birmingham, AL 35242, USA.
* Corresponding author: kent.hoekman@dri.edu

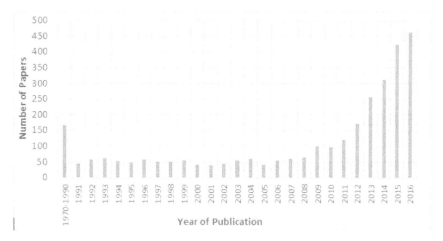

Fig. 1. Number of publications on the topic of HTC. Derived from Web of Science, using "hydrothermal carbonization" or "HTC" as search terms.

these other names, the total number of publications shown in Fig. 1 would increase by about 25%.)

Friedrich Bergius, a Nobel Prize winning German chemist, is generally credited with having first investigated the HTC process as part of his efforts to understand the natural transformation of biomass into coal (Bergius, 1913). Other significant early work in this area was conducted by Ruyter (1982); van Krevelen (1982); van Krevlen et al. (1993); Bobleter (1994); and others. In the past decade, several reviews have been published, which provide useful overviews of the numerous aspects of HTC (Peterson et al., 2008; Yu et al., 2008; Funke and Ziegler, 2010; Libra et al., 2011; Titirici and Antonietti, 2010; Reza et al., 2014a).

In some respects, HTC can be considered as a distinct subset of the broad continuum of thermochemical processes that can be applied to biomass feedstocks. This perspective was discussed by Libra et al. (2011) who compared HTC reaction conditions and product yields with those from other thermochemical processes that also produced solid char products. This comparison is summarized in Table 1 which also includes information about additional forms of thermochemical treatment. Hydrothermal liquefaction (HTL) and hydrothermal gasification (HTG) can be considered more severe forms of HTC. All three forms of hydrothermal treatment are conducted in hot, pressurized, liquid water, but with the objective of providing different products. The dominant intended products of HTC, HTL, and HTG are solids, liquids, and gases, respectively.

Table 1. Typical reaction conditions and yields of thermochemical treatment of biomass.*

Process	Process conditions		Product Distribution, wt.%		
	Temp.,°C	Time	Char	Liquid	Gas
Slow Pyrolysis	350–500	hours–days	25–35	20–30	25–35
Intermediate Pyrolysis	~500	10–20 sec	15–25	30–50	20–30
Fast Pyrolysis	~500	1–2 sec	10–15	60–75	10–15
Gasification	~800	10–20 sec	~10	~5	80–90
Torrefaction	200–300	0.5–4 hr	70–90	5–10	10–30
Hydrothermal carbonization (HTC)	175–275	5 min–8 hr	50–80	5–20	2–5
Hydrothermal liquefaction (HTL)	300–375	5 min–4 hr	5–20	30–60	10–30
Hydrothermal gasification (HTG)	375–500	5 min–2 hr	0–5	5–10	30–80

*Data adapted from Funke and Ziegler (2010), Libra et al. (2011), Eseyin et al. (2015), Elliott (2011), Kambo and Dutta (2015a), Toor et al. (2011), Kambo and Dutta (2015b), and Aller (2016).

HTC resembles the mild pyrolysis process called torrefaction in that both processes are conducted over a similar temperature range and are used to convert biomass into an energy-densified solid char with similar, but distinctively different physical and chemical properties. That is, HTC and torrefaction both produce a friable, 'coal-like' product having lower oxygen content, higher energy density, and higher hydrophobicity as compared to raw biomass feedstocks. Since HTC and torrefaction produce products that appear similar, HTC has often been called "wet torrefaction" (Yan et al., 2009; Yan et al., 2010; Bach et al., 2013; Bach and Skreiberg, 2016), although this terminology is chemically inappropriate as the chemical processes involved are quite different. Some authors further differentiate HTC and wet torrefaction on the basis of the intended uses of their respective char products (Eseyin et al., 2015; Bach et al., 2013). However, this distinction has not found traction in the literature. Torrefaction involves heating of biomass in an inert atmosphere (limited oxygen supply) at temperatures of 200–300°C. Under these conditions, all free water is driven off, along with volatile organic compounds (VOCs) originally contained in the feedstock. Typically, 15–30% of the feedstock mass (dry basis) is lost during torrefaction, but only about one-half as much of the energy content is removed; thus, the remaining char has a higher energy content (MJ/kg) than the feedstock (Kambo and Dutta, 2015a; Tumuluru et al., 2010; Tumuluru et al., 2011; Agar and Wihersaari, 2011).

In contrast, HTC is carried out in a liquid, aqueous environment. Under the typical temperature range of 175–275°C, autogenous pressure ranges from approximately 2 to 6 MPa. While these conditions are well below the critical point of water (374°C and 22.1 MPa), it should be noted that water has distinctly different characteristics under HTC conditions as compared to ambient conditions (Kruse and Dahmen, 2015; Brunner, 2009). This is shown in Table 2, which compares the properties of water under three conditions (1) ambient (25°C), (2) sub-critical hot compressed (250°C), and (3) super-critical hot compressed (400°C). As temperature and pressure increase, the density and viscosity of water decrease, which increases diffusion and leads to higher reaction rates. The decrease in dielectric constant at elevated temperature changes the solvent characteristics of water, so that under HTC conditions, it behaves in a fashion similar to a non-polar solvent, thereby enhancing the solubility of organic constituents such as biomass and its degradation products. Finally, under HTC conditions, water has a higher degree of ionization, with an enhanced dissociation into acidic hydronium (H_3O^+) and basic hydroxyl (OH^-) ions. This ionization makes sub-critical water an effective medium for carrying out acid hydrolysis.

A significant advantage of HTC is its ability to utilize wet feedstocks and avoid the complexity, cost, and energy requirements of pre-drying. It should be noted that hydrochar is obtained from the HTC process as a wet product that still requires drying. However, due to its hydrophobic character, it is much easier and less costly to dry hydrochar than to dry a wet biomass feedstock. Although considerable energy is needed to heat the process water used in HTC, this expenditure is small compared to the energy required to evaporate an equivalent mass of water (Berge et al., 2011). As HTC is conducted in a closed system, the large phase-change energy requirement of converting from liquid water to steam is avoided. Because of this feedstock

Table 2. Properties of water as a function of temperature and pressure.*

Property	Ambient water	Sub-critical hot, compressed water	Super-critical hot, compressed water
Temperature, °C	25	250	400
Pressure, MPa	0.1	5	25
Density, g/cm^3	0.997	0.80	0.17
Dielectric constant, ε	78.5	27.1	5.9
Heat capacity, KJ/kg°K	4.22	4.86	13
Viscosity, mPa s	0.89	0.11	0.03
Ionic product, pK$_w$	14.0	11.2	19.4
Thermal Conductivity, mW/m°K	608	620	160

*Data adapted from Tekin et al. (2014), Onwudilli and Williams (2008), Kruse and Dahmen (2015), and Bach and Skreiberg (2016).

flexibility, HTC has been applied to a wide range of wet and low-value materials. Some examples reported in the literature include food wastes (Sabio et al., 2016; Parshetti et al., 2014; Berge et al., 2015), sewage sludge (Peng et al., 2016; Danso-Boateng et al., 2015; Buttmann, 2011; Parshetti et al., 2013; He et al., 2014; He et al., 2013), municipal waste streams (Berge et al., 2011; Burguete et al., 2016; Hwang et al., 2012; Lu et al., 2012; Reza et al., 2016a), pulp mill streams (Wikberg et al., 2016; Makela et al., 2016; Lin et al., 2015; Areeprasert et al., 2014; Kang et al., 2012), anaerobic digester streams (Mumme et al., 2011), wet distillers/brewers grains (Heilmann et al., 2011; Poerschmann et al., 2014), animal manure (Dong et al., 2009; Cao et al., 2012), raw peat (Mursito et al., 2010), bamboo (Yang et al., 2016), aquatic plants (Heilmann et al., 2010; Savage, 2012; Duan et al., 2013; Broch et al., 2014; Gao et al., 2016), human faecal sludge (Danso-Boateng et al., 2015; Fakkaew et al., 2015), and other feedstocks (Poerschmann et al., 2013).

Although HTC has proven to be an effective process for treating diverse, wet biomass waste streams, woody materials and agricultural residues that are available in sufficient amounts to support commercial operations may be more suitable feedstocks for large-scale energy applications. While wood chips and wood waste have received the greatest attention, the literature also reports many examples of agricultural products and residues being suitable for HTC. These include coconut fiber (Liu et al., 2013; Liu et al., 2014a), palm empty fruit bunches (Parshetti et al., 2013), sugarcane bagasse (Hoekman et al., 2013; Chen et al., 2012), corn cobs/stover (Hoekman et al., 2013; Zhang et al., 2015), rice hulls (Hoekman et al., 2013), barley straw (Sevilla et al., 2011), arid land plants (Reza et al., 2016b), grape pomace (Pala et al., 2014), walnut shells (Roman et al., 2012), oil palm shell (Nizamuddin et al. 2015), and tobacco stalk (Cai et al., 2016).

The primary solid product of HTC is an energy-dense hydrochar; the nature and uses of these chars are quite diverse. We use the term 'hydrochar' to denote chars produced by the HTC process. This is to distinguish from the term 'biochar,' which is commonly used to define a product of pyrolysis, frequently intended for improving soils. According to Lehmann and Joseph, "Biochar is the product of heating biomass in the absence of or with limited air to above 250°C, a process called charring or pyrolysis" (Lehman et al., 2015). In addition to fuel, hydrochar has been used in a wide range of applications, including soil amendment (Lehman et al., 2006; Fowles, 2007; Rilling et al., 2010), water purification (Hu et al., 2010; Liu and Zhang, 2009; Mochidzuki et al., 2005), carbon-based microspheres and nanocomposites (Hu et al., 2010; Sevilla and Fuertes, 2009; Kruse et al., 2013), carbon sequestration (Titirici et al., 2007; Naisse et al., 2015), and catalysts (Titirici and Antonietti, 2010). (Biochar has also been used in most of these similar applications.) However, in this paper, we focus on the generation and use of hydrochar as a solid, renewable fuel.

2. HTC Process Applied to Lignocellulosic Biomass

Lignocellulosic biomass consists primarily of three polymeric constituents—cellulose, hemicellulose, and lignin. While composition varies from one source to another (hardwoods, softwoods, agricultural wastes, etc.) most biomass contains approximately 40–50 wt.% cellulose, 25–30 wt.% hemicellulose, and 15–20 wt.% lignin (Bobleter, 1994; Kambo and Dutta, 2015b; Nizamuddin et al., 2017). In addition, small amounts of extractable organic materials (terpenes, waxes, tannins, tall oils, etc.) and inorganic ash constituents are present. Occasionally, the term holocellulose is used, meaning the sum of cellulose and hemicellulose. For most biomass materials, the sum of holocellulose, lignin, and ash approaches 100% when expressed on a dry basis.

Cellulose is generally regarded as the world's most abundant organic chemical product. It consists of long, linear chains of glucose molecules that are linked with β-(1-4)-D-glucosidic ether bonds. The number of repeating glucose units in cellulose typically varies from a few thousand to 15,000. Hemicellulose is also a polymeric material consisting of linked sugar units, but with much less uniformity than cellulose. Whereas cellulose consists exclusively of a repeating 6-carbon sugar (glucose), hemicellulose polymers consist of both 5- and 6-carbon sugars. In addition, hemicellulose is not linear, like cellulose, but contains considerable branching with side chain groups. In contrast, lignin has a very different structure, consisting primarily of linked aromatic structures. The three principal aromatic building blocks within lignin are p-coumaryl alcohol, coniferyl alcohol, and sinapyl alcohol. An illustration showing the chemical components and macroscopic structures of cellulose, hemicellulose, and lignin are provided in Fig. 2.

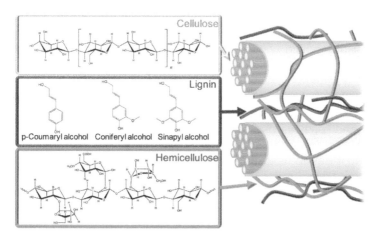

Fig. 2. Structural representation of lignocellulosic biomass showing components of cellulose, hemicellulose, and lignin. From Alonso, Wettstein, and Dumesic (Alonso et al., 2012) (Used with permission).

Due to its aromatic structure—and hence, higher C/O ratio, lignin has a higher energy content than cellulose or hemicellulose. This is an important consideration when preparing energy dense solid biofuels as it is advantageous to retain as much lignin as possible. Fortunately, this is frequently the case, as lignin tends to be much more resistant to acid hydrolysis and decomposition in HCW than holocellulose. The higher heating value (HHV) of lignin is approximately 26 MJ/kg, while the HHVs of cellulose and hemicellulose are 17–18 MJ/kg (Demirbas, 2005). The HHV of raw biomass materials is typically in the range of 16–20 MJ/kg (on a dry basis) (Nizamuddin et al., 2017).

2.1 HTC chemical processes

The chemical processes involved in HTC treatment of biomass are complex and variable, depending upon feedstock and reaction conditions. Because of this complexity, a clear chemical mechanism for HTC is difficult to define. Nevertheless, there has been a considerable mechanistic discussion in the literature and a consensus of views has emerged (Funke and Ziegler, 2010; Libra et al., 2011; Titirici and Antoniette, 2010; Reza et al., 2014a; Kruse et al., 2013; Coronella et al., 2014; Reza et al., 2014b). Most researchers agree that HTC involves the simultaneous occurrence of five chemical processes: (1) hydrolysis, (2) dehydration, (3) decarboxylation, (4) condensation polymerization, and (5) aromatization. An illustration that shows the complexity and inter-relatedness of these chemical processes are given in Fig. 3. Additional comments about each of these processes are provided below.

2.1.1 Hydrolysis

Generally, it is accepted that the first step in the HTC process involves hydrolysis, by which water reacts with ether and ester bonds within the three classes of biomacromolecules (cellulose, hemicellulose, and lignin) to create fragments. These fragments—oligosaccharides from cellulose and hemicellulose, and phenolic species from lignin—then continue to react through a variety of other chemical processes. Due to the fragility of its ether and ester bonds, hemicellulose is the least stable fraction and is readily hydrolyzed at temperatures as low as 180°C. Because cellulose is more robust due to its uniform, linear structure, higher temperatures are required to hydrolyze this component—typically around 230°C. Lignin is even more resistant to such reactions. While lignin may begin to react at temperatures as low as 230°C, much of it remains unreacted, even at considerably higher temperature, and is thus retained as an identifiable

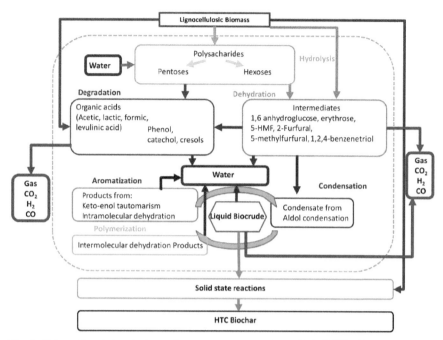

Fig. 3. Schematic of chemical reaction processes occurring within HTC of biomass. (From Reza et al. (2014). Used with permission.)

component of the hydrochar product of HTC. For use as a solid biofuel, inclusion of lignin is beneficial due to its high energy content and good binding behavior (Kaliyan and Morey, 2009; Kaliyan and Morey, 2010).

Hydrolysis of cellulose and hemicellulose leads to the production of smaller fragments, including molecules of simple mono- and disaccharides, such as glucose, fructose, galactose, xylose, etc. However, under typical HTC reaction conditions, these saccharides are not stable but react further to produce other water-soluble products, including acetic acid, furans, furfurals, and other species. Because some of these products are valuable chemicals, there is interest in recovering them, although economical isolation of discrete products of interest is challenging. Due to the production of organic acids (especially acetic acid), the pH level of HTC process water is lower than the initial level. While initial pH may be 6–7, the final pH level is generally 3–4. The issue of pH during the HTC process is more complex, as water itself becomes quite acidic under hot, pressurized conditions (Zeitsch, 2000).

2.1.2 Dehydration

Dehydration of biomass during HTC occurs by two different processes. Physical dehydration (also called dewatering) results from the exclusion of water due to the increasingly hydrophobic character of the hydrochar product (Coronella et al., 2014; Acharjee et al., 2011). Dewatering does not change the chemical composition of HTC products. Chemical dehydration results from elimination of hydroxyl groups within biomass and its degradation products, thereby changing the chemical composition (Funke and Ziegler, 2010; Sevilla and Fuertes, 2009). Chemical dehydration results in loss of oxygen and hydrogen, thereby decreasing both H/C and O/C ratios while increasing carbon content. While dehydration can involve several other chemical processes, a very common process is loss of water from 5- and 6-carbon sugar units to produce furfural and 5-HMF, respectively. A simple reaction pathway illustrating this is shown in Fig. 4, whereby holocellulose is first hydrolyzed, leading to the production of simple sugars, which then undergo dehydration to produce furfural and 5-HMF.

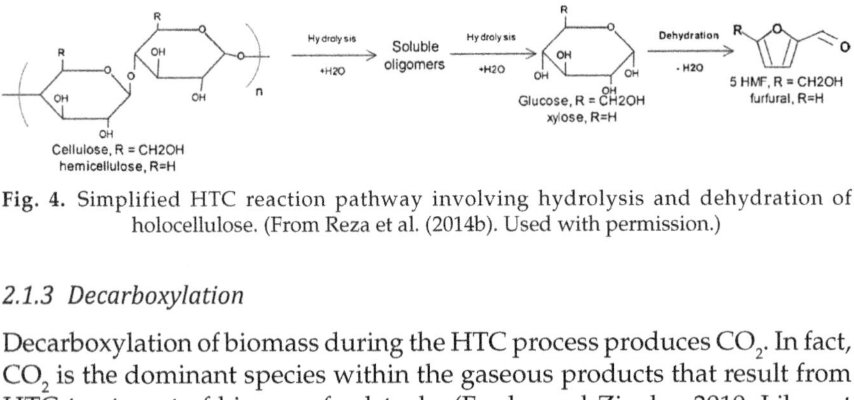

Fig. 4. Simplified HTC reaction pathway involving hydrolysis and dehydration of holocellulose. (From Reza et al. (2014b). Used with permission.)

2.1.3 Decarboxylation

Decarboxylation of biomass during the HTC process produces CO_2. In fact, CO_2 is the dominant species within the gaseous products that result from HTC treatment of biomass feedstocks (Funke and Ziegler, 2010; Libra et al., 2011; Hoekman et al., 2013; Hoekman et al., 2011). Formation of CO_2 from thermal degradation of carboxyl groups is well-known. However, because the amount of CO_2 produced by HTC is more than can be explained by simple elimination of carboxyl groups within the biomass, other CO_2-forming mechanisms must also be involved. One possible contributing factor is formic acid, which is formed in the HTC process and is known to produce CO_2 under hydrothermal conditions (Yu and Savage, 1998). In addition, some condensation reactions may produce CO_2.

The hydrothermal treatment of biomass results in simultaneous dehydration and decarboxylation processes. Decarboxylation produces CO_2, while dehydration produces H_2O. Thus, the ratio of decarboxylation to dehydration can be expressed as $r = (mol\ CO_2)/(mol\ H_2O)$. The value of r varies with feedstock and reaction conditions but generally is in the range of 0.2–1.0 (Funke and Ziegler, 2010; Reza et al., 2014b). It is thought that under most HTC conditions, the rate of dehydration is much faster than the rate of decarboxylation. However, as HTC process conditions become

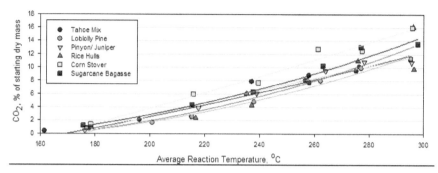

Fig. 5. Production of CO_2 from HTC treatment of various biomass feedstocks. (From Hoekman et al. (2013). Used with permission.)

more severe (higher temperature and longer reaction time), considerable CO_2 can be produced. This is illustrated in Fig. 5, which shows that for several biomass feedstocks, the yield of CO_2 increases from about 2% at an HTC temperature of 200°C to over 10% at an HTC temperature of 275°C.

2.1.4 Condensation-polymerization

Many of the initially-produced HTC products are quite reactive and continue to react through condensation and polymerization processes (Funke and Ziegler, 2010; Coronella et al., 2014; Reza et al., 2014b). Aldol condensation is a common example, by which new C-C bonds are formed, along with the elimination of water. Multiple condensation steps result in polymerization of smaller entities into larger structures. Condensation reactions can also result in cross-linking between larger fragments originating from the decomposition of cellulose and lignin. By means of such condensation-polymerization processes, water-soluble fragments originally produced by HTC can be combined to form water-insoluble, solid products, or be joined to existing solid materials. The evidence for this is provided from HTC experiments in which a range of temperature/time conditions was explored (Hoekman et al., 2013). It was observed that under relatively mild HTC conditions, up to 20% of the initial feedstock was converted into water-soluble but non-volatile material [sometimes called non-volatile residue (NVR)]. However, with more severe reaction conditions, the amount of NVR material was found to decrease as it reacted further to contribute to the formation of solids and gases.

2.1.5 Aromatization

Even though cellulose and hemicellulose contain no aromatic structures, such structures can be produced during the HTC process (Funke and Ziegler, 2010; Sevilla and Fuertes, 2009). In addition, existing aromatic fragments from lignin are incorporated into hydrochar. Because of their high chemical stability, such aromatic structures are considered to be the basic building blocks of solid hydrochar, just as they are of natural coal (Funke and Ziegler, 2010). Within the HTC process, aromatization requires a relatively high process temperature. Several researchers have investigated this using infrared spectroscopy (FTIR) to analyze hydrochars produced over a range of process conditions (Liu et al., 2014a; Coronella et al., 2014; Reza et al., 2014b; Guo et al., 2015). An example is given in Fig. 6, which shows FTIR spectra for raw loblolly pine and hydrochars produced from it. As explained by the authors, the increases and decreases of absorption bands at different process temperatures can be used to indicate the progression of specific chemical processes. This type of analysis shows that hemicellulose degrades at the lower temperatures, followed by cellulose degradation at higher temperatures, with aromatization occurring at even higher temperatures. Further evidence of increased aromatic content at higher process temperature is provided by researchers who used ¹³C NMR spectroscopy to characterize hydrochar produced from HTC treatment of cellulose (Liu et al., 2013; Kim et al., 2015).

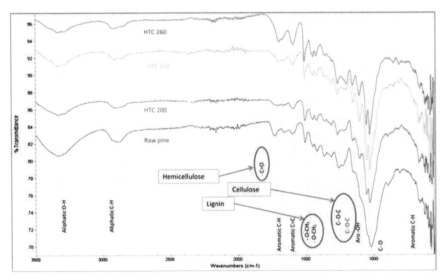

Fig. 6. FTIR spectra of loblolly pine and hydrochars produced at 200, 230, and 260°C. (From Coronella et al. (2014). Used with permission.)

2.1.6 Catalysis of HTC

Several researchers have explored the use of catalysts to accelerate or, in some other way, improve the HTC process. Minowa et al. (1998) reported that use of an alkali catalyst, sodium carbonate (Na_2CO_3), lowered the decomposition temperature of cellulose and inhibited char formation, resulting in greater oil formation. However, these effects were seen mainly at higher process temperatures (300–350°C), where conditions are more suitable for HTL than for HTC. Also at these high temperatures, the same researchers observed that use of a nickel catalyst promoted steam-reforming reactions, leading to the formation of methane. The work of this group was confirmed and expanded in a more recent paper (Fang et al., 2004). Kumar and Gupta (2008) explored the use of potassium carbonate (K_2CO_3) as a catalyst in the HTC treatment of cellulose and found similar results. When using HTC at milder conditions to pre-treat switchgrass and corn stover prior to enzymatic ethanol production, the inclusion of K_2CO_3 was found to be beneficial in reducing the pre-treatment temperature (Kumar et al., 2011). Poerschmann et al. (2014) used citric acid as a catalyst in the HTC treatment of brewers' spent grain, although the effectiveness of this was not discussed. Bach et al. (2015) explored the use of CO_2 as a catalyst by conducting HTC treatment of spruce and birch woods in an atmosphere of pressurized CO_2 (70 bar), as compared to an atmosphere of pressurized N_2. They found that CO_2 was beneficial in reducing the process severity required to conduct HTC, and the resulting hydrochar had slightly improved grindability (up to 7% lower specific grinding energy) and hydrophobicity (1–2% reduction in equilibrium moisture content), along with substantially lower ash content.

Toor et al. (2011) and Nizamuddin et al. (2017) have summarized the use of various HTC catalysts as reported in the literature. While some effects have been observed in certain cases, it seems likely that the overall benefits of catalysts for producing hydrochars as solid fuels are too small to be pursued in commercial applications. Although no thorough economic analysis has been performed to demonstrate this, there are several considerations that lead us to this conclusion. First, many benefits attributed to catalyst-use occur only at HTC conditions that are more severe than required to produce energy-densified hydrochar fuels. Second, hydrolysis—which is the key initial step in the HTC process—is likely to be catalyzed already by the action of HCW and by acetic acid which is produced during the process of HTC. Third, on a commercial scale, biomass feedstocks are unlikely to be as clean as those used in laboratory experiments. In particular, commercial feedstocks contain ash constituents, some of which may be problematic with respect to combustion (see Section 4.1.2). In such cases, use of additional ash-forming catalysts could add to this concern. Because of this situation, the recent report that adding CO_2 to the HTC process may be helpful in reducing ash is intriguing and should be explored further (Bach et al., 2015).

2.2 HTC process conditions and yields

A defining feature of the HTC process is that it is carried out in HCW. While carbonization is possible in oil as well as water, the process and products are quite different (Funke and Ziegler, 2010; Heilmann, 2011). The presence of liquid water promotes hydrolysis and ionic chemical reactions but suppresses pyrolysis—presumably due to its good heat transfer behavior which minimizes local temperature peaks. Broadly speaking, HTC treatment of any biomass feedstock produces three product streams: solid hydrochar, aqueous products, and gaseous products. As described in later sections, the partitioning of products among these three types varies considerably based on feedstock and process conditions. Under reasonably severe operating conditions, a typical partitioning of solid, aqueous, and gaseous products is approximately 50/40/10, as shown in Fig. 7.

In most situations, the HTC process is conducted under autogenous pressure, meaning that the pressure is dictated by the saturated vapor pressure of water at a given temperature. However, several researchers reported over-pressurizing the reactor vessel (usually with N_2) to ensure

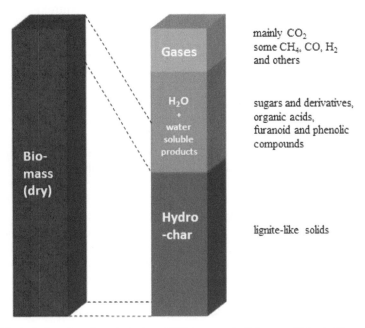

Fig. 7. Typical allocation of products from HTC treatment of biomass. (Adopted from Funke and Ziegler (2010).)

that the water remains in a liquid state. Bach et al. (2013) reported that over-pressurization slightly enhanced the carbonization rate of Norway spruce and birch woods under HTC conditions. However, this effect seems too small to justify on a commercial scale, where higher pressure operations would entail higher capital expenses.

2.2.1 Effect of process conditions on yield and energy content of hydrochar

In utilizing HTC to create an energy-dense solid fuel, the product of greatest interest is hydrochar. In fact, one of the most positive attributes of HTC is its ability to increase the energy content of hydrochar, as compared to the initial feedstock. Therefore, the yield of hydrochar under different process conditions is a critical parameter. The total energy yield is another critical parameter when evaluating the effects of process conditions. The third metric of interest is the energy densification ratio, which expresses the degree to which the hydrochar energy content exceeds the feedstock energy content. Equations for these three inter-related yield parameters are given below. In all cases, these metrics are expressed on a dry mass basis.

Mass yield = (Mass of hydrochar)/(Mass of biomass) x 100%   (Eq. 1)

Energy densification ratio = (HHV of hydrochar)/(HHV of biomass) (Eq. 2)

Energy yield = mass yield x energy densification ratio   (Eq. 3)

Optimization of HTC process conditions depends upon which metrics are being used to judge the outcome. For fuel purposes, maximizing energy densification is an important objective, but this must be balanced by also achieving a reasonable mass yield, from a process economics perspective. Thus, total energy yield, which is a function of mass yield and energy densification, is also an important factor. In addition, physical handling properties (ease of drying, pelletization behavior, grindability, etc.), combustion behavior, and process economics are important in determining optimized HTC process conditions.

Numerous process parameters have been shown to influence hydrochar yield and energy content. These include temperature, residence time, solids loading (ratio of water/biomass), biomass particle size, and pH. Of these, process temperature and retention time have been shown to be most important. In general, as process temperature and time increase, the extent of carbonization increases. This results in reduced hydrochar yield, but increased energy density (MJ/kg) of the hydrochar. The effect of temperature on the physical appearance of hydrochar produced from coniferous wood chips is shown in Fig. 8. As the process temperature was increased, the hydrochar product became darker in color, less fibrous/more friable, and more 'coal-like' in appearance. Also, hydrochar produced

Fig. 8. The effect of reaction temperature on hydrochar products from HTC treatment of coniferous wood chips. Reaction time was constant at 6 hours (From Sermyagina et al. (2015). Used with permission.)

from many feedstocks has a distinctive, but pleasant odor, reminiscent of barbeque cooking.

In most reported cases where both temperature and time have been varied, temperature has been shown to have the dominant effect on hydrochar yield and energy content (Reza et al., 2014a; Bach et al., 2013; Hoekman et al., 2011; Sermyagina et al., 2015; Yan et al., 2014; Funke and Ziegler, 2009; Bach et al., 2012; Nizamuddin et al., 2016; Lu et al., 2013; Kleinert and Wittman, 2009). Sermyagina et al. (2015) investigated HTC of raw coniferous wood chips and determined that the mass and energy yields of hydrochar could be expressed by mathematical formulas including terms for temperature (T,°C), reaction time (t, hr), and water/biomass ratio (r), as shown in Eqs. 4 and 5.

$$\text{Mass yield} = 1 - 0.04079 \, (T-150)^{0.337} \cdot t^{0.2142} \cdot r^{0.3055} \tag{Eq. 4}$$

$$\text{Energy yield} = 1 - 0.05632 \, (T-150)^{0.062} \cdot t^{0.2846} \cdot r^{0.4405} \tag{Eq. 5}$$

Equations 4 and 5 were developed for particular feedstocks treated under a narrow range of process conditions. It is not known how broadly applicable they may be for a wider range of feedstocks and process conditions.

The effect of reaction temperature upon energy content of hydrochars derived from a variety of feedstocks is shown in Fig. 9 (Hoekman et al., 2013). All experiments represented in this figure were conducted with a 30-min reaction time and an 8/1 ratio of water/biomass. In all cases, very little energy densification occurred at temperatures below 200°C. A rapid increase in energy densification occurred as the temperature was increased from 200 to 275°C, with little additional densification at higher temperatures.

Figure 9 shows differences in behavior of woody and herbaceous feedstocks. The three woody feedstocks (Tahoe Mix, Loblolly Pine, and Pinyon/Juniper) have higher initial energy contents (HHV) at around 20–21 MJ/kg and produce hydrochars with energy contents as high as 28–30 MJ/kg, which is within the range typically found for lignite and sub-bituminous coal. Two of the herbaceous feedstocks (corn stover and sugarcane bagasse)

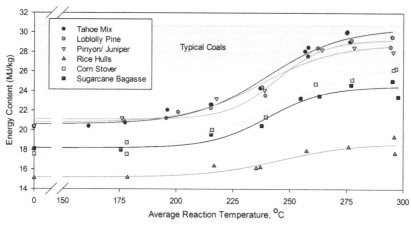

Fig. 9. The effect of HTC process temperature on energy content (HHV) of hydrochar product. Reaction time was constant at 30 min. (From Hoekman et al. (2013). Used with permission).

produced a similar degree of energy densification (30–40%) at the highest process temperatures, although the absolute energy content values of the feedstocks and hydrochars were lower than those from the woody biomass. A third herbaceous feedstock, rice hulls, appears to be an outlier, having much lower energy content and producing hydrochar products with much less energy densification than the other biomass materials. However, most of this difference can be explained by varying ash contents among the feedstocks. All three woody materials had ash contents of < 1%; corn stover and sugarcane bagasse had approximately 6% ash; rice hulls had 20% ash. Much of this ash (particularly silica in the case of rice hulls) is retained in the hydrochar, thereby 'diluting' the energy content of the product.

2.2.2 Reaction severity

To characterize reaction processes in which both temperature and time are important parameters, it is useful to combine these parameters into a single factor that defines overall reaction severity. Overend and Chornet (1987) defined a reactivity factor to characterize hydrolytic depolymerization processes during wood pulping and similar operations using the following equation:

$$R_0 = t * e^{[(T-100)/14.75]} \qquad \text{(Eq. 6)}$$

In this expression, R_0 (also called the reaction ordinate) has units of time (min). The variable "t" is the experimental reaction time (min); the variable "T" is the reaction temperature (°C). This expression assumes the hydrolytic process involves first-order reactions having an Arrhenius dependence upon temperature. A reference temperature of 100°C is used. When graphically portraying experimental data in terms of process severity, it is common to plot the logarithm of the reaction ordinate ($\log R_0$), which is called the "severity factor" (SF). This severity factor equation has been used by several researchers in describing the HTC process as applied to various biomass feedstocks (Garrote et al., 1999; Roos et al., 2009; Peterson et al., 2009; Kim et al., 2013; Hoekman et al., 2017).

Hoekman et al. (2017) showed that by utilizing SF as an HTC reaction parameter, hydrochar energy content results from many different feedstocks, reactor systems, and process conditions became more consistent. To investigate whether this approach is useful across a broader range of situations, we have used Eq. 6 to compute SF values for a large number of literature-reported HTC experiments. These data are compiled in Table 3, which is divided into two sections for woody (W) and herbaceous (H) feedstocks. Most values shown in Table 3 for hydrochar yield and properties were taken directly from the individual literature reports, although some were calculated using the relationships shown in Equations 1–3. Figure 10 shows energy contents and O/C ratios of hydrochar produced from these feedstocks as a function of SF. Although there is considerable scatter in the data, especially for the herbaceous feedstocks, it is clear that hydrochar energy content increases with increasing SF. The effect of increasing SF upon energy content is similar for woody and herbaceous materials, although, at a given SF value, the woody feedstocks produce hydrochars having higher energy content.

While Fig. 10 includes linear "best fit lines" to help illustrate the overall data trends, it is unlikely that energy content and O/C ratio vary linearly with SF for any given feedstock. As was shown with a more limited dataset of biomass feedstocks, energy content exhibited an "S-shaped" relationship with SF (Hoekman et al., 2017). Very little increase in energy content was seen up to an SF value near 4.5, then rapid increase occurred between SF of 4.5 and 6.5 with no further increase in higher SF values. Such S-shaped behavior is difficult to see in Fig. 10 due to a large number of feedstocks and variable process conditions (water/biomass ratio, reaction times, heating rate, etc.) that are included in this dataset.

2.2.3 Mass and water balances

Despite dozens of literature reports of HTC studies, rigorous determinations of mass balance are quite rare. One of the reasons for this is that in many cases, gaseous products are simply vented, without being captured and measured. The second problem is that water is both a reactant and a product of the HTC process, and it is very difficult to determine accurately small changes in water that occur within an aqueous reaction environment. Water is consumed in hydrolysis reactions, but is produced in dehydration and

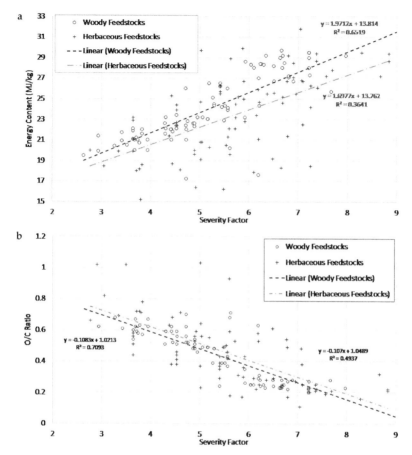

Fig. 10. The effect of HTC process severity on energy content (top) and O/C ratio of hydrochar product (bottom). Various feedstocks and conditions as reported in the literature. Data were taken from Table 3.

condensation reactions. Through a series of careful laboratory experiments, it was shown that when conducting HTC with loblolly pine as feedstock, water production changed from being negative at a process temperature of 200°C to positive at 260°C (Cornella et al., 2014; Reza et al., 2014b). On a severity factor basis, this crossover from water consumption to water production occurred near an SF value of 4.5.

Hoekman et al. (2011, 2013) conducted a large number of HTC experiments over a range of reaction temperatures using six different biomass feedstocks. In these experiments, gaseous products were captured and characterized, and produced water was determined by careful measurement of all water inputs and outputs. A significant fraction of initial feedstock mass is converted to water-soluble organic products, some of

Table 3. Hydrothermal carbonization of woody and herbaceous biomass reported in the literature.

Pub. date	Feedstock Name	Type	Temp.,°C	Hold time	W/B ratio	Severity factor	Mass yield, %	Energy yield, %	HHV, MJ/kg	Energy densification ratio	Atomic O/C	Ref.
2002	Konera wood	W	200	1 min	6/1	2.944	62.2	67.8	17.2	1.09	0.70	Inoue et al., 2002
			250			4.417	56.6	62.8	17.6	1.11	0.63	
			300			5.889	38.3	62.4	25.7	1.63	0.28	
			350			7.361	39.8	68.8	27.3	1.73	0.24	
2009	Loblolly pine	W	200	5 min	5/1	3.643	88.7	95.8	21.1	1.08	0.54	Yan et al., 2009
			230			4.527	70.6	79.8	22.1	1.13	0.51	
			260			5.410	57.0	77.5	26.5	1.36	0.24	
2011	Tahoe mix	W	215	30 min	8/1	4.863	69.1	76.7	22.6	1.11	0.48	Hoekman et al., 2011
			235	30 min		5.452	63.7	76.1	24.2	1.19	0.39	
			255	30 min		6.041	50.3	70.0	28.3	1.39	0.25	
			275	30 min		6.630	50.9	72.7	29.0	1.43	0.23	
			295	30 min		7.219	50.1	72.8	29.5	1.45	0.20	
			255	5 min		5.263	57.7	71.2	25.1	1.23	0.39	
			255	10 min		5.564	55.5	71.1	26.0	1.28	0.37	
			255	30 min		6.041	50.3	70.0	28.3	1.39	0.25	
			255	60 min		6.342	52.1	74.8	29.2	1.43	0.23	
2012	Beech wood chips	W	210	3 hr	7/1	5.494	68	81	22.9	1.19	0.49	Tremel et al. 2012
2012–2013	Spruce	W	175	30 min	5/1	3.685	88	89	21	1.02	0.62	Bach et al., 2013; Bach et al., 2012
			200	10 min		3.944	82	85	21	1.02	0.62	
			200	30 min		4.421	79	82	22	1.07	0.59	
			200	60 min		4.723	73	80	22	1.07	0.56	
			225	30 min		5.158	70	80	23	1.12	0.49	

2012–2013	Birch	W	175	30 min	5/1	3.685	80	81	20	1.01	0.67	Bach et al., 2013; Bach et al., 2012
			200	10 min		3.944	66	69	20	1.01	0.67	
			200	30 min		4.421	64	66	21	1.05	0.62	
			200	60 min		4.723	62	64	21	1.05	0.62	
			225	30 min		5.158	58	60	23	1.15	0.49	
2013	Loblolly pine	W	177	30 min	8/1	3.744	77.7	80.5	21.0	1.04	0.58	Hoekman et al., 2013
			201			4.451	70.1	75.4	21.8	1.08	0.52	
			216			4.893	72.4	79.4	22.3	1.10	0.54	
			239			5.570	63.0	73.9	23.9	1.17	0.43	
			262			6.247	50.1	70.2	28.4	1.40	0.24	
			277			6.689	48.2	69.3	29.2	1.44	0.24	
			295			7.219	52.4	74.5	29.0	1.42	0.22	
2013	Tahoe mix	W	161	30 min	8/1	3.273	85.3	85.5	20.4	1.00	0.68	Hoekman et al., 2013
			178			3.774	78.8	80.4	20.7	1.02	0.63	
			196			4.304	73.4	78.7	21.6	1.07	0.59	
			216			4.893	68.7	76.4	22.6	1.11	0.48	
			237			5.511	63.7	76.1	24.3	1.19	0.39	
			257			6.100	48.5	67.0	28.0	1.38	0.25	
			276			6.659	51.8	75.2	29.7	1.46	0.23	
			295			7.219	50.1	72.8	29.5	1.45	0.20	
2013	Loblolly pine	W	200	0.5 min	30/1	2.643	85.0	86.4	19.5	1.02		Reza et al., 2013a
				1 min		2.944	81.4	84.3	19.9	1.03		
				3 min		3.421	74.5	81.0	20.9	1.09		
				5 min		3.643	63.9	72.8	21.9	1.14		
2013	Loblolly pine	W	230	0.5 min	30/1	3.527	76.5	81.5	20.5	1.07		Reza et al., 2013a
				1 min		3.828	66.5	76.0	22.0	1.14		
				3 min		4.305	62.7	75.6	23.2	1.21		
				5 min		4.527	58.0	70.8	23.4	1.22		
2013	Loblolly pine	W	260	0.5 min	30/1	4.410	73.8	84.3	22.0	1.14		Reza et al., 2013a
				1 min		4.711	55.9	74.4	26.0	1.35		
				3 min		5.188	54.7	76.3	26.5	1.38		
				5 min		5.410	54.3	74.3	26.2	1.36		

Table 3 contd. ....

...Table 3 contd.

Pub. date	Feedstock Name	Type	HTC Reaction conditions Temp.,°C	Hold time	W/B ratio	Severity factor	Mass yield, %	Hydrochar yield and properties Energy yield, %	HHV, MJ/kg	Energy densification ratio	Atomic O/C	Ref.
2013	Pinyon/Juniper	W	176	30 min	8/1	3.715	77.4	81.1	21.2	1.05	0.59	Hoekman et al., 2013
			218			4.951	71.5	80.8	23.2	1.13	0.46	
			239			5.570	62.7	74.6	24.1	1.19	0.42	
			264			6.306	50.6	69.6	28.3	1.38	0.25	
			278			6.718	49.6	69.8	28.5	1.41	0.25	
			295			7.219	48.6	66.3	28.0	1.36	0.23	
2014	Loblolly pine	W	200	1 min	30/1	2.944	79.1	84.8	21.4	1.07	0.62	Yan et al., 2014
				3 min		3.421	73.7	79.9	21.7	1.08	0.61	
				5 min		3.643	72.9	80.5	22.1	1.10	0.60	
2014	Loblolly pine	W	230	1 min	30/1	3.828	68.6	74.3	21.6	1.08	0.57	Yan et al., 2014
				3 min		4.305	64.8	71.8	22.1	1.11	0.59	
				5 min		4.527	62.4	70.0	22.4	1.21	0.59	
2014	Slash pine	W	235	6 min		4.753	60.9	68.8	22.8	1.13	0.52	Hoekman et al., 2014
			235	12 min		5.054	69.7	80.1	23.2	1.15	0.49	
			275	6 min		5.931	54.2	76.4	28.6	1.41	0.26	
			275	12 min		6.232	53.6	75.6	28.4	1.41	0.28	
2015	Umbila wood	W	180	150 min	16/1	4.532	87				0.56	Cuvilas et al., 2015
			180	350 min		4.900	82				0.51	
2016	Eucalyptus bark	W	220	2 hr	10/1	5.612	46.4	50.1	20.2	1.09	0.68	Gao et al., 2016
			240	2 hr		6.201	41.2	60.2	27.0	1.46	0.29	
			275	2 hr		7.232	42.2	62.5	27.3	1.48	0.27	
			300	2 hr		7.968	40.0	63.2	29.2	1.58	0.23	
			240	4 hr		6.502	41.5	63.1	28.2	1.52	0.28	
			240	6 hr		6.678	40.9	60.1	28.0	1.51	0.28	
			240	8 hr		6.803	40.1	59.3	27.1	1.46	0.27	
			240	10 hr		6.900	40.3	59.6	26.7	1.44	0.28	

Year	Biomass	W/H	Temp	Time	Ratio							Reference
2016	Norway spruce	W	210 222	30 min 5 min		4.716 4.291	74.1 73.8	80.3 80.7	22.4 22.6	1.08 1.09	0.52 0.51	Bach and Skreiberg, 2016
2017	Loblolly pine	W	290 290	20–30 s 20–30 s	3–4 3–4	5.293 5.293	83.5 83.5	101 100	24.5 24.3	1.21 1.20	0.48 0.48	Hoekman et al., 2017
2017	Holocellulose from wood fibers	W	200 210 220 230 240	8 hr		5.626 5.920 6.214 6.509 6.803			– 23.2 24.3 26.7 27.2		– 0.35 0.31 0.24 0.22	Liu et al., 2017
2011	Distillers grains	H	200	2 hr	4/1	5.024	38.9	57.2	29.7	1.47	0.23	Heilmann et al., 2011
2012	Corn stalk	H	250	4 hr	10/1	6.797	36	60.1	29.2	1.67	0.17	Xiao et al., 2012
2012	Walnut shells	H	190 230	20 hr	20/1	5.729 6.907	48.6 36.3	57.5 50.2	23.2 27.1	1.18 1.38		Roman et al., 2012
2012	Sunflower stalk	H	190 230	20 hr	20/1	5.729 6.907	39.7 29.2	58.9 50.4	24.5 28.5	1.49 1.74		Roman et al., 2012
2012	Sugarcane bagasse	H	180	5 min 30 min	10/1 10/1	3.054 3.833	70.4 60.8	75.9 66.1	18.5 18.6	1.08 1.09	0.82 0.78	Chen et al., 2012
2013	Coconut fiber	H	220 250 300 350	30 min	10/1	5.010 5.894 7.366 8.838	57 45 40 37	76.7 65.7 65.0 55.8	24.7 26.7 29.4 28.7	1.34 1.45 1.60 1.56	0.37 0.30 0.21 0.21	Liu et al., 2013
2013	Eucalyptus leaves	H	200 250 300 350	30 min	10/1	4.421 5.894 7.366 8.838	65 46 40 32	87.3 61.1 61.3 47.8	25.3 25.0 28.7 29.4	1.33 1.32 1.51 1.55	0.38 0.36 0.25 0.22	Liu et al., 2013

Table 3 contd. ...

...*Table 3 contd.*

Pub. date	Feedstock		HTC reaction conditions				Hydrochar yield and properties					Ref.
	Name	Type	Temp.,°C	Hold time	W/B ratio	Severity factor	Mass yield, %	Energy yield, %	HHV, MJ/kg	Energy densification ratio	Atomic O/C	
2013	Palm empty fruit bunches	H	150 250 350	20 min	10/1	2.773 5.718 8.662	76 62 49	78.2 70.4 68.5	20.0 22.1 27.2	1.02 1.13 1.40	0.66 0.53 0.32	Parshetti et al., 2013
2013	Sugarcane bagasse	H	176 216 238 259 277 296	30 min	8/1	3.715 4.893 5.540 6.159 6.689 7.248	69.6 63.8 59.1 45.0 44.6 42.7	67.9 67.7 65.5 57.6 59.5 56.5	18.0 19.6 20.5 23.4 24.6 24.2	0.98 1.06 1.11 1.28 1.34 1.32	0.72 0.56 0.50 0.22 0.25 0.23	Hoekman et al., 2013
2013	Corn stover	H	178 216 239 261 277 296	30 min	8/1	3.774 4.893 5.570 6.218 6.689 7.248	67.0 56.6 56.4 40.0 42.4 38.0	68.5 64.6 68.5 56.3 60.7 56.2	18.2 20.0 21.3 24.7 25.2 26.3	1.02 1.14 1.21 1.41 1.43 1.48	0.72 0.53 0.46 0.25 0.23 0.20	Hoekman et al., 2013
2013	Rice hulls	H	179 217 236 257 276 296	30 min	8/1	3.803 4.922 5.481 6.100 6.659 7.248	78.5 72.6 64.3 55.4 54.6 52.2	78.5 78.7 68.6 65.8 65.8 63.7	15.2 16.4 16.2 17.9 18.3 18.5	1.00 1.08 1.07 1.19 1.21 1.22	0.71 0.59 0.49 0.29 0.24 0.23	Hoekman et al., 2013
2014	Spent brewery grain	H	210	4 hr		5.619	67	101	28.9	1.51	0.18	Gunarathne et al., 2014
2014	Spent brewery grain	H	200 240	14 hr 14 hr	1/1 1/1	5.869 7.046	51.1 47.5	68.5 67.9	29.9 31.8	1.34 1.43	0.17 0.11	Poerschmann et al., 2014

Year	Biomass	Type	Temp	Time	Ratio							Reference
2014	Maize silage	H	200	20 min	12/1	4.245	51.8	71.7	21.6	1.38		Reza et al., 2014c
			200	60 min		4.723	49.2	72.5	23.0	1.47		
			200	180 min		5.200	47.4	75.6	24.9	1.60		
			200	360 min		5.501	45.4	74.5	25.6	1.64		
			250	20 min		5.718	41.0	58.9	22.4	1.44		
			250	60 min		6.195	39.4	70.2	27.8	1.78		
			250	180 min		6.672	39.3	71.5	28.4	1.82		
			250	360 min		6.973	39.6	76.9	30.3	1.94		
			230	60 min		5.606	45.3	64.7	22.3	1.43		
2015	Corn cobs	H	175	5 min	9/1	2.907	77.7				1.02	Zheng et al., 2015
			195	5 min	9/1	3.496	53.0				1.02	
2015	Corn cob residues	H	190	1.5 hr	10/1	4.604	66	69	18.1	1.05	0.71	Zhang et al., 2015
			210			5.193	58	68	20.3	1.18	0.55	
			230			5.782	48	70	25.2	1.46	0.31	
			250			6.371	45	67	25.5	1.48	0.28	
			270			6.960	40	59	25.5	1.48	0.27	
2015	Oil palm shell	H	220	30 min	7/1	5.010	62.4		20.8		1.03	Nizamuddin et al., 2015
			240			5.599	54.4		22.1		0.93	
			260			6.188	48.6		25.9		0.58	
			280			6.777	44.8		26.7		0.54	
			290			7.071	43.0		26.8		0.48	
2015	Corn stalk	H	180	8 hr	10/1	5.037	70	81	20.8	1.16	0.56	Guo et al., 2015
			210			5.920	59	75	22.8	1.27	0.43	
			230			6.509	48	68	25.6	1.42	0.29	
			250			7.098	43	64	26.7	1.48	0.25	
			270			7.687	40	62	27.8	1.54	0.21	
			290			8.276	36	59	29.5	1.64	0.16	
2015	Miscanthus	H	190	5 min	6/1	3.349	83.5	90.2	19.9	1.08	0.69	Kambo and Dutta, 2015a
			225			4.379	66.8	77.6	21.4	1.16	0.63	
			260			5.410	47.8	66.5	25.7	1.39	0.39	

Table 3 contd. ....

...Table 3 contd.

| Pub. date | Feedstock | | HTC Reaction Conditions | | | | Hydrochar Yield and Properties | | | | | Ref. |
	Name	Type	Temp.,°C	Hold time	W/B ratio	Severity factor	Mass yield, %	Energy yield, %	HHV, MJ/kg	Energy densification ratio	Atomic O/C	
2016	Tobacco stalk	H	180	2 hr	10/1	4.435	80	79.7	18.7	1.00	0.68	Cai et al., 2016
			200	2 hr		5.024	67	67.7	19.1	1.02	0.64	
			220	2 hr		5.612	61	64.2	19.6	1.05	0.59	
			240	2 hr		6.201	63	71.2	21.1	1.12	0.53	
			260	2 hr		6.790	59	66.9	21.4	1.14	0.49	
			260	1 hr		6.489	62	63.8	19.2	1.02	0.61	
			260	4 hr		7.091	53	63.6	22.5	1.20	0.38	
			260	8 he		7.392	43	60.2	26.1	1.40	0.22	
			260	12 hr		7.568	41	59.8	27.2	1.45	0.19	
2016	Grindelia (gum weed)	H	200	5 min	5/1	3.643	59	64	19.9	1.08	0.51	Reza et al., 2016b
			230			4.527	57	69	22.4	1.22	0.46	
			260			5.410	50	71	26.2	1.42	0.35	
2016	Grindelia bagasse	H	200	5 min	5/1	3.643	60	67	21.1	1.11	0.44	Reza et al., 2016b
			230			4.527	51	61	22.7	1.19	0.41	
			260			5.410	44	59	25.5	1.34	0.29	
2016	Rabbitbrush	H	200	5 min	5/1	3.643	79	90	23.2	1.14	0.58	Reza et al., 2016b
			230			4.527	70	84	24.4	1.20	0.38	
			260			5.410	62	81	26.3	1.30	0.37	
2016	Rabbitbrush bagasse	H	200	5 min	5/1	3.643	69	75	22.9	1.09	0.56	Reza et al., 2016b
			230			4.527	62	71	23.9	1.14	0.44	
			260			5.410	57	71	26.3	1.25	0.43	

which are volatile compounds that are lost during the evaporative process used to determine the non-volatile residue (NVR) fraction of the aqueous products. Thus, specific volatile species (e.g., furfural, acetic acid, formic acid, and 5-HMF) were measured independently and added to the NVR for mass balance determinations. Even when exercising great care, it proved difficult to account for over 90–95% of the total feedstock mass.

To avoid the problems introduced by water, it may be preferable to express mass balance on the basis of carbon mass, rather than total mass. Figure 11 provides a carbon balance for the set of HTC experiments mentioned above (Hoekman et al., 2013). This figure shows that under all conditions, a significant fraction of feedstock carbon is converted into aqueous phase products. Furthermore, as the process severity increases,

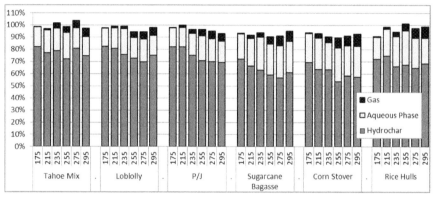

Fig. 11. Carbon balance of HTC process for six biomass feedstocks treated at six different temperatures (°C). Reaction time was constant at 30 min. (From Hoekman et al. (2013). Used with permission.)

the fraction of carbon in the hydrochar product decreases, while increased carbon is found in the gaseous and aqueous phase products.

2.3 HTC process energetics

The overall energy balance of the HTC process is an important consideration in evaluating the commercial viability of solid biofuels produced by the process. However, the energetics of biomass carbonization is difficult to determine due to the complex chemical nature of the feedstocks and the multitude of reaction processes that occur. Based on simple stoichiometry, Titirici et al. (2007) reported that treatment of carbohydrates by HTC is more highly exothermic than fermentation of anaerobic digestion. Also using theoretical considerations, Libra et al. (2011) determined that HTC of cellulose was mildly exothermic, with a heat of reaction of –1.6 MJ/kg. However, these authors also pointed out that their analysis ignored formation of water-soluble organic products, which constitute an important byproduct stream from HTC processing. In addition, they suggested that initial stages of HTC are likely to require energy inputs as hydrolysis of cellulose is known to be endothermic.

Yan et al. (2010) investigated the heat of reaction from HTC treatment of loblolly pine at process temperatures from 200 to 260°C. Based on energy balances determined from product characterization, they postulated that the process was slightly endothermic, with a heat of reaction of 0.56 MJ/kg at 200°C, and 0.25 MJ/kg at 260°C. However, these authors also performed a rigorous uncertainty analysis of their methodology and determined that the heats of reaction at all three process temperatures investigated (200, 230, and 260°C) were close to zero and could not be differentiated with a high degree of confidence. A more recent assessment of this work pointed out that Yan et al. (2010) failed to account for the production of water, which is known to occur in the HTC process (Reza et al., 2014a). By including the exothermic formation of water, the overall HTC process would become slightly exothermic rather than slightly endothermic.

Funke and Ziegler (2011) used differential scanning calorimetry (DSC) in an isothermal mode to investigate the heat of reaction from HTC conducted at 240°C when using a water/solid ratio of 4/1. They observed slightly exothermic conditions for glucose, cellulose, and poplar wood, with ΔH values of −1.06, −1.07, and −0.76 MJ/kg, respectively. Rebling et al. (2015) utilized a heat flow DSC operated in a temperature scanning mode to measure the heat of reaction from HTC treatment of undefined biomass obtained from a nature protection area. They confirmed an exothermic process and measured heat release of 0.72 MJ/kg. Although direct comparison is difficult due to different feedstocks and measurement methods, this value is in reasonable agreement with Funke and Ziegler (2011). Such a small heat of reaction is expected to have minimal influence on the overall energy balance of the HTC process and is certainly insufficient to provide the energy required to operate a commercial process.

2.4 Kinetics of HTC

The kinetics of biomass degradation by HTC treatment is complex and difficult to ascertain. As with the process energetics discussed above, HTC kinetics is complicated by the variable chemical structures within biomass feedstocks and the multitude of reaction processes that occur. Several researchers have attempted to simplify this by focusing on the separate behaviors of the main biomass constituents: hemicellulose, cellulose, and lignin. A good background discussion of this is provided by Coronella et al. (2014).

Mochidzuki et al. (2003) utilized a liquid-phase thermogravimetric method to determine activation energy values of 100 and 96 kJ/mol for HTC treatment of rice husks and spent malt, respectively. Luo et al. (2011)

investigated HTC treatment of water hyacinth over two temperature ranges, 150–210°C and 200–280°C, and determined activation energy values of 147 and 91 kJ/mol, respectively. Liu et al. (2014a) investigated the kinetics of HTC treatment of coconut fibers and eucalyptus leaves and also found distinctly different behaviors at lower temperatures (220–300°C) and higher temperatures (300–375°C). In the lower temperature regime, which is more representative of HTC conditions for producing solid fuels, they observed first-order reaction kinetics and determined activation energies of 67.5 kJ/mol and 59.2 kJ/mol for coconut fibers and eucalyptus leaves, respectively. They also treated the same feedstocks under torrefaction conditions and found the activation Energies to be slightly higher than under HTC conditions.

Reza et al. (2013a) utilized a novel reactor system to enable experimental determination of mass loss kinetics of loblolly pine under HTC process temperatures of 200–260°C. This reactor system, shown in Fig. 12, allowed for rapid and controlled heating of the biomass feedstock and cooling of the product mixture, such that reaction times as short as 15 sec could be investigated. They found that the experimental results could be explained by two parallel first order reactions—one representing hemicellulose, the other representing cellulose. In this kinetic model, lignin was considered to be inert. The activation energies determined for hemicellulose and cellulose were 30 and 73 kJ/mol, respectively. The higher activation energy for cellulose compared to hemicellulose is consistent with other reports, and confirms the higher degree of the recalcitrance of cellulose. As pointed out by the authors, these activation energies are considerably lower than

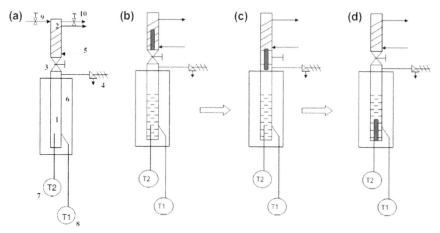

Fig. 12. Schematic showing operation of two-chamber reactor system used to investigate the kinetics of HTC process. Biomass sample in top chamber is dropped into pre-heated water in the bottom chamber by means of a ball valve. (Taken from Reza et al. (2013a). Used with permission.)

those reported in the literature for pure hemicellulose and cellulose. It is believed that when treating the whole lignocellulosic material, interactions of various chemical processes result in lowering the activation energy, as compared to the treatment of individual components (Cornella et al., 2014; Reza et al., 2013a).

2.5 Product characterization

HTC treatment of biomass produces gases, aqueous-phase, and solid products—all of which are of interest and have been characterized by numerous researchers. As described above, the allocation of products depends upon the biomass feedstock and the HTC process conditions, particularly the severity of the reaction. As process severity increases, the amounts of gaseous and aqueous-phase products also increase, while the amount of solid hydrochar decreases, although the energy density of the hydrochar increases. A brief summary of the three products streams is provided below.

2.5.1 Characterization of hydrochar

Generally, hydrochar is the desired solid product from HTC treatment of biomass. Thus, much work has been done to characterize hydrochar produced from different feedstocks under various process conditions. Perhaps the most common means of characterizing hydrochar involves proximate and ultimate analyses, which are analytical tests originally developed to characterize coals and other solid fuels. Proximate analysis is used to determine the amount of water, volatile carbon, fixed carbon, and ash in a solid fuel. Ultimate analysis is used to determine the elemental composition of fuel in terms of mass percent of carbon, hydrogen, nitrogen, sulfur, and oxygen. Another important characteristic of solid fuels is their energy density, usually expressed as a higher heating value (HHV) on a dry basis. Alternatively, HHV can be calculated from the elemental analysis results through a variety of predictive equations developed for biomass materials, although there is some question about the appropriateness of such equations when applied to hydrochar products.

   The van Krevelen diagram is a useful way to illustrate the degree of "coalification" that has occurred through HTC processing, and hence, the fuel value of the produced hydrochar. Originally proposed by van Krevelen and co-workers in 1960 (Schuhmacher et al., 1960), the diagram plots atomic H/C ratios vs. atomic O/C ratios. An example diagram is provided in Fig. 13, where it is used to show the increasing coal-like behavior of hydrochar from loblolly pine, as it is produced under increasingly severe process conditions. The colored shapes indicate regions where typical

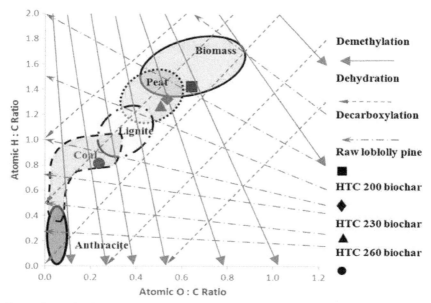

Fig. 13. Example of a Van Krevelen diagram showing increased carbonization at higher HTC process temperature. (Taken from Coronella et al. (2014). Used with permission.)

biomass, peat, lignite, and coal appear. Thus, Fig. 13 illustrates how hydrochar changes from being 'peat-like' to being 'coal-like' as the HTC process severity increases. The solid and dashed guidelines shown on this van Krevelen diagram indicate directional changes that biomass and its chars would undergo as a consequence of three chemical processes: dehydration, decarboxylation, and demethylation. An inspection of Fig. 13 indicates that HTC treatment of loblolly pine is dominated by dehydration with a lesser contribution from decarboxylation. The general behaviors seen in Fig. 13 are typical of many biomass feedstocks.

Information about the molecular composition of hydrochar can be obtained by infrared (IR) spectroscopy. An example of FTIR spectra was shown in Fig. 6 to illustrate changes that occur in hydrochar produced from loblolly pine under increasingly severe process conditions (Cornella et al., 2014; Reza et al., 2014b; Reza et al., 2014d). Numerous other FTIR examples are available in the literature to illustrate the effects of HTC process conditions on hydrochar composition when treating coconut fibers (Liu et al., 2014a), palm empty fruit bunches (Parshetti et al., 2013), Konera wood (Inoue et al., 2002), landscape management biomass (Rohrdanz et al., 2016), and arid land plants (grindelia and rabbitbrush) (Reza et al., 2016b). In general, these spectroscopic analyses indicate a sequential reduction in hydroxyl groups (OH), carbonyl groups (C=O), and ether groups (C-O-C), as well as increased aromatic character. These changes are consistent with the initial reaction of hemicellulose followed by cellulose and lignin conversion by means of the chemical processes described in Section 2.1.

Another method used to characterize both raw biomass and hydrochars is the so-called 'fiber analysis', which is based on a modified van Soest method (Goering and van Soest, 1970; Reza et al., 2013b). This method uses the ANKOM filter bag technique (FBT) for the determination of extractives, hemicellulose, cellulose, and lignin plus ash. This fractionation is based on sequential dissolution in a series of solvents, with mass losses determined at each step. Although lignin content is not measured directly, it can be calculated by subtracting ash content, which is determined as part of the proximate analysis. Several researchers have used this fiber analysis method to confirm that hemicellulose is the most reactive component under HTC conditions, followed by cellulose and finally lignin (Kambo and Dutta, 2015a; Reza et al., 2014b; Reza et al., 2013a; Reza et al., 2014c; Uddin et al., 2014).

A major benefit of hydrochar is its friability and ease of grinding, as compared to raw biomass. This is critical when considering the use of hydrochar in a co-firing operation with pulverized coal. In fact, the difficulty and expense of pulverizing raw biomass make such co-firing applications commercially unattractive. Bach et al. (2012) measured the specific energy required to grind raw spruce and birch, as well as hydrochars produced from these woods. They found that the grinding energy required to pulverize hydrochar was reduced as the HTC process severity increased (temperature range of 175–225°C). When produced at 225°C, the specific grinding energy of the hydrochars was less than 10 kWh/t, as compared to 130–150 kWh/t for the raw wood feedstocks.

Kambo and Dutta (2015a) showed that hydrochar produced from miscanthus at 260°C, when pulverized in a planetary ball mill, gave a particle size distribution very similar to that of pulverized coal. This was in contrast to torrefied miscanthus which produced larger size fractions when pulverized. Tremel et al. (2012) subjected beech wood chips to the HTC process at 210°C in a pilot-scale operation. They pulverized the hydrochar in a pin mill and measured a specific energy consumption of 135 kJ/kg, which was comparable to the value of pulverizing torrefied biomass but was much lower than the 700 kJ/kg required to pulverize raw biomass. When considering the use of hydrochar in combination with coal, the Hardgrove grindability index (HGI) is a useful metric.

Numerous researchers have used microscopy to investigate the morphology of hydrochar; a few examples of published micrographs are (Liu et al., 2014a; Sevilla et al., 2011; Roman et al., 2012; Nizamuddin et al., 2015; Titirici et al., 2007; Liu et al., 2017). Micrographs of pinewood and its hydrochars (produced at process temperatures of 250–350°C) are shown in Fig. 14. These micrographs are typical in showing a coarser surface for

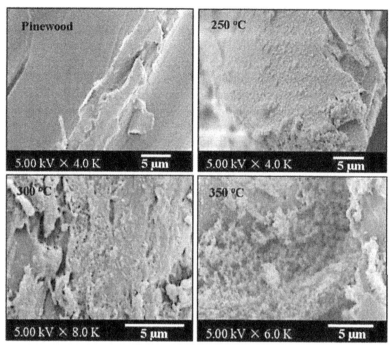

Fig. 14. Micrographs of pinewood and hydrochar produced from HTC treatment at different temperatures. (Taken from Liu et al. (2014a). Used with permission.)

hydrochar as compared to the raw biomass. As HTC process temperature increases, the surface becomes rougher and pitted. This is because aqueous reactions of the hemicellulose and cellulose constituents occur, decomposing these materials into smaller fragments, which exit the solid scaffold structure of the biomass.

While HTC treatment causes increased surface roughness of hydrochar, and hence, increased surface area compared to raw biomass, an extensive pore structure apparently does not develop. This is based on the observations of several researchers who have measured porosity and surface area of hydrochars, along with other solid materials. Surface area determinations are generally based upon N_2 adsorption measurements according to the methods of Brunauer, Emmitt, and Teller (BET). As an example, hydrochar produced from barley straw at 250°C was found to have a surface area of 8.3 m^2/g, while a pyrolysis char produced from the same feedstock (at 400°C) had a surface area of 150 m^2/g (Sevilla et al., 2011).

Roman et al. (2012) measured the surface area of hydrochars produced from sunflower stem and walnut shell under a range of HTC conditions and found them all to be within 25–30 m^2/g. Fang et al. (2015) reported surface areas of 10.7 m^2/g and below for hydrochars produced from sugarcane bagasse, hickory wood, and peanut shell. They also observed that surface area decreased as the process temperature to produce the hydrochar increased. This was attributed to collapse of pore wall structure as the reaction severity increased. Parshetti et al. (2013) reported similar behaviors for hydrochar produced from palm empty fruit bunches, with BET surface areas of 6.08, 8.03, and 2.04 m^2/g for materials produced at 150°C, 250°C, and 350°C, respectively. Liu et al. (2014a) determined that the surface area of pinewood-derived hydrochar reached a maximum of 20.4 m^2/g at a process temperature of 300°C, but decreased at higher temperatures. Nizamuddin et al. (2015) measured a surface area of 12.6 m^2/g for hydrochar produced from oil palm shell at 260°C, while the surface area of the raw biomass was only 0.31 m^2/g. Similar results were reported by Cai et al. (2016) for the surface area of raw tobacco stalk and hydrochars produced at various temperatures. Additional discussion of hydrochar surface area and the factors that influence this is provided by Kambo and Dutta (2015b). While the relatively low surface area of hydrochars can be increased by additional chemical and/or thermal activation, this is not necessary when used in fuel applications.

Another important characteristic of hydrochar is its resistance to water, or hydrophobicity, as compared to raw biomass. High hydrophobicity is critical for economically attractive applications of solid biofuels as this reduces transportation costs due to shipping less water. In addition, water resistance simplifies fuel storage and handling operations. In contrast, raw biomass is typically protected from the elements by use of indoor storage to prevent damage resulting from water absorption.

A useful measure of hydrophobicity is equilibrium moisture content (EMC) which is defined as the moisture content of a material that is in thermodynamic equilibrium with the moisture in the surrounding atmosphere under prescribed conditions (Coronella et al., 2014). Acharjee et al. (2011) showed that the EMC of hydrochar produced from loblolly pine was lower than that from raw feedstock over a relative humidity range of 11–85%. Furthermore, the EMC of hydrochar was reduced as the HTC process temperature was increased from 200°C to 260°C. The EMC of dry torrefied loblolly pine was also lower than that of raw biomass, but it was not as hydrophobic as hydrochar produced under similar severity (Yan et al., 2009; Acharjee et al., 2011). For example, at a relative humidity of 84%, the measured EMC of loblolly pine raw, torrefied, and hydrochar were 15.6%, 8.7%, and 5.3%, respectively. Bach et al. (2013) demonstrated similar effects with hydrochars produced from spruce wood which had much higher hydrophobicity when prepared at 225°C as compared to 175°C. Finally, Kambo and Dutta (2015a) showed that the hydrophobicity of miscanthus-derived hydrochar increased with HTC temperature and that this hydrochar was more hydrophobic than torrefied miscanthus at similar process temperatures.

2.5.2 Characterization of HTC process water

Because HTC is conducted in an aqueous environment and numerous water-soluble products are produced in the process, the final process water generally contains a complex mixture of organic compounds. This aqueous process stream is sometimes referred to as "liquid products", which is not strictly correct, as many of these products are actually water-soluble solids. Another term used to describe these process waters is "liquors".

Various approaches have been used to characterize these aqueous products. Typically, pH is measured and is found to be in the range of 3–4. The aqueous solutions are acidic due to the formation of organic acids— particularly acetic acid—in the HTC process. A simple, non-volatile residue (NVR) analysis can be used to determine the mass of products that are not volatilized upon complete drying in an oven at 105°C (Hoekman et al., 2013). However, this drying method also drives off volatile organic compounds (VOCs) contained in the process water, such as acetic acid and furfural. A method that directly measures organic carbon in the aqueous products (both volatile and non-volatile) is total organic carbon (TOC) analysis. This instrumental analysis involves catalytic oxidation of all organic materials to CO_2, which are then quantified by non-dispersive infrared (NDIR) detection (Hoekman et al., 2013; Hoekman et al., 2011).

Various methods have been used to measure individual organic species within the HTC aqueous products. Xiao et al. (2012) used GC-MS to analyze organic extracts of aqueous products obtained from HTC treatment of corn stalks and forest waste. The numerous compounds identified were classified as being sugar-derived (organic acids, furfural, and other furans) or lignin-derived (phenolic compounds). Reza et al. (2014b) used GC-MS to characterize organic compounds in the aqueous products from HTC treatment of loblolly pine, as well as the same compounds which remained adsorbed on the solid hydrochar. To do this, the aqueous products were evaporated to dryness, then reconstituted in an organic solvent and derivatized to improve the GC-MS analysis. The solid hydrochar was extracted with organic solvent, evaporated, reconstituted, derivatized, and analyzed by the same GC-MS method. Results showed that all the major monosaccharide products were partitioned between the solid hydrochar and the aqueous product stream. These analyses were performed on products of HTC treatment conducted at temperatures between 200°C and 260°C. It was observed that several simple sugars (such as arabinose, xylose, and galactose) were present in higher amounts under low-temperature conditions (200°C), but were diminished at 230°C and above. Glucose concentrations were maximized at 230°C but diminished at 260°C.

Further investigation into the effects of process temperature upon yields of specific compounds within the aqueous products was conducted by Hoekman et al. (2013), who treated six biomass feedstocks under HTC conditions over a temperature range of 175–295°C. Sugars and other carbohydrates were measured using a high-performance liquid chromatography (HPLC) method, with detection by refractive index (RI). The results, summarized in Fig. 15, show that yields of individual sugars varied substantially among the feedstocks and process temperatures. Some simple sugars, such as galactose, xylose, and mannose, had significant yields only at low process temperatures (175–215°C) but were not seen at higher temperatures. Glucose was slightly more stable, with measurable yields at temperatures up to 235°C.

Two very significant species found in the HTC aqueous products are furfural and 5-HMF. As shown in Fig. 15, the yields of these species were maximized at temperatures of 215–255°C, which corresponds to severity factors of approximately 4.8–5.6. Becker et al. (2014) reported very similar behaviors for furfural and 5-HMF obtained from HTC treatment of various other biomass feedstocks. These two species are valuable chemicals that can serve as precursors to numerous other organic products (Tekin et al., 2014; Alonso et al., 2012; Rosatella et al., 2011). Thus, being able to isolate these species from the HTC process water would likely improve the overall economics of the process. Isolation of simple sugar products could also be

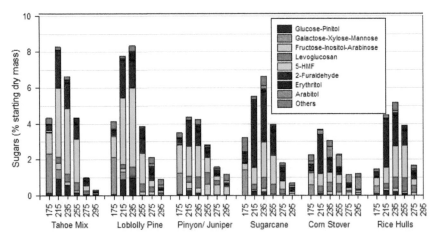

Fig. 15. Sugars measured by HPLC-RI in aqueous products from HTC treatment of six biomass feedstocks at six different temperatures. All reaction times were 30-min. (Taken from Hoekman et al. (2013). Used with permission.)

beneficial, but is more challenging, given the low and narrow temperature range under which these compounds are stable.

Organic acids are also major constituents of HTC process waters. The dominant product is acetic acid, although formic, lactic, levulinic, and other organic acids are also formed (Hoekman et al., 2013; Becker et al., 2014). Unlike the sugar products, total acid concentrations continue to increase as the process temperature (severity) increases. This is illustrated in Fig. 16 which shows that the total yield of acids and sugars remains fairly constant at 10–15% over a wide range of feedstock types and process conditions. Consequently, TOC yields remain relatively constant over a wide range of conditions. This observation was also made by Becker et al. (2014).

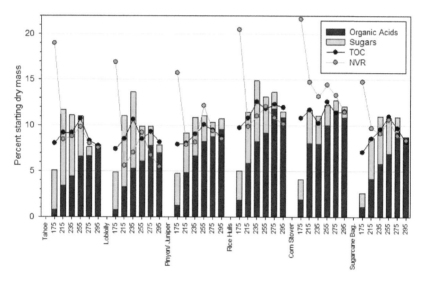

Fig. 16. Total sugars and organic acids compared to TOC and NVR contents of HTC aqueous products from HTC treatment of six biomass feedstocks at six different temperatures. All reaction times were 30-min. (Taken from Hoekman et al. 2013. Used with permission.)

2.5.3 Characterization of gaseous products

As already discussed, CO_2 is the dominant gaseous product from the HTC process, generally representing > 90% of total gaseous products (Eterson et al., 2008; Funke and Zieger, 2010; Libra et al., 2011; Hoekman et al., 2011). CO_2 is formed through decarboxylation, which increases with higher process severity. As shown in Fig. 5, the yield of CO_2 can vary from less than 1% to over 10% (expressed as wt. % of initial feedstock), depending on process severity. Carbon monoxide (CO) is likely the second most dominant gaseous product, although its concentration is typically only 3–6% that of CO_2 (Hoekman et al., 2013). Trace levels of several other gaseous products —including hydrogen, methane, and other light hydrocarbons—are also produced. However, the total energy value of the produced gases is very low, making it uneconomical to use as a fuel.

While the permanent gases from HTC processing are dominated by CO_2, the headspace vapors in any reactor system will also contain VOCs at concentrations that vary depending upon the pressure and temperature conditions. Detailed GC-MS analysis of headspace vapors produced from HTC treatment of pinyon-juniper woods, collected at room temperature, revealed significant concentrations of furan and methyl furans, along with trace amounts of many other C_2–C_{10} VOCs (Hoekman et al., 2014). Similar analyses were conducted on process gases produced in a continuous HTC reactor being used to treat loblolly pine feedstock (Hoekman et al., 2017). Both hot gas (produced from rapid decompression of the hot product stream) and cold gas (after passing through a condenser) were analyzed. Results showed these gases to contain furan, furfural, terpenes, and a wide range of other VOCs, with most species having a much higher concentration in the hot gas as compared to the cold gas.

3. Pelletization of Hydrochar

An important advantage of hydrochar compared to raw biomass or dry torrefied biomass is its ability to form robust, water-resistant pellets (or briquettes). For commercial applications of solid biofuels, this is critically important, as it enables conversion of the fuel material into a form that is much more easily handled, transported, stored, and integrated with existing infrastructure used for coal. Compression into a compact, solid form reduces dust problems and increases the energy density of bulk shipments, thereby reducing transportation costs. Additionally, flowable materials having a uniform size and shape are easier to feed into process units.

While torrefied biomass can also be pelletized, this often requires the addition of an external binder, and the resulting pellets are not robust (Reza et al., 2014d; Reza et al., 2012). A common way to assess the stability of pellets is through use of a modified Micum rotating drum test, which was originally developed for characterizing coal (Gil et al., 2010; Stelte et al., 2011; Stelte et al., 2012). In this test, pellets are tumbled under prescribed conditions. During this tumbling, small particles are lost from the pellets due to abrasion. The abrasion index is defined as the mass percentage of these fine particles compared to the initial mass of the pellets. The durability is defined as the mass percentage of remaining pellets compared to the starting mass. Thus, robust or highly stable pellets have a high durability value (> 95%) and a low abrasion index (< 5%).

The mechanical strength of pellets is commonly measured by applying an increasing compressive force on the side of a horizontally-oriented pellet until the pellet breaks (Liu et al., 2016; Liu et al., 2014b; Kambo and Dutta, 2014). The tensile strength is related to the compressive force applied and the geometric dimensions of the pellet. Liu et al. (2014b) reported that the tensile strength of pinewood sawdust pellets was 3.91 MPa, while the strength of pellets produced from hydrochar of pinewood sawdust was 7.10 MPa.

When dry, raw wood pellets are generally quite stable, as evidenced by their high durability values (> 95%) and low abrasive index values. However, as shown in Fig. 17, exposure to water dramatically reduces pellet stability. This figure includes photographs of different pelletized materials before and after being immersed in water for 60-min. Also shown are the photos of the same water-immersed pellets after they had undergone a tumbler durability test. Clearly, raw wood pellets cannot survive exposure to such aggressive test conditions. Pellets made from torrefied wood are noticeably more stable, although they still swell significantly upon water immersion, and disintegrate completely upon tumbling. In contrast, pellets made from hydrochar are extremely stable, even when immersed in water and subsequently tumbled. Numerous researchers have commented on the superiority of hydrochar pellets as compared to torrefied wood pellets concerning dimensional stability and durability against water damage (Reza et al., 2014d; Reza et al., 2012; Liu et al., 2016; Kambo and Dutta, 2014; Jose-Vicente et al., 2016).

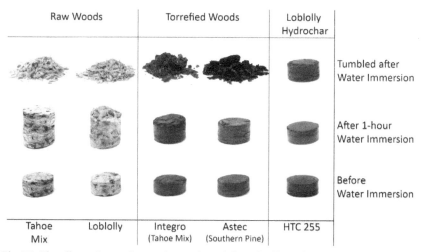

Fig. 17. The effects of water immersion on the stability of pelletized woody biomass materials. (Taken from Hoekman et al. (2015). Used with permission.)

The fact that raw wood pellets have excellent durability (when dry) is usually attributed to the presence of lignin and its ability to flow when heated to the glass transition temperature. Several researchers have investigated the glass transition temperature of hydrochar materials and have used microscopy to observe how hydrochar reduces cracks and surface roughness, compared to dry torrefied material (Liu et al., 2014a; Reza et al., 2014d; Reza et al., 2012; Liu et al., 2016). This work has shown that hydrochar can form bridges that connect adjacent particles and thus serve as a binder that strengthens pellets. Reza et al. (2014d) have utilized the strong binding behavior of hydrochar in mixing with torrefied materials to produce "engineered pellets" having acceptable durability. Others have also shown that hydrochar can be an effective binder when mixed with torrefied wood as well as with coal (Liu et al., 2016; Felix et al., 2015; Hoekman et al., 2015).

While lignin undoubtedly contributes to the binding behavior of hydrochar, it is likely that other factors are also involved. It is of note that chemical reactions occurring in the HTC process (e.g., hydrolysis, dehydration, and condensation) lead to the production of chemical species not originally present in raw biomass. For example, from detailed GC-MS analysis of organic extracts from hydrochar we have detected significant levels of abietic acids (Hoekman et al., 2015). These species are major constituents of resin acids (or tall oils) which are obtained commercially from woods and are used in various applications including coatings and adhesives (Coll et al., 2001; Axelsson et al., 2011). Also, HTC process conditions are similar to various commercial wood pretreatment methods used to produce particle board, fiberboard, strand board, and other wood composite materials (Palaez-Samaniego et al., 2013). Generally, these pretreatment methods involve exposure to steam or hot water, which removes all hemicellulose, resulting in a remaining solid that has improved strength and water resistance compared to the raw wood feedstock. A commercial steam-based process is used to produce Zilkha Black® Pellets (Zilkha).

As is well known, products of the HTC process include furan derivatives, such as furfural and 5-HMF. These materials are capable of reacting to produce cross-linked oligomers and polymers, known as furan resins, which are also used as adhesives (Pfriem et al., 2012). Therefore, it is likely that during the HTC process, organic materials such as abietic acids, tall oils, furan resins, phenolic resins, and others are produced and remain within the solid hydrochar, thereby imparting it with highly effective binding characteristics.

The effects of HTC process conditions upon properties of hydrochar pellets have been investigated. Reza et al. (2012) showed that both the mass density (kg/m³) and volumetric energy density (GJ/m³) of pellets made from loblolly pine hydrochar increased as the HTC process temperature increased from 200 to 260°C. Compared to pellets of raw loblolly pine, the 260°C hydrochar pellets had 33% higher mass density and 70% higher volumetric energy density. Very similar percentage increases in mass and energy densities were reported for pellets produced from miscanthus hydrochar, although the absolute values were lower for this herbaceous feedstock (Kambo and Dutta, 2014). In contrast, pellets produced from torrefied loblolly pine and torrefied miscanthus had greatly reduced mass and energy densities. With the combination of hydrochar's higher energy content (MJ/kg) and higher mass density (kg/m³), hydrochar pellets have nearly double the volumetric energy density (GJ/m³) of pellets produced from torrefied biomass. These results are summarized in Table 4.

Hoekman et al. (2015) also investigated properties and behaviors of pellets produced from hydrochar of loblolly pine. They explored a wide range of HTC temperatures, from 175°C to 295°C. Their results, shown graphically in Fig. 18, indicate that while a slight increase in pellet mass density occurs between 175°C and 215°C, the density falls off sharply as the process temperature is increased further, particularly beyond 250°C. This figure also shows that pellet volumetric energy density reaches a maximum near a process temperature of 250°C, and then decreases at higher temperatures. It was also shown that pellet robustness—as measured by swelling upon water immersion and tumbler durability values—varied with HTC process temperature. The pellets made from hydrochar produced at 175°C were quite unstable, while those made from hydrochar produced at 215°C were much more stable. Pellet stability increased only slightly at higher process temperatures and then decreased at temperatures above 275°C. Thus, it appears that the ideal HTC temperature range for acceptable

Table 4. Mass and energy densities of pellets.

Biomass material	Mass density, kg/m³	Vol. Energy density, GJ/m³	HHV, MJ/kg	Ref.
Raw loblolly pine	1102.4	22.8	20.7	Reza et al., 2012
HTC-200	1125.8	24.3	21.6	
HTC-230	1331.5	30.0	22.6	
HTC-260	1468.2	38.8	26.4	
Torrefied-300	805.5	19.0	23.6	
Raw miscanthus	834.1	15.7	18.8	Kambo and Dutta, 2014
HTC-190	886.9	17.9	20.2	
HTC-205	959.4	20.7	21.6	
HTC-260	1035.9	26.9	25.9	
Torrefied-260	819.6	16.7	20.3	

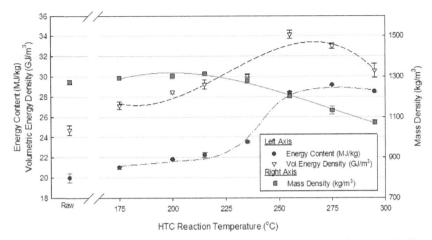

Fig. 18. The effect of HTC process temperature upon mass and volumetric densities of pellets. Hydrochar was produced by treatment of Loblolly Pine at indicated temperatures for 30-min. In a 2-L stirred Parr reactor. (Taken from Hoekman et al. (2015). Used with permission.)

pellet formation is 200–250°C, which corresponds to a severity factor range of about 4.4–5.9.

4. Commercial Considerations

When considering application of the HTC process to treat biomass on a commercial scale, several additional factors become important. For example, the size, configuration, and throughput of reactor vessels are critical. To date, no commercial-scale deployment of HTC has occurred. Most of the work involving HTC of biomass has been conducted at small-scale, in R&D facilities. The most commonly used equipment includes batch-type Parr reactors and other autoclave systems. Other simple laboratory apparatus has also been deployed, including sealed capillary tubes (Knezevic et al., 2009; Knezevic et al., 2010; Potic et al., 2004) and small, flow-through reactors in which liquids are pumped through a fixed bed of biomass (Ando et al., 2000; Sasaki et al., 2003). In another flow-through configuration, slurries of biomass in water are pumped through tubular reactor systems (Kumar and Gupta, 2008; Kumar et al., 2011). On a somewhat larger scale, Tremel et al. (2012) utilized a 250-L batch reactor for HTC treatment of 2.5–3.5 mm wood chips. By employing such a configuration in a semi-continuous batch mode, commercial-scale operation would be possible. A pilot plant operated by the Spanish company, Ingelia may function in this manner, although very few details have been provided (Hitzl et al., 2015). Additionally, Erlach et al. have described modeling simulations of a commercial-scale continuous reactor system, whose design was based on a plant previously used for upgrading peat (Erlach et al., 2011; Erlach et al., 2012). An auger-based continuous process for pre-treating wheat straw was described by Petersen et al. (2009). A laboratory version of such an auger-driven continuous process development unit (PDU) was described by Hoekman et al. (2015). Several other pilot-scale operations are mentioned in company website pages by Avalon Industries, TerraNova Energy, Artec Biotechnologie, and Grenol (Avalon Industries; TerraNova Energy; Artec Biotechnologies; Grenol GmbH). In all these cases, insufficient details are provided to fully understand the process configurations.

Very recently, a fully continuous process utilizing a twin-screw extruder (TSE) reactor system was described (Hoekman et al., 2017). While this TSE was designed for laboratory applications, it could be scaled-up for commercial operation (Liu et al., 2017). Due to the extreme mastication and controlled de-pressurization that occurs within the TSE process, the resulting hydrochar has significantly different physical characteristics than hydrochar produced in batch or auger-based reactors. A photograph of hydrochar produced from TSE treatment of loblolly pine at 250–260°C is shown in Fig. 19. Compared to conventional hydrochar, the TSE product is much lighter in color and has a smooth, non-fibrous, paste-like texture.

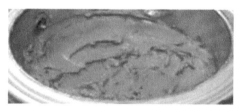

Fig. 19. Photo of hydrochar produced from loblolly pine using a twin-screw extruder reactor system. (From Hoekman et al. (2017). Used with permission.)

4.1 Combustion behavior

For hydrochar utilized as a solid fuel, a critical factor is its combustion behavior. Because the HTC process is not commercially deployed at present, only limited supplies of hydrochar have been produced. To our knowledge, no full-scale test burn of hydrochar has been conducted. However, combustion properties of hydrochar have been investigated by many researchers using surrogate analyses, with use of thermogravimetric analysis (TGA) in the presence of air being most common. In this analysis, a small amount of sample is heated under controlled conditions while carefully monitoring weight loss as the temperature increases. Typically, both thermogravimetric (TG) and derivative thermogravimetric (DTG) profiles are generated and analyzed. Examples of TG and DTG profiles are shown in Fig. 20.

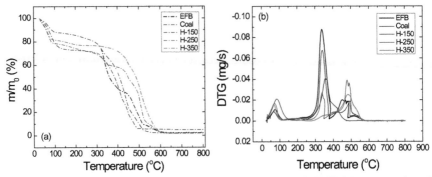

Fig. 20. TG (left) and DTG (right) curves for empty fruit bunch (EFB) biomass, coal, and hydrochar. All samples were heated in air at 10°C/min. (From Parshetti et al. (2013). Used with permission.)

From the analysis of such TG and DTG profiles, various combustion characteristics can be defined, including ignition temperature, maximum mass loss rate, peak temperature, burnout temperature, ignition index, and others. Such analyses have been reported for TGA of hydrochars produced from coconut fiber, coconut shell, and eucalyptus leaves (Bach and Skreiberg, 2016; Liu et al., 2013; Liu et al., 2014a; Liu et al., 2016), palm empty fruit bunches (Parshetti et al., 2013), palm shell (Nizamuddin et al., 2016), Norway spruce and birch (Bach et al., 2014), and others. In general, such analyses show that upon HTC processing, the biomass is converted to a material having less volatile matter, higher ignition temperature, and combustion over a wider temperature range. In all these ways, the hydrochar has become more coal-like and better suited for combustion either alone, or co-fired with coal.

Liu et al. (2016) have shown that pellets produced from blends of hydrochar (from coconut fiber and shell) and lignite have combustion properties very similar to those of lignite alone, although the blended pellets combust with slightly higher thermal efficiency. Bach et al. (2017) compared combustion properties of hydrochar and dry torrefied biomass from Norway spruce and found slightly different behaviors between the two. The dry torrefied material gave a sharper combustion peak at a lower temperature, whereas the hydrochar sustained combustion over a wider temperature range.

Another fuel use of hydrochar is as a feedstock for gasification. Tremel et al. (2012) investigated the treatment of beech wood-derived hydrochar in a laboratory-scale entrained-flow gasifier. They demonstrated high carbon conversion efficiency over a gasification temperature range of 1000–1400°C, with residence times as short as 1 sec. Comparable experiments using lignite as feedstock had similar or slightly lower carbon conversions. The authors concluded that the combination of HTC and entrained flow gasification shows high potential as a substitute for lignite gasification.

Gunarathne et al. (2014) investigated gasification of pellets produced from HTC treatment of spent Brewer's grains using a fixed-bed, pilot scale, updraft gasifier with an oxygen equivalence ratio of 0.20–0.22. They obtained the highest efficiency (80%) and highest syngas energy content (LHV of 7.9 MJ/m³) at an equivalence ratio of 0.20. Gasifying raw biomass pellets under these conditions gave lower efficiency (76%) and lower syngas energy content (6 MJ/m³).

4.2 Ash considerations

While biomass feedstocks generally have lower ash content than most coals, the different nature of the ash constituents can be of concern. In particular, the relatively high levels of alkali and alkaline earth metals (AAEM), especially in some agricultural wastes, lead to concerns about slagging and fouling on boiler heat transfer surfaces (Liu et al., 2014a). To relate ash's chemical composition with such problems and to maintain satisfactory operational performance, a set of indices has been established, along with acceptable ranges for these indices. These include a slagging index, fouling index, alkali index, and slag viscosity index (Coronella et al., 2014; Reza et al., 2013b; Liu et al., 2016; Masia et al., 2007).

Several literature reports indicate that HTC treatment of biomass reduces the ash content of the resulting hydrochar, primarily by leaching of AAEM into the aqueous product stream. Reza et al. (2013b) conducted HTC treatment of several herbaceous feedstocks (miscanthus, corn stover, switchgrass, and rice hulls) and reported up to 90% removal of Ca, Mg, P, and K. In addition, the resulting hydrochar products had improved slagging and fouling indices compared to the untreated biomass materials.

Liu et al. investigated HTC treatment of coconut fiber and eucalyptus leaves and reported approximately 90% reductions in K, Na, and Mg, and 60–70% reduction in Ca (Liu et al., 2014a; Liu and Balasubramanian, 2014). They also determined that use of mixed hydrochar/lignite pellets decreased slagging inclination as compared to lignite alone, and concluded that such co-firing would not adversely affect subsequent disposal or re-utilization of the ash (Liu et al., 2016). Beneficial ash reduction has also been demonstrated by HTC treatment of spruce and birch woods (Bach et al., 2013; Bach et al., 2012), miscanthus (Kambo and Dutta, 2015a), and other feedstocks (Burguete et al., 2016).

4.3 Techno-economic considerations

Although economic considerations are critical in determining the commercial viability of HTC processes, there is very little information available in this area. Erlach et al. (2011) used *Aspen Plus* simulation modeling to compare the economics of integrating an HTC process with a wood-fired combined heat and power (CHP) system and integrating a wood pellet process with the CHP system. Due to its complexity, the total equipment costs of the HTC plant are about twice as high as for the pellet plant. However, the overall economics depend largely upon the cost of biomass feedstock. Using poplar wood chips as feedstock, the HTC pathway costs are 33% higher than the pellet pathway (€12.94/GJ vs. €9.73/GJ; based on 2009 values of the Euro). If a biodegradable organic waste material (available at no cost) is used as HTC feedstock, the energy-specific costs of the two pathways are comparable. These researchers also computed the costs of avoided CO_2 emissions from the use of raw wood pellets to be €82/MT. The comparative costs for the use of hydrochar were €116/MT if raw wood is used as feedstock and €81/MT if organic waste materials are used as feedstock. This research group later extended their modeling study to compare the overall energy efficiency of gasification when using raw wood or hydrochar (Erlach et al., 2012). They determined that the gasification process itself was more efficient when using hydrochar, but this was offset by the energy required to operate the HTC process. Thus, the overall energy efficiency of producing syngas was slightly lower when using hydrochar.

Felix et al. (2015) performed a techno-economic modeling study to estimate the costs of producing fuel briquettes from blends of coal fines and hydrochar that was produced in a twin-screw extruder (TSE) reactor system. Yield and product compositional information about the hydrochar were derived from laboratory experiments in which loblolly pine was processed in a TSE reactor (Hoekman, 2017). *Aspen Plus* simulations were performed to represent HTC production plants of two sizes: 120 MT/day and 2400 MT/day. The minimum selling price (MSP) of the blended pellets was found to vary with plant size but depended more strongly on the ratio of hydrochar/coal fines comprising the briquettes. Hydrochar has a high cost relative to coal fines; thus, increasing the blend ratio of hydrochar from 10% to 30% increased the MSP of briquettes from the 2400 MT/day plant from $22.5/MT to $46.1/MT.

4.4 Process optimization and integration

As with any process, the overall practicality and economics of producing fuels via HTC can be expected to improve as individual steps in the process are optimized and the process is fully integrated into a broader supply chain. An obvious area of improvement involves recycling of the HTC process water. Not only does this reduce total water demand and minimize costs of treatment and disposal, but it also presents opportunities for improved heat recovery, which lowers the overall energy requirements (Kambo and Dutta, 2015b). Stemann et al. (2011; 2013) investigated stepwise recycling of process water for HTC treatment of poplar wood using a small laboratory-scale reactor. In each step, process water from the previous step was used to treat a fresh batch of poplar feedstock. The concentrations of organic acids and TOC in the recycled process water increased sequentially, reaching maximum levels after about 5–8 steps. It was also observed that this recycling process slightly increased the mass, energy content, and dewaterability of the hydrochar product, presumably because some of the components from the process water became incorporated into the hydrochar.

Similar experiments were reported by Uddin et al. (2014) who recycled process water from HTC treatment of loblolly pine at several reaction temperatures. They too observed an increase in TOC with successive recycle steps, reaching a plateau after a few steps. In addition, they reported a 5–10% increase in hydrochar yield upon the use of water recycling, although the energy content of the hydrochar was unchanged.

Several researchers have commented on the possibility of improving the versatility and economics of the HTC process by recovering valuable chemicals from the process water such as furfural, 5-HMF, acetic acid, formic acid, glycolic acid, and levulinic acid (Kambo and Dutta, 2015b). However, efficient recovery of these byproducts in high purity becomes more difficult in recirculation situations, in which the process water becomes increasingly rich in TOC but also more chemically complex as initial products react further to produce secondary and tertiary products. In this regard, the TSE-based system for producing hydrochar may be advantageous, as a very short reaction time arrests the chemical processes, thus reducing production of secondary products (Hoekman et al., 2017). Also, the rapid de-pressurization that occurs in the TSE process causes 'flashing' of volatile products, resulting in a relatively clean process water stream containing only highly volatile organic compounds, such as furfural and acetic acid, perhaps making recovery of these chemicals more attractive. An additional benefit of the TSE process is the increased mass and energy yields of hydrochar as much of the organic material that would be lost into the process water using conventional HTC processing is now retained as part of the hydrochar.

Integration of the HTC process into a broader set of processes can also improve the overall economics. For example, based on a techno-economic modeling analysis, Erlach et al. (2011) showed that the costs of producing hydrochar could be reduced by integrating an HTC plant with a CHP plant as this simplifies heat recovery schemes and takes advantage of other facility/utility overlap. Similarly, Saari et al. (2016) modeled several scenarios in which an HTC plant was integrated with a wood-fired CHP plant. The results showed considerable economic benefit in integrating these two, particularly in cases where hot, pressurized water from the CHP plant boiler was used as process water for the HTC plant. Although this approach had the poorest overall energy efficiency, it resulted in the greatest cost reductions in the HTC process and the longest CHP operating times.

The concept of utilizing HTC as part of a biorefinery has been promoted by Hitzl et al. (2015). These authors suggested that HTC can be an effective way to convert wet biomass grown locally into acceptable solid biofuels while closing the nutrient cycle by using the process water for crop irrigation and applying the combustion ash as fertilizer. While appealing in its simplicity, this closed-loop biorefinery concept certainly has considerable logistical and economic challenges. In addition, the efficacy of using hydrochar or its process waters as a soil treatment is a very complex area, with reported results being highly variable. A useful introduction to this area is provided by Reza et al. (2014a).

A more attractive way to handle HTC process water involves anaerobic digestion (AD). Due to their high levels of TOC and chemical oxygen demand (COD), HTC process waters are potentially good candidates for AD treatment. The resulting methane product can be captured and used in a variety of ways, thereby improving the overall economics of the operation. Wirth and Mumme (2013) investigated AD treatment of process water from HTC of corn silage. The feedstock had a TOC level of about 16 g/L and a COD level of 41 g/L. While initial results showed good removal efficiency for TOC and COD, along with expected methane production, there were concerns about lack of certain trace elements that promote AD and the presence of some inhibiting organic species. Similar AD investigations were conducted using process waters from HTC treatment of sewage sludge (Wirth et al., 2015; Wirth and Reza, 2016). Here too, initial results look promising. However, this area of AD treatment of HTC process waters is clearly in its infancy and much more work is required to develop fully optimized systems.

4.5 Environmental considerations

A major driver for the use of biofuels is a reduction in the life-cycle emissions of greenhouse gases (GHGs), particularly CO_2. In most life-cycle assessment (LCA) studies, the CO_2 emissions associated with combustion of biofuels are ignored, assuming that they simply represent recycling of contemporary carbon that recently was extracted from the atmosphere to grow the biogenic feedstock. However, other life-cycle impacts are also of interest.

Berge et al. (2015) conducted an LCA study to investigate the environmental impacts associated with HTC treatment of food waste and the subsequent combustion of the hydrochar product to generate electricity. Besides global warming potential (GWP), several other environmental impact categories were assessed—including human toxicity, photochemical ozone formation, terrestrial acidification, eutrophication, freshwater ecotoxicity, and others. As the hydrochar combustion was used to displace coal combustion, clear benefits were seen in terms of GWP and acidification. However, this study also highlighted potential adverse impacts of discharging HTC process waters and pointed out the need to treat these materials prior to discharge. Other concerns were raised about adverse impacts of certain metals (Cr, As, Ni, Hg, Cd, Zn, Pb, and Cu), although some of this may be attributed to the presence of packaging materials included with the food waste, and thus may not be relevant to hydrochar produced from woody biomass.

A somewhat similar LCA study was recently published by Owsianiak et al. (2016), who considered HTC treatment of green waste, food waste, and organic fraction of municipal solid waste (MSW). This differed from the above-mentioned Berge et al. (2015) study in using more realistic pilot plant data to parameterize the LCA model, including additional processes within the system boundaries, and considering a larger set of environmental impact factors. In the Owsianiak et al. (2016) modeling study, the hydrochar was produced and pelletized in Spain, then transported to the U.K. where it was combusted. Life-cycle environmental impacts varied by feedstock, but in general, the largest benefits were seen in the categories of climate change, non-cancer human toxicity, particulate matter emissions, and acidification. The authors concluded that HTC of biowaste, coupled with the use of hydrochar as a solid fuel, may be an attractive option, depending upon the specific conditions.

Finally, Liu et al. (2017) conducted an LCA study to quantify the GHG impacts of a system in which hydrochar produced from loblolly pine is mixed with coal fines, and the resulting co-formed solid fuel is burned in a coal-fired power plant to generate electricity. The size of the simulated HTC plant ranged from 120 to 2400 mt/d of raw loblolly pine feedstock. The HTC process utilized a TSE reactor system, and the amount of hydrochar in the co-formed fuel ranged from 0% to 100%, with emphasis on 10% and 30% blends. Clear benefits in GHG reductions were demonstrated from the use of hydrochar as a fuel. For example, generation and use of the 30/70 hydrochar/coal fines blend reduced GHG emissions by 24.1% compared to the baseline case in which coal alone was used. However, as pointed out by the authors, the HTC process is relatively energy- and cost-intensive. Thus, economically-favorable commercial applications may require the use of fuel blends in which the hydrochar component is minimized, using only enough to ensure good binding behavior with less expensive components—such as coal fines or torrefied biomass.

5. Conclusions and Future Developments

HTC has become a widely investigated process for converting raw, lignocellulosic biomass into an energy-densified hydrochar, suitable for use as a solid biofuel. In a reaction medium consisting of hot, pressurized water, the three main components of biomass—hemicellulose, cellulose, and lignin—degrade through a series of complex chemical reactions. Initial hydrolysis is followed by dehydration, decarboxylation, condensation-polymerization, and aromatization reactions. Besides hydrochar, HTC processing of biomass produces gaseous products (primarily CO_2) and a complex mixture of water-soluble organics—including sugars, organic acids, furans, furfurals, phenols, and other constituents. A major advantage of HTC is its ability to utilize wet feedstocks and avoid the complexity, cost, and energy requirements of pre-drying.

On an elemental basis, HTC results in preferential loss of oxygen, thereby increasing the carbon content and decreasing O/C ratios of hydrochar. Compared to raw feedstocks, hydrochar has higher heating value, better grindability, greater hydrophobicity, and improved pelletization behavior. The mass and energy yields of hydrochar, as well as its physical and chemical properties, depend upon the process conditions employed—particularly reaction temperature and time. Wide ranges of process conditions have been utilized, with temperatures usually in the range of 200–300°C and time between a few minutes and several hours. The concept of reaction severity is a useful way to characterize process conditions as this combines reaction temperature and time into a single metric, called severity factor (SF). The use of this metric enables easier comparison of HTC process conditions across a range of feedstocks and reactor systems.

An SF range of 4.5–7.0 is suitable for effecting significant carbonization of most feedstocks. (For a 1-hour reaction time, this SF range corresponds to a temperature range of approximately 175–275°C.) Near the low end of this range, the hydrochar product may have only 5–10% higher energy density than the starting feedstock, although it exhibits significant improvement in friability, hydrophobicity, and pelletization. Near the high end of this severity range, hydrochar's energy density may increase by 30–40% and be comparable or higher than that of lignite. Increasing the process severity factor above 7.0 provides no significant additional benefits in terms of hydrochar energy content or other fuel characteristics.

A significant advantage of hydrochar over-torrefied biomass (and other pyrolytically-produced biofuels) is its ability to form robust, water-resistant pellets (or briquettes). This improves the handling, transport, and storage of hydrochar as compared to other fuels, and makes hydrochar pellets fully compatible with the existing infrastructure used for coal. The extremely effective binding behavior of hydrochar is attributed not only to the presence of natural binders, such as lignin but also to the chemical production of additional binders that occur during the HTC process. Due to its higher energy content and higher mass density, pellets produced from hydrochar have volumetric energy density values (GJ/m³) approximately twice as large as pellets produced from torrefied biomass.

Most literature reports of HTC refer to laboratory-scale applications, usually involving small batch reactors. However, successful commercial deployment of HTC will likely require the use of continuous process systems. To date, full commercial-scale deployment of HTC has not yet occurred. There are reports of several pilot-scale applications, although insufficient details are provided to fully understand the process configurations and reaction conditions. Other challenges and limitations for HTC at commercial scale include the use of high-pressure vessels and other equipment, the need for a complex high-pressure water handling system, the need for hydrochar drying and briquetting processes, and the requirement for wastewater treatment and disposal.

An interesting continuous process that has been described in some detail, although only at a laboratory scale, involves the use of a twin-screw extruder (TSE) reactor system. This system enables very short reaction times (< 30 sec) and produces a hydrochar having significantly different physical characteristics than conventional hydrochar, while still possessing the benefits of energy densification, hydrophobicity, and good binding behavior.

Due to limited availability, no full-scale test burn of hydrochar has been conducted. Nevertheless, based on surrogate analyses and small-scale demonstrations, hydrochar appears to be an acceptable combustion fuel and gasifier feedstock to replace or supplement coal. Techno-economic analyses have shown the production and use of hydrochar as a fuel to be relatively expensive, as compared to raw biomass fuels. However, by fully optimizing the HTC process and integrating it with other processes (such as CHP), there may be economically attractive options for HTC. Recovery of valuable chemicals from HTC process water could also improve the economics, although this appears to be quite challenging. Treatment of HTC process water by anaerobic digestion (AD) to produce methane may also improve the overall economics.

The use of hydrochar as a biofuel offers clear environmental benefits in the form of reduced greenhouse gas (GHG) emissions. Life cycle assessment (LCA) studies have been performed to quantify these benefits, as well as other environmental benefits resulting from the displacement of coal. Considering the high cost of hydrochar as compared to coal, an attractive approach may be to use a relatively small amount of hydrochar as a binder for less expensive solid fuels—such as coal fines. This creates a partially-renewable coal substitute that can be utilized in a conventional pulverized coal-fired power plant. Considerable additional work is necessary to demonstrate the feasibility of this approach.

References

Acharjee, T. C., C. J. Coronella and V. R. Vasquez. 2011. Effect of thermal pretreatment on equilibrium moisture content of lignocellulosic biomass. Bioresource Technology 102: 4849–4854.

Agar, D. and M. Wihersaari. 2011. Torrefaction technology for solid fuel production. Global Change Biology 1–4.

Aller, M. 2016. Biochar properties: transport, fate, and impact. Critical Reviews in Environ. Sci. Technol. 46: 1183–1296.

Alonso, D. M., S. G. Wettstein and J. A. Dumesic. 2012. Bimetallic catalysts for upgrading of biomass to fuels and chemicals. Chem. Soc. Rev. 41: 7965–8216.

Ando, H., T. Sakaki, T. Kokusho, M. Shibata, Y. Uemura and Y. Hatate. 2000. Decomposition behavior of plant biomass in hot-compressed water. Industrial & Engineering Chemistry Research 39: 3688–3693.

Areeprasert, C., P. Zhao, D. Ma, Y. Shen and K. Yoshikawar. 2014. Alternative solid fuel production from paper sludge employing hydrothermal treatment. Energy & Fuels 28: 1198–1206.

Artec Biotechnologie GmbH. Artec Biotechnologie GmbH Website [Internet]. Available from: http: //artec-biotechnologie.com/.

Avalon Industries. Avalon Industries Website [Internet]. Available from: http: //www.avalon-industries.com/web/pages/en/home.php.

Axelsson, S., K. Eriksson and U. Nilsson. 2011. Determination of resin acids during production of wood pellets-a comparison of HPLC/ESI-MS With the GC/FID MDHS 83/2 method. Journal of Environmental Monitoring 13: 2940–2945.

Bach, Q.-V., K.-Q. Tran and Ø. Skreiberg. 2012. Wet Torrefaction of Norwegian Biomass Fuels. 20th European Biomass Conference and Exhibition, 18–22 June 2012, Milan, Italy, pp. 1755–1763.

Bach, Q.-V., K.-Q. Tran, R. A. Khalil and Ø. Skreiberg. 2013. Comparative assessment of wet torrefaction. Energy & Fuels 27: 6743–6753.

Bach, Q. V., K.-Q. Tran, O. Skreiberg, R. A. Khalil and A. N. Phan. 2014. Effects of wet torrefaction on reactivity and kinetics of wood under air combustion conditions. Fuel 137: 375–383.

Bach, Q.-V., K.-Q. Tran and Ø. Skreiberg. 2015. Accelerating wet torrefaction rate and ash removal by carbon dioxide addition. Fuel Processing Technology 140: 297–303.

Bach, Q. V. and O. Skreiberg. 2016. Upgrading biomass fuels via wet torrefaction: a review and comparison with dry torrefaction. Renewable & Sustainable Energy Reviews 54: 665–677.

Bach, Q.-V., K.-Q. Tran and Ø. Skreiberg. 2017. Comparative study on the thermal degradation of dry- and wet-torrefied woods. Applied Energy 185: 1051–1058.

Becker, R., U. Dorgerloh, E. Paulke, J. Mumme and I. Nehls. 2014. Hydrothermal carbonization of biomass: major organic components of the aqueous phase. Chem. Eng. Technol. 37: 511–518.

Berge, N. D., K. S. Ro, J. Mao, J. R. V. Flora, M. A. Chappell and S. Bae. 2011. Hydrothermal carbonization of municipal waste streams. Environ. Sci. Technol. 45: 5696–5703.

Berge, N. D., L. Li, J. R. V. Flora and K. S. Ro. 2015. Assessing the environmental impact of energy production from hydrochar generated via hydrothermal carbonization of food wastes. Waste Management 43: 203–217.

Bergius, F. 1913. Die Anwendung Hoher Drucke Bei Chemischen Vorgangenund Eine Nachbil Dung Des Entstehungsprozesses Der Steinkohle. Wilhelm Knapp, Halle A.S., Druck and Verlag Von Wilhelm Knapp, pp. 41–58.

Bobleter, O. 1994. Hydrothermal degradation of polymers dervied from plants. Progress in Polymer Science 19: 797–841.

Broch, A., U. Jena, S. K. Hoekman and J. Langford. 2014. Analysis of solid and aqueous phase products from hydrothermal carbonization of whole and lipid-extracted algae. Energies 7: 62–79.

Brunner, G. 2009. Near critical and supercritical water. Part I. Hydrolytic and hydrothermal processes. J. Supercritical Fluids 47: 373–381.

Burguete, P., A. Corma, M. Hitzl, R. Modrego, E. Ponce and M. Renz. 2016. Fuel and chemicals from wet lignocellulosic biomass waste streams by hydrothermal carbonization. Green Chemistry 18: 1051–1060.

Buttmann, M. 2011. Climate friendly coal from hydrothermal carbonization of biomass. Chemie Ingenieur Technik 83: 1890–1896.

Cai, J., B. Li, C. Chen, J. Wang, M. Zhao and K. Zhang. 2016. Hydrothermal carbonization of tobacco stalk for fuel application. Bioresource Technology 220: 305–311.

Cao, X., J. J. Pignatello, Y. Li, C. Lattao, M. A. Chappell and N. Chen. 2012. Characterization of wood chars produced at different temperatures using advanced solid-state ^{13}C NMR spectroscopic techniques. Energy & Fuels 26: 5983–5991.

Chen, W.-H., S.-C. Ye and H.-K. Sheen. 2012. Hydrothermal carbonization of sugarcane bagasse via wet torrefaction in association with microwave heating. Bioresource Technology 118: 195–203.

Coll, R., S. Udas and W. A. Jacoby. 2001. Conversion of the rosin acid fraction of crude tall oil into fuels and chemicals. Energy & Fuels 15: 1166–1172.

Coronella, C. J., J. G. Lynam, M. T. Reza and M. H. Uddin. 2014. Hydrothermal carbonization of lignocellulosic biomass. pp. 275–311. In: Jin, F. (ed.). Application of Hydrothermal Reactions to Biomass Conversion, Green Chemistry and Sustainable Technology, Springer-Verlag Berlin, Heidelberg.

Cuvilas, C. A., E. Kantarelis and W. Yang. 2015. The impact of a mild sub-critical hydrothermal carbonization pretreatment on umbila wood. A mass and energy balance perspective. Energies 8: 2165–2175.

Danso-Boateng, E., R. G. Holdich, A. D. Wheatley, S. J. Martin and G. Shama. 2015. Hydrothermal carbonization of primary sewage sludge and synthetic faeces: effect of reaction temperature and time on filterability. Environ. Prog. Sustainable Energy 34: 1279–1290.

Demirbas, A. 2005. Estimating of structural composition of wood and non-wood biomass samples. Energy Sources 27: 761–767.

Dong, R., Y. Zhang, L. L. Christianson, T. L. Funk, X. Wang and Z. Wang. 2009. Product distribution and implication of hydrothermal conversion of swine manure at low temperatures. Transaction of the ASABE 52: 1239–1248.

Duan, P., Z. Chang, Y. Xu, B. Xiujun, F. Wang and L. Zhang. 2013. Hydrothermal processing of duckweed: effect of reaction conditions on product distribution and composition. Bioresource Technology 135: 710–719.

Elliott, D. 2011. Hydrothermal processing. pp. 200–231. In: Brown, R. (ed.). Thermochemical Processing of Biomass: Conversion into Fuels, Chemicals, and Power, John Wiley & Sons.

Erlach, B., B. Wirth and G. Tsatsaronis. 2011. Co-Production of Electricity, Heat, and Biocoal Pellets From Biomass: a Techno-Economic Comparison With Wood Pelletizing. World Renewable Energy Conference 2011 - Sweden: World Renewable Energy Congress 2011 - Sweden; 2011 May; Linkoping, Sweden.

Erlach, B., B. Harder and G. Tsatsaronis. 2012. Combined hydrothermal carbonization and gasification of biomass with carbon capture. Energy 45: 329–338.

Eseyin, A. E., P. H. Steele and C. U. Pittman. 2015. Current trends in the production and applications of torrefied wood/biomass—A review. BioResources 10: 8812–8858.

Fakkaew, K., T. Koottatep, T. Pussayanavin and C. Polprasert. 2015. Hydrochar production by hydrothermal carbonization of faecal sludge. Journal of Water Sanitation and Hygiene for Development 5: 439–447.

Fang, J., B. Gao, J. Chen and A. R. Zimmerman. 2015. Hydrochars derived from plant biomass under various conditions: characterization and potential applications and impacts. Chemical Engineering Journal 267: 253–259.

Fang, Z., T. Minowa, R. L. Smith, T. Ogi and J. A. Kozinski. 2004. Liquefaction and gasification of cellulose with Na_2CO_3 and Ni in subcritical water at 350 degrees C. Industrial & Engineering Chemistry Research 43: 2454–2463.

Felix, L. G., W. E. Farthing and S. K. Hoekman. 2015. Research and Development to Prepare and Characterize Robust Coal/Biomass Mixtures for Direct Co-Feeding into Gasification Systems. 2015 Mar. Report No.: Final Scientific/Technical Report under DOE Award No. DE-FE0005349; OSTI ID No. 1176858.

Fowles, M. 2007. Black carbon sequestration as an alternative to bioenergy. Biomass & BioEnergy 31: 426–432.

Funke, A. and F. Ziegler. 2009. Hydrothermal Carbonization of Biomass: A Literature Survey Focussing on Its Technical Application and Prospects. 17th European Biomass Conference and Exhibition, Hamburg, Germany: Proceedings of the 17th European Biomass Conference, Technische Universitat Berlin, Institute of Energy Engineering, Hamburg, Germany, 29 June–3 July, 2009; Hamburg, Germany. Technische Universitat Berlin, Institute of Energy Engineering.

Funke, A. and F. Ziegler. 2010. Hydrothermal carbonization of biomass: a summary and discussion of chemical mechanisms for process engineering. Biofuels Bioproducts & Biorefining 4: 160–177.

Funke, A. and F. Ziegler. 2011. Heat of reaction measurements for hydrothermal carbonization of biomass. Bioresource Technology 102: 7595–7598.

Gao, P., Y. Zhou, F. Meng, Y. Zhang, Z. Liu, W. Zhang et al. 2016. Preparation and characterization of hydrochar from waste eucalyptus bark by hydrothermal carbonization. Energy 97: 238–245.

Gao, Y., B. Yu, K. Wu, Q. Yuan, X. Wang and H. Chen. 2016. Physicochemical, pyrolytic, and combustion characteristics of hydrochar obtained by hydrothermal carbonization of biomass. BioResources 11: 4113–4133.

Garrote, G., H. Dominguez and J. C. Parajo. 1999. Hydrothermal processing of lignocellulosic materials. Holz Als Roh- Und Werkstoff 57: 191–202.

Gil, M. V., P. Oulego, M. D. Casal, C. Pevida, J. J. Pis and F. Rubiera. 2010. Mechanical durability and combustion characteristics of pellets from biomass blends. Bioresource Technology 101: 8859–8867.

Goering, H. K. and P. J. Van Soest. 1970. Forage fiber analyses (apparatus, reagents, procedures, and some applications). Agriculture Handbook 379: 1–20.

Grenol GmbH. Grenol Website [Internet]. Available from: http://www.grenol.org/index.php?id=8&L=1.

Gunarathne, D. S., A. Mueller, S. Fleck, T. Kolb, J. K. Chmielewski and W. Yang. 2014. Gasification characteristics of steam exploded biomass in an updraft pilot scale gasifier. Energy 71: 496–506.

Gunarathne, D. S., A. Mueller, S. Fleck, T. Kolb, J. K. Chmielewski, W. H. Yang et al. 2014. Gasification characteristics of hydrothermal carbonized biomass in an updraft pilot-scale gasifier. Energy & Fuels 28: 1992–2002.

Guo, S., X. Dong, T. Wu, F. Shi and C. Zhu. 2015. Characteristic evolution of hydrochar from hydrothermal carbonization of corn stalk. Journal of Analytical and Applied Pyrolysis 116: 1–9.

He, C., A. Giannis and J. Y. Wang. 2013. Conversion of sewage sludge to clean solid fuel using hydrothermal carbonization: hydrochar fuel characteristics and combustion behavior. Applied Energy 111: 257–266.

He, C., K. Wang, Y. Yang and J.-Y. Wang. 2014. Utilization of sewage-sludge-derived hydrochars toward efficient cocombustion with different-rank coals: effects of subcritical water conversion and blending scenarios. Energy & Fuels 28: 6140–6150.

Heilmann, S. M., H. T. Davis, L. R. Jader, P. A. Lefebvre, M. J. Sadowsky, F. J. Schendel et al. 2010. Hydrothermal carbonization of microalgae. Biomass & Bioenergy 34: 875–882.

Heilmann, S. M., L. R. Jader, M. J. Sadowsky, F. J. Schendel, M. G. von Keitz and K. J. Valentas. 2011. Hydrothermal carbonization of distiller's grains. Biomass and Bioenergy 35: 2526–2533.

Hitzl, M., A. Corma, F. Pomares and M. Renz. 2015. The hydrothermal carbonization (HTC) plant as a decentral biorefinery for wet biomass. Catalysis Today 257: 154–159.

Hoekman, S. K., A. Broch and C. Robbins. 2011. Hydrothermal carbonization (HTC) of lignocellulosic biomass. Energy & Fuels 25: 1802–1810.

Hoekman, S. K., A. Broch, A. Warren, L. Felix and J. Irvin. 2015. Laboratory pelletization of hydrochar from woody biomass. Biofuels 5: 651–666.

Hoekman, S. K., A. Broch, C. Robbins, B. Zielinska and L. G. Felix. 2013. Hydrothermal carbonization (HTC) of selected woody and herbaceous biomass feedstocks. Biomass Conversion and Biorefinery 3: 113–126.

Hoekman, S. K., A. Broch, C. Robbins, R. Purcell, B. Zielinska and L. Felix. 2014. Process development unit (PDU) for hydrothermal carbonization (HTC) of lignocellulosic biomass. Waste Biomass Valor 5: 669–678.

Hoekman, S. K., A. Broch, L. Felix and W. Farthing. 2017. Hydrothermal carbonization (HTC) of loblolly pine using a continuous, reactive twin-screw extruder. Energy Conversion and Management 134: 247–259.

Hu, B., K. Wang, L. Wu, S.-H. Yu, M. Antonietti and M.-M. Titirici. 2010. Engineering carbon materials from the hydrothermal carbonization process of biomass. Advanced Materials 22: 813–828.

Hwang, I.-H., H. Aoyama, T. Matsuto, T. Nakagishi and T. Matsuo. 2012. Recovery of solid fuel from municipal solid waste by hydrothermal treatment using subcritical water. Waste Management 32: 410–416.

Inoue, S., T. Hanaoka and T. Minowa. 2002. Hot compressed water treatment for production of charcoal from wood. Journal of Chemical Engineering of Japan 35: 1020–1023.

Jose-Vicente, O.-V., G.-A. Enrique and G.-G. Patricia. 2016. Analysis of durability and dimensional stability of hydrothermal carbonized wooden pellets. Wood Research 61: 321–330.

Kaliyan, N. and R. V. Morey. 2009. Factors affecting strength and durability of densified biomass products. Biomass & Bioenergy 33: 337–359.

Kaliyan, N. and R. V. Morey. 2010. Natural binders and solid bridge type binding mechanisms in briquettes and pellets made from corn stover and switchgrass. Bioresource Technology 101: 1082–1090.

Kambo, H. S. and A. Dutta. 2014. Strength, storage, and combustion characteristics of densified lignocellulosic biomass produced via torrefaction and hydrothermal carbonization. Applied Energy 135: 182–191.

Kambo, H. S. and A. Dutta. 2015a. Comparative evaluation of torrefaction and hydrothermal carbonization of lignocellulosic biomass for the production of solid biofuel. Energy Conversion and Management 105: 746–755.

Kambo, H. S. and A. Dutta. 2015b. A comparative review of biochar and hydrochar in terms of production, physico-chemical properties and applications. Renewable & Sustainable Energy Reviews 45: 359–378.

Kang, S., X. Li, J. Fan and J. Chang. 2012. Solid fuel production by hydrothermal carbonization of black liquor. Bioresource Technology 110: 710–718.

Kim, D. S., A. A. Myint, H. W. Lee, J. Yoon and Y.-W. Lee. 2013. Evaluation of hot compressed water pretreatment and enzymatic saccharification of tulip tree sawdust using severity factors. Bioresource Technology 144: 460–466.

Kim, D., K. Yoshikawa and K. Y. Park. 2015. Characteristics of biochar obtained by hydrothermal carbonization of cellulose for renewable energy. Energies 8: 14040–14048.

Kleinert, M. and T. Wittman. 2009. Carbonisation of Biomass Using a Hydrothermal Approach: State-of-the-Art and Recent Developments. 17th European Biomass Conference and Exhibition, Hamburg, Germany: Proceedings of the 17th European Biomass Conference; 2009; Hamburg, Germany. Department of Chemistry, University of Bergen, Norway, pp. 1683–1687.

Knezevic, D., W. P. M. van Swaaij and S. R. A. Kersten. 2009. Hydrothermal conversion of biomass: I. glucose conversion in hot compressed water. Industrial & Engineering Chemistry Research 48: 4731–4743.

Knezevic, D., W. Van Swaaij and S. Kersten. 2010. Hydrothermal conversion of biomass. II. Conversion of wood, pyrolysis oil, and glucose in hot compressed water. Industrial & Engineering Chemistry Research 49: 104–112.

Kruse, A., A. Funke and M.-M. Titirici. 2013. Hydrothermal conversion of biomass to fuels and energetic materials. Current Opinion in Chemical Biology 17: 515–521.

Kruse, A. and N. Dahmen. 2015. Water—a magic solvent for biomass conversion. Journal of Supercritical Fluids 96: 36–45.

Kumar, S. and R. B. Gupta. 2008. Hydrolysis of microcrystalline cellulose in subcritical and supercritical water in a continuous flow reactor. Ind. Eng. Chem. Res. 47: 9321–9329.

Kumar, S., U. Kothari, L. Kong, Y. Y. Lee and R. B. Gupta. 2011. Hydrothermal pretreatment of switchgrass and corn stover for production of ethanol and carbon microspheres. Biomass & Bioenergy 35: 956–968.

Lehmann, J. and S. Joseph. 2015. Biochar for environmental management: an introduction. pp. 1–14. *In*: Lehmann, J. and S. Joseph (eds.). Biochar for Environmental Management, Routledge.

Lehmann, J., J. Gaunt and M. Rondon. 2006. Bio-char sequestration in terrestrial ecosystems—a review. Mitigation and Adaptation Strategies for Global Change 11: 403–427.

Libra, J. A., K. S. Ro, C. Kammann, A. Funke, N. D. Berge and Y. Neubauer. 2011. Hydrothermal carbonization of biomass residuals: a comparative review of the chemistry, processes and applications of wet and dry pyrolysis. Biofuels 2: 89–124.

Lin, Y., X. Ma, X. Peng, S. Hu, Z. Yu and S. Fang. 2015. Effect of hydrothermal carbonization temperature on combustion behavior of hydrochar fuel from paper sludge. Applied Thermal Engineering 91: 574–582.

Liu, F., R. Yu and M. Guo. 2017. Hydrothermal carbonization of forestry residues: influence of reaction temperature on holocellulose derived hydrochar properties. J. Mater. Sci. 52: 1736–1746.

Liu, X., S. K. Hoekman, W. E. Farthing and L. G. Felix. 2017. TC2015: life cycle analysis of co-formed coal fines and hydrochar produced in twin-screw extruder (TSE). Environ. Prog. Sustainable Energy 36: 668–676.

Liu, Z., A. Quek, S. K. Hoekman and R. Balasubramanian. 2013. Production of solid biochar fuel from waste biomass by hydrothermal carbonization. Fuel 103: 943–949.

Liu, Z. and F.-S. Zhang. 2009. Removal of lead from water using biochars prepared from hydrothermal liquefaction of biomass. J. Hazardous Materials 167: 933–939.

Liu, Z. G. and R. Balasubramanian. 2014. Upgrading of waste biomass by hydrothermal carbonization (HTC) and low temperature pyrolysis (LTP): a comparative evaluation. Applied Energy 114: 857–864.

Liu, Z., R. Balasubramanian and S. K. Hoekman. 2014a. Production of renewable solid fuel hydrochar from waste biomass by sub- and supercritical water treatment. pp. 231–260. *In*: Fang, Z. and C. Xu (eds.). Near-Critical and Supercritical Water and Their Applications for Biorefineries, Springer Netherlands+Business Media Dordrecht 2014.

Liu, Z., A. Quek and R. Balasubramanian. 2014b. Preparation and characterization of fuel pellets from woody biomass, agro-residues and their corresponding hydrochars. Applied Energy 113: 1315–1322.

Liu, Z., Y. Guo, R. Balasubramanian and S. K. Hoekman. 2016. Mechanical stability and combustion characteristics of hydrochar/lignite blend pellets. Fuel 164: 59–65.

Lu, X. W., B. Jordan and N. D. Berge. 2012. Thermal conversion of municipal solid waste via hydrothermal carbonization: comparison of carbonization products to products from current waste management techniques. Waste Management 32: 1353–1365.

Lu, X., P. J. Pellechia, J. R. V. Flora and N. D. Berge. 2013. Influence of reaction time and temperature on product formation and characteristics associated with the hydrothermal carbonization of cellulose. Bioresource Technology 138: 180–190.

Luo, G. E., P. J. Strong, H. L. Wang, W. Z. Ni and W. Y. Shi. 2011. Kinetics of the pyrolytic and hydrothermal decomposition of water hyacinth. Bioresource Technology 102: 6990–6994.

Makela, M., V. Benavente and A. Fullana. 2016. Hydrothermal carbonization of industrial mixed sludge from a pulp and paper mill. Bioresource Technology 200: 444–450.

Masia, A. A. T., B. J. P. Buhre, R. P. Gupta and T. F. Wall. 2007. Characterising ash of biomass and waste. Fuel Processing Technology 88: 1071–1081.

Minowa, T., F. Zhen and T. Ogi. 1998. Cellulose decomposition in hot-compressed water with alkali or nickel catalyst. J. Supercritical Fluids 13: 253–259.

Mochidzuki, K., A. Sakoda and M. Suzuki. 2003. Liquid-phase thermogravimetric measurement of reaction kinetics of the conversion of biomass wastes in pressurized hot water: a kinetic-study. Advances in Environmental Research 7: 421–428.

Mochidzuki, K., N. Sato and A. Sakoda. 2005. Production and characterization of carbonaceous adsorbents from biomass wastes by aqueous phase carbonization. Adsorption 11: 669–673.

Mumme, J., L. Eckervogt, J. Pielert, M. Diakite, F. Rupp and J. Kern. 2011. Hydrothermal carbonization of anaerobically digested maize silage. Bioresource Technology 102: 9255–9260.

Mursito, A. T., T. Hirajima and K. Sasaki. 2010. Upgrading and dewatering of raw tropical peat by hydrothermal treatment. Fuel 89: 635–641.

Naisse, C., C. Girardin, R. Lefevre, A. Pozzi, R. Maas and A. Stark. 2015. Effect of physical weathering on the carbon sequestration potential of biochars and hydrochars in soil. Global Change Biology BioEnergy 7: 488–496.

Nizamuddin, S., N. S. J. Kumar, J. N. Sahu, P. Ganesan, N. M. Mubarak and S. A. Mazari. 2015. Synthesis and characterization of hydrochars produced by hydrothermal carbonization of oil palm shell. Canadian Journal of Chemical Engineering 93: 1916–1921.

Nizamuddin, S., N. Mubarak, M. Tiripathi, N. Jayakumar, J. Sahu and P. Ganesan. 2016. Chemical, dielectric and structural characterization of optimized hydrochar produced from hydrothermal carbonization of palm shell. Fuel 163: 88–97.

Nizamuddin, S., H. A. Baloch, G. J. Griffin, N. M. Mubarak, A. W. Bhutto and R. Abro. 2017. An overview of effect of process parameters on hydrothermal carbonization of biomass. Renewable & Sustainable Energy Reviews 73: 1289–1299.

Onwudili, J. A. and P. T. Williams. 2008. Hydrothermal gasification and oxidation as effective flameless conversion technologies for organic wastes. J. Energy Institute 81: 102–109.

Overend, R. P. and E. Chornet. 1987. Fractionation of lignocellulosics by steam-aqueous pretreatments. Phil. Trans. R. Soc. Lond. A 321: 523–536.

Owsianiak, M., M. W. Ryberg, M. Renz, M. Hitzl and M. Hauschild. 2016. Environmental performance of hydrothermal carbonization of four wet biomass waste streams at industry-relevant scales. ACS Sustainable Chem. Eng. 4: 6783–6791.

Pala, M., I. C. Kantarli, H. B. Buyukisik and J. Yanik. 2014. Hydrothermal carbonization and torrefaction of grape pomace: a comparative evaluation. Bioresource Technology 161: 255–262.

Parshetti, G. K., S. K. Hoekman and R. Balasubramanian. 2013. Chemical, structural and combustion characteristics of carbonaceous products obtained by hydrothermal carbonization of palm empty fruit bunches. Bioresource Technology 135: 683–689.

Parshetti, G. K., Z. Liu, A. Jain, M. P. Srinivasan and R. Balasubramanian. 2013. Hydrothermal carbonization of sewage sludge for energy production with coal. Fuel 111: 201–210.

Parshetti, G. K., S. Chowdhury and R. Balasubramanian. 2014. Hydrothermal conversion of urban food waste to chars for removal of textile dyes from contaminated waters. Bioresource Technology 161: 310–319.

Pelaez-Samaniego, M. R., V. Yadama, E. Lowell and R. Espinoza-Herrera. 2013. A review of wood thermal pretreatments to improve wood composite properties. Wood Sci. Technol. 47: 1285–1319.

Peng, C., Y. Zhai, Y. Zhu, B. Xu, T. Wang and C. Li. 2016. Production of char from sewage sludge employing hydrothermal carbonization: char properties, combustion behavior and thermal characteristics. Fuel 176: 110–118.

Petersen, M. O., J. Larsen and M. H. Thomsen. 2009. Optimization of hydrothermal pretreatment of wheat straw for production of bioethanol at low water consumption without addition of chemicals. Biomass & Bioenergy 33: 834–840.

Peterson, A. A., F. Vogel, R. P. Lachance, M. Froling, M. J. Antal and J. W. Tester. 2008. Thermochemical biofuel production in hydrothermal media: a review of sub- and supercritical water technologies. Energy Environ. Sci. 1: 32–65.

Pfriem, A., T. Dietrich and B. Buchelt. 2012. Furfuryl alcohol impregnation for improved plasticization and fixation during the densification of wood. Holzforschung 66: 215–218.

Poerschmann, J., I. Baskyr, B. Weiner, R. Koehler, H. Wedwitschka and F. Kopinke. 2013. Hydrothermal carbonization of olive mill wastewater. Bioresource Technology 133: 581–588.

Poerschmann, J., B. Weiner, H. Wedwitschka, I. Baskyr, R. Koehler and F. D. Kopinke. 2014. Characterization of biocoals and dissolved organic matter phases obtained upon hydrothermal carbonization of brewer's spent grain. Bioresource Technology 164: 162–169.

Potic, B., S. R. A. Kersten, W. Prins and W. P. M. van Swaaij. 2004. A high-throughput screening technique for conversion in hot compressed water. Industrial & Engineering Chemistry Research 43: 4580–4584.

Rebling, T., P. von Frieling, J. Buchholz and T. Greve. 2015. Hydrothermal carbonization: combination of heat of reaction measurements and theoretical estimations. J. Therm. Anal. Calorim. 119: 1941–1953.

Reza, M. T., J. G. Lynam, V. R. Vasquez and C. J. Coronella. 2012. Pelletization of biochar from hydrothermally carbonized wood. Environ. Prog. Sustainable Energy 31: 225–234.

Reza, M. T., W. Yan, M. H. Uddin, J. G. Lynam, S. K. Hoekman and C. J. Coronella. 2013a. Reaction kinetics of hydrothermal carbonization of loblolly pine. Bioresource Technology 139: 161–169.

Reza, M. T., J. G. Lynam, M. H. Uddin and C. J. Coronella. 2013b. Hydrothermal carbonization: fate of inorganics. Biomass & Bioenergy 49: 86–94.

Reza, M. T., J. Andert, B. Wirth, D. Busch, J. Pielert and J. G. Lynam. 2014a. Hydrothermal carbonization of biomass for energy and crop production. Applied Bioenergy 1: 11–29.

Reza, M. T., M. H. Uddin, J. G. Lynam, S. K. Hoekman and C. J. Coronella. 2014b. Hyrothermal carbonization of loblolly pine: reaction chemistry and water balance. Biomass Conversion and Biorefinery 4: 311–321.

Reza, M. T., W. Becker, K. Sachsenheimer and J. Mumme. 2014c. Hydrothermal carbonization (HTC): near infrared spectroscopy and partial least-squares regression for determination of selective components in HTC solid and liquid products derived from maize silage. Bioresource Technology 161: 91–101.

Reza, M. T., M. H. Uddin, J. G. Lynam and C. J. Coronella. 2014d. Engineered pellets from dry torrefied and HTC biochar blends. Biomass and Bioenergy 63: 229–238.

Reza, M. T., C. Coronella, K. M. Holtman, D. Franqui-Villanueva and S. R. Poulson. 2016a. Hydrothermal carbonization of autoclaved municipal solid waste pulp and anaerobically treated pulp digestate. ACS Sustainable Chemistry & Engineering 4: 3649–3658.

Reza, M., X. Yang, C. J. Coronella, H. Lin, U. Hathwaik and D. Shintani. 2016b. Hydrothermal carbonization (HTC) and pelletization of two arid land plants bagasse for energy densification. ACS Sustainable Chemistry & Engineering 4: 1106–1114.

Rillig, M. C., M. Wagner, M. Salem, P. M. Antunes, C. George and H.-G. Ramke. 2010. Material derived from hydrothermal carbonization: effects on plant growth and arbuscular mycorrhiza. Applied Soil Ecology 45: 238–242.

Röhrdanz, M., T. Rebling, J. Ohlert, J. Jasper, T. Greve and R. Buchwald. 2016. Hydrothermal carbonization of biomass from landscape management-influence of process parameters on soil properties of hydrochars. Journal of Environmental Management 173: 72–78.

Román, S., J. M. V. Nabais, C. Laginhas, B. Ledesma and J. F. González. 2012. Hydrothermal carbonization as an effective way of densifying the energy content of biomass. Fuel Processing Technology 103: 78–83.

Roos, A. A., T. Persson, H. Krawczyk, G. Zacchi and H. Stalbrand. 2009. Extraction of water-soluble hemicelluloses from barley husks. Bioresource Technology 100: 763–769.

Rosatella, A. A., S. P. Simeonov, R. F. M. Frade and C. A. M. Afonso. 2011. 5-Hydroxymethylfurfural (HMF) as a building block platform: biological properties, synthesis and synthetic applications. Green Chem. 13: 754–793.

Ruyter, H. P. 1982. Coalification model. Fuel 61: 1182–1187.

Saari, J., E. Sermyagina, J. Kaikko, E. Vakkilainen and V. Sergeev. 2016. Integration of hydrothermal carbonization and a CHP plant: Part 2 - operational and economic analysis. Energy 113: 574–585.

Sabio, E., A. Alvarez-Murillo, S. Roman and B. Ledesma. 2016. Conversion of tomato-peel waste into solid fuel by hydrothermal carbonization: influence of the processing variables. Waste Management 47: 122–132.

Sasaki, M., T. Adschiri and K. Arai. 2003. Fractionation of sugarcane bagasse by hydrothermal treatment. Bioresource Technology 86: 301–304.

Savage, P. E. 2012. Algae under pressure and in hot water. Science 338: 1039–1040.

Schuhmacher, J. P., F. J. Huntjens and D. W. van Krevelen. 1960. Chemical structure and properties of coal XXVI—studies on artificial coalification. Fuel 39: 223–234.

Sermyagina, E., J. Saari, J. Kaikko and E. Vakkilainen. 2015. Hydrothermal carbonization of coniferous biomass: effect of process parameters on mass and energy yields. Journal of Analytical and Applied Pyrolysis 113: 551–556.

Sevilla, M. and A. B. Fuertes. 2009. Chemical and structural properties of carbonaceous products obtained by hydrothermal carbonization of saccharides. Chem. Eur. J. 15: 4195–4203.

Sevilla, M. and A. B. Fuertes. 2009. The production of carbon materials by hydrothermal carbonization of cellulose. Carbon 47: 2281–2289.

Sevilla, M., J. A. Marciá-Agulló and A. B. Fuertes. 2011. Hydrothermal carbonization of biomass as a route for the sequestration of CO_2: chemical and structural properties of the carbonized products. Biomass and Bioenergy 35: 3152–3159.

Stelte, W., C. Clemons, J. K. Holm, A. R. Sanadi, J. Ahrenfeldt, L. Shang et al. 2011. Pelletizing properties of torrefied spruce. Biomass & Bioenergy 35: 4690–4698.

Stelte, W., A. R. Sanadi, L. Shang, J. K. Holm, J. Ahrenfeldt and U. B. Henriksen. 2012. Recent developments in biomass pelletization—a review. BioResources 7: 4451–4490.

Stemann, J. and F. Ziegler. 2011. Hydrothermal Carbonisation (HTC): Recycling of Process Water. 19th European Biomass Conference and Exhibition Berlin, Germany, pp. 1894–1899.

Stemann, J., A. Putschew and F. Ziegler. 2013. Hydrothermal carbonization: process water characterization and effects of water recirculation. Bioresource Technology 143: 139–146.

Tekin, K., S. Karagöz and S. Bektas. 2014. A review of hydrothermal biomass processing. Renewable & Sustainable Energy Reviews 40: 673–687.

TerraNova Energy. TerraNova Energy Website [Internet]. Available from: http: //terranova-Energy.com/en/.

Titirici, M.-M., A. Thomas and M. Antonietti. 2007. Back in the black: hydrothermal carbonization of plant material as an efficient chemical process to treat the CO_2 problem? New Journal of Chemistry 31: 787–789.

Titirici, M. M. and M. Antonietti. 2010. Chemistry and materials options of sustainable carbon materials made by hydrothermal carbonization. Chemical Society Reviews 39: 103–116.

Toor, S. S., L. Rosendahl and A. Rudolf. 2011. Hydrothermal liquefaction of biomass: a review of subcritical water technologies. Energy 36: 2328–2342.

Tremel, A., J. Stemann, M. Herrmann, B. Erlach and H. Spliethoff. 2012. Entrained flow gasification of biocoal from hydrothermal carbonization. Fuel 102: 396–403.

Tumuluru, J. S., C. T. Wright, K. L. Kenney and J. R. Hess. 2010. A Technical Review on Biomass Processing: Densification, Preprocessing, Modeling, and Optimization. 2010 ASABE Annual International Meeting; 2010 Jun; Pittsburgh, PA.

Tumuluru, J., S. Sokhansanj, C. Wright, J. Hess and R. Boardman. 2011. A Review on Biomass Torrefaction Process and Product Properties. Idaho National Laboratory; 2011 Aug. Report No.: INL/CON-11-22634.

Uddin, M. H., M. T. Reza, J. G. Lynam and C. J. Coronella. 2014. Effects of water recycling in hydrothermal carbonization of loblolly pine. Environ. Prog. Sustainable Energy 33: 1309–1315.

van Krevelen, D. W. 1982. Development of coal research—a review. Fuel 61: 786–790.

van Krevelen, D. W. 1993. Coalification Revisited. Coal: Typology, Physics, Chemistry, Constitution. Third, Completely Revised Edition, Amsterdam: Elsevier Science Publishers B.V., pp. 837–846.

Wikberg, H., T. Ohra-aho, M. Honkanen, H. Kanerva, A. Harlin and M. Vippola. 2016. Hydrothermal carbonization of pulp mill streams. Bioresource Technology 212: 236–244.

Wirth, B. and J. Mumme. 2013. Anaerobic digestion of waste water from hydrothermal carbonization of corn silage. Applied Bioenergy 1–10.

Wirth, B., T. Reza and J. Mumme. 2015. Influence of digestion temperature and organic loading rate on the continuous anaerobic treatment of process liquor from hydrothermal carbonization of sewage sludge. Bioresource Technology 198: 215–222.

Wirth, B. and M. Reza. 2016. Continuous anaerobic degradation of liquid condensate from steam-derived hydrothermal carbonization of sewage sludge. ACS Sustainable Chemistry & Engineering 4: 1673–1678.

Xiao, L.-P., Z.-J. Shi, F. Xu and R.-C. Sun. 2012. Hydrothermal carbonization of lignocellulosic biomass. Bioresource Technology 118: 619–623.

Yan, W., Acharjee, T. C., C. J. Coronella and V. R. Vasquez. 2009. Thermal pretreatment of lignocellulosic biomass. Environ. Progress & Sustainable Energy 28: 435–440.

Yan, W., J. T. Hastings, T. C. Acharjee, C. J. Coronella and V. R. Vasquez. 2010. Mass and energy balances of wet torrefaction of lignocellulosic biomass. Energy & Fuels 24: 4738–4742.

Yan, W., S. K. Hoekman, A. Broch and C. J. Coronella. 2014. Effect of hydrothermal carbonization reaction parameters on the properties of hydrochar and pellets. Environ. Prog. Sustainable Energy 33: 676–680.

Yang, W., H. Wang, M. Zhang, J. Zhu, J. Zhou and S. Wu. 2016. Fuel properties and combustion kinetics of hydrochar prepared by hydrothermal carbonization of bamboo. Bioresource Technology 205: 199–204.

Yu, J. L. and P. E. Savage. 1998. Decomposition of formic acid under hydrothermal conditions. Industrial & Engineering Chemistry Research 37: 2–10.

Yu, Y., X. Lou and H. Wu. 2008. Some recent advances in hydrolysis of biomass in hot-compressed water and its comparisons with other hydrolysis methods. Energy & Fuels 22: 46–60.

Zeitsch, K. J. 2000. The Chemistry and Technology of Furfural and Its Many by-Products. 1st Edition. Elsevier.

Zhang, L., Q. Wang, B. Wang, G. Yang, L. A. Lucia and J. Chen. 2015. Hydrothermal carbonization of corncob residues for hydrochar production. Energy Fuels 29: 872–876.

Zheng, A. Q., Z. L. Zhao, S. Chang, Z. Huang, K. Zhao and G. Q. Wei. 2015. Comparison of the effect of wet and dry torrefaction on chemical structure and pyrolysis behavior of corncobs. Bioresource Technology 176: 15–22.

Zilkha Biomass Energy Website. Zilkha Biomass Energy [Internet]. Available from: http: // zilkha.com/.

CHAPTER 9

Thermal Pretreatment of Biomass to make it Suitable for Biopower Application

Jaya Shankar Tumuluru

1. Introduction

If grown sustainably, lignocellulosic biomass is considered a carbon-neutral energy source, meaning that the greenhouse gas (GHG) emissions, namely carbon dioxide (CO_2), released from converting biomass to energy are equivalent to the amount of CO_2 absorbed by biomass plants during their growing cycles. According to a 2011 study conducted by the Organisation for Economic Cooperation and Development/International Energy Agency (OECD/IEA, 2011), biomass can be net carbon negative if it is coupled with carbon capture and storage (CCS) technology. The primary lignocellulosic biomass resources are forest- and agriculture-derived biomass such as logging residues, fuel treatment thinnings, crop residues, and perennially grown grasses and trees. The updated Billion-Ton Report published by the United States (U.S.) Department of Energy Office of Energy Efficiency (DOE/EE) has divided biomass into two parts: forest-derived resources and agriculture-derived resources (DOE/EE, 2016). Thus, biomass fuels

750 University Blvd, Energy Systems Laboratory, Idaho National Laboratory, Idaho Falls, Idaho, 83415-3570.
Email: JayaShankar.Tumuluru@inl.gov

"recycle" atmospheric carbon and may reduce global warming impacts. Figure 1 indicates the different fuel sources used for power generation in 2015 in the U.S.

Figure 2 indicates the total energy consumption in the U.S., as well as the contribution of various energy resources including the renewable resources, which equaled about 10 percent in 2015. About 49% of the total

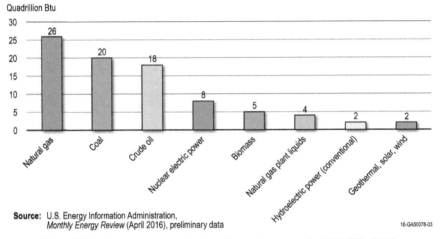

Source: U.S. Energy Information Administration, *Monthly Energy Review* (April 2016), preliminary data

18-GA50078-03

Fig. 1. U.S. primary energy production by a major source in 2015 (EIA, 2016).

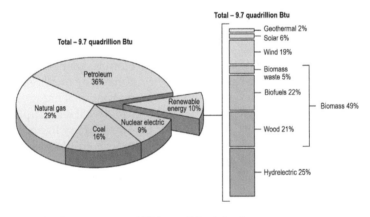

Note: Sum of components may not equal 100% because of independent rounding

Source: U.S. Energy Information Administration, *Monthly Energy Review* Table 1.3 and 10.1 (April 2016), preliminary data

18-GA50078-04

Fig. 2. Total energy consumption in the U.S. in 2015 via renewable energy production (EIA, 2016).

energy produced from renewable sources comes from biomass (waste, biofuels, and wood).

2. Biopower

In general, biopower—the production of electricity from biomass—holds significant potential as a major renewable energy source in a low-carbon energy future. Globally, an estimated 72 gigawatts (GW) of biopower capacity was in operation at the end of 2011, representing a 9 percent increase from 2010 (REN 21, 2012). Biomass for biopower is produced from mill residues, urban wood waste, forest harvesting residues, agricultural waste material, dedicated herbaceous crops, and specified woody crops (Tumuluru et al., 2012). The major feedstocks used for commercial-scale electricity and heat generation are residuals from timber harvesting, saw milling, and pulp and paper production. Currently, biopower generation has increased by using wood pellets. The world's largest biopower plant in the United Kingdom (UK) that has a capacity of 750 megawatts (MW) (REN 21, 2012) is currently using wood pellets. In the U.S., more than 7,000 MW of biopower capacity are installed at more than 350 plants. Various biopower producers include electric utilities, independent power producers, and the pulp and paper industry. Biomass provides 10.2 percent of total global energy consumption; 61 percent of which is attributed to traditional uses of biomass such as domestic cooking, lighting, and heating in the developing world (Chum et al., 2011). Other application examples of biomass use include combined heat and power (CHP) as well as the production of liquid transportation fuels. Figure 3 shows the generation of biopower in the U.S. since 1980. In 2011, the U.S. generated 57 terawatt hours (tWhs) of biopower

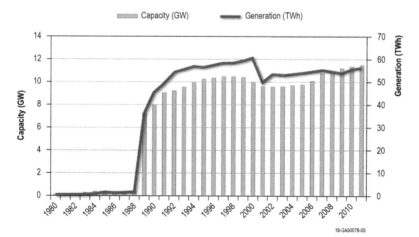

18-GA50078-05

Fig. 3. Biopower capacity and generation in the U.S., 1980–2011 (EIA/AER, 2012).

(see Fig. 3). In the U.S., 67 percent of biopower uses woody biomass as a feedstock, whereas the rest comes from MSW, landfill gas, and agricultural and other byproducts (EIA/AER, 2012).

According to EIA/AER (2012), there is about 3.7 GW of biogenic municipal waste capacity in the U.S., which can potentially be used for biopower production. Landfill gas produced 14.3 billion kWhs of electricity in 2011. In 2010, only 12 percent of solid biogenic U.S. trash was diverted from the waste stream and combusted for energy (EPA/CHP, 2010). To help expand opportunities for biopower production, DOE established the Biopower Program (originally called the Biomass Power Program) and sponsored efforts to increase the productivity of dedicated energy crops (Tumuluru et al., 2012). Figure 4 provides the potential biopower plant expansion in the U.S. Currently, utility-scale co-firing in coal-fired power plants is becoming commercially feasible. The major challenge in increasing the co-firing of biomass in power plants is in retrofitting the plants, as well as getting the feedstock to desired specifications. Approximately 300 commercial-scale power plants around the world were modified to incorporate biomass to diversify fuel sources (REN 21, 2012). In Europe, several utility-scale power plants have been retrofitted to burn 100 percent

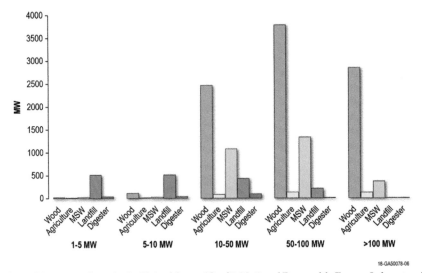

18-GA50078-06

Fig. 4. Bio-power plants in the United States (Credit: National Renewable Energy Laboratory).

biomass in response to climate change legislation restricting coal use and the ability to pay higher costs for low-GHG electricity and obtain carbon credits.

2.1 Advantages of biopower

Biomass can be an excellent solution to address the GHG emissions challenges associated with fossil fuel usage. In general energy activities that release carbon into the atmosphere are carbon-positive (such as burning coal) while energy activities that remove carbon from the atmosphere are carbon-negative. Biomass burning like CCS. Biopower also emits CO_2 directly but is considered carbon-neutral because the released carbon is already a part of the carbon cycle.

Biomass, which is low in sulfur, can result in lower sulfur dioxide emissions, which result in acid rain and health problems (Tumuluru et al., 2012; Mann and Spath, 2001). A DOE study (2004) suggested that co-firing biomass by 10 percent reduces SO_2 and CO_2 emissions by about 10 percent. Spath and Mann (2004) indicated that co-firing 5 percent urban waste biomass with coal could reduce overall GHG emissions by 19 percent due to the diversion of methane, which would have been released due to natural decomposition of the organic matter. The major limitation of co-firing biomass is the emission of particulate matter, carbon monoxide, volatile organic compounds, and nitrogen oxide emissions. Studies conducted by Bracmort (2012) and Spath and Mann (2004), indicated that the biopower generated can result in lower carbon emissions than that coming from traditional fossil fuels.

Understanding the life cycle emissions from biopower can help to understand its sustainability. Life cycle emissions depend on many factors such as the type of biomass, technology used, feedstock production (cultivating and harvesting), transportation, power plant operating standards, global land use change, and the sustainability of biomass sources (Manomet, 2010). A study conducted by the U.S. National Renewable Energy Laboratory found that GHG emissions per kWh were lower for biopower when compared with fossil-based systems (see Table 1). It is clear that life cycle GHG emissions of electricity generation using biomass have lower plant average as compared to the fossil fuels such as coal and natural gas.

Table 1. Life cycle GHG emissions of electricity generation technologies (g CO_2e/kWh).

Energy source	Emissions		
	Lowest 25% of plants	Plant average	Highest 25% of plants
Coal	883	1001	1141
Natural gas	427	477	543
Biopower (average)	−3	52	69
Co-firing*	9	49	65
Direct combustion	7	56	84
Gasification	16	49	63
Gasification Engine	85	81	100

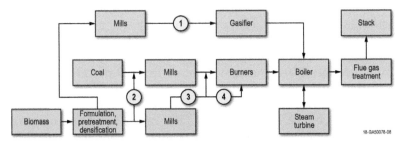

Fig. 5. Various co-firing approaches (Tumuluru et al., 2012).

2.2 Biopower generation methods

Combustion is commonly used for biopower generation. Different commercial stoker/grate and fluid-bed combustion technologies are being used to burn various forms of biomass. The heat produced via combustion is converted into steam, which is further used to generate electrical power. Various combustion technologies include direct firing, repowering, co-firing, and combined heat and power (CHP). In direct firing, biomass is the only fuel used in a power plant. The feedstock is burned in a boiler to create steam, and this steam is further used to power a steam turbine to produce electricity. In a co-fired system, biomass substitutes a portion of the fossil fuel used in a power plant. Cofiring technology is widely used today for power generation.

In co-firing applications, up to 20 percent of the fossil fuel can be substituted with biomass (Tumuluru et al., 2012). Various co-firing approaches are given in Fig. 5. During co-firing, feedstock can be blended with coal before entering the boiler using a separate feed system. The efficiencies of co-firing systems are closer to coal than dedicated biomass systems. In the case of CHP, both electricity and useful heat are generated from various fuels. CHP combusts fuel to produce steam that powers a turbine generator, and the CHP plant exhaust is then used for electricity generation or for industrial applications. Other biopower generation technologies are gasification and pyrolysis. In gasification, biomass is exposed to high temperatures and an oxygen-deficient environment to produce synthesis gas, which consists of carbon monoxide and hydrogen. The syngas can then be used to fuel the gas turbine to produce electricity. In the pyrolysis process, biomass is converted into a liquid product by thermal decomposition. The bio-oil can be used as fuel oil in furnaces, turbines, and engines.

2.2.1 Limitations in using biomass for biopower applications

Traditionally, heat or electricity is generated from biomass through the direct combustion of forest and wood products, residues, yard clippings, construction and demolition debris, and municipal solid waste. There are challenges in using biomass for power generation; the diversity of biomass resource types and production conditions cause variability in biomass composition and physical properties. Technical limitations, such as inherent low heating values and low bulk densities of biomass relative to coal thereby necessitate transporting larger units of biomass, pose further challenges to use biomass as the sole feedstock for bioenergy production (Tumuluru et al., 2011). In addition, inherent physical (e.g., particle size, density) and chemical (e.g., proximate, ultimate, energy properties) characteristics of raw biomass restrict its use in higher percentages for direct combustion application. Furthermore, grinding raw biomass is very challenging due to its fibrous nature. High moisture in biomass increases the grinding energy and negatively impacts the particles size distribution (Tumuluru et al., 2014). High moisture in the biomass also results in plugging the grinder screens and reactor and bridge the particles in the conveyors. In terms of chemical composition, the raw biomass has higher oxygen and hydrogen content and lower carbon and calorific value as compared to fossil fuels (Tumuluru, 2015; 2016). The limitations of both the physical and chemical properties do not make biomass a great candidate for solid and liquid fuels production using thermochemical conversion pathways. Table 2 indicates chemical (proximate and ultimate) composition of woody and herbaceous biomass in comparison to coal.

Table 3 indicates the physical and chemical properties of biomass and their limitations for co-firing with coal. Recent studies by Tumuluru et al. (2016) suggest novel harvesting, mechanical preprocessing, and thermal

Table 2. Typical coal, wood, and herbaceous biomass composition.

Parameter	Central Appalachian (Long Fork)	Wood	Switchgrass	Corn stover
Proximate Analysis (wt. %)				
Moisture	7.16	42.00	9.84	8.00
Ash	11.52	2.31	8.09	6.90
Volatile matter	31.23	47.79	69.14	69.74
Fixed Carbon	50.09	7.90	12.93	15.36
Ultimate Analysis (wt. %)				
Carbon	66.93	29.16	42	42.60
Hydrogen	4.43	2.67	5.24	5.06
Oxygen	7.55	23.19	33.97	36.52
Nitrogen	1.34	0.60	0.69	0.83
Sulfur	1.07	0.07	0.17	0.09
Moisture	7.16	42	9.84	8.00
Ash	11.52	2.31	8.09	6.90
Chlorine (%)	0.12	0.01	0.18	0.24
Higher Heating Value (MJ/kg)	28.17	11.86	16.28	16.28

pretreatments can help to improve the physical and thermal properties of biomass. These improvements will make biomass to meet the specifications in terms of density, particle size, ash composition, and carbohydrate content for both biochemical and thermochemical conversion application.

2.2.2 Thermal pretreatment to improve the physical properties and chemical composition of the biomass

According to Tumuluru et al. (2016), the thermal pretreatment of biomass in the range of 180–300°C makes it suitable for co-firing applications and can help to co-fire higher percentages with coal. The common thermal pretreatment methods are deep drying and torrefaction, which significantly alter the physical and chemical composition of the biomass (Tumuluru, 2015; 2016). Deep drying and torrefaction are defined as slowly heating biomass in an inert environment and temperature range of 150–300°C. This process improves the physical properties and chemical composition of the biomass, making it perform better for co-firing, gasification, and pyrolysis applications. Figure 6 indicates the different drying zones (Tumuluru et al., 2011). At drying temperatures of 50–150°C (temperature regime A), most

Table 3. Co-firing issues in terms of physical properties and chemical composition of the biomass (Tumuluru et al., 2012).

Chemical composition	Issues
Moisture content	Dry-matter losses, spontaneous combustion, calorific value
Volatile content	Local fuel-air mixing and combustion considerations, soot formation and radiation phenomenon impacts
Ash content	Particle emissions, ash utilization and disposal, heat transfer surface fouling, combustion chemistry interactions, SCR catalyst dusting, and poisoning
Carbon and Hydrogen	Positive effect on gross calorific value
Oxygen	Negative effect on gross calorific value, peak flame temperature impacts
Chlorine	Corrosion, enhanced mercury capture (conversion of elemental mercury to particulate chlorides)
Nitrogen	NO_x, N_2O, and HCN emissions (potentially reduces fuel/prompt NO_x)
Sulfur	Sox emissions and corrosion generally reduced
Fluorine	HF emissions and corrosion
Potassium & sodium	Corrosion, lower ash melting temperature, aerosol formation, catalyze coal char reactions
Magnesium	Hotter ash melting temperature
Calcium	Ash utilization, catalyzes coal char reactions
Phosphorus	Hotter ash melting temperature
Heavy metals	Emission of pollutants, aerosol formation, ash utilization, and disposal issues
Physical properties	
Bulk density	Fuel logistics (e.g., storage, transport, handling costs)
Particle size/shape	Fuel feeding system, combustion chemistry, drying properties, dust formation, safety (e.g., explosion considerations)
Flow characteristics	Plugging, segregation, obstructed or limited discharge, erratic flow, sudden uncontrollable flow, and sticking of material to container walls which causes spoilage and structural damage

of the moisture in the biomass is lost and results in shrinkage. This region is also called the non-reactive drying zone, where most of the chemical constituents of the biomass remain intact. Temperature regime C, also

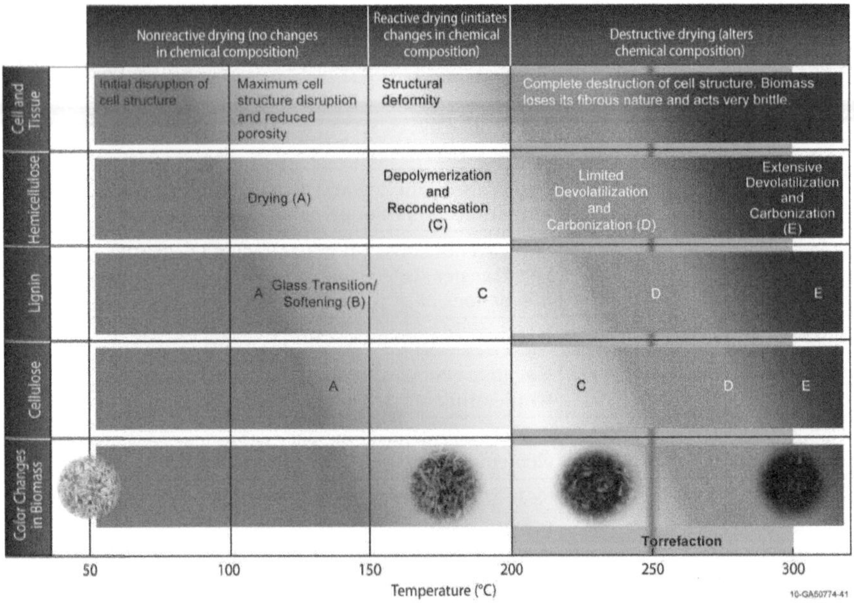

Fig. 6. Impact of thermal pretreatment on the biomass components (Tumuluru et al., 2011).

called the reactive drying range (e.g., deep drying temperature), is between 150–200°C. At these temperatures, breakage of hydrogen and carbon bonds will occur and result in the emission of lipophilic extractives. Finally, temperature regime D is the destructive drying (200–300°C) temperature zone, which is also the torrefaction temperature zone. At these temperatures, most inter- and intra-molecular hydrogen bonds and C-C and C-O bonds result in the formation of hydrophilic extractives, carboxylic acids, alcohols, aldehydes, ether, and gases like CO, CO_2, and CH_4. At these temperatures, the cell structure is completely destroyed; biomass loses its fibrous nature and becomes brittle.

2.2.3 Impact of thermal pretreatment on physical properties

2.2.3.1 Grinding properties

Grinding the high moisture and fibrous, elastic biomass of different sizes suitable for co-firing is an energy-intensive and difficult exercise in a standard coal mill. As a result, almost all power plants engaged in co-firing have established separate biomass (e.g., pellet) milling and a following burner feed-in system. Having separate biomass milling and burner feed systems is very cost-intensive. Biomass is thermally pretreated to remove most of the moisture and create brittle and easy-to-grind feedstock. This material can be milled in existing coal mills without any modifications,

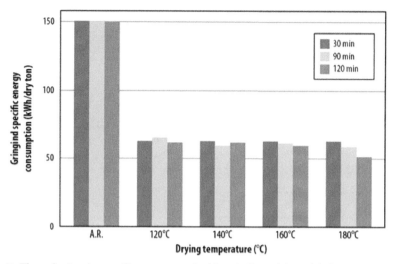

Fig. 7. The reduction in specific energy required in grinding of deep dried corn stover. A.R. indicates the corn stover with moisture content as received (10.7% wb) (INL research data, August 2011).

thereby reducing cost challenges. They could easily switch between coal and biomass to realize short-term production optimums. According to Tumuluru et al. (2011), the deep drying and torrefaction process makes biomass lose most of its moisture and tenacious nature, which is mainly coupled to the breakdown of the hemicellulose matrix and depolymerization of the cellulose, resulting in the decrease of fiber length (Bergman and Kiel, 2005; Bergman et al., 2005). There is a significant decrease in particle length, but not the diameter, per se, resulting in better grindability, handling characteristics, and biomass ability to flow unhindered through processing and transportation systems. Also during the torrefaction process, the biomass tends to shrink becoming lightweight, flaky, and fragile. It also loses its mechanical strength, making it easier to grind and pulverize (Arias et al., 2008).

Figure 7 indicates the grinding tests conducted on as-received material and further drying to the temperature range of 120–180°C. The raw and dried samples were tested in an instrumented laboratory knife mill, where the grinding energy was recorded. Results showed that grindability was significantly improved by drying over the entire temperature range when compared to the raw sample. The reduction in grinding energy ranged from 58–65 percent with little variation over the range of thermal treatment process parameters. Notably, grindability was mostly unaffected by drying time, with only incremental improvements at the highest temperature (e.g., 180°C).

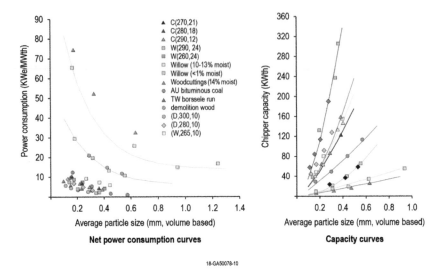

18-GA50078-10

Fig. 8. Grinding power consumption and mill capacity, torrefaction for biomass co-firing in existing coal-fired power stations (Bergman et al., 2005).

Bergman and Kiel (2005) conducted studies on the energy requirements for grinding raw and torrefied biomass such as willow, woodcuttings, demolition wood, and coal using a heavy-duty cutting mill. They concluded that power consumption reduces dramatically when biomass is first torrefied. This reduction ranges from 70–90 percent based on the conditions under which the material is torrefied. Bergman and Kiel (2005) also found that the capacity of the mill increases by a factor of 7.5–15 percent. The most important phenomenon observed by these authors was that size-reduction characteristics of torrefied biomass were greatly similar to those of coal. Studies conducted by Phanphanich and Mani (2011) on the grinding characteristics of torrefied biomass indicated a significant decrease in the grinding energy from 237.7–37.6 kWh/t from raw to torrefied forest biomass at 300°C for 30 min. Figure 7 indicates the typical grinding energy reduction for torrefied wood as compared to untreated wood. Figure 8 depicts the differences between coal and biomass grindability. It is clear from the figure that raw biomass takes more energy to grind as compared to torrefied biomass. Easier grindability of torrefied biomass offers an economic and capacity advantage for co-firing (Wild et al., 2016). Figure 9 indicates the comparison of grinding energy for raw and biomass torrefied at different temperatures.

2.2.3.2 Particle size distribution

Particle-size distribution, sphericity, and surface area are important parameters for understanding flowability and combustion behavior during

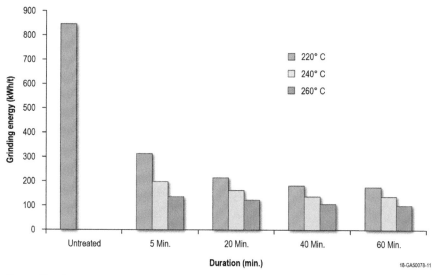

Fig. 9. Grinding energy of beech as a function of torrefaction duration, obtained with a Retsch ZM1 ultracentrifugal mill equipped with a 500-mm grid (Tumuluru et al., 2012).

co-firing. As suggested by Tumuluru et al. (2012) and Tumuluru (2015; 2016), the major limitation of ground raw biomass is the wide range of particle size distribution and irregular particle size and shape that makes it difficult to feed. Ground torrefied biomass produces a narrower, more uniform particle size as compared to untreated biomass due to its brittle nature. Phanphanich and Mani (2010) studied the particle size and shape of torrefied pine chips and logging residues and concluded that smaller particle sizes are produced as compared to untreated biomass. In addition, they observed that the particle distribution curve was skewed towards smaller particle sizes with increased torrefaction temperatures. The same authors also indicated that the sphericity and particle surface area increased as the torrefaction temperature was increased up to 300°C. Further Idaho National Laboratory (INL) studies also indicated that mean particle size did not change significantly at 180°C, but decreased significantly at the torrefaction temperature (see Fig. 10). Wild et al. (2016) discovered that it is not only the reduction of grinding energy but more so the shape of the resulting particles that is of importance to the quality of combustion in co-firing. Research done by the Energy Research Centre of the Netherlands (ECN) showed that the "sphericity" of ground coal is best matched by ground torrefied biomass pellets. The shape of the particles of ground torrefied biomass pellets is very different to the one achieved when only grinding wood or wood pellets (see Fig. 11) (Carbo et al., 2015).

The Hardgrove Index (HGI) usually expresses grindability of coal in the power sector. In general, power plants expect coal with an HGI from 50–80

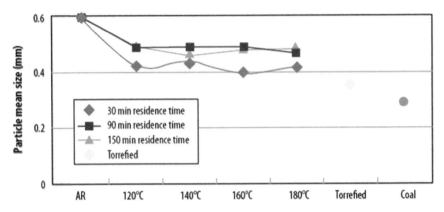

Fig. 10. Mean particle sizes of deep dried corn stover grinds as determined by Camsizer™ Digital Image Processing System. AR is the corn stover with moisture content as received (10.7% wb) (INL research data, August 2011).

Fig. 11. The shape of milled raw spruce, torrefied spruce chips, torrefied spruce pellets, and coal (Carbo et al., 2015).

(Wild et al., 2016), as coals with higher HGI values require less energy for grinding. Generally, wood pellets have HGI values of about 20, whereas torrefied biomass has HGI values in the low to mid-50s, making them behave like coal in power plants (Wild et al., 2016). Therefore, torrefaction helps to improve the particle surface area or decreases particle size, which increases the HGI of the biomass making it more suitable for combustion and cofiring applications.

2.2.3.3 Moisture content

The moisture content of biomass is influenced by thermal pretreatment process parameters such as temperature and residence time. The raw material loses most of its moisture content during deep drying and torrefaction temperatures. Studies conducted by Tumuluru (2016) indicated that deep drying and torrefaction of lodgepole pine grind decreased the moisture content from an initial value of 4.2 to a final value of 1.15 percent (w.b.) (see Fig. 12). His studies indicate the decrease in moisture content from 160, 180, 230, and 270°C and 30 min residence time was about 29.04 percent, 43.80 percent, 56.90 percent, and 68.57 percent, respectively. Similar studies conducted by Tumuluru (2015) on switchgrass and corn stover confirmed these findings regarding significant moisture reduction at the torrefaction regime (see Fig. 13). Both of these studies indicate that if moisture content reduction is the main objective, then deep drying at 160 and 180°C for < 30 min will help to reduce most of the moisture in both woody and herbaceous biomass. The loss of moisture due to thermal pretreatment is mainly due to evaporation and dehydration reactions and due to the release of organic and inorganic products.

2.2.3.4 Bulk density

Torrefaction of biomass loses mass in the form of solids, liquids, and gases, thereby making it more porous. Research conducted by Bergman and Kiel (2005) indicated there will be a reduction in the volumetric density of biomass after torrefaction. This reduction density is dependent on torrefaction process conditions such as temperature and time. Table 4 provides the loss of bulk density of *Eucalyptus grandis* wood at different torrefaction temperatures ranging from 220–280°C (Oliveira-Rodrigues and Rousset, 2009). Their studies show there is about a 14.12 percent loss of bulk density when torrefied at 280°C for 30 minutes. Loss of density during torrefaction will have an impact on transportation costs. One way to overcome the density limitation of torrefied biomass is to compress it using mechanical densification systems.

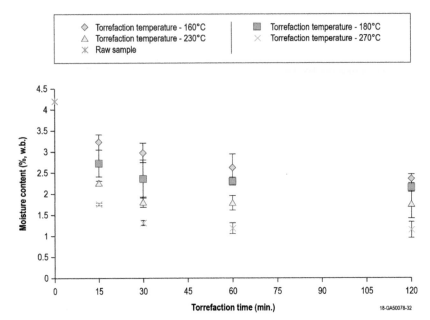

Fig. 12. Lodgepole pine grind moisture content at deep drying and torrefaction temperatures (Tumuluru, 2016).

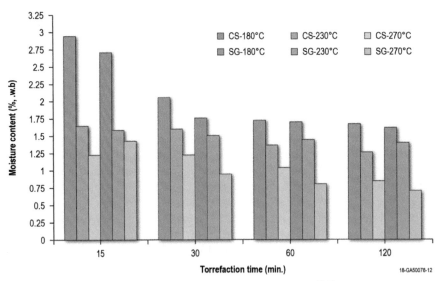

Fig. 13. Switchgrass and corn stover moisture content at different temperatures (Tumuluru, 2015).

Table 4. Bulk density (dry basis) of *Eucalyptus grandis* wood in three different treatments.

Treatment	Bulk density (g/cm³)	Percentage loss
Control*	0.85a	—
T2-220°C	0.83a	2.35
T3-250°C	0.79b	7.06
T4-280°C	0.73c	14.12

Note: Means followed by the same letter are statistically similar to the 5% probability level.
*Average moisture content of control treatment = 15%.

18-GA50078-13

Fig. 14. Flow diagram of production of torrefied wood pellets.

3. Densification

Biomass densification depends upon the densification system used, process conditions (e.g., feedstock moisture content, particle size), and composition of the biomass (mainly lignin). Biomass with higher lignin content can be compressed with less energy. Lignin in the biomass is the major binding component. According to Lipinsky et al. (2002), Bergman et al. (2005), Reed and Bryant (1978), and Koukios (1993), the torrefaction process breaks down the hemicellulose matrix forming fatty unsaturated structures and opens up more lignin-active sites resulting in better binding. Bergman and Kiel (2005) discovered that densification of hot torrefied biomass using Pronto-Press produced torrefied pellets with densities in the range of 750–850 kg/m³ with significantly higher calorific values. Figure 14 shows the flow diagram developed by Bergman (2005) for pelleting of the torrefied biomass. The major challenge associated in densification of torrefied material is the combustion issue during compression and extrusion in a pellet die. This can be overcome by cooling the torrefied materials and pelleting them by adding moisture and binder. Studies conducted by Sarkar et al. (2014) on gasification of torrefied and densified switchgrass has corroborated this observation.

4. Impact of Thermal Pretreatment on Chemical Composition

4.1 Proximate and ultimate composition

According to Tumuluru et al. (2011), torrefaction of biomass has a significant impact on the chemical composition of both herbaceous and woody biomass. Further studies conducted by Tumuluru (2015; 2016) indicate that both herbaceous and woody biomass' chemical composition in terms of proximate and ultimate composition changes with respect to torrefaction temperature and residence time. These studies also show that by increasing torrefaction temperature and residence time, the proximate composition, such as fixed carbon and ash content, increased while volatile content and moisture content decreased. The ultimate composition of hydrogen, oxygen, sulfur, and nitrogen decreased whereas the carbon content and calorific value increased. Figures 15, 16, 17, 18, and 19 indicate the changes in lodgepole

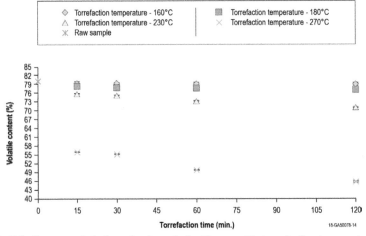

Fig. 15. Volatile content in lodgepole pine grind with respect to torrefaction temperature and time (Tumuluru, 2016).

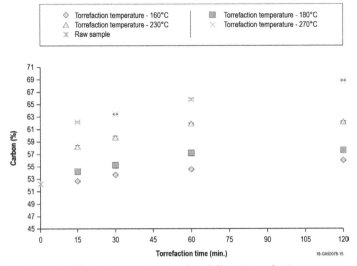

Fig. 16. Carbon content in lodgepole pine grind at different torrefaction process conditions (Tumuluru, 2016).

pine volatile content, carbon content, hydrogen content, oxygen content, and calorific value, respectively. In a review on the torrefaction of biomass, Tumuluru et al. (2011) discovered that at > 250°C, the biomass undergoes

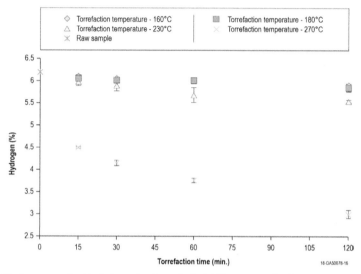

Fig. 17. Hydrogen content in lodgepole pine grind at different torrefaction process conditions (Tumuluru, 2016).

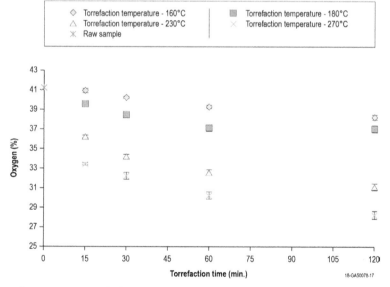

Fig. 18. Oxygen content in lodgepole pine grind at different torrefaction process conditions (Tumuluru, 2016).

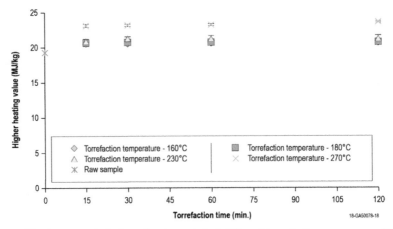

Fig. 19. Heating value in lodgepole pine grind at different torrefaction process conditions (Tumuluru, 2016).

extensive de-volatilization and carbonization. These reactions result in a loss of volatiles and an increase in carbon content. Bates and Ghoniem (2012) indicate that at a lower temperature of 200°C, the compositional changes are minimal in willow, whereas at higher temperatures the mass fraction of carbon is increased, while those of hydrogen and oxygen decrease. The decrease of oxygen content is mainly due to the dehydration reaction that produces water vapor and releases CO and CO_2. During torrefaction, the loss of volatiles, gases, and water also results in a decrease in hydrogen content. Table 5 indicates the ultimate analysis and elemental composition of torrefied woody and lignocellulosic biomass properties as compared to the original raw materials.

The other property that changes due to torrefaction is the calorific value. The heating value of biomass is an important quality attribute for biopower generation. Torrefaction of biomass reduces the volatiles and increases the carbon content, and in turn, the calorific value. Zanzi et al. (2002) and Nimlos et al. (2003) confirm that increasing torrefaction temperature increases the heating value. During torrefaction, biomass loses relatively more oxygen and hydrogen than carbon, which in turn increases the calorific value (Uslu et al., 2008). Sadaka and Negi (2009) also observed that the highest heating value of 22.75 MJ/kg was achievable at torrefaction conditions of 315°C and 3 hours using agricultural straws. Table 5 also shows the calorific value of the varied torrefied woody and herbaceous biomass.

The van Krevelen diagram (see Fig. 20) is used to understand the effect of torrefaction on the atomic ratio of carbon, hydrogen, and oxygen. Higher temperature torrefaction helps atomic ratios of the biomass to move closer

Table 5. Ultimate analysis, HHV (dry ash free basis), and moisture content of untreated and torrefied biomass (Bridgeman et al., 2008).

	Raw	**Torrefaction temperature (°K)**			
Reed Canary Grass		503	523	543	563
C (%)	48.6	49.3	50.3	52.2	54.3
H (%)	6.8	6.5	6.3	6.0	6.1
N (%)	0.3	0.1	0.0	0.1	0.1
O (%)	37.3	—	37.0	37.3	36.3
Moisture (%)	4.7	2.5	1.9	1.3	1.2
CV (kJ/kg)	19,500	—	20,000	20,800	21,800
Wheat Straw					
C (%)	47.3	48.7	49.6	51.9	5.6
H (%)	6.8	6.3	6.1	5.9	1.0
N (%)	0.8	0.7	0.9	0.8	27.6
O (%)	37.7	—	35.6	33.2	0.8
Moisture (%)	4.1	1.5	0.9	0.3	0.8
CV (kJ/kg)	18,900	19,400	19,800	20,700	22,600
Willow					
C (%)	49.9	50.7	51.7	53.4	54.7
H (%)	6.5	6.2	6.1	6.1	6.0
N (%)	0.2	0.2	0.2	0.2	0.1
O (%)	39.9	39.5	38.7	37.2	36.4
Moisture (%)	2.8	0.5	0.1	0.1	0.0
CV (kJ/kg)	20,000	20,600	20,600	21,400	21,900

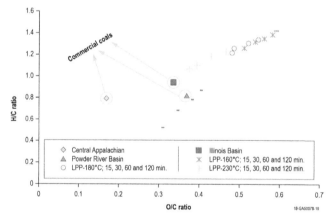

Fig. 20. van Krevelen diagram for raw, torrefied lodgepole pine grind, and commercial coals (Tumuluru, 2016).

to some of the commercially available coals (Tumuluru, 2015; 2016). H/C and O/C ratios are good indicators to characterize biomass for suitability in co-firing applications. The van Krevelen diagram drawn for the torrefied H/C and O/C ratios indicated that these values are closer to some commercially available coals (see Fig. 20). At higher torrefaction temperatures, lower O/C ratios are due to the release of oxygen-rich compounds (e.g., CO, CO_2, H_2O), whereas lower H/C ratios are due to the formation of CH_4 and C_2H_6 during torrefaction. Lower H/C and O/C ratio fuels also produce less smoke, less water-vapor formation, and lower energy loss during combustion.

Tumuluru (2015; 2016) compared the H/C and O/C ratios of woody and herbaceous biomass with respect to various commercially available coals (e.g., Central Appalachian, Illinois Basin, Powder River Basin) (see Fig. 20). In general, high-quality coals have a lower ratio of H/C to O/C mainly due to a lower oxygen content and a higher carbon content as compared to lodgepole pine grind. At the torrefaction temperature of 270°C and a 15–120 min residence time, H/C and O/C values of lodgepole pine grind moved it closer to Power River Basin and Illinois Basin coal. The shift is majorly due to a significant increase in carbon content and a steep decrease in oxygen and hydrogen content. As detailed by Tumuluru et al. (2011), at higher torrefaction temperatures of > 250°C, there will be breakage of inter- and intra-molecular hydrogen and carbon-oxygen and carbon-carbon bonds resulting in the emission of extractives and oxygenated compounds, resulting in lower H/C and O/C values.

4.2 Impact of thermal pretreatment on storage properties

Moisture uptake by biomass during storage is due to the presence of OH groups. Torrefaction of biomass destroys the OH groups causing the biomass to lose the capacity to form hydrogen bonds (Tumuluru et al., 2011). Bergman (2005) conducted studies on the hydrophobicity of raw and torrefied pellets, which indicated that the raw pellets swelled rapidly and disintegrated into original particles, whereas torrefied pellets did not disintegrate and showed little water uptake (7–20 percent on a mass basis). Sokhansanj et al. (2010) reported that torrefied pellets had a 25 percent decrease in water uptake when compared to the control. ECN did a long-term test on outdoor storage and handling of torrefied biomass and related this to both coal and white wood pellets (Wild et al., 2016). The results indicated that torrefied pellets are better than white pellets (see Fig. 21). Storage tests conducted in Europe for a period of 10 months on 3–4 metric tonnes of torrefied wood pellets, indicated that outdoor storage does not reduce the quality of the product, such as density and durability, significantly. Similar results were

Fig. 21. White wood pellets and torrefied wood pellets stored outside after 12 days of storage (Wild et al., 2016).

reported by Andritz testing torrefied wood briquettes at their ACB plant in Frohnleiten, Austria, where the quality of the torrefied briquettes did not change significantly after 43 days of outdoor storage (Wild et al., 2016).

Emission of storage off-gases such as CO and CO_2 from woody and herbaceous biomass is a major issue in safe transportation (Tumuluru et al., 2015). The off-gas emissions are mainly due to chemical oxidation and microbial activity. Many fatal accidents were reported during the transportation of white pellets. Tumuluru et al. (2011) suggested that biomass moisture content coupled with high storage temperatures can cause severe off-gassing and self-heating from biomass-based fuels. Tumuluru et al. (2015) also reported that off-gassing in torrefied wood chips emits less as compared to regular wood chips and white pellets (see Fig. 22). Table 6 summarizes the findings on biomass torrefaction and suitability for co-firing applications.

4.3 Techno-economic analysis of thermal pretreatment process

The torrefaction thermal pretreatment process upgrades the quality of the biomass and provides a great economic potential for the whole biomass supply chain. For raw biomass, the grinding energy is very energy-intensive. According to Phanphanich and Mani (2011), grinding energy reduces to 24 kW h/t (at torrefaction temperature of 300°C), which is 1/10 of the energy

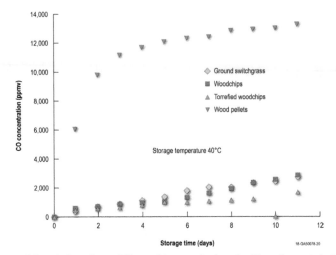

Fig. 22. CO emissions from different biomass feedstocks (Tumuluru et al., 2015).

needed for grinding raw biomass. The major reason for the reduction of grinding energy of torrefied biomass is due to its brittle nature (Tumuluru et al., 2011). Arias et al. (2008) reports that hammer mills typically used for raw biomass can be replaced by simple cutting mills. According to Thek and Obernberger (2004), pelletization torrefied biomass saves about 20% of the energy required for conventional pelletization by increasing energy efficiency. Torrefaction reduces the bulk density of the biomass but increases in energy density.

Xu et al. (2014) conducted a techno-economic analysis of the torrefaction process. The economic analysis was conducted by total feedstock cost ($/ ton feedstock). The capital costs in their studies were taken from the work conducted by Bergman et al. (2005). The cost estimation was calculated for an annual capacity of 100,000 tons with an operation window of six months per year. These authors concluded that besides saving the production cost torrefaction also reduce the transportation cost. According to these authors, the main reason for the reduction of the transportation cost is due to increase in the energy density of the torrefied biomass. Also, these authors have concluded that torrefaction combined with pelletization decreases the transportation cost by about 30% compared to conventional pelleting process. Table 7 indicates the cost of torrefaction and the torrefaction and pelletization process (TOP).

Table 6. Summary of torrefaction effects on the biomass quality attributes (Tumuluru and Hess, 2015).

RESEARCH FINDINGS

IMPROVED STORABILITY
Hydrophobicity studies (temperature zone D) conducted by Idaho National Laboratory (INL) and Oak Ridge National Laboratory (ORNL) found that torrefaction dramatically reduces the need for expensive covered storage and helps to retain energy value during storage and long-distance transportation.

IMPROVED FLOWABILITY
Grinding studies (temperature zones D and E) at INL have found that torrefied biomass has better particle size and shape sphericity after grinding, which makes it easier to handle in existing high-volume transportation systems and more suitable for thermochemical applications, such as gasification, cofiring and pyrolysis.

IMPROVED ENERGY VALUE
Torrefaction studies (temperature zones D and E) conducted by INL indicated that biomass retains most of its energy content while giving up moisture and low-energy-content volatiles, which increases heating and improves combustion efficiency.

REDUCED LOGISTICS COSTS
The Energy Research Centre of the Netherlands (ECN) reported that torrefaction and densification increase bulk density nearly four-fold and can reduce logistics costs by 30%.

REDUCED PREPROCESSING COSTS
INL grinding studies on deep dired (temperature zone C) and torrefied biomass (temperature zones D and E) found significant reduction in grinding energy required (50-70% less) compared with raw biomass.

REDUCED VARIABILITY
The ECN has found that torrefied biomass (temperature zones D and E) has more consistent moisture content (less than 30%) and pulverises more evenly than untreated biomass, resulting in better blending of varying plant fractions.

REDUCED STORAGE OFF-GASES
INL and University of British Columbia studies indicated torrefied woody biomass (temperature zone D) emits lower (CO, CO_2 and CH_4 off-gases compared with non-torrefied wood chips and commercial wood pellets.

18-GA50078-07

4.4 *Torrefaction technical feasibility and initiatives in Europe and North America*

Koppejan et al. (2012) summarize different torrefaction technologies based on six criteria: (1) heating medium; (2) quality of particle mixing; (3) quality of heat exchange; (4) product uniformity; (5) scalability; and (6) flexibility in using different input materials (see Table 8). Most of the technologies use gas as the heat-carrying medium, except for the screw-type reactors, which use hot oil instead.

Table 7. Techno-economic analysis of torrefaction and pelletization process (Xu et al., 2014).

Items[a]	Torrefaction	Torrefaction and pelletization (TOP)[b]
A. Capital costs		
Capital investment, $ (in millions)	8	10
Annual capacity (tons/year)	100,000	100,000
Financing (8%) ($/ton)	1.6	2
Construction and management (C&M) ($/ton)[c]	1	1.2
Subtotal ($/ton)	6.6	8.2
B. Operation costs		
Fuel gas cost ($/ton)	0.76	1.63
Utility ($/ton)	0.2	0.3
Labor ($/ton)	2.00	2.80
Maintenance ($/ton)	0.20	0.24
Depreciation ($/ton)	5.5	6.8
Plant overhead ($/ton)	1.0	1.5
Subtotal ($/ton)	9.66	13.27
C. Transportation		
Energy density (GJ/ton)	21	21
Cost in 100 km ($/GJ)	1	0.28

[a]The analysis was based on raw feedstock (ton). References: Bergman (2005), EIA (2013), and Van der Stelt et al. (2011).
[b]TOP, combined unit of torrefaction and pelletization.
[c]C&M, construction and management.

Table 8. Comparison of various torrefaction technologies (Koppejan et al., 2012).

Technology	Heating medium	Mixing	Heat exchange	Uniformity	Scalability	Input flexibility
Rotating drum	Gas	+	+	+/−	−	+
Screw type	Hot oil	−	+ +	+/−	−	−
Multiple furnace	Gas	+	+	+/−	+	+
Torbed reactor	Gas	+	+ +	+/−	−	−
Moving bed	Gas	−	−	−	−	−
Belt dryer	Gas	+ +	+ +	+	−	+
Vibrating Electrical Elevator and Reactor (REVE)	Electricity	+ +	+ +	+	+	+

Product uniformity, process scalability, and flexibility are currently the main issues to overcome with these technologies. Additional criteria needing to be addressed include energy efficiency and power consumption. According to Doassans-Carrère et al. (2014), the Vibrating Electrical Elevator and Reactor (REVE) is a new industrial technology for biomass torrefaction that helps to overcome some of the product uniformity and energy consumption limitations. According to the REVTECH Research and Development Department, REVE uses electricity as its heating medium, which is one of the main differences with other technologies. One of the major variable which influences the energy consumption of the torrefaction process is the initial moisture content. According to Ohliger et al. (2013), the energy required to torrefy biomass having a moisture content of 10–15% at 230 and 250°C is about 215 kW h/ton. A number of companies in Europe, North America, and Asia have currently initiated and commissioned torrefaction pilot and commercial scale plants. Table 9 indicates the list of pilot and commercial scale initiatives to produce torrefied biomass (Wild et al., 2016).

4.5 Other pretreatment methods to improve biomass specifications for co-firing

4.5.1 Formulation/Blending

In general, formulation can include blending (biomass feedstocks from multiple sources of the same biomass resource to average out compositional and moisture variations), aggregation (combination of different raw or preprocessed biomass resources to produce a single, consistent feedstock with desirable properties), and amendment (combination of raw or preprocessed biomass resources with non-biomass additives to produce a consistent, on-spec feedstock) (Tumuluru et al., 2012). Both woody and herbaceous have great variability in their chemical composition. Uniform biomass feedstock properties are needed to successfully co-fire higher percentages of biomass with coal. Formulation helps to develop customized feedstock by blending two or more biomass feedstocks to achieve the desired chemical composition (e.g., volatiles, oxygen, hydrogen, nitrogen, chlorine, sulfur, and nitrogen contents), physical properties (e.g., particle size distribution, density), and calorific value (Tumuluru et al., 2011; 2012). Grasses that are higher in ash content can be blended with woody biomass, which is lower in ash content to meet the ash specification desirable for

Table 9. Biomass torrefaction initiatives in North America and Europe (Wild et al., 2016).

Developer	Technology	Location(s)	Production capacity (ton/a)	Scale and status Pilot scale: ~500 kg/h (Demo scale: > 0,5 t/h–2 ton/h Commercial scale: > 2 ton/h)	Full integration (pre-treatment, torrefaction, combustion, heat cycle, densification)	Status
Clean Electricity Generation (BV, UK) Solvay (FR)/New Biomass Energy (USA)	Oscillating belt Screw reactor	Derby (UK) Quitman (USA/MS)	30,000 80,000	Commercial scale Commercial scale	Yes Yes	Available/operational Available/operational
Topell Energy (NL)/Blackwood	Fluidized bed	Duiven (NL)	60,000	Commercial scale	Yes	taken over by Blackwood
Arigna Fuels (IR)	Screw conveyor	County Roscommon (IR)	20,000	Commercial scale	Yes	Available/operational
Torr-Coal B.V. (NL)	Rotary drum	Dilsen-Stokkem (BE)	30,000	Commercial scale	Yes	Available/operational
Airex (CAN/QC)	Cyclonic bed	Bécancour (CAN/QC)	16,000	Demonstration scale		Available
Andritz (AT)	Rotary drum	Frohnleiten (AT)	8,000	Demonstration scale	Yes	New ownership
Andritz (DK)/ECN (NL) BioEndev (SWE)	Moving bed Dedicated screw reactor	Stenderup (DK) Holmsund, Umea (SWE)	10,000 16,000	Demonstration scale Commercial demo	Yes Yes	stand by Available
CMI NESA (BE)	Multiple hearths	Seraing (BE)	Undefined	Demonstration scale		Unknown
Earth Care Products (USA)	Rotary drum	Independence (USA/KS)	20,000	Demonstration scale		Available/operational
Grupo Lantec (SP)	Moving bed	Urnieta (SP)	20,000	Demonstration scale		Unknown
Integro Earth Fuels, LLC (USA)	Multiple hearths	Greenville (USA/SC)	11,000	Demonstration scale		Unknown

LMK Energy (FR)	Moving bed	Mazingarbe (FR)	20,000	Demonstration scale		Unknown
Konza Renewable Fuels (USA)	Rotary drum	Healy (USA/KS)	5,000	Demonstration scale		Unknown
River Basin Energy (USA)	Fluidized bed (Aerobic)	Rotterdam	7000	Demonstration scale		In commission
Teal Sales, Inc. (USA)	Rotary drum	White Castle (USA/LA)	15,000	Demonstration scale		Available/operational
Agri-Tech Producers LLC (US/SC)	Screw conveyor	Raleigh (USA/NC)	Undefined	Pilot stage		Available/operational
Airex (CAN/QC)	Cyclonic bed	Rouyn-Noranda (CAN/QC)	Undefined	Pilot stage		Available/operational
Airex (CAN/QC)	Cyclonic bed	Trois-Rivières (CAN/QC)	Undefined	Pilot stage		Available/operational
CENER (SP)	Rotary drum	Aoiz (SP)	Undefined	Pilot scale		Available/operational
Terra Green Energy (USA)	Multiple hearth	McKean County (USA/PA)	Undefined	Pilot scale		Available/operational
Wyssmont (USA)	Multiple hearth	Fort Lee (USA/NJ)	Undefined	Pilot scale		Unknown
CEA (FR)	Multiple hearth	Paris (FR)	Undefined	Laboratory scale		Available/operational
Rotawave, Ltd. (UK)	Microwave	Chester (UK)	Undefined	Laboratory scale		Probably closed
Bio Energy Development & Production (CAN)	Fluidized bed	Nova Scotia (CAN/NS)	Undefined	Unknown		Unknown
Horizon Bioenergy (NL)	Oscillating belt	Steenwijk (NL)	45,000	Commercial scale	Yes	Dismantled

co-firing applications. Boavida et al. (2004), Shih and Frey (1995), and Sami et al. (2001) indicate that different grades of coals are blended to reduce the sulfur, nitrogen, and ash content. These researchers suggest that different low-grade coals can be mixed with high-grade coals to meet the ash, sulfur, and nitrogen specifications.

4.5.2 Washing/Leaching

The presence of metals (e.g., Cl, Na, K) in biomass results in ash deposition, slag formation, corrosion, sintering, and agglomeration. Washing and leaching of biomass can help to reduce these metals. Davidsson et al. (2002), Jenkins (1996), and Jensen (2001) suggest that washing herbaceous biomass reduces ash content. Reduction of alkali compounds in biomass feedstock before co-firing can save costs involved in maintenance (i.e., superheaters would become more durable for longer periods). Leaching biomass with water will also help to remove alkali sulfates, carbonates, and chlorites. Ammonia can help to leach out magnesium, calcium, sodium, and potassium. Leaching with hydrochloric acid removes carbonates and sulfates of alkaline earth and other metals. Studies conducted by Cerezo (2011) on leaching biomass with de-ionized water and organic solvents helps to modify ash composition, especially reducing the alkali index (K_2O+Na_2O) and ash melting temperatures. The other methods to increase ash melting temperature and reduce slagging and fouling is to add dolomite and kaolin additives. Studies by Thompson et al. (2003) and Lacey et al. (2015) indicate that selecting proper harvesting methods and mechanical fractionation of the harvested materials help to reduce the silica content in the harvested biomass.

4.5.3 Steam explosion

According to Marchessauk et al. (1980; 1981; 1982), DeLong (1981), and Foody (1980), a steam explosion breaks lignin into low-molecular weight products but retains its basic structure and remains reactive due to extensive depolymerization. Hemicelluloses are partially broken down and become predominantly soluble in water as well. This makes biomass easier to pellet under milder process conditions. Shaw et al. (2009) and Lam et al. (2011) found steam exploded pellets made denser pellets with a higher calorific value. Lam et al. (2011), Bruno et al. (2007), and Ding et al. (2011) indicate that pellets made from steam-exploded woody biomass are more stable during storage due to a lack of OH group after the steam explosion.

4.5.4 Hydrothermal carbonization

Hydrothermal carbonization is a wet process, where hot compressed water is used for carbonization. According to Yan et al. (2009), utilizing hydrothermal carbonization produces a solid product with about 55–90 percent mass and 80–95 percent fuel value of the original feedstock. They also report that the biomass after hydrothermal carbonization acted hydrophobic. Pelleting studies conducted by Reza et al. (2011) indicate that higher mass and energy density is achievable (e.g., 1400 kg/m^3 and 38–40 GJ/m^3). Table 10 provides the hurdles and remedies that can help to co-fire higher percentages of biomass with coal. Table 11 indicates the properties of the pretreated and densified biomass. It is clear from Table 11

Table 10. Biomass co-firing hurdles, desired attributes, and remedies.

Hurdles	Desired attributes	Remedies
Feed Milling, Entrainment, Classification	(a) Hydrophobic; (b) brittle in nature; (c) low in moisture; (d) particle density equivalence (particle size/momentum) and flowability	Torrefaction, hydrothermal, and steam explosion followed by densification reduce moisture, make biomass brittle, increase density, particle hardness and flowability
Feed System	(a) Dense-phase particle pneumatic transport; (b) uniform particle size; (c) low explosivity index; and (d) particle density equivalence	Torrefaction or hydrothermal carbonization, steam explosion combined with densification
Burner Performance	(a) Flame properties match fuel-oxidant mixing and reactions design parameters; (b) volatiles behavior similarity; (c) soot formation & radiation similarity; and (d) acceptable pollutant formation/control	(a) Torrefaction, hydrothermal, and steam explosion to adjust fuel heat content (b) Feedstock selection (nitrogen, sulfur, chlorine, etc.)
Slagging/Fly Ash	(a) Mineral composition affecting slag formation temperature, viscosity, and corrosivity; (b) elemental ash composition; and (c) softening/ash fusion temperature	(a) Washing and leaching or formulation and blending of different feedstocks. (b) Cleaning of deposits by soot blowing or exchange of agglomerated bed material, adding chemicals to reduce corrosion, and increasing the ash melting point in order to avoid agglomeration and deposit formation
Post Combustion Processes	(a) Minimal impact on SCR catalyst bed poisoning; (b) minimal impact on precipitators and scrubbers; and (c) minimal impact on baghouse particle collectors	Managing fuel properties by pre-treatment processes—i.e., torrefaction, washing/leaching Feedstock selection/blending

Table 11. Properties indicative of co-firing performance for different pretreated and densified biomass and coal-based fuels (Tumuluru et al., 2012).

	Wood chips	Briquettes	Wood pellets	Extruded logs	Formulated pellets	SEP	HTP	TOP	Coal
Moisture content (% wt)	30–45	<10	5–7	<8	4–7	5–7	3–6	1–5	10–15
Calorific value (MJ/kg)	9–12	17–19	17–19	19–20	17	20–21	20–25	20–24	23–28
Volatiles (% bd)	70–75	70–75	70–75	<60	65	~75–77	70–75	55–65	15–30
Elemental carbon (% bd)	30–35	45	45	45–50	42	53–55	60	55–60	80–90
Bulk density (g/cm³)	0.2–0.25	0.45–0.55	0.6–0.7	0.3–0.4	0.6–0.7	0.7–0.8	0.75–0.8	0.75–0.8	0.8–0.85
Unit density (g/cm³)	N/A	0.8–1.0	1.0–1.1	1.3–1.4	1.1–1.2	1.1–1.3	1.1–1.4	1.1–1.4	N/A
Dust	Average	Average	Limited	Limited	Limited	Limited	Limited	Limited	Limited
Hygroscopic properties	HPL	HPL	HPL	HPL	HPL	HPB	HPB	HPB	HPB
Biological degradation	Yes	Yes	Yes	Low	Yes	No	No	No	No
Milling requirement	Special	Special	Special	Special	Special	Classic	Classic	Classic	Classic
Handling requirements	Special	Special	Easy	Special	Easy	Easy	Easy	Easy	Easy
Product consistency	Limited	Limited	High	High	High	High	High	High	High
Transport cost	High	Average	Average	Average	Average	Low	Low	Low	Low

Note: SEP: steam-exploded pellets
HTTP: hydrothermal pellets
TOP: torrefied pellets
HPL: hydrophilic
HPB: hydrophobic, formulated pellets—equal proportions of corn stover, switchgrass, eucalyptus, and pine.

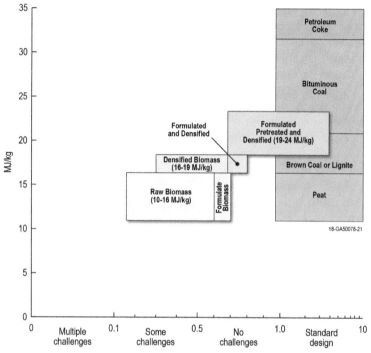

Fig. 23. Raw, formulated, pretreated, and densified biomass fuel ranking with respect to available coals.

that many co-firing biomass hurdles can be overcome by pretreating and densifying. Figure 23 indicates how thermal pretreatment and other methods can help biomass move into the category of high-quality coal characteristics.

5. Conclusions

The main constraints for co-firing biomass with coal—especially in more common direct co-firing operations—are handling, storage, milling, and feeding problems. Torrefaction of biomass helps to overcome these technical constraints. Torrefaction within a temperature range of 200–300°C makes biomass brittle, helps to reduce the grinding energy by about 50–70 percent, and increases the throughput of the mill by three to four times. Torrefaction not only improves the grinding performance of biomass, but also helps to improve the size and shape of the ground biomass, making it look more like coal. The major disadvantage of torrefaction is a loss of bulk density. The bulk density limitation can be overcome by compressing the torrefied biomass using a pellet mill or briquette press. Torrefied pellets have a bulk density of about 800–850 kg/m^3. Torrefaction also results in significant changes in biomass chemical composition, reducing hydrogen, oxygen, and volatile matter, while increasing carbon content and calorific values. The H/C and O/C ratio of torrefied woody and herbaceous biomass are closer to commercially available coals such as lignite and Powder River basin. In addition, torrefied biomass is hydrophobic in nature, making it suitable for long-term storage without any degradation. The ash-related challenges in terms of deposit formation (e.g., slagging and fouling), agglomeration, corrosion, or erosion can be overcome by formulation, washing, and leaching. Other thermal pretreatment methods that can help make biomass more suitable for biopower generation are steam explosion and hydrothermal carbonization. Integrating torrefaction and densification help to make biomass suitable for biopower generation with little modification to coal power plants. Formulation and pretreatment followed by densification can also help to modify the physical properties, chemical composition, and storage behavior of the biomass for improved co-firing with coal.

References

Arias, B. R., C. G. Pevida, J. D. Fermoso, M. G. Plaza, F. G. Rubiera and J. J. Pis-Martinez. 2008. Influence of torrefaction on the grindability and reactivity of woody biomass. Fuel Process. Technol. 89: 169–175.

Bates, R. B. and A. F. Ghoniem. 2012. Biomass torrefaction: Modeling of volatile and solid product evolution kinetics. Bioresour. Technol. 124: 460–469.

Bergman, P. C. A. 2005. Combined torrefaction and pelletization: The TOP process. Report ECN-C-05-073, ECN, Petten.

Bergman, P. C. A., A. R. Boersma, R. W. H. Zwart and J. H. A. Kiel. 2005. Torrefaction for biomass co-firing in existing coal-fired power stations. Report ECN-C-05-013, ECN, Petten.

Bergman, P. C. A. and J. H. A. Kiel. 2005. Torrefaction for biomass upgrading. *In*: Proceedings of the 14th European Biomass Conference & Exhibition. Paris, France.

Boavida, D., P. Abelha and I. Gulyurtlu. 2004. A study on coal blending for reducing NO$_x$ and N$_2$O levels during fluidized bed combustion, Clean Air 5. Accessible at: http://bdigital. ufp.pt/bitstream/10284/295/1/cleanair.pdf.

Bracmort, K. 2012. Is Biopower Carbon Neutral? Congressional Research Service: R41603. Accessible at: http://www.fas.org/sgp/crs/misc/R41603.pdf.

Bridgeman, T. G., J. M. Jones, I. Shield and P. T. Williams. 2008. Torrefaction of reed canary grass, wheat straw, and willow to enhance solid fuel qualities and combustion properties. Fuel 87: 844–856.

Bruno, E. A., V. Marques, I. Domingos and H. Pereira. 2007. Influence of steam heating on the properties of pine (*Pinus pinaster*) and eucalypt (*Eucalyptus globulus*) wood. Wood Sci. Technol. 41: 193–207.

Carbo, M., P. Abelha, M. Cieplik, P. Kroon, C. Mourão and J. Kiel. 2015. Handling, storage, and large-scale co-firing of torrefied biomass pellets. Fifth IEA CCC Workshop on Co-firing Biomass with Coal, Drax, UK.

Cerezo, L. 2011. Issues associated with trace contaminants in biomass co-firing: Feedstock upgrading leaching solution. *In*: Biomass Densification Workshop, Idaho National Laboratory (presentation). August 22–24, 2011. Idaho Falls, ID, USA.

Chum, H., A. Faaij, J. Moreira, G. Berndes, P. Dhamija, H. Dong, B. Gabrielle, A. Goss Eng, W. Lucht, M. Mapako, O. Masera Cerutti, T. McIntyre, T. Minowa and K. Pingoud. 2011. Bioenergy. *In*: O. Edenhofer, R. Pichs-Madruga, Y. Sokona, K. Seyboth, P. Matschoss, S. Kadner, T. Zwickel, P. Eickemeier, G. Hansen, S. Schlömer and C. von Stechow (eds.). IPCC Special Report on Renewable Energy Sources and Climate Change Mitigation. Cambridge University Press, Cambridge, United Kingdom and New York, NY, USA. Accessible at: http://srren.ipcc-wg3.de/report/IPCC_SRREN_Ch02.pdf.

Davidsson, K. O., J. G. Korsgren, J. B. C. Pettersson and U. Jäglid. 2002. The effects of fuel washing techniques on alkali release from biomass. Fuel 81: 137–142.

DeLong, E. A. 1981. Canadian Patent 1096374.

Ding, T., L. Gu and T. Li. 2011. Influence of steam pressure on physical and mechanical properties of heat treated Mongolian pine lumber. Eur. J. Wood Prod. 69(1): 121–126.

Doassans-Carrère, N., S. Muller and M. Mitzkat. 2014. REVE—a new industrial technology for biomass torrefaction: pilot studies. Fuel Process Technol. 126: 155–162.

Foody, P. 1980. Optimization of Steam Explosion Pretreatment. Final Report to DOE. Contract AC02-79ET23050. 1, B80.

Jenkins, B. M., R. R. Bakker and J. B. Wei. 1996. On the properties of washed straw. Biomass Bioenergy 10: 177–200.

Jensen, P. A., B. Sander and K. Dam-Johansen. 2001. Pretreatment of straw for power production by pyrolysis and char wash. Biomass Bioenergy 20: 431–446.

Koppejan, J., S. Sokhansanj, S. Melin and S. Madrali. 2012. Status overview of torrefaction technologies, 2012. IEA Bioenergy Task 32 report (Enschede).

Koukios, E. G. 1993. Progress in thermochemical, solid state refining of biofuels: From research to commercialization. Advanced Thermochem. Biomass Conversion, 2.

Lacey, A. L., J. E. Aston, T. L. Westover, R. S. Cherry and D. N. Thompson. 2015. Removal of introduced inorganic content from the chipped forest residues via air classification. Fuel 160: 265–273.

Lam, P. S., S. Sokhansanj, X. Bi et al. 2011. Energy input and quality of pellets made from steam-exploded Douglas fir (*Pseudotsuga menziesii*). Energy Fuels 25: 1521–1528.

Lipinsky, E. S., J. R. Arcate and T. B. Reed. 2002. Enhanced wood fuels via torrefaction. Fuel Chemistry Division Preprints 47: 408–410.

Mann, M. K. and P. L. Spath. 2001. A life cycle assessment of biomass co-firing in a coal-fired power plant. Clean Products and Processes 3: 81–91.

Manomet Center for Conservation Sciences. 2010. Massachusetts Biomass Sustainability and Carbon Policy Study: Report to Massachusetts Department of Energy Resources. Contributors: Cardellichio, P., A. Colnes, J. Gunn, B. Kittler, R. Perschel, C. Recchia, D. Saah and T. Walker. Natural Capital Initiative Report NCI-2010-03.

Marchessauk, R. H., S. Coulombe, T. Hanal and H. Morikawa. 1980. Monomers and oligomers from wood. Pulp Paper Mag. Can. Trans. 6: TR52-56.

Marchessauk, R. H., S. Coulombe, T. Hanai and H. Morikawa. 1981. Vinylic Monomers from Bioconversion of Wood. *In*: 181st National Meeting of the American Chemical Society, Atlanta, Georgia, USA.

Marchessauk, R. H., S. Coulombe, H. Morikawa and D. Robert. 1982. Characterization of aspen exploded wood lignin. Can. J. Chem. 1082: 2372.

Nimlos, N. M., B. Emily, J. L. Michael and J. E. Robert. 2003. Biomass torrefaction studies with a molecular beam mass spectrophotometer, National Bioenergy Center, Presentation of Paper—American Chemical Society, Division Fuel Chemistry 48: 590–591.

Ohliger, A., M. Förster and R. Kneer. 2013. Torrefaction of beechwood: a parametric study including heat of reaction and grindability. Fuel 104: 607–613.

Oliveira-Rodrigues, T. and P. L. A. Rousset. 2009. Effects of torrefaction on energy properties of eucalyptus grandis wood. Cerne 15: 446–452.

Organisation for Economic Cooperation and Development (OECD)/International Energy Agency (IEA). 2011. Combining Bioenergy with CCS: Reporting and Accounting of

Negative Emissions under UNFCC and Kyoto Protocol. Accessible at: http://www.iea. org/publications/freepublications/publication/bioenergy_ccs.pdf.

Phanphanich, M. and S. Mani. 2011. Impact of torrefaction on the grindability and fuel characteristics of forest biomass. Bioresour. Technol. 102: 1246–1253.

Reed, T. B. and B. Bryant. 1978. Densified biomass: A new form of solid fuel. Solar Energy Research Institute Report #SERI–35. U.S. Department of Energy. Division of Solar Technology. Golden, Colorado, USA.

REN 21. 2012. Renewables 2012, Global Status Report. Accessible at: http://www.ren21.net/ Portals/0/documents/Resources/GSR2012_low%20res_FINAL.pdf.

Reza, M. T., J. Lynam, C. Coronella and R. Vasquez. 2011. Pellets from HTC (hydrothermal carbonization) biochar. *In*: TC Biomass 2011: The International Conference on Thermochemical Conversion Science (presentation). September 27–30, 2011. Chicago, IL, USA.

Sadaka, S. and S. Negi. 2009. Improvements of biomass physical and thermochemical characteristics via torrefaction process. Environ. Prog. Sustainable Energy 28(3): 427–434.

Sami, M., K. Annamalai and M. Wooldridge. 2001. Co-firing of coal and biomass fuel blend. Prog. Energy Combust. Sci. 27: 171–214.

Sarkar, M., A. Kumar, J. S. Tumuluru, K. N. Patil and D. D. Bellmer. 2014. Gasification performance of switchgrass pretreated with torrefaction and densification. Appl. Energy. 127: 194–201.

Searcy, E., P. Flynn, E. Ghafoori and A. Kumar. 2007. The relative cost of biomass energy transport. Appl. Biochem. Biotechnol. 137(1): 639–652.

Shaw, M. D., C. Karunakaran and L.G. Tabil. 2009. Physicochemical characteristics of densified untreated and steam exploded poplar wood and wheat straw grinds. Biosyst. Eng. 103: 198–207.

Shih, J. S. and H. C. Frey. 1995. Coal blending optimization under uncertainty. Eur. J. Oper. Res. 83: 452–465.

Sokhansanj, S., J. Peng, J. Lim, X. Bi, L. Wang, P. Lam, J. Hoi, S. Melin, J. Tumuluru and C. Wright. 2010. Optimum torrefaction and pelletization of biomass feedstock. TCS 2010 Symposium on Thermal and Catalytic Sciences for Biofuels and Biobased Products. Iowa State University. Ames, IA, USA.

Spath, P. and M. Mann. 2004. Biomass Power and Conventional Fossil Systems with and without CO_2 Sequestration—Comparing the Energy Balance, Greenhouse Gas Emissions and Economics, U.S. DOE NREL/TP-510-32575. Accessible at: http://www.nrel.gov/ docs/fy04osti/32575.pdf.

Thompson, D. N., J. A. Lacey and P. G. Shaw. 2003. Post-harvest processing methods for reduction of silica and alkali metals in wheat straw. Appl. Biochem. Biotechnol. 105: 205–218.

Tumuluru, J. S., S. Sokhansanj, J. R. Hess, C. T. Wright and R. D. Boardman. 2011. A review on biomass torrefaction process and product properties for energy applications. Ind. Biotechnol. 7: 384–401.

Tumuluru, J. S., J. R. Hess, R. D. Boardman, C. T. Wright and T. L. Westover. 2012. Formulation, pretreatment, and densification options to improve biomass specifications for co-firing high percentages with coal. Ind. Biotechnol. 8: 113–132.

Tumuluru, J. S., L.G. Tabil, Y. Song, K. L. Iroba and V. Meda. 2014. Grinding energy and physical properties of chopped and hammer-milled barley, wheat, oat and canola straws. Biomass Bioenergy 60: 58–67.

Tumuluru, J. S. and R. J. Hess. 2015. New market potential: Torrefaction of woody biomass. Mater. World 06: 41–43.

Tumuluru, J. S. 2015. Comparison of chemical composition and energy property of torrefied switchgrass and corn stover. Front. Energy Res. 3: 1–11.

Tumuluru, J. S., C. J. Lim, X. T. Bi, X. Kuang, S. Melin, F. Yazdanpanah and S. Sokhansanj. 2015. Analysis on storage off-gas emissions from woody, herbaceous, and torrefied biomass. Energies 8: 1745–1759.

Tumuluru, J. S. 2016. Effect of deep drying and torrefaction temperature on proximate, ultimate composition, and heating value of 2-mm lodgepole pine (*Pinus contorta*) grind. J. Bioeng. 3: 16.

Tumuluru, J. S., E. Searcy, K. L. Kenney, W. A. Smith, G. L. Gresham and N. A. Yancey. 2016. Impact of Feedstock Supply Systems Unit Operations on Feedstock Cost and Quality for Bioenergy Applications. Nova Publishers. USA.

U.S. Department of Energy (DOE). 2004. Biomass Cofiring in Coal-Fired Boilers. Federal Energy Management Program: Federal Technology Alert. DOE/EE-0288. Accessible at: http://www.nrel.gov/docs/fy04osti/33811.pdf.

U.S. Department of Energy, Office of Energy Efficiency (DOE/EE). 2016. Billion-Ton Report: Advancing Domestic Resources for a Thriving Bioeconomy. DOE/EE-1440. Accessible at: https://energy.gov/sites/prod/files/2016/12/f34/2016_billion_ton_report_12.2.16_0. pdf.

U.S. Energy Information Administration (EIA). 2016. Accessible at: https://www.eia.gov/.

U.S. Energy Information Administration, Annual Energy Review (EIA/AER). 2012. Accessible at: http://www.eia.gov/totalenergy/data/annual/pdf/aer.pdf.

U.S. Environmental Protection Agency, Combined Heat and Power Partnership (EPA/ CHP). 2010. Biomass Combined Heat and Power Catalog of Technologies, Vol. 5. Biomass Conversion Technologies. Accessible at: https://www.epa.gov/sites/ production/files/2015-07/documents/biomass_combined_heat_and_power_catalog_ of_technologies_5._biomass_conversion_technologies.pdf.

Uslu, A., A. Faaij and P. C. A. Bergman. 2008. Pre-treatment technologies, and their effect on international bioenergy supply chain logistics: Techno-economic evaluation of torrefaction, fast pyrolysis, and pelletisation. Energy 33: 1206–1223.

Wild, M., M. Deutmeyerm. D. Bradley, B. Hektor, J. R. Hess, L. Nikolaisen, W. Stelte, J. S. Tumuluru, P. Lamers, S. Prosukurina, E. Vakkilainen and J. Heinimö. 2016. Possible effects of torrefaction on biomass trade. IEA Bioenergy Task 40. April 2016.

Xu, F., K. Linnebur and D. Wang. 2014. Torrefaction of conservation reserve program biomass: a techno-economic evaluation. Ind. Crops Prod. 61: 382–387.

Yan, W., T. C. Acharjee, C. J. Coronella and V. R. Vasquez. 2009. Thermal pretreatment of lignocellulosic biomass. Environ. Prog. Sustainable Energy 28: 435–440.

Zanzi, R., D. T. Ferro, A. Torres, P. B. Soler and E. Bjornbom. 2002. Biomass torrefaction. Sixth Asia-Pacific International Symposium on Combustion and Energy Utilization, Kuala Lumpur.

CHAPTER 10

The Impacts of Thermal Pretreatments on Biomass Gasification and Pyrolysis Processes

Zixu Yang[1,2] and *Ajay Kumar*[2,*]

1. Introduction

Diminishing fossil fuel reserves, significant growth in demand for fossil fuels driven by emerging economies, and environmental and political concerns with the utilization of fossil fuels have prompted considerable research efforts to utilize renewable and sustainable energy resources such as solar, wind, biomass, tide wave, and geothermal. Among these renewable and sustainable energy sources under development, biomass is the only source of sustainable carbon. The utilization of biomass for power production can significantly reduce the emissions of NO_x and SO_x due to negligible contents of nitrogen and sulfur. Furthermore, biomass is typically acclaimed as a carbon neutral fuel since the carbon dioxide released from combustion or biomass-based fuels is consumed through photosynthesis.

[1] Institution changed into: State Key Laboratory of Chemical Engineering, Department of Chemical Engineering, East China University of Science and Technology, Shanghai 200237, China.
[2] Department of Biosystems and Agricultural Engineering, Oklahoma State University, Stillwater, OK, USA.
* Corresponding author: ajay.kumar@okstate.edu

As a result, biomass and biomass-based fuels can mitigate greenhouse gas (GHG) emissions by replacing use of fossil fuels (Bhattacharya et al., 2003).

Lignocellulosic biomass sources are available in a wide range of species. Generally, lignocellulosic biomass can be divided into two types, namely organic wastes/residues and purposely grown energy crops (e.g., switchgrass, miscanthus, etc.), oil crops (e.g., oil palm), and sugar crops (e.g., corn, sorghum, etc.). The organic wastes/residues mainly include wastes from agro-forestry industries, residues from paper and food industries, municipal green wastes, animal waste, and sewage sludge. These materials can be converted into more convenient solid, liquid and gaseous fuels which have the potential to substitute fossil fuels. Figure 1 (IE, 2007) summarizes the technologies of biomass conversion. Biomass conversion technologies mainly involve biochemical, thermochemical, and mechanical processes. Biochemical conversions utilize microorganisms and/or enzymes to decompose polymers in biomass structure into bio-alcohols through fermentation or biogas through anaerobic digestion. Mechanical conversion of biomass primarily involves extracting oils by physical extrusion of fruit or kernel of oil crops. The details on these two types of biomass conversions have been discussed in the other chapters of the book, whereas this chapter mainly focuses on thermochemical conversions of biomass.

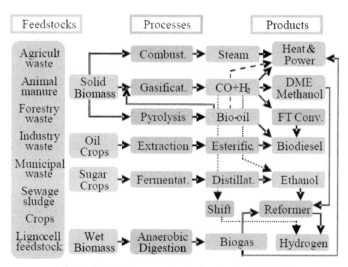

Fig. 1. Biomass conversion technologies (IE, 2007).

1.1 Thermochemical conversions of biomass

1.1.1 Thermochemical conversion technologies

Thermochemical conversion is the controlled heating or oxidation of feedstocks to produce energy carriers or heat (Tanger et al., 2013). Due to its high conversion efficiency, flexibility in feedstock varieties and fast conversion rate, thermochemical conversions are more attractive than other biomass conversion technologies (Tanger et al., 2013). Conversion techniques can be mainly categorized as combustion, gasification, and pyrolysis as shown in Fig. 1. Combustion of biomass directly releases the chemical energy stored in biomass in the form of heat through complete oxidization and is the most direct way of converting biomass to energy. Technically, combustion does not convert biomass into any advanced energy carriers that can be reserved for future use or further refined into versatile fuel products. Gasification and pyrolysis, however, offer efficient ways to produce gaseous, liquid, or solid products that are capable of being energy carriers or intermediate platforms for valuable chemicals (Patel et al., 2016). Gasification is the conversion of carbonaceous materials, such as biomass, coal, and tars into a gaseous mixture product called synthesis gas (or syngas) that contains primarily H_2, CO, CH_4, and CO_2 under the oxygen-constrained environment (Kumar et al., 2009). Syngas can be used directly as a fuel in combustion engines or gas turbines for power generation. Biomass integrated gasification combined cycle (IGCC) is a high energy efficiency process (35–60%) (Molino et al., 2016; Bhattacharya et al., 2011). Synthesis of fuel products such as alcohols production through syngas fermentation and hydrocarbon production through Fischer-Tropsch Synthesis (FTS) are also attractive options of syngas utilization, but the performance of the conversion is highly dependent on the quality of syngas. Pyrolysis, unlike gasification, thermally cracks organic materials by heating in an oxygen-free environment. The primary product of pyrolysis, char, liquid, or gas can be selected by manipulating the operation conditions such as heating rate, temperature, and residence time. Generally, a high yield of char is favored by low temperatures (~400°C) and long residence times; high temperatures (> 600°C) and long residence times result in high gas yield; while short residence times (less than 5 s) at moderate temperatures (400–600°C) lead to high liquid yield (Yang et al., 2015). The liquid product of pyrolysis, also known as bio-oil or bio-crude, is obtained by rapid quenching of vapors and aerosols generated from the decomposition of biomass. Some considerable research efforts have focused on fast pyrolysis of biomass due to its liquid yield, which can go as high as 90 wt.% depending on the type of feedstock and reaction conditions (Mohan et al., 2006). Bio-oil is considered to be a clean and renewable liquid fuel that can potentially substitute fossil fuels for transportation and power generation. However, the composition and fuel properties of bio-oil are substantially different from petroleum. Due to high oxygen content, bio-oil has many undesirable attributes such as immiscibility with hydrocarbon fuels, high acidity, and high viscosity. These attributes cause issues and challenges during the end-use, refining, storage, and transportation of the bio-oil.

1.1.2 Biomass feedstocks for thermochemical conversions

The intrinsic biomass feedstock properties including texture, bulk density, particle size/shape, moisture content, chemical compositions, heating value, etc., can vary significantly across different biomass resources. The change in properties can severely affect the operation, efficiency, and economics of the selected conversion processes. The variability of feedstock properties is a result of multiple factors such as feedstock types, agronomic conditions, harvest practices, and storage conditions (Williams et al., 2016).

The composition of biomass can be conceptualized in three ways, namely biochemical, proximate, and ultimate analyses. The biochemical analysis includes relative portions of biopolymers (e.g., cellulose, hemicellulose, lignin, and extractives) in the biomass. Proximate analysis, adapted from the characterization of coal (e.g., ASTM D3172), describes the fuel properties in terms of moisture content, volatile matter, fixed carbon, and ash contents (Tanger et al., 2013). Biomass feedstock composition can also be described in terms of ultimate analysis, which includes relative percentages of individual elements (C, H, N, O, and S). In practice, stoichiometric ratios of H/C and O/C are also commonly used to present the ultimate analysis. Feedstock chemical properties related to the performance of thermochemical conversions mainly include H/C and O/C ratios, moisture content, mineral content and cellulose/lignin ratio (Tanger et al., 2013; Williams et al., 2016).

Stoichiometric ratios of H/C and O/C directly relate to the energy content of the feedstock. In general, high H/C ratio and low O/C ratio implies high energy content. The energy content (heating value) is obtained by measuring the heat released during complete combustion, where C and H are oxidized into CO_2 and H_2O, respectively, and minerals (K, Ca, Na, P, Si, etc.) remain as ash. The moisture content is determined by measuring water available in the feedstock through drying. The moisture content impacts transportation, drying, pretreatment, and feeding processes. A high moisture content decreases the energy efficiencies of most thermochemical processes as more energy input is needed to remove moisture (Tumuluru et al., 2012). It should be noted that the tolerance on acceptable moisture content is different for different thermochemical processes. For example, biomass combustion or co-combustion has the strictest requirement for moisture content (< 5 wt.% in most cases) (Khan et al., 2009). During pyrolysis, the moisture contributes to the water content of bio-oil. Thus, high moisture content (> 10 wt.%) is also not recommended (Babu, 2008). For gasification, a high moisture content (up to 30 wt.%) is acceptable since water can be involved in the gasification reactions to increase H_2 yield (Martinez et al., 2012). The minerals do not convert into fuels or energy carrier but end up with ash during thermochemical conversions. Thus, high mineral content leads to a decrease in energy content of the products. Biomass ash contains alkali and alkaline earth metals (AAEM), which are proved to have catalytic effects on decomposition of biomass, changing decomposition pathways, product yield, and selectivity (Carpenter et al., 2014; Bridgwater, 2012). In addition, AAEMs have a low melting point (600–700°C), above which these behave as a liquid (at high temperature) and deposit as solid slag when temperature drops, causing the fouling and slagging of equipment (Du et al., 2014; Turn et al., 1997). Cellulose and lignin have different polymer structures, which results in different decomposition pathways, leading to different product distributions (Yang et al., 2007; Hosoya et al., 2007). The impacts of key feedstock chemical properties on gasification and pyrolysis are summarized in Table 1.

Table 1. Impacts of feedstock chemical properties on thermochemical conversions.

Feedstock chemical properties	Process	Impacts
H/C and O/C ratios	Gasification	Increase in O/C and/or decrease in H/C ratios results in decrease in H_2/CO molar ratio and the heating value of the syngas.
	Pyrolysis	Biomass has a low H/C ratio (0.7 to 2.8) as compared to that of the desired liquid fuel products (2–4 for alcohols and alkanes) (Tanger et al., 2013). Therefore, oxygen must be removed either in the form of water via hydrodeoxygenation or CO_2 via catalytic cracking.
Moisture content	Gasification	Biomass moisture content within 30 wt% is acceptable. Increase in moisture level reduces the temperature of the oxidation zone, leading to an increase in tar yield (McKendry, 2002).
	Pyrolysis	Increase in biomass moisture content reduces the heating rates and increases the energy input as more heat is needed for drying. Biomass moisture collected in bio-oil, affecting its stability, viscosity, acidity, heating values, and other properties.
Mineral content	Gasification	High mineral content causes ash deposition, leading to slagging and blockage problems.
	Pyrolysis	Trace amount (1 wt%) minerals have significant catalytic effects on depolymerization, resulting in a change in product distribution and bio-oil yield (Carpenter et al., 2014).
Cellulose/ lignin ratio	Gasification	The conversion of cellulose (98%) is higher than that of lignin (57%). However, the H_2/CO molar ratio of lignin (1.2) is higher than that of cellulose (0.8) (Claude et al., 2016).
	Pyrolysis	The bio-oil compounds obtained from cellulose are primarily anhydrosugar and its decomposition products (furans, ketones, aldehydes, and acids) (Carpenter et al., 2014). While lignin generates mainly phenolics compounds, a decrease in cellulose/ lignin ratio leads to an increase in bio-oil average molecular weight (Fahmi et al., 2008).

1.2 Challenges associated with biomass

The successful long-term sustainability of biomass-to-liquid fuels (BTL) technology depends on the effective conversion of biomass into products that are compatible with existing transportation and refinery infrastructure. However, the variability in biomass feedstock properties makes conversion processes more complex and challenging as compared to fossil energy sources. Apart from the key chemical properties related to thermochemical conversions (shown in Table 1), physical properties such as hygroscopicity, bulk density, and grindability are also important to the storage, transportation, and handling of biomass. Hygroscopicity is the capability of absorbing water from surroundings. Most biomass plants are hygroscopic, making biomass vulnerable to fungal attack and biodegradation during storage and transportation (Chew and Doshi, 2011). The low bulk density of biomass requires a large space for collection and transportation, leading to increase in storage and transportation costs. The fibrous and tenacious structure of biomass makes it difficult to grind and process in the existing systems. To achieve optimal performance, the physical and chemical properties related to thermochemical processes must be improved via pretreatments. Chemical pretreatments and thermal pretreatments are the two types of pretreatments that have been studied. Chemical pretreatments alter the biomass chemical compositions (cellulose, hemicellulose, lignin, extractives, and minerals) by using solvents such as hot water, acids, bases, or organic solvents (Gao et al., 2016). Most minerals can be efficiently removed by acid washing/leaching. Thermal pretreatment techniques, such as torrefaction and densification, have shown to improve the suitability of biomass as a feedstock for power generation and fuel production. Much recent research and development efforts have focused on investigating the properties of pretreated biomass and its impact on thermochemical conversions. Torrefaction technology has been extensively reviewed with a focus on torrefaction chemistry, kinetics, and product properties (Chew and Doshi, 2011; Van der Stelt et al., 2011; Nunes et al., 2014; Chen et al., 2015; Ciolkosz and Wallace, 2011; Bach and Skreiberg, 2016). This review aims to discuss the impacts of thermal pretreatments (torrefaction and densification) on thermochemical conversions of the pre-processed biomass.

2. Thermal Pretreatments

2.1 Torrefaction

The word "torrefaction" is the French translation for "roasting" because the studies on biomass torrefaction were pioneered in France in 1930s (Van der Stelt et al., 2011). Recently, torrefaction technology has diverted into two different types, namely dry torrefaction and wet torrefaction (also known as hydrothermal carbonization). Dry torrefaction is a thermal pretreatment that processes biomass feedstock at a mild temperature (200–300°C) under an inert environment at atmospheric pressure. In comparison to dry torrefaction that is conducted in a gaseous phase, wet torrefaction processes biomass in a liquid phase, usually hot compressed water at a temperature range of 180–260°C (Bach and Skreiberg, 2016). The objectives of both torrefaction technologies are the same, which is to produce a solid fuel with improved energy content and improved physical and chemical properties. Torrefaction yields a product that is superior to raw biomass in the following aspects: (1) higher energy content and energy density; (2) lower moisture content; (3) lower O/C and H/C ratios; (4) higher hydrophobicity and improved fungal durability; (5) improved grindability. Temperature and residence time are the two operation parameters in dry torrefaction that influence the properties of torrefied products. Generally, the higher temperature and longer residence time, the higher is the degree of carbonization of the torrefied biomass.

2.2 Densification

The purpose of densification is to apply mechanical force to compact the selected materials into uniformly shaped and standardized commodities such as pellets and briquettes. Densification provides multiple benefits including (1) higher biomass density and uniformity; (2) better handling and storage efficiencies; (3) improved flow properties (Faizal et al., 2016; Kaliyan and Morey, 2010). Lignocellulosic biomass must be dried and size-reduced before densification. For this reason, torrefaction is usually combined with densification, since torrefaction can significantly reduce the energy consumption needed for densification. The performance of conventional densification process and a combination of torrefaction with pelletization (TOP) was compared in a case study using two biomass feedstocks, sawdust, and green wood chips with the same initial moisture content (57 wt.%), at a capacity of 170 kton/h (Bergman, 2005). Results showed that the energy consumption of TOP was 0.83 MW for sawdust and 1.01 MW for green wood chips, compared with an energy consumption of 1.26 MW for sawdust and 1.84 MW for green wood chips using conventional densification process. The combination of torrefaction and densification results in a torrefied biomass pellet or briquette, which is a solid fuel resembling coal with high bulk and energy density, low moisture content and hydrophobicity, and improved grindability compared with raw biomass. In addition, torrefied biomass is more likely to cause dust formation than raw biomass. This causes a respiratory or explosion hazard during storage, and densification of torrefied biomass solves this issue (Ciolkosz and Wallace, 2011).

3. Impacts of Thermal Pretreatments on Thermochemical Conversions

The overall conversion efficiency, especially through thermochemical conversion, are expected to enhance due to improved fuel properties such as high energy content, low moisture and oxygen content, and high flow properties due to thermal pretreatments of the raw biomass. Due to this reason, many studies have focused on the comparison of non-pretreated (raw) and pretreated biomass during thermochemical conversions. However, since biomass pretreatment requires energy and cost, the operation conditions of pretreatment are closely scrutinized to obtain a feedstock that will achieve the performance of the overall conversion processes.

3.1 Gasification

3.1.1 Challenges in biomass gasification

Although gasification offers efficient conversion of biomass to energy, commercialization of this technology has been obstructed due to several technical challenges. First of all, the product gas contains mostly syngas but also significant impurities such as unconverted tars, char, and ash. These impurities must be removed before the downstream processing operations. Tars mainly consist of high molecular weight aromatic compounds which condense and cause blocking of pipes and catalyst coking in downstream syngas upgrading (Kumar et al., 2009). Gasification efficiencies of the raw biomass are low because of the high O/C ratios and moisture content of raw biomass (as compared to coal) (Prins et al., 2006). Thermal pretreatment provides a solution to these issues by improving the fuel properties of the biomass.

3.1.2 Impacts of torrefaction on biomass gasification

The studies using numerical modeling and experiments to investigate the impacts of torrefaction pretreatment on the performance of biomass gasification were initiated relatively recently (Pinto et al., 2017; Prando et al., 2014; Fisher et al., 2012; Sermyagina et al., 2015; Couhert et al., 2009; Chen et al., 2011; Yang et al., 2013; Deng et al., 2009; Yang et al., 2014). In this section, the impacts of torrefaction on the gasification efficiencies, syngas yield and quality and the behavior of torrefied biomass during co-gasification are discussed.

3.1.2.1 Gasification efficiencies

The evaluation of gasification efficiencies is based on the thermodynamic analysis. There are two types of thermodynamic analysis frequently used

—energy analysis, such as cold gas efficiency (CGE, defined as the ratio of chemical energy of the syngas and total energy in the input) and hot gas efficiency (HGE, defined as the ratio of sum of chemical energy and sensible heat in syngas and total energy in the input); and exergy analysis, such as energetic efficiency (defined as the ratio of exergy in syngas and fuel). The energy analysis focuses on the energy conversion efficiency, whereas the exergy analysis focuses on evaluating the energy losses that occur during conversion processes (Ptasinski, 2008). Thermodynamic efficiencies have been widely used to assess the impact of torrefaction on gasification. Many studies on gasification efficiency analysis have been performed using process simulation. The composition of gasification products (syngas, char, and tars) at thermodynamic equilibrium are predicted based on the Gibbs free energy minimization approach. The fuel properties, especially fuel oxygen and moisture content are greatly related to gasification efficiencies. In a study by Prins et al. (2007), gasification efficiencies of different fuels (coal, woody biomass, sludge, and manure) were compared. The chemical exergy efficiency (the ratio of chemical exergy in the fuel and syngas) was around 75% for coal as compared to 70% for woody and herbaceous biomass (Ptasinski et al., 2007), and 60% for wet biomass such as sludge and manure. They found that higher O/C atomic ratio and moisture content of the fuel led to higher exergy loss, implying that the highly oxygenated and wet fuels (e.g., untreated biomass) are not ideal fuels for gasification. Later, they evaluated efficiencies of biomass gasification with torrefaction in three different systems; air-blown circulating fluidized bed gasification of raw biomass, biomass torrefaction followed by circulating fluidized bed gasification of torrefied biomass, and biomass torrefaction integrated with oxygen-blown entrained flow gasification of torrefied biomass (Prins et al., 2006). The exergetic efficiency of air-blown gasification of torrefied biomass was found to be even lower than that of the raw biomass because the volatiles from the torrefaction step were not utilized. However, the exergetic efficiency of torrefaction combined with entrained flow gasification of torrefied biomass (72.6%) was higher than that of direct circulating fluidized bed gasification of raw biomass (68.6%). This improvement was because the heat produced from the entrained flow gasifier was used to drive the torrefaction process and both the solid and volatiles from biomass torrefaction were introduced to the gasification process. Based on the results from Prins et al. (2007), Clausen (2014) proposed the concept of integrated torrefaction, that is, torrefaction integrated with entrained flow gasification and external torrefaction, which is defined as the decentralized production of torrefied biomass and centralized conversion of torrefied biomass through entrained flow gasification. In integrated torrefaction, the volatiles from torrefaction are further converted to syngas, whereas the volatiles are combusted for heat supply for torrefaction in external torrefaction. A thermodynamic analysis (Clausen, 2014) was applied to compare the energy efficiency of these two torrefaction-gasification processes, and the results showed that integrated torrefaction at a temperature of 250°C achieved CGE of 81%, which was higher than CGE of 63% obtained from external torrefaction. It should be noted that the torrefied biomass obtained from the integrated process at 250°C had similar composition to that obtained from external torrefaction. Although the integrated torrefaction process has the potential to increase energy efficiency, the more complex design and no electricity generation credit will add up to the overall cost. Therefore, an economic analysis is needed to determine which process can bring more economic benefits. The thermodynamic analysis on the gasification of raw and torrefied bamboo in a downdraft gasifier from Kuo et al. (2014) indicated that CGE of torrefied bamboo was lower than that of raw bamboo due to the higher HHV of torrefied bamboo. In another simulation study, Tapasvi et al. (2015) found that the CGE and exergy efficiencies were 76.1–97.9% and 68.3–85.8%, respectively for torrefied wood, compared with 67.9–91.0% and 60.7–79.4%, respectively, for untreated wood. Apart from these gasification efficiencies evaluated under thermodynamic equilibrium, numerical simulations based on gasification reaction kinetics are also used to analyze the benefits of torrefaction during gasification in the most recent years. Chen et al. (2013) simulated the gasification performance of raw bamboo, torrefied bamboo (280°C for 1 h), and high-volatile bituminous coal in an oxygen-blown entrained flow gasifier. The numerical model was established based on a two-dimensional steady flow with turbulence and kinetically limited chemical reactions (devolatilization, char gasification reactions, and gas phase reactions). The results indicated that the gasification performance was highly dependent on the HHV of fuels. In detail, the maximum gasification temperature of raw bamboo, torrefied bamboo, and bituminous coal was 1348, 2268 and 2840 K, respectively, showing a positive correlation to the HHV of these fuels. Similarly, CGE was reported as 29.0, 49.8 and 69.1% for raw bamboo, torrefied bamboo, and bituminous coal, respectively.

The experimental results on gasification of torrefied biomass have also been reported. The reported gasification efficiencies, such as CGE varied significantly depending on the gasifier design, biomass feedstock, and torrefaction conditions (Table 2). As a result, the impacts of torrefaction on the gasification efficiencies become ambiguous. Two studies from oxygen-blown entrained flow gasification (Chen et al., 2011; Weiland et al., 2014) confirmed that gasification CGE of torrefied biomass was higher as compared to that of raw biomass, and the optimum gasification CGE can be achieved at mild torrefaction conditions. However, the studies from the fluidized bed and fixed bed gasification (Kwapinska et al., 2015; Kulkarni et al., 2016; Sarkar et al., 2014; Dudyński et al., 2015) revealed that gasification CGE of raw biomass was lower than that of torrefied biomass. Kwapinska

Table 2. Summary of lab-scale experimental studies on gasification of torrefied biomass.

Gasifier design	Biomass type	Torrefaction conditions	Gasification conditions	CGE raw vs. torrefied biomass (%)	Ref.
Entrained flow gasifier	Sawdust	230, 250, 270 and 290°C, residence time: 20 or 30 min	1200°C, oxygen equivalence ratio = 0.3	76, 80–87	(Chen et al., 2011)
Entrained flow gasifier	Wood residues and stem wood pellets	Wood residues torrefied at 300 or 340°C for 4.5 min. Wood pellets torrefied at 245°C for 45–60 min	1210–1250°C, oxygen equivalence ratio = 0.44	54, 61–76	(Weiland et al., 2015)
Fluidized bed gasifier	Pinewood	Not given	935°C, air equivalence ratio = 0.25	52, 66	(Kulkarni et al., 2016)
Fluidized bed gasifier	Miscanthus x gigantenus	250°C for 4 h	660–800°C, air equivalence ratio = 0.18–0.32	50–75, 37–65	(Kwapinska et al., 2015)
Fixed bed gasifier	Wood pellets	250–270°C	470–960°C, air equivalence ratio = 0.14–0.31	77, 75	(Dudyński et al., 2015)
Fixed bed gasifier	Switchgrass	230 and 270°C for 30 min	700–900°C	58, 33–43	(Sarkar et al., 2014)
Auger gasifier	Cotton gin wastes	280°C for 1–2 h	750–950°C, air equivalence ratio = 0.09–0.1	30–37, 40–43	(Sadaka, 2013)

et al. (2015) attributed the decreased CGE to higher char elutriation rate and poor reactivity of torrefied biomass. The study from an auger gasifier (Sadaka, 2013) showed that the gasification CGE of torrefied biomass was 40–43%, which was higher than that of raw biomass (30–37%), but these values were much lower as compared to that reported from other gasifier designs. One of the reasons for the diversity of the impacts of torrefaction on gasification efficiency is that several of the gasifier designs are not optimized. On comparing the experimental data, it can be seen that gasification efficiency is partially affected by the fuel properties, but also affected by the gasifier designs and operating conditions. Therefore, based on the reported results, the improvement of fuel properties such as higher H/C ratios, lower moisture, and oxygen content do not necessarily lead to the improvement of the gasification efficiencies. For the long-term success of bioenergy industry, the research and development should not simply focus on the impacts of torrefaction on the fuel properties, but on the structure and chemistry changes during the torrefaction and the impact of these changes on its reactivity subject to thermal decomposition.

3.1.2.2 Syngas composition and quality

The composition and quality of syngas produced are directly related to biomass properties, gasifier design, and operating conditions and downstream syngas cleaning and conditioning. Torrefaction has been utilized to modify the compositional and structural properties of biomass, thus making it more suitable to produce high-quality syngas, which is usually characterized by high concentrations of H_2 and CO and low concentrations of CO_2, CH_4, and other light hydrocarbons. Most studies confirmed that the content of CO_2 in syngas from torrefied biomass gasification is generally lower than that from gasification of raw biomass due to the lower oxygen content of the torrefied biomass (Chen et al., 2011; Kuo et al., 2014; Chen et al., 2013; Weiland et al., 2014; Kwapinska et al., 2015; Kulkarni et al., 2016; Sarkar et al., 2014; Dudyński et al., 2015). Kwapinska et al. (2015) and Kuo et al. (2014) concluded that the release of chemically bound hydrogen, such as H_2O, CH_4, and condensable organics (acetic acid, hydroxyacetone, and methanol) during torrefaction process reduced the H_2 produced from the reactions (e.g., water gas shift reaction and steam-carbon reaction) involved with these compounds. However, many studies have also found that H_2 content did not significantly change after torrefaction (Chen et al., 2011; Weiland et al., 2014; Sarkar et al., 2014; Couhert et al., 2009). The use of steam as a gasifying agent can significantly enhance the production of H_2 through water-gas reaction. Some studies (Tapasvi et al., 2015; Kuo et al., 2014) reported that torrefied biomass yielded a higher content of H_2 than raw biomass during steam gasification. The content of CO is also expected to be higher for torrefied biomass due to the higher fixed carbon content in the torrefied biomass.

The concentration of tars in the syngas stream is another important index to evaluate the quality of syngas. Tars are mainly composed of condensable heavy aromatic compounds, which tend to condense and cause fouling issues in engines and coolers. Torrefaction was reported to effectively inhibit the production of tars during gasification (Kulkarni et al., 2016; Dudyński et al., 2015). This can be explained by the fact that torrefaction releases part of volatiles that can end up as tars during gasification. In addition, torrefied biomass enhances gasification temperature above 1000°C resulting in cracking of the tars formed. Tars obtained from gasification of torrefied biomass mainly are composed of acids (70%) and contain significantly less amount of aromatic and cyclic compounds than that from gasification of raw biomass (Table 3).

Table 3. Chemical composition of tar from gasification of raw and torrefied biomass (wt.%) (Dudynski, 2015).

Compound groups	Polish pellets	Torrefied pellets	South Africa pellets	Polish sawdust
Aliphatic	0.0	0.0	0.0	0.4
Acids	11.7	70.0	26.0	23.1
Aliphatic esters	0.2	0.0	1.4	0.0
Aliphatic aldehydes and ketones	1.6	0.8	3.5	5.8
Aliphatic alcohol	1.3	7.7	1.6	0.7
Alkylbenzenes	0.8	0.4	0.6	5.2
Alkylphenols	21.9	11.8	26.8	34.1
Furan	0.1	1.2	0.0	7.9
Furan (polyfunctional oxygen)	5.2	0.0	4.4	1.1
Linear and cyclic aliphatic oxygenates	27.0	1.7	5.8	5.2
Aromatic oxygenates	25.0	2.2	29.7	15.2
Nitrogen and sulfur heteroatoms	4.2	4.0	0.0	1.2

3.1.2.3 Co-gasification

Co-utilization (co-firing and co-gasification) of biomass and coal has become more attractive in recent years due to multiple environmental benefits, such as reduction of GHG emissions and pollutant gases (NO_x, SO_x, NH_3 and HCN) emissions (Tchapda and Pisupati, 2014). The high concentrations of alkali (K and Na) and alkali earth metals (Ca) in biomass are also expected to exert catalytic behavior when blended with coal during co-gasification (Habibi et al., 2012). With all of these inherent benefits, torrefied biomass renders more advantages than raw biomass during co-gasification due to its improved energy density, hydrophobic nature, and improved grindability. These improved fuel properties can significantly decrease the energy requirement for size reduction (up to 85%), and also offer a solution to biomass feeding issues such as bridging and blockage (Tchapda and Pisupati, 2014). Hence, improved fuel properties of torrefied biomass make it more attractive for co-gasification with the potential to use high mix ratios in existing coal-fired boilers without major modifications. The co-feed ratio of torrefied biomass, especially torrefied biomass pellet, in existing coal-fired boilers, can be as high as 50%, which is much higher than the co-feed ratio of raw biomass (10–15%) (Nunes et al., 2014). A case study (Li et al., 2012) on co-firing pulverized boilers indicated that the net CO_2 and NO_x emissions significantly reduced with the increase in mixing ratios of torrefied biomass. The loss of boiler load was only 9.52% when coal was replaced completely by torrefied biomass.

Co-utilization of torrefied biomass and coal, because of the potential benefits discussed above, has been given close scrutiny in recent years. However, the research and development of co-gasification using torrefied biomass are still at a proof-of-concept stage. Therefore only a limited number of studies are reported. Chen et al. (2013) employed the Taguchi method to numerically optimize the performance of co-gasification of torrefied eucalyptus and coal in an entrained flow gasifier and found that torrefaction temperature was a significant factor in optimizing the gasification performance. Kuo et al. (2016) conducted a thermodynamic analysis of a hybrid power plant that consisted of an integrated torrefied biomass co-gasification, a solid oxide fuel cell, and two calcium looping CO_2 capture configurations. Co-gasification using torrefied biomass appeared to provide a promising approach for clean and efficient utilization of coal and biomass. However, additional research, especially experimental work, is needed to clarify the impacts of key parameters (e.g., torrefaction conditions and fuel mixing ratios) on the co-gasification efficiency, syngas quality, and economics. Moreover, the slagging, fouling and corrosion issues related to ash-melting of biomass need to be concerned.

3.1.3 Impacts of densification on biomass gasification

Pelletized or otherwise densified biomass exhibits better fuel operability (handling, transportation, storage, and feeding) than raw biomass. It can also be easily processed by coal pulverization and fed into the boiler or gasifier. Due to these benefits, pelletized biomass is frequently used in gasification, especially in fixed bed gasifiers where mechanically stable fuel particles of limited size are required for successful operation (Dai et al., 2008). Densification also improves the particle density favoring mixing of materials in fluidized bed gasifier (Bronson et al., 2012). Pelletization renders biomass particles with uniform size, which offers a solution to the bridging or slugging problems during fluidization. In addition, the permeability and thermal conductivity of biomass change during densification. Pelletized biomass has been extensively studied as a co-feed in gasification or combustion system; however, a few studies have compared the gasification performance of pelletized biomass with its parent biomass.

Yoon et al. (2012) found that pelletization changed the chemical composition (due to the addition of binder, ten wt.% of Larch sawdust) and flow characteristics of biomass particles, leading to an increased calorific value and improved breathability in the gasifier. Accordingly, the gasification performance of rice husk pellet (CGE: 70%) was higher than that of rice husk (CGE: 60%). In another study by Sakar et al. (2014), gasification of untreated switchgrass, torrefied switchgrass (230 and 270°C), switchgrass pellet, and torrefied switchgrass pellet was conducted in a fixed-bed gasifier. The results showed that the CGE followed the order of torrefied switchgrass pellet>switchgrass pellet>untreated switchgrass>torrefied switchgrass. It should also be noted that the percentage of binder (corn starch) addition was only two wt.%, which was much lower than that reported in other studies (Yoon et al., 2012; Pirraglia et al., 2013). The analysis (Sarkar et al., 2014) of microstructure revealed that the pelletized switchgrass presented a severely disintegrated fibrous structure, whereas the fibrous matrix was still visible in unpelletized switchgrass. This structural variation may lead to the difference in reactivity during gasification. The study on fragmentation and attrition of wood chips and wood pellets indicated that higher mechanical strength of wood pellets inhibited primary fragmentation and also reduced carbon elutriation during gasification, thus leading to a higher carbon conversion (Ammendola et al., 2013; Ammendola et al., 2011). The investigation (Erlich and Fransson, 2011) on char bed properties, such as porosity and pressure drop from three pellets (wood, palm-oil residues, and bagasse) indicated that pellets with higher reactivity (e.g., wood and bagasse) created denser char bed, leading to longer residence time for air and gases in the bed, and thus increased char conversion. Teixeira et al. (2012) compared the char bed properties of wood pellets and wood chips during a continuous fixed-bed gasification. It was found that the reactive zone of char bed was three times longer for wood chips than that for pellets. In addition, the bed compaction, defined as the ratio between initial and in-progress char particle velocities, was similar for wood chips and pellets, indicating that pelletization did not significantly change the char bed properties. Using a commercialized down-draft gasifier, Gautam et al. (2011) reported a total syngas tar concentration of 340–680 mg/Nm^3 from wood pellets and 54 mg/Nm^3 from wood chips. They attributed this high tar concentration to the poor heat transfer efficiency of the denser wood pellet. Prasad et al. (2015) also pointed out that the size of pellets should be optimized to avoid incomplete gasification, leading to a high concentration of tars in the syngas.

3.2 Pyrolysis

Pyrolysis can be divided into different modes (e.g., fast pyrolysis and conventional pyrolysis) depending on heating rates and residence time. Currently, extensive research has focused on fast pyrolysis of biomass due to its high liquid yield and fast conversion rate. The liquid product, bio-oil, has shown the potential to serve as either an alternative to fossil fuels or an ideal source for chemical synthesis. However, as mentioned earlier, the commercialization of bio-oil has been challenged by its poor qualities, such as high oxygen content, high viscosity, acidity, and low stability. Thermal pretreatment such as torrefaction is seen as a potential pretreatment method prior to fast pyrolysis to mitigate these issues because torrefaction lowers the oxygen and water content of biomass feedstock. In this section below, a discussion on how thermal pretreatments affect the pyrolytic behavior and chemical compositions of bio-oil is provided.

3.2.1 Impacts of torrefaction on fast pyrolysis

3.2.1.1 Pyrolytic behavior

The thermal decomposition behavior of torrefied biomass during pyrolysis has been analyzed by several researchers in order to understand the change of reaction mechanism due to torrefaction pretreatment. Biomass pyrolysis is usually characterized by a three-stage thermal decomposition, where the first stage is dehydration process (~120°C) with a weight loss up to 10%, followed by the active pyrolysis (220–400°C) accompanied by massive weight loss up to 70%, and then a final stage, also known as passive pyrolysis (> 450°C), which incurs minimal weight loss. The decomposition of hemicellulose and cellulose mainly occurs at the active pyrolysis stage. Decomposition of hemicellulose involves exothermic reactions which mainly take place in the temperature range of 220–315°C, and decomposition of cellulose is endothermic occurring in the temperature range of 315–400°C (Yang et al., 2007). Studies on pyrolysis kinetics (Chen and Kuo, 2011) reveal that torrefaction reactions in the temperature range of 200–300°C mainly include faster decomposition (pre-exponential factor = $4.13*10^{16}$ min^{-1}) of hemicellulose starting from 225°C, followed by a slower decomposition of cellulose (pre-exponential factor = $2.86*10^9$ min^{-1}) starting from 275°C. Because hemicellulose is decomposed during torrefaction, the devolatilization of torrefied hemicellulose in biomass was significantly reduced during pyrolysis and even eliminated at high torrefaction temperature (Ren et al., 2013; Sarkar et al., 2013; Tapasvi et al., 2013; Cao et al., 2016). The impact of torrefaction on activation energy (E) was studied by Zhang et al. (2016). As shown in Fig. 2, the activation energy showed an increasing trend in the initial pyrolysis stage (conversion rate < 0.4) which is mainly due to transition from lower energy barrier reactions such as weak-linkage cleavage in hemicellulose to higher energy barrier reactions such as chain scission and depolymerization of cellulose molecules. After that, cellulose converts into active cellulose with a lower degree of polymerization and molecular weight, leading to the decrease in activation energy. Torrefaction leads to decomposition of hemicellulose and formation of active cellulose, which are supposed to occur at an initial stage of pyrolysis. Hence, the energy barrier for pyrolysis reactions is significantly decreased, and the change of activation energy becomes minimal.

The structural changes of biomass during torrefaction are revealed using analytical techniques, such as solid-state ^{13}C NMR, FTIR spectroscopy, and XRD (Zheng et al., 2013; Neupane et al., 2015; Ru et al., 2015; Zheng et al., 2015). Torrefaction severity, the factor controlling the degree of weight loss

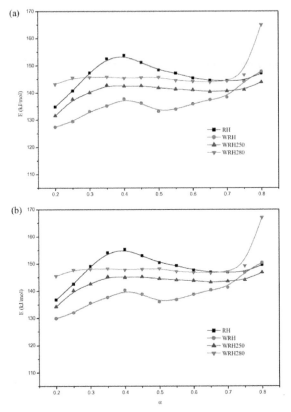

Fig. 2. The activation energy (E) as a function of conversion during pyrolysis (Zhang et al., 2016), RH = untreated rice husk, WRH = rice husk pretreated by water washing, WRH250/280 = water washed rice husk followed by torrefaction at 250/280°C.

of biomass during torrefaction, is mainly related to torrefaction temperature and residence time (Li et al., 2012; Peng et al., 2013). Chen et al. (Chen et al., 2014a) proposed a normalized parameter "torrefaction severity index" (TSI) to determine the torrefaction severity. TSI is defined as (Chen et al., 2014a):

$$TS\,I = \frac{\Delta WI}{\Delta WI_{ref}} \tag{1}$$

where ΔWI and ΔWI_{ref} refer to the weight loss at a given torrefaction condition and a reference condition, respectively. Typically, the reference condition is selected as the combination of torrefaction temperature and residence time (e.g., 300°C and one hr) that gives the highest degree of weight loss during torrefaction. Hence, the TSI ranges from 0 to 1, corresponding to the onset of torrefaction and the reference condition, respectively.

Zheng et al. (2015) characterized the structural changes of hemicellulose, cellulose, and lignin during torrefaction at different torrefaction temperatures (210–300°C) and residence times (20–60 min). Both FTIR and ^{13}C NMR analysis confirmed depolymerization, ring breakage, and fragmentation of hemicellulose during torrefaction. Additionally, polycondensation reactions and charring of hemicellulose were also observed. Lignin underwent a series of β-O-4 scission, demethoxylation, and polycondensation, whereas the structure of cellulose changed slightly during torrefaction (Zheng et al., 2015). Ru et al. (2015) concluded that the dehydration of hydroxyls, dissociation of O-acetyl branches, and breakage of weak ether linkages (e.g., β-O-1,4-glycosidic bond, β-O-4 aryl ether linkage in lignin) were the main reactions during torrefaction. A few studies also proposed that the polymerization and recondensation of primary decomposition products of hemicellulose and lignin result in aromatic and aliphatic C-C and C-H bonds (Zheng et al., 2013; Neupane et al., 2015). XRD analysis (Zheng et al., 2013; Neupane et al., 2015) indicated that mild torrefaction (~250°C) increased the crystallinity due to the recrystallized amorphous cellulose and decomposition of hemicellulose. As the torrefaction severity further increased, the crystallinity decreased significantly due to decomposition of crystalline cellulose. The effect of torrefaction on the pyrolysis of cellulose is illustrated in Fig. 3. Cellulose first depolymerizes to form active cellulose, and then active cellulose undergoes cross-linking and polycondensation to form char. The cross-linking and charring of cellulose enhances with an increasing severity of torrefaction, thus leading to an increase in char yield during pyrolysis.

3.2.1.2 Bio-oil properties

Torrefaction parameters (temperature and residence time) have significant impacts on the yield, physical properties, and chemical compositions of bio-oil derived from fast pyrolysis of torrefied biomass. The yield of bio-oil from torrefied biomass is expected to decrease due to following reasons. First of all, part of volatiles is released during torrefaction, leading to the reduction of total condensable volatiles for pyrolysis. Secondly, cross-linking and charring of cellulose during torrefaction lead to the increase in char yield and decrease in bio-oil yield (as shown in Fig. 3). Thirdly, torrefaction concentrates lignin, which tends to undergo carbonization and polycondensation to form char during pyrolysis. All of these impacts are enhanced with the increase in the torrefaction severity. Therefore, the bio-oil yield decreases as the torrefaction severity increases. Figure 4 illustrates the impacts of torrefaction temperature on the pyrolysis products distribution of rice husk.

Torrefaction removes some water and oxygenates from biomass, therefore leading to the increase in the calorific value and decrease in the water content of the resulting bio-oil (Table 4). The pH value of bio-oil is not significantly affected by torrefaction (Chen et al., 2015). Viscosity and a solid

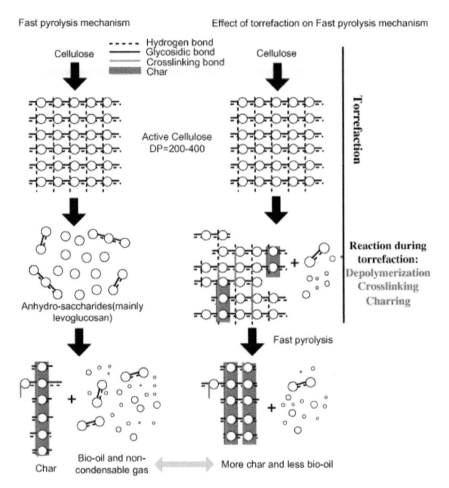

Fig. 3. Effect of torrefaction on cellulose pyrolysis mechanisms (Zheng et al., 2013).

content of bio-oil from torrefied biomass were found to increase significantly as the torrefaction temperature increased (Zheng et al., 2012). The variation in the chemical composition of bio-oil due to torrefaction has been analyzed extensively. Bio-oil is a complex mixture of oxygenated organics. Depending on functional groups and corresponding biomass constituent origin, these compounds can be roughly categorized into hemicellulose-derived compounds (e.g., acids, ketones, aldehydes, furans, and pyrans), cellulose derived compounds (e.g., furans and anhydrosugars), lignin-derived compounds (e.g., unbranched phenols, guaiacols and syringols), extractives (e.g., lipids and N-containing compounds). Torrefaction alters relative concentrations of hemicellulose, cellulose and lignin in the biomass, thus leads to changes in the contents of their corresponding pyrolysis product.

As expected, the content of hemicellulose-derived compounds significantly decreases as the torrefaction severity increases, whereas the content of lignin-derived compounds significantly increases as the torrefaction severity increases. A few studies (Zhang et al., 2016; Ren et al., 2013; Yang et al., 2014; Meng et al., 2012) reported that the major pyrolysis product of cellulose, anhydrosugars, increased significantly as the torrefaction severity increased. This can be explained by the interaction between hemicellulose and cellulose that negatively affects the production of sugars during pyrolysis (Wang et al., 2011). However, this interaction effect mitigates due to massive decomposition of hemicellulose during torrefaction pretreatment. Due to the variation among pyrolysis reactor systems, biomass feedstock and torrefaction parameters, the reports of bio-oil composition vary significantly.

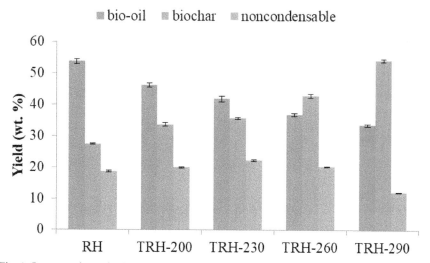

Fig. 4. Impacts of torrefaction temperature on the pyrolysis products distribution (RH = rice husk, TRH-200/230/260/290 = rice husk torrefied at 200/230/260/290°C, data adapted from (Chen et al., 2014b)).

Table 4. Effects of torrefaction temperature on HHV and water content of bio-oil (Adapted from (Meng et al., 2012). LP = Loblolly pine, torrefaction condition: 270, 330, and 330°C for 2.5 min).

Sample	HHV (MJ/kg dry basis)	Water in BOC[a] (wt.%)	Water in BOP[b] (wt.%)
LP-Raw	21	44.4	14.3
LP-T270	21.2	36.7	8.46
LP-T300	23.5	41.3	9.04
LP-T330	28.7	35.7	6.29

[a] BOC: bio-oil collected from condensers.
[b] BOP: bio-oil collected from electrostatic precipitator.

Table 5 summarizes the impacts of torrefaction on representative bio-oil compounds as reported in the literature. Catalytic fast pyrolysis of

Table 5. Summary of impacts of torrefaction on selected compounds in bio-oil (+/–: increasing/decreasing).

Compound	Group	Impacts of torrefaction (+/–)	Reference
Hemicellulose/cellulose derived compounds			
Acetic acid	Acids	–	Zhang et al., 2016; Ru et al., 2015; Chen et al., 2014b; Chen et al., 2015; Zheng et al., 2012
Propanoic acid	Acids	–	Chen et al., 2015
1,2-Cyclopentanedione	Ketones	–	Zhang et al., 2016
2-Cyclopenten-1-one, 2-hydroxy-3-methyl-	Ketones	+	Zhang et al., 2016; Ru et al., 2015
Acetaldehyde, hydroxy-	Aldehydes	–	Zhang et al., 2016; Chen et al., 2014b; Zheng et al., 2012
Furfural	Furans	–	Zhang et al., 2016; Ru et al., 2015; Chen et al., 2014b; Zheng et al., 2012; Meng et al., 2012
2-furanmethanol	Furan	No effects	Ru et al., 2015; Zheng et al., 2012
2(5H)-furanone	Furans	No effects	Ru et al., 2015; Chen et al., 2014b
Benzofuran, 2,3-dihydro-	Furans	–	Zhang et al., 2016; Chen et al., 2014
Levoglucose	Anhydrosugars	+	Zhang et al., 2016; Chen et al., 2014b; Zheng et al., 2012
Lignin-derived compounds			
Phenol	Phenols	+	Chen et al., 2014b; Chen et al., 2015; Zheng et al., 2012
Methylphenol	Phenols	+	Ru et al., 2015; Chen et al., 2015; Zheng et al., 2012
1,2-benzendiol	Phenols	+	Chen et al., 2015
Methoxyphenol	Phenols	+	Zhang et al., 2016; Chen et al., 2014b
Phenol, 2-methoxy-4-methyl-	Phenols	+	Zhang et al., 2016; Chen et al., 2014b; Chen et al., 2014
2-Methoxy-4-vinylphenol	Phenols	–	Ru et al., 2015; Chen et al., 2014b
2-methoxy-4-(1-propenyl)-phenol	Phenols	–	Zhang et al., 2016; Ru et al., 2015; Chen et al., 2014b
Extractives derived compounds			
Octadecanoic acid	Lipids	–	Zhang et al., 2016
n-Hexadecanoic acid	Lipids	–	Zhang et al., 2016

biomass, a promising method to produce renewable hydrocarbons, has received particular interest in recent years. In this process, the biomass-derived vapors are upgraded *in situ* into deoxygenated products in the presence of zeolites and other catalysts through enhanced dehydration, decarbonylation, and decarboxylation reactions. Yields of hydrocarbons increase and coke production decrease with increasing H/C_{eff} ratios of feedstock ($H/C_{eff} = H-2*O/C$) (Dickerson and Soria, 2013). In this regard, catalytic pyrolysis of torrefied biomass is more attractive due to increased H/C ratio and decreased O/C ratio in the feedstock. A few studies confirmed that torrefaction could improve the hydrocarbon yield (Neupane et al., 2015; Zheng et al., 2015; Zheng et al., 2014; Srinivasan et al., 2012; Srinivasan et al., 2014; Adhikari et al., 2014; Hilten et al., 2013; Hilten et al., 2012). It is also reported that carbohydrate portions of biomass (hemicellulose and cellulose) are the major contributor to hydrocarbons, whereas lignin negatively affects the hydrocarbon yields because lignin predominantly ends up with char and coke during catalytic pyrolysis (Wang et al., 2014). The ability of biomass components to producing hydrocarbons is ranked as follows: cellulose > hemicellulose > lignin. Therefore, mild torrefaction parameters are recommended to minimize the loss of cellulose in order to maximize the hydrocarbon yields from catalytic pyrolysis.

4. Techno-Economic Prospects of Thermal Pretreatment

It is well-known that thermal pretreatment (torrefaction and densification) improves energy density, hydrophobic properties, uniformity, and grindability. Pretreatment means adding a process in the whole bioenergy production process chain. The net benefit of thermal pretreatment is established based on the assumption that the additional capital and operational cost of thermal pretreatment is compensated by the reduced cost or improved efficiencies in other processing units. Transportation and logistics steps benefit from the higher energy density of the fuel, allowing for decreased costs per energy unit of fuel transported or stored. In practice, torrefied biomass is usually required to be densified due to increased volumetric density and mechanical strength of the product, especially for long-distance transportation. Hence, the combination of torrefaction and densification offers significant benefits to logistics. The improvement in grindability can reduce the energy consumption required for size reduction by 70–90% compared with untreated biomass (Bergman, 2005). This is especially important for the co-fired power plant. Pretreated biomass is expected to provide benefits during energy utilization segments, especially thermochemical applications due to higher energy content, lower oxygen content and lower moisture content.

Techno-economic assessment (TEA) provides an evaluation of a specific process from an economic perspective through a comprehensive analysis of mass and energy balance, process efficiency, and cost. TEA was applied to examine the economics of bioenergy production chains with a focus on the impacts of pretreatments (torrefaction, densification, and combined torrefaction and densification) on the cost of bioenergy production (Uslu et al., 2008). The study was based on the assumption that the biomass feedstock was harvested in Brazil and converted through entrained flow gasification and Fisher-Tropsch liquid production-biomass integrated gasification combined cycle (BIGCC), fluidized bed combustion, and co-firing in existing power plant in Europe. The analysis indicated that torrefaction combined with densification provided the lowest cost of fuels (50 €/ton), followed by densification (54 €/ton) and torrefaction (58 €/ton). Table 6 compared the cost of fuel production, indicating that combined torrefaction and densification was superior to densification alone to reduce the overall production cost.

Table 6. Costs of bioenergy production (Uslu et al., 2008).

	Intermediate delivered to harbor (€/GJ$_{HHV}$)	FT-liquid fuel (€/ GJ$_{HHV}$)	Power (BIGCC) (€/kWh)	Power (combustion) (€/kWh)	Power (co-firing) (€/kWh)
Torrefied biomass pellet	3.3	7.4	4.6	7.7	4.6
Pellet	3.9	7.9	5.5	8.2	4.8

The economic analysis of an integrated compact moving-bed reactor based torrefaction plant (Fig. 5) indicated that the production costs for woody biomass torrefied pellets were between 3.3 and 4.8 $/GJ$_{LHV}$ for the short term production when the plant's capacity varied from 50 to 250 ton/year (Batidzirai et al., 2013). The production cost was estimated to further decrease from 2.1 to 5.1 $/GJ$_{LHV}$ in the long-term production at a much higher capacity (500 ton/year) due to scaling up and technical learning. The produced torrefied biomass pellet was claimed to be economically competitive with traditional pellets and could satisfy the commercial quality standards. In another study, an economic evaluation of full-scale production of torrefied biomass pellets was modeled based on a case study with 100,000 metric tons/year (Pirraglia et al., 2013). The production costs distribution indicated that delivered cost of biomass accounted for the largest portion (29%) of the production cost, followed by the depreciation of the equipment (25%), and the cost of binder (13%). The selection of binder materials had a significant impact on the economics. Among four binders investigated, distillers dried grains (DDGs) and feed corns were superior to steam and soybeans in improving the profitability of the process. The overall cost (production costs + transportation costs) of torrefied biomass pellets was estimated to be $282/metric ton, which is comparable to the average price of regular biomass pellets. Winjobi et al. (2016) had compared the bio-oil

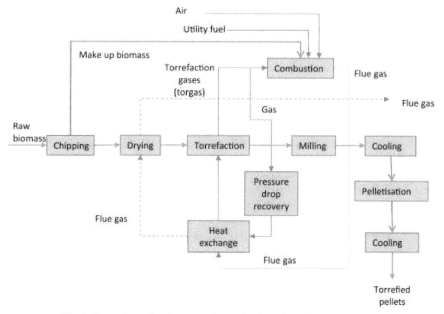

Fig. 5. Overview of an integrated torrefaction plant (Bergman, 2005).

produced from a two-step pyrolysis (torrefaction followed by pyrolysis) and a one-step fast pyrolysis. The economic analysis showed that the addition of condensable volatiles from torrefaction during two-step pyrolysis increased the overall yield of bio-oil, thus resulting in a lower price of bio-oil on a volume basis ($1.04–$1.27/gal) as compared to bio-oil produced from one-step fast pyrolysis ($1.32/gal). However, the bio-oil energy content of the two-step pyrolysis was lower than that of the one-step pyrolysis. Chai et al. (2016) conducted an economic analysis on the production of green BTEX (benzene, toluene, ethylbenzene, and xylenes) from decentralized torrefaction combined with centralized catalytic pyrolysis. The sensitivity analysis revealed that torrefaction temperature and residence time significantly affect the torrefied biomass yield, transportation cost, BTEX yield, and BTEX production cost. The minimum production cost of BTEX was $1271/tonne from torrefied biomass (239°C for 34 min) as compared to the cost of $1423/tonne from untreated biomass. The cost saving was a result of reduced transportation cost and increased BTEX yield due to the reduced oxygen content in the torrefied biomass. Torrefaction also reduced the net production cost when the distance of transporting feedstock increased. However, it should be noted that thermal pretreatment is still an emerging technology that has not been fully commercialized. Hence, cost estimates have high uncertainties due to lack of actual commercial-scale operation data and other unexpected technical challenges. Additional analysis and performance data from the large-scale operation are needed to provide more information for more accurate evaluation of economic benefits of biomass thermal pretreatments.

In a recent study by Winjobi et al. (2016; 2017a; 2017b), the effects of torrefaction on fast pyrolysis and subsequent bio-oil upgrading were investigated using 1000 dry metric tons/day of loblolly pine through a fast pyrolysis unit operating at 530°C and three torrefaction temperatures of 290, 310, and 330°C. Three scenarios with different combinations of the primary heat input resources (fossil fuels, pyrolysis char, and condensable liquid from torrefaction) and production goals (maximizing bio-oil yield or bio-oil heating value) were also investigated (Table 7). The lowest minimum selling price (MSP) of bio-oil on a volume basis is 1.04 $/gal obtained from Scenario 2 of the two-step pyrolysis at the torrefaction temperature of 330°C, compared to 1.32 $/gal for Scenario 1 of the one-step pyrolysis without torrefaction. When total bio-oil yield is maximized, that is, the condensable liquid from torrefaction is blended with bio-oil produced from pyrolysis, the use of *in situ* produced char for heat input is more profitable than turning it into revenue credits (Winjobi et al., 2016). Another interesting finding was that the heating value of the bio-oil under Scenario 3 (without torrefaction condensable blends) was always higher than that of the bio-oil under Scenario 2 (with torrefaction condensable blends), indicating that there is a trade-off between the bio-oil selling price and bio-oil quality. The minimum bio-oil selling price was 16.41 $/GJ obtained in Scenario 2 of the two-step pyrolysis at the torrefaction temperature of 330°C. The similar bio-oil selling price of 16.89 $/GJ was obtained in Scenario 2 of the one-step pyrolysis. In a follow-up economic assessment study (Winjobi et al., 2017b), torrefaction temperature was found to have a dominant effect on the yield of hydrocarbon fuels (ranging from 25.5 to 11.1 dry wt%). High torrefaction temperature decreased the yield of total liquid intermediates (condensable liquids from torrefaction plus bio-oil from pyrolysis). Accordingly, the MSP of hydrocarbon fuels increased significantly as the torrefaction increased, ranging from 4.71 $/gal (Scenario 2, 290°C) to 6.84 $/gal (Scenario 3, 330°C). However, the lowest MSP obtained with the inclusion of torrefaction was almost the same as that obtained with one-step pyrolysis. The environmental impact study of torrefaction (Winjobi et al., 2017a) revealed that torrefaction helped reduce the global warming potential (GWP). As shown in Fig. 7, the GWP for heat integrated hydrocarbon fuel process decreased dramatically when torrefaction was included. Further analysis indicated that the reduction of GWP was a result of a reduction in energy supply for size reduction due to the improvement of grindability. The GWP also reduced increasing torrefaction temperature due to the increase in credits from char and condensable liquids, and more significant reductions can be achieved with the heat integrated processes. The lowest GWP

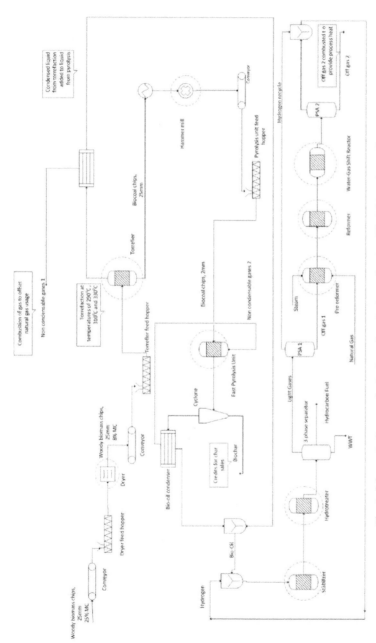

Fig. 6. Schematic diagram for hydrocarbon fuels production (Winjobi et al., 2017b).

Table 7. Summary of three pyrolysis scenarios (Winjobi et al., 2017b).

Scenarios	Primary heat input source	Production Objective
Scenario 1[a]	Fossil fuel	Maximizing bio-oil yield
Scenario 2	Pyrolysis char	Maximizing bio-oil yield
Scenario 3	Condensable liquid from torrefaction	Maximizing bio-oil heating value

[a]The unused char is sold as a byproduct of revenue credit.

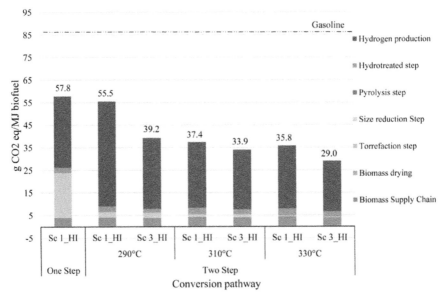

Fig. 7. GWP for heat integrated hydrocarbon fuel production pathways (energy allocation) (Winjobi et al., 2017a).

(29 g CO_2 equiv/MJ biofuel) was achieved via a two-step heat integrated pathway with torrefaction at 330°C. From economic and environmental perspectives, a two-step heat integrated hydrocarbon fuel production process, which includes a low temperature (290°C) torrefaction, fast pyrolysis, and bio-oil upgrading was recommended.

5. Concluding Remarks

Thermal pretreatments of biomass (such as torrefaction) and densification have been studied extensively in recent years to investigate their potential of improving the performance of bioenergy and fuel production via thermochemical conversions. Although torrefied biomass exhibits enhanced fuel properties, such as higher energy density, lower moisture, and oxygen content, the impacts of torrefaction on gasification performance in terms of energy efficiencies and syngas quality are not consistent, since gasification performance is also highly dependent on other operation parameters, such as gasifier design, equivalence ratio, and gasifying agent. Integrated biomass torrefaction with entrained-flow gasification seems to be a promising technology due to minimum theoretical exergy loss, but its high energy input hampers its economic feasibility. Densified biomass improves bulk density, uniformity, and mechanical strength, making size reductions and feeding of fuels more efficient and less energy-intensive. The uniform particle size and shape improves the flow characteristics of fuel in the gasifier, resulting in enhanced gasification performance. However, the increased particle density of torrefied biomass also decreases heat transfer rate, possibly leading to incomplete gasification. Therefore, the pellet size needs to be carefully selected to optimize net process efficiency.

Torrefaction also varies the chemical structure and devolatilization kinetics of biomass and hence affects pyrolysis. Torrefaction process is dominated by thermal decomposition of hemicellulose, activation of cellulose, and depolymerization of lignin. Increasing severity of torrefaction also favors the repolymerization and cross-linking of primary pyrolysis products of cellulose and lignin, resulting in more char production. The bio-oil produced from torrefied biomass is characterized by higher contents of anhydrosugars, phenols and guaiaols, and lower contents of furans, ketones, acids, and lignin monomers.

Thermal pretreatment also offers potential to save production cost by improved handling, transportation, and storage properties. Torrefied biomass pellets are commercially attractive commodities that have potential to substitute regular pellets for bioenergy production. Additional operational data from pilot and demonstration scale tests are needed to accurately evaluate the economic impacts of torrefaction and densification on bioenergy production chains.

References

Adhikari, S., V. Srinivasan and O. Fasina. 2014. Catalytic pyrolysis of raw and thermally treated lignin using different acidic zeolites. Energy & Fuels.

Ammendola, P., R. Chirone, F. Miccio, G. Ruoppolo and F. Scala. 2011. Devolatilization and attrition behavior of fuel pellets during fluidized-bed gasification. carbon (wt%), 49(55.1): 58.7.

Ammendola, P., R. Chirone, G. Ruoppolo and F. Scala. 2013. The effect of pelletization on the attrition of wood under fluidized bed combustion and gasification conditions. Proceedings of the Combustion Institute 34(2): 2735–2740.

Babu, B. V. 2008. Biomass pyrolysis: a state-of-the-art review. Biofuels Bioproducts & Biorefining-Biofpr. 2(5): 393–414.

Bach, Q.-V. and Ø. Skreiberg. 2016. Upgrading biomass fuels via wet torrefaction: A review and comparison with dry torrefaction. Renewable and Sustainable Energy Reviews 54: 665–677.

Batidzirai, B., A. Mignot, W. Schakel, H. Junginger and A. Faaij. 2013. Biomass torrefaction technology: Techno-economic status and future prospects. Energy 62: 196–214.

Bergman, P. 2005. Combined torrefaction and pelletisation. The TOP process.

Bergman, P. C. and J. H. Kiel. 2005. Torrefaction for biomass upgrading. In Proc. 14th European Biomass Conference, Paris, France.

Bhattacharya, A., D. Manna, B. Paul and A. Datta. 2011. Biomass integrated gasification combined cycle power generation with supplementary biomass firing: Energy and exergy based performance analysis. Energy 36(5): 2599–2610.

Bhattacharya, S. C., P. A. Salam, H. L. Pham and N. H. Ravindranath. 2003. Sustainable biomass production for energy in selected Asian countries. Biomass and Bioenergy 25(5): 471–482.

Bridgwater, A.V. 2012. Review of fast pyrolysis of biomass and product upgrading. Biomass and Bioenergy 38: 68–94.

Bronson, B., F. Preto and P. Mehrani. 2012. Effect of pretreatment on the physical properties of biomass and its relation to fluidized bed gasification. Environmental Progress & Sustainable Energy 31(3): 335–339.

Cao, L., X. Yuan, L. Jiang, C. Li, Z. Xiao, Z. Huang, X. Chen, G. Zeng and H. Li. 2016. Thermogravimetric characteristics and kinetics analysis of oil cake and torrefied biomass blends. Fuel 175: 129–136.

Carpenter, D., T. L. Westover, S. Czernik and W. Jablonski. 2014. Biomass feedstocks for renewable fuel production: a review of the impacts of feedstock and pretreatment on the yield and product distribution of fast pyrolysis bio-oils and vapors. Green Chemistry 16(2): 384–406.

Chai, L., C. M. Saffron, Y. Yang, Z. Zhang, R. W. Munro and R. M. Kriegel. 2016. Integration of decentralized torrefaction with centralized catalytic pyrolysis to produce green aromatics from coffee grounds. Bioresource Technology 201: 287–292.

Chen, D., J. Zhou and Q. Zhang. 2014b. Effects of torrefaction on the pyrolysis behavior and bio-oil properties of rice husk by using TG-FTIR and Py-GC/MS. Energy & Fuels 28(9): 5857–5863.

Chen, D., Z. Zheng, K. Fu, Z. Zeng, J. Wang and M. Lu. 2015. Torrefaction of biomass stalk and its effect on the yield and quality of pyrolysis products. Fuel 159: 27–32.

Chen, Q., J. S. Zhou, B. J. Liu, Q. F. Mei and Z. Y. Luo. 2011. Influence of torrefaction pretreatment on biomass gasification technology. Chinese Science Bulletin 56(14): 1449–1456.

Chen, W.-H. and P.-C. Kuo. 2011. Isothermal torrefaction kinetics of hemicellulose, cellulose, lignin and xylan using thermogravimetric analysis. Energy 36(11): 6451–6460.

Chen, W.-H., C.-J. Chen, C.-I. Hung, C.-H. Shen and H.-W. Hsu. 2013. A comparison of gasification phenomena among raw biomass, torrefied biomass and coal in an entrained-flow reactor. Applied Energy 112: 421–430.

Chen, W. H., C. J. Chen and C. I. Hung. 2013. Taguchi approach for co-gasification optimization of torrefied biomass and coal. Bioresource Technology 144: 615–622.

Chen, W.-H., M.-Y. Huang, J.-S. Chang and C.-Y. Chen. 2014. Thermal decomposition dynamics and severity of microalgae residues in torrefaction. Bioresource Technology 169: 258–264.

Chen, W.-H., J. Peng and X. T. Bi. 2015. A state-of-the-art review of biomass torrefaction, densification and applications. Renewable and Sustainable Energy Reviews 44: 847–866.

Chew, J. J. and V. Doshi. 2011. Recent advances in biomass pretreatment—Torrefaction fundamentals and technology. Renewable and Sustainable Energy Reviews 15(8): 4212–4222.

Ciolkosz, D. and R. Wallace. 2015. A review of torrefaction for bioenergy feedstock production. Biofuels, Bioproducts and Biorefining 5(3): 317–329.

Claude, V., C. Courson, M. Köhler and S. p. D. Lambert. Overview and essentials of biomass gasification technologies and their catalytic cleaning methods. Energy & Fuels 30(11): 8791–8814.

Clausen, L. R. 2014. Integrated torrefaction vs. external torrefaction—A thermodynamic analysis for the case of a thermochemical biorefinery. Energy 7: 597–607.

Couhert, C., S. Salvador and J. M. Commandre. 2009. Impact of torrefaction on syngas production from wood. Fuel 88(11): 2286–2290.

Dai, J., S. Sokhansanj, J. R. Grace, X. Bi, C. J. Lim and S. Melin. 2008. Overview and some issues related to co-firing biomass and coal. The Canadian Journal of Chemical Engineering 86(3): 367–386.

Deng, J., G. J. Wang, J. H. Kuang, Y. L. Zhang and Y. H. Luo. 2009. Pretreatment of agricultural residues for co-gasification via torrefaction. Journal of Analytical and Applied Pyrolysis 86(2): 331–337.

Dickerson, T. and J. Soria. 2013. Catalytic fast pyrolysis: a rev. Energies 6(1): 514–538.

Du, S. L., H. P. Yang, K. Z. Qian, X. H. Wang and H. P. Chen. 2014. Fusion and transformation properties of the inorganic components in biomass ash. Fuel 117: 1281–1287.

Dudynski, M., J. C. van Dyk, K. Kwiatkowski and M. Sosnowska. 2015. Biomass gasification: Influence of torrefaction on syngas production and tar formation. Fuel Processing Technology 131: 203–212.

Erlich, C. and T. H. Fransson. 2011. Downdraft gasification of pellets made of wood, palm-oil residues respective bagasse: experimental study. Applied Energy 88(3): 899–908.

Fahmi, R., A. V. Bridgwater, I. Donnison, N. Yates and J. Jones. 2008. The effect of lignin and inorganic species in biomass on pyrolysis oil yields, quality and stability. Fuel 87(7): 1230–1240.

Faizal, H. M., M. R. A. Rahman and Z. A. Latiff. 2016. Review on densification of palm residues as a technique for biomass energy utilization. Jurnal Teknologi 78(9-2): 9–18.

Fisher, E. M., C. Dupont, L. I. Darvell, J. M. Commandre, A. Saddawi, J. M. Jones, M. Grateau, T. Nocquet and S. Salvador. 2012. Combustion and gasification characteristics of chars from raw and torrefied biomass. Bioresource Technology 119: 157–165.

Gao, W., R. F. Zhao, D. J. Liu and I. Destech Publications. 2016. Literature Review on Pretreatments of Lignocellulosic Biomass. International Conference on Material Science and Civil Engineering, Msce 2016, 205–212.

Gautam, G., S. Adhikari, S. Thangalazhy-Gopakumar, C. Brodbeck, S. Bhavnani and S. Taylor. 2011. Tar analysis in syngas derived from pelletized biomass in a commercial stratified downdraft gasifier. BioResources 6(4): 4653–4661.

Habibi, R., J. Kopyscinski, M. S. Masnadi, J. Lam, J. R. Grace, C. A. Mims and J. M. Hill. 2012. Co-gasification of biomass and non-biomass feedstocks: synergistic and inhibition effects of switchgrass mixed with sub-bituminous coal and fluid coke during CO_2 gasification. Energy & Fuels 27(1): 494–500.

Hilten, R. N., R. A. Speir, J. R. Kastner, S. Mani and K. Das. 2012. Effect of torrefaction on bio-oil upgrading over HZSM-5. Part 2: Byproduct formation and catalyst properties and function. Energy & Fuels 27(2): 844–856.

Hilten, R. N., R. A. Speir, J. R. Kastner, S. Mani and K. Das. 2013. Effect of torrefaction on bio-oil upgrading over HZSM-5. Part 1: Product yield, product quality, and catalyst effectiveness for benzene, toluene, ethylbenzene, and xylene production. Energy & Fuels 27(2): 830–843.

Hosoya, T., H. Kawamoto and S. Saka. 2007. Cellulose–hemicellulose and cellulose–lignin interactions in wood pyrolysis at gasification temperature. Journal of Analytical and Applied Pyrolysis 80(1): 118–125.

International Energy Agency. 2007. IEA Energy Technology Essentials: Biomass for Power Generation and CHP. Available from: https://www.iea.org/publications/freepublications/publication/essentials3.pdf.

Kaliyan, N. and R. V. 2010. Morey, densification characteristics of corn cobs. Fuel Processing Technology 91(5): 559–565.

Khan, A. A., W. de Jong, P. J. Jansens and H. Spliethoff. 2009. Biomass combustion in fluidized bed boilers: Potential problems and remedies. Fuel Processing Technology 90(1): 21–50.

Kulkarni, A., R. Baker, N. Abdoulmomine, S. Adhikari and S. Bhavnani. 2016. Experimental study of torrefied pine as a gasification fuel using a bubbling fluidized bed gasifier. Renewable Energy 93: 460–468.

Kumar, A., D. D. Jones and M. A. Hanna. 2009. Thermochemical biomass gasification: a review of the current status of the technology. Energies 2(3): 556–581.

Kuo, P.-C. and W. Wu. 2014. Design, optimization and energetic efficiency of producing hydrogen-rich gas from biomass steam gasification. Energies 8(1): 94–110.

Kuo, P.-C., W. Wu and W.-H. Chen. 2014. Gasification performances of raw and torrefied biomass in a downdraft fixed bed gasifier using thermodynamic analysis. Part B: Fuel 117: 1231–1241.

Kuo, P.-C. and W. Wu. 2016. Design and thermodynamic analysis of a hybrid power plant using torrefied biomass and coal blends. Energy Conversion and Management 111: 15–26.

Kwapinska, M., G. Xue, A. Horvat, L. P. Rabou, S. Dooley, W. Kwapinski and J. J. Leahy. 2015. Fluidized bed gasification of torrefied and raw grassy biomass (Miscanthus × gigantenus). The effect of operating conditions on process performance. Energy & Fuels 29(11): 7290–7300.

Li, H., X. Liu, R. Legros, X. T. Bi, C. Lim and S. Sokhansanj. 2012. Torrefaction of sawdust in a fluidized bed reactor. Bioresource Technology 103(1): 453–458.

Li, J., A. Brzdekiewicz, W. Yang and W. Blasiak. 2012. Co-firing based on biomass torrefaction in a pulverized coal boiler with aim of 100% fuel switching. Applied Energy 99: 344–354.

Martinez, J. D., K. Mahkamov, R. V. Andrade and E. E. S. Lora. 2012. Syngas production in downdraft biomass gasifiers and its application using internal combustion engines. Renewable Energy 38(1): 1–9.

McKendry, P. 2002. Energy production from biomass (part 3): gasification technologies. Bioresource Technology 83(1): 55–63.

Meng, J., J. Park, D. Tilotta and S. Park. 2012. The effect of torrefaction on the chemistry of fast-pyrolysis bio-oil. Bioresource Technology 111: 439–446.

Mohan, D., C. U. Pittman and P. H. Steele. 2006. Pyrolysis of wood/biomass for bio-oil: a critical review. Energy & Fuels 20(3): 848–889.

Molino, A., S. Chianese and D. Musmarra. 2016. Biomass gasification technology: The state of the art overview. Journal of Energy Chemistry 25(1): 10–25.

Neupane, S., S. Adhikari, Z. Wang, A. Ragauskas and Y. Pu. 2015. Effect of torrefaction on biomass structure and hydrocarbon production from fast pyrolysis. Green Chemistry 17(4): 2406–2417.

Nunes, L., J. Matias and J. Catalão. 2014. A review on torrefied biomass pellets as a sustainable alternative to coal in power generation. Renewable and Sustainable Energy Reviews 40: 153–160.

Patel, M., X. Zhang and A. Kumar. 2016. Techno-economic and life cycle assessment on lignocellulosic biomass thermochemical conversion technologies: a review. Renewable and Sustainable Energy Reviews 53: 1486–1499.

Peng, J., X. Bi, S. Sokhansanj and C. Lim. 2013. Torrefaction and densification of different species of softwood residues. Fuel 111: 411–421.

Pinto, F., J. Gominho, R. N. Andre, D. Goncalves, M. Miranda, F. Varela, D. Neves, J. Santos, A. Lourenco and H. Pereira. 2017. Effect of rice husk torrefaction on syngas production and quality. Energy & Fuels 31(5): 5183–5192.

Pirraglia, A., R. Gonzalez, D. Saloni and J. Denig. 2013. Technical and economic assessment for the production of torrefied ligno-cellulosic biomass pellets in the US. Energy Conversion and Management 66: 153–164.

Prando, D., F. Patuzzi, P. Baggio and M. Baratieri. 2014. CHP gasification systems fed by torrefied biomass: assessment of the energy performance. Waste and Biomass Valorization 5(2): 147–155.

Prasad, L., P. Subbarao and J. Subrahmanyam. 2015. Experimental investigation on gasification characteristic of high lignin biomass (Pongamia shells). Renewable Energy 80: 415–423.

Prins, M. J., K. J. Ptasinski and F. J. J. G. Janssen. 2006. More efficient biomass gasification via torrefaction. Energy 31(15): 3458–3470.

Prins, M. J., K. J. Ptasinski and F. J. Janssen. 2007. From coal to biomass gasification: Comparison of thermodynamic efficiency. Energy 32(7): 1248–1259.

Ptasinski, K. J., M. J. Prins and A. Pierik. 2007. Exergetic evaluation of biomass gasification. Energy 32(4): 568–574.

Ptasinski, K. J. 2008. Thermodynamic efficiency of biomass gasification and biofuels conversion. Biofuels, Bioproducts and Biorefining 2(3): 239–253.

Ren, S., H. Lei, L. Wang, Q. Bu, S. Chen, J. Wu, J. Julson and R. Ruan. 2013. The effects of torrefaction on compositions of bio-oil and syngas from biomass pyrolysis by microwave heating. Bioresource Technology 135: 659–664.

Ren, S., H. Lei, L. Wang, Q. Bu, S. Chen and J. Wu. 2013. Thermal behaviour and kinetic study for woody biomass torrefaction and torrefied biomass pyrolysis by TGA. Biosystems Engineering 116(4): 420–426.

Ru, B., S. Wang, G. Dai and L. Zhang. 2015. Effect of torrefaction on biomass physicochemical characteristics and the resulting pyrolysis behavior. Energy & Fuels 29(9): 5865–5874.

Sadaka, S. S. 2013. Gasification of raw and torrefied cotton gin wastes in an auger system. Applied Engineering in Agriculture 29(3): 405–414.

Sarkar, M., A. Kumar, J. S. Tumuluru, K. N. Patil and D. D. Bellmer. 2013. Thermal devolatilization kinetics of switchgrass pretreated with torrefaction and densification. Transactions of The ASABE 57(1): 1–12.

Sarkar, M., A. Kumar, J. S. Tumuluru, K. N. Patil and D. D. Bellmer. 2014. Gasification performance of switchgrass pretreated with torrefaction and densification. Applied Energy 127: 194–201.

Sermyagina, E., J. Saari, B. Zakeri, J. Kaikko and E. Vakkilainen. 2015. Effect of heat integration method and torrefaction temperature on the performance of an integrated CHP-torrefaction plant. Applied Energy 149: 24–34.

Srinivasan, V., S. Adhikari, S. A. Chattanathan and S. Park. 2012. Catalytic pyrolysis of torrefied biomass for hydrocarbons production. Energy & Fuels 26(12): 7347–7353.

Srinivasan, V., S. Adhikari, S. A. Chattanathan, M. Tu and S. Park. 2014. Catalytic pyrolysis of raw and thermally treated cellulose using different acidic zeolites. BioEnergy Research 1–9.

Tanger, P., J. L. Field, C. E. Jahn, M. W. DeFoort and J. E. Leach. 2013. Biomass for thermochemical conversion: targets and challenges. Frontiers in Plant Science 4: 218.

Tapasvi, D., R. Khalil, G. Várhegyi, K.-Q. Tran, M. Grønli and Ø. Skreiberg. 2013. Thermal decomposition kinetics of woods with an emphasis on torrefaction. Energy & Fuels 27(10): 6134–6145.

Tapasvi, D., R. S. Kempegowda, K.-Q. Tran, Ø. Skreiberg and M. Grønli. 2015. A simulation study on the torrefied biomass gasification. Energy Conversion and Management 90: 446–457.

Tchapda, A. H. and S. V. Pisupati. 2014. A review of thermal co-conversion of coal and biomass/waste. Energies 7(3): 1098–1148.

Teixeira, G., L. Van de Steene, E. Martin, F. Gelix and S. Salvador. 2012. Gasification of char from wood pellets and from wood chips: Textural properties and thermochemical conversion along a continuous fixed bed. Fuel 102: 514–524.

Tumuluru, J. S., J. R. Hess, R. D. Boardman, C. T. Wright and T. L. Westover. 2012. Formulation, pretreatment, and densification options to improve biomass specifications for co-firing high percentages with coal. Industrial Biotechnology 8(3): 113–132.

Turn, S. Q., C. M. Kinoshita and D. M. Ishimura. 1997. Removal of inorganic constituents of biomass feedstocks by mechanical dewatering and leaching. Biomass & Bioenergy 12(4): 241–252.

Uslu, A., A. P. Faaij and P. C. Bergman. 2008. Pre-treatment technologies, and their effect on international bioenergy supply chain logistics. Techno-economic evaluation of torrefaction, fast pyrolysis and pelletisation. Energy 33(8): 1206–1223.

Van der Stelt, M., H. Gerhauser, J. Kiel and K. Ptasinski. 2011. Biomass upgrading by torrefaction for the production of biofuels: a review. Biomass and Bioenergy 35(9): 3748–3762.

Wang, K., K. H. Kim and R. C. Brown. 2014. Catalytic pyrolysis of individual components of lignocellulosic biomass. Green Chem. 16(2): 727–735.

Wang, S., X. Guo, K. Wang and Z. Luo. 2011. Influence of the interaction of components on the pyrolysis behavior of biomass. Journal of Analytical and Applied Pyrolysis 91(1): 183–189.

Weiland, F., M. Nordwaeger, I. Olofsson, H. Wiinikka and A. Nordin. 2014. Entrained flow gasification of torrefied wood residues. Fuel Processing Technology 125: 51–58.

Williams, C. L., T. L. Westover, R. M. Emerson, J. S. Tumuluru and C. Li. 2016. Sources of biomass feedstock variability and the potential impact on biofuels production. BioEnergy Research 9(1): 1–14.

Winjobi, O., D. R. Shonnard, E. Bar-Ziv and W. Zhou. 2016. Techno-economic assessment of the effect of torrefaction on fast pyrolysis of pine. Biofuels, Bioproducts and Biorefining 10(2): 117–128.

Winjobi, O., D. R. Shonnard and W. Zhou. 2017a. Production of hydrocarbon fuel using two-step torrefaction and fast pyrolysis of pine. Part 2: Life-cycle carbon footprint. ACS Sustainable Chemistry & Engineering.

Winjobi, O., D. R. Shonnard and W. Zhou. 2017b. Production of hydrocarbon fuel using two-step torrefaction and fast pyrolysis of pine. Part 1: Techno-economic analysis. ACS Sustainable Chemistry & Engineering 5(6): 4529–4540.

Yang, H., R. Yan, H. Chen, D. H. Lee and C. Zheng. 2007. Characteristics of hemicellulose, cellulose and lignin pyrolysis. Fuel 86(12-13): 1781–1788.

Yang, K. C., K. T. Wu, M. H. Hsieh, H. T. Hsu, C. S. Chen and H. W. Chen. 2013. Co-gasification of woody biomass and microalgae in a fluidized bed. Journal of the Taiwan Institute of Chemical Engineers 44(6): 1027–1033.

Yang, X. Q., X. J. Liu, H. X. Liu, X. M. Yue, J. P. Cao and M. Zhou. 2014. Synergy effect in co-gasification of lignite and char of pine sawdust. Acta Physico-Chimica Sinica 30(10): 1794–1800.

Yang, Z., M. Sarkar, A. Kumar, J. S. Tumuluru and R. L. Huhnke. 2014. Effects of torrefaction and densification on switchgrass pyrolysis products. Bioresource Technology 174(0): 266–273.

Yang, Z., A. Kumar and R. L. Huhnke. 2015. Review of recent developments to improve storage and transportation stability of bio-oil. Renewable and Sustainable Energy Reviews pp. 859–870.

Yoon, S. J., Y.-I. Son, Y.-K. Kim and J.-G. Lee. 2012. Gasification and power generation characteristics of rice husk and rice husk pellet using a downdraft fixed-bed gasifier. Renewable Energy 42: 163–167.

Zhang, S., Q. Dong, L. Zhang and Y. Xiong. 2016. Effects of water washing and torrefaction on the pyrolysis behavior and kinetics of rice husk through TGA and Py-GC/MS. Bioresource Technology 199: 352–361.

Zheng, A., Z. Zhao, S. Chang, Z. Huang, F. He and H. Li. 2012. Effect of torrefaction temperature on product distribution from two-staged pyrolysis of biomass. Energy & Fuels 26(5): 2968–2974.

Zheng, A., Z. Zhao, S. Chang, Z. Huang, X. Wang, F. He and H. Li. 2013. Effect of torrefaction on structure and fast pyrolysis behavior of corncobs. Bioresource Technology 128: 370–377.

Zheng, A., Z. Zhao, Z. Huang, K. Zhao, G. Wei, X. Wang, F. He and H. Li. 2014. Catalytic fast pyrolysis of biomass pretreated by torrefaction with varying severity. Energy & Fuels 28(9): 5804–5811.

Zheng, A., L. Jiang, Z. Zhao, Z. Huang, K. Zhao, G. Wei, X. Wang, F. He and H. Li. 2015. Impact of torrefaction on the chemical structure and catalytic fast pyrolysis behavior of hemicellulose, lignin, and cellulose. Energy & Fuels 29(12): 8027–8034.

CHAPTER 11

Hydrothermal Liquefaction
A Promising Technology for High Moisture Biomass Conversion

Ankita Juneja,[1,] Deepak Kumar[1] and Jaya Shankar Tumuluru[2]*

1. Introduction

A steady increase in energy demand, limited fossil fuel reserves, and environmental concerns from fossil fuel burning necessitates exploring an alternative renewable and environmentally friendly energy source. More than 80% of global energy is derived from fossil fuels and at the current rate of consumption, petroleum reserves will be depleted in the next few decades (Brennan and Owende, 2010; Guo et al., 2015a). Biofuels derived from biomass such as agricultural waste, dedicated energy crops, aquatic plants, forestry material, municipal waste, sewage sludge, animal waste, are promising and sustainable alternative energy sources with less environmental burdens. Being non-food materials, biofuel produced from these feedstocks also address the issue of "food vs. fuel", a major concern with first generation fuels (e.g., corn ethanol, soybean biodiesel). Several regions of the world which lack fossil fuel reserves have an abundant supply

[1] Agricultural and Biological Engineering, University of Illinois at Urbana–Champaign, Urbana - 61801.
[2] Biofuels Department, 750 University Blvd., Energy Systems Laboratory, Idaho National Laboratory, Idaho Falls, Idaho - 83415-3570.
* Corresponding author: ajuneja@illinois.edu

of biomass and energy produced from biomass can help in fulfilling their energy demand and boost their economy (Akhtar and Amin, 2011). Plant biomass, also called lignocellulosic feedstocks (e.g., agricultural residues, forestry waste, energy crops, etc.), are abundant in supply, renewable, and fix CO_2 through photosynthesis. The fuel produced from lignocellulosic biomass conversion are considered to have a high net energy ratio (ratio of energy in fuel and fossil energy used to produce fuel) and to produce significantly less greenhouse gas emissions than those produced from petroleum (Kumar and Murthy, 2012; Spatari et al., 2005). Aquatic plants such as microalgae provide advantages of high photosynthetic efficiency, high biomass yields, and faster growth, which make them an attractive source of biofuel production (Zou et al., 2009).

Several processes have been developed for biomass energy conversion and can be broadly classified into biochemical (such as alcoholic fermentation, anaerobic digestion, etc.) and thermochemical conversion processes (such as combustion, gasification, pyrolysis, and liquefaction). Although thermochemical conversions are more energy intensive due to their requirement of high temperatures and pressure, however, provide the advantage of higher rates of reaction. Most of the thermochemical processes involve high temperature (> 200°C) and pressure (> 5 MPa) treatment under oxygen-deprived environment. A large range of biomass with high moisture such as algal biomass, tropical grasses, municipal waste, and sewage sludge are not suitable feedstocks for gasification and pyrolysis, because of the requirement of dry feedstock or high energy penalties for water vaporization. Pre-drying using solar energy or mechanical dryers is one option. However, it adds up extra time and cost in the process (Akhtar and Amin, 2011). Hydrothermal technologies that process biomass at sub or supercritical water are the most feasible option for processing high moisture feedstocks. Based on temperature and pressure conditions, the hydrothermal processing can be divided into three broad regions: carbonization, liquefaction, catalytic gasification, and high-temperature gasification as shown in the phase diagram of water (Fig. 1). This chapter will focus only on the hydrothermal liquefaction process.

The hydrothermal liquefaction (HTL) involves biomass treatment in water at moderate temperatures (200–400°C) under pressure (5–25 MPa) for about 10–60 minutes, with or without a catalyst (Akhtar and Amin, 2011; Elliott et al., 2015; Peterson et al., 2008). Water at elevated temperature and pressure becomes non-polar, highly reactive, and miscible for organic components; thereby works as a catalyst for hydrolytic reactions. High pressure keeps water in the liquid state and the temperature is kept high to imitate the pyrolytic mechanism and produce gas, aqueous, and oily products. Due to change in dielectric properties of water at high-temperature water or supercritical water, insoluble components of biomass such as cellulose, lignin become solubilized which improve the hydrolytic

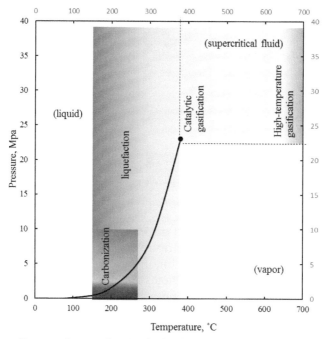

Fig. 1. Phase diagram of water showing hydrothermal processing regions (adapted from Peterson et al., 2008).

decomposition and fragmentation (Guo et al., 2015b; Kumar and Gupta, 2008). Detailed chemistry and reaction pathways during HTL are explained in the later sections of the chapter. Due to the ability to process wet biomass, a wide range of feedstocks (aquatic plants, lignocellulosic biomass, and waste) can be efficiently converted via HTL. For example, the most studied route for oil production from algae is lipid extraction with hexane but this process for biodiesel production requires the use of dry algae (85–95% solids), and only the drying step can add up to 30% of the total process cost (Becker, 1994). In contrast, HTL can use wet algae, saving the drying cost, avoiding a separate cell disruption process, and reducing about 50% energy requirements of the overall process. Other advantages of HTL over traditional lipid extraction process are the potential use of low lipid algae and less freshwater consumption (Guo et al., 2015b; Yu et al., 2011).

Products of HTL process include a high heating value biocrude or bio-oil, an aqueous fraction, gases, and solid residues (Fig. 2). Biocrude, with reduced oxygen content (10–20%), contains a mixture of a wide range of molecular weight compounds depending on the feedstock composition and process conditions (Vardon et al., 2011). Due to lower O:C and H:C ratios, the biocrude obtained after liquefaction is of superior quality and has relatively very high heating value (35–40 MJ/kg) compared to raw biomass

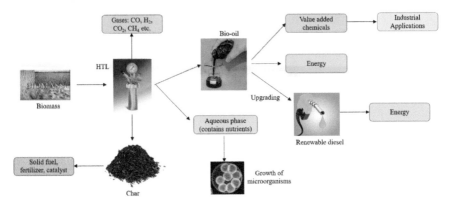

Fig. 2. Product and byproducts after liquefaction of biomass.

(15–25 MJ/kg). Bio-oil can be used directly to co-fire with coal or can be upgraded for high-quality fuels (Barreiro et al., 2013). For example, bio-oil can be hydrotreated to produce a highly usable form of fuel, renewable diesel, or jet fuel. Some of the advantages of renewable diesel over biodiesel and petroleum-based diesel are (1) direct use in diesel-powered vehicles without modifications, (2) compatibility with current diesel distribution infrastructure, (3) production using existing oil refinery capacity, (4) use of advance emission control devices due to ultra-low sulfur content and (5) better performance than diesel (Tier, 2010). Aqueous phase after HTL process contains a high amount of organic matter and nutrients. Minerals such as nitrates, phosphorus, iron, and potassium can be separated from the aqueous phase and reused for various purposes such as algae growth (Biller et al., 2012; Jena et al., 2011b; Zhou et al., 2013). The produced gases could be recirculated for various purposes such as CO_2 for microalgae production and H_2 for biocrude upgradation (Barreiro et al., 2013). Solid residues, also called as biochar, contain nutrients and can be used as a soil enhancer.

2. Role of Hydro-Media in HTL

Water plays a critical role in HTL. At normal temperature, water is a polar liquid and does not react with organic compounds. Elevated temperature (superheated water) leads to the breaking of strong hydrogen bonds which makes water less polar and behave more like an organic solvent and, consequently, highly reactive and miscible with organic components. Under these conditions, the dielectric constant (relative permittivity), ε_r, of water decreases substantially. For example, dielectric constant reduces from 78.85 to 19.66 with a temperature increase from 25°C to 300°C. Secondly, an increase in temperature increases the self-ionization (dissociation) of water. Water dissociates into hydronium ion (H_3O^+) and hydroxide ions (OH^-) as a reversible reaction. The ionic product of water (K_W), written as molar concentrations of the ionic products of water increases considerably ($1*10^{-14}$ to $1*10^{-12}$) with an increase in temperature from normal to subcritical range.

$$KW = [H_3O^+] [OH^-] \tag{1}$$

With the increase in the dissociation constant, water acts as a catalyst and accelerates the rate of both acid and base-catalyzed reactions. These two factors make water, at an elevated temperature, a good solvent for non-polar hydrophobic hydrocarbons and lead to fast and efficient reactions.

3. Effect of Feedstock on HTL Process and Products

3.1 Particle size

Particle size is considered to be an important parameter in any biochemical or thermochemical conversion. The purpose of reducing the particle size is to increase the surface area of the biomass and increase its accessibility for conversion. Although grinding is an energy-intensive step, making this process expensive, the cost incurred is justified with the increase in yield of the product due to the increased surface area. However, in case of hydrothermal liquefaction, the reduction of particle size is not observed to impact bio-oil yields. Zhang et al. (2009) did not observe any change in bio-oil yield with a change in particle size of biomass at 350°C. Moreover, the oil yield was observed to decrease from 14% to 7–9% with a reduction in particle size at supercritical water conditions. The reason for this observation could be that sub/super-critical water helps to overcome heat transfer limitations in HTL as it acts as an extractant and a heat transfer agent. It indicates that excessive grinding of biomass is unnecessary in case of hydrothermal liquefaction and the cost incurred for grinding is not justified.

3.2 Moisture content

As discussed earlier in this chapter, water plays a very important role in biomass conversion. High solid contents can result in capital cost savings, reduce downstream process (product recovery and wastewater treatment) cost, and improve energy efficiency (less material to heat), however, they have a negative effect on the biomass conversion efficiency. The high solids content is unfavorable for proper mixing in the HTL reactor and reduces the heat and mass transfer, which results in lower conversion and bio-oil yields. Karagoz et al. (2006) reported about 25% lower oil yield at water to solid ratio of 3 as compared to that at water to solid ratio 6 during liquefaction of pine sawdust. The effect concentration of solids is highly dependent on the type of biomass. In case of HTL of *Spirulina*, about 22% higher biocrude yield was observed at 20% solids than that at 10% solids, however, no further significant change was observed with an increase in solid content (up to 50%) (Jena et al., 2011a).

3.3 Composition

A range of biomass types can be used as a feedstock for hydrothermal liquefaction for biocrude production, including bio-waste, lignocellulosic biomass, and microalgae. Lignocellulosic biomass contains cellulose, hemicellulose, and lignin as their primary constituents, whereas microalgae are low in cellulose but mainly are composed of lipids, proteins, and other carbohydrates. Although the basic reaction mechanism of HTL process can be described as shown in Fig. 3, however, the actual mechanism and reaction pathway is very complex due to different behavior of individual biomass components. This section of the chapter describes the basic structure, reaction pathway, and kinetics of breakdown of major components in the feedstock during HTL process.

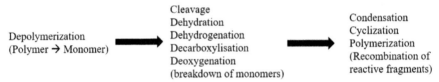

Fig. 3. Reaction mechanism for HTL of biomass (Demirbaş, 2000).

3.3.1 Lipids

Lipids are non-polar organic compounds insoluble in water at room temperature. However, as mentioned earlier, water under HTL conditions becomes non-polar and therefore miscible with these organic compounds. Triacylglycerides (TAGs), triesters of fatty acids and glycerol, are the most common form of lipids in biological systems. TAGs are readily hydrolyzed in hot compressed water without the use of a catalyst (Toor et al., 2011). The reaction pathway for TAG degradation is shown in Fig. 4. The bio-oil yield from HTL increases directly with the increase in lipid content of the biomass. TAGs undergo rapid hydrolysis and the highest fatty acid yields obtained with vegetable oil and soybean oil under hydrothermal conditions were close to 100% (Holliday et al., 1997; King et al., 1999). The kinetics of fatty acid esters hydrolysis under hydrothermal conditions follow the first-order reaction rate with rate constant $4.03*10^{-4}$ s^{-1} at 350°C, following the Arrhenius equation with an activation energy of 79 kJ mol^{-1} for palmitic acid (Fu et al., 2010).

The main component of lipid hydrolysis is fatty acids, which are highly stable under thermal conditions (Holliday et al., 1997). However, they can be partially degraded in supercritical conditions with long chain hydrocarbons. Watanable et al. (2006) investigated the decomposition of stearic acid under supercritical conditions and observed it be highly stable without the use

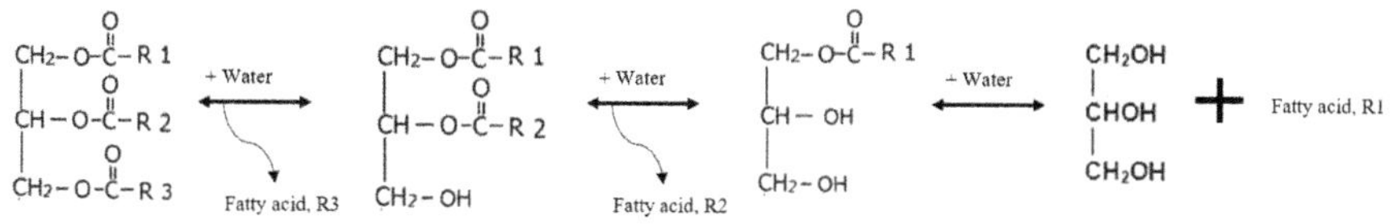

$$
\begin{array}{c}
\overset{O}{\underset{|}{CH_2-O-\overset{\|}{C}-R\,1}} \\
\overset{O}{\underset{|}{CH-O-\overset{\|}{C}-R\,2}} \\
\overset{O}{\underset{}{CH_2-O-\overset{\|}{C}-R\,3}}
\end{array}
\;\underset{\text{Fatty acid, R3}}{\overset{+\,\text{Water}}{\rightleftharpoons}}\;
\begin{array}{c}
\overset{O}{\underset{|}{CH_2-O-\overset{\|}{C}-R\,1}} \\
\overset{O}{\underset{|}{CH-O-\overset{\|}{C}-R\,2}} \\
CH_2-OH
\end{array}
\;\underset{\text{Fatty acid, R2}}{\overset{+\,\text{Water}}{\rightleftharpoons}}\;
\begin{array}{c}
\overset{O}{\underset{|}{CH_2-O-\overset{\|}{C}-R\,1}} \\
CH-OH \\
CH_2-OH
\end{array}
\;\overset{+\,\text{Water}}{\rightleftharpoons}\;
\begin{array}{c}
CH_2OH \\
CHOH \\
CH_2OH
\end{array}
\;+\;\text{Fatty acid, R1}
$$

Fig. 4. Breakdown mechanism of triglycerides during HTL process (Changi et al., 2015).

of a catalyst. However, the addition of KOH led to 32% decomposition yield. The increase in yield was attributed to the promotion of dissociation of CH_3COOH into CH_3COO^- and H^+ which made the decarboxylation faster (Watanabe et al., 2006). Another main component of lipids degradation is glycerol which is also considered to be an important energy source. Further degradation of glycerol in subcritical water (360°C, 34 MPa, 0–180 s) produces acrolein, therefore glycerol alone is not a suitable feedstock for HTL (Lehr et al., 2007).

3.3.2 Carbohydrates

Carbohydrates are the collective name for monosaccharides, oligosaccharides, and polysaccharides, with cellulose, hemicellulose, and starch being the most abundant carbohydrates. Under hydrothermal conditions, carbohydrates undergo fast hydrolysis resulting in the formation of smaller molecules such as glucose, which are further decomposed drastically. The overall reaction mechanism for carbohydrate degradation under high pressure and temperature is shown in Fig. 5. Kabyemela et al. (1999) observed 55% conversion of glucose after 2 seconds at 300°C and 90% conversion at 350°C. Glucose degradation undergoes 1st-order kinetics with Arrhenius activation energy of 88 kJ mol^{-1} (Kabyemela et al., 1997). Degradation of carbohydrates does not directly produce biocrude oil; rather liquefaction breaks the carbohydrates to polar water-soluble organics such as organic acids. The aldehyde and benzene type structures may produce larger hydrocarbons which are then part of biocrude fraction (Barreiro et al., 2013). The overall reaction mechanism of hydrolysis of carbohydrates to glucose and further degradation products are summarized in Fig. 5. The hydrothermal degradation of individual carbohydrates is explained below.

Starch

Starch is a polysaccharide of amylose (glucose molecules connected with α-1,4 bonds) and amylopectin (highly branched polymer with glucose molecules connected by α-1,4 bonds and α-1,6 glycoside branches). Although starches can be rapidly broken down to glucose monomers under elevated

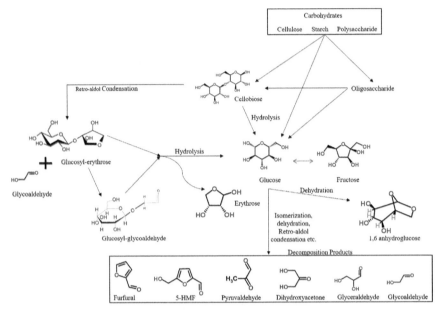

Fig. 5. Decomposition mechanism of carbohydrate under high temperature and pressure (Cantero et al., 2013; Yu et al., 2013).

conditions, the glucose yield is lower as compared to enzymatic hydrolysis, presumably due to further decomposition of glucose (Peterson et al., 2008). A 10 to 30-minute residence time at a temperature of 200°C gave a glucose yield of 63% from hydrothermal hydrolysis of starch from sweet potato. Increasing the severity of conditions and residence times, the yields of glucose decreased with an increase in 5-hydroxymethylfurfural (5-HMF) (Nagamori and Funazukuri, 2004). Li et al. (2016) investigated the effect of HTL process conditions on the conversion of starch from three sources (rice, potato, and sweet potato). Due to variation in the type of starches, reducing sugar yields (15–33%), and optimum process conditions were observed different for three starch sources (Li et al., 2016). A kinetic study on starch hydrolysis in subcritical water found starch decomposition as the first order Arrhenius equation with an activation energy of 147.9 kJ/mol, which was found to be less than that of cellulose degradation due to the fact that starch is less crystalline than glucose and glycosidic linkages in starch are much easier to hydrolyze than the β-1,4-glycosidic linkages in cellulose (Rogalinski et al., 2008).

Cellulose

Cellulose, the most abundant polymer on earth, is made up of a straight chain glucose units linked by β,1-4 glycosidic bonds. Anhydroglucose unit of cellulose contains three free hydroxyl groups that form numerous intra- and inter-molecule hydrogen bonds. Glucose chains bounded by these hydrogen bonds and van der Waal's forces make a highly crystalline structure which is insoluble in water and resistant to degradation. However, at near critical conditions, cellulose is rapidly hydrolyzed to its constituents. The breaking of the crystalline structure of cellulose under sub- and super-critical conditions is shown as a reaction pathway in Fig. 6.

The products obtained in bio-oil from cellulose HTL are different under acidic, neutral, and alkaline conditions (Yin and Tan, 2012). It was observed that under acidic and neutral conditions, 5-HMF was the main component of bio-oil, whereas under alkaline conditions C_{2-5} carboxylic acids seemed dominant. Under all the three conditions, the bio-oil yield increased up to 300°C but reduced drastically after that, however with different mechanisms. Under acidic conditions (pH 3), the bio-oil yield reduced from 34% at 300°C to 6.1% at 320°C with a corresponding increase in solids from 17.4% to 50%. A similar reduction was observed under neutral and alkaline conditions with a corresponding increase in their degradation products. The bio-oil degradation products for these conditions are shown in Table 1.

Rogalinski et al. (2008) investigated the hydrolysis of cellulose and observed almost 100% conversion being achieved within 2 minutes at

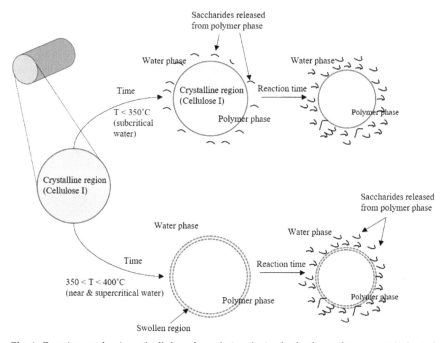

Fig. 6. Reaction mechanism of cellulose degradation during hydrothermal treatment (Adapted from Saski et al., 2004).

Table 1. Bio-oil components and degradation products from cellulose HTL (Yin and Tan, 2012).

Condition	Main component of bio-oil	Degradation product after further increase in temperature (> 300°C)
Acidic	5-HMF	Solid residue/char by polymerization
Neutral	5-HMF and carboxylic acids	Residual solids and gaseous products
Alkaline	C_{2-5} carboxylic acids	Gaseous products

280°C. However, between 250°C to 270°C, the rate of glucose decomposition exceeded the rate of glucose release by cellulose hydrolysis. Cellulose decomposition also shows a first-order kinetic reaction followed by Arrhenius equation with activation energy of 163.9 kJ/mol (Rogalinski et al., 2008).

Hemicellulose

Hemicellulose is an amorphous heterogeneous short chain polymer of sugars (xylose, arabinose, glucose, mannose, galactose) and uronic acids (4-O-methyl-glucuronic, galacturonic acids). Xylose is the dominant sugar in hemicellulose for most of the biomass. Due to amorphous nature, hemicellulose gets relatively easily hydrolyzed in water at high temperatures. Hemicellulose degradation leads to the formation of oligomers that are further broken down into monomers (mainly xylose) almost instantly. The monomers are further degraded to form smaller compounds. The activation energy of hemicellulose hydrolysis to xylose was estimated as 41 kJ/mol and for xylose degradation to smaller compounds as 32 kJ/mol, irrespective of the acid concentration used in hydrolysis (Bhandari et al., 1984). At 230°C, 34.5 MPa and 2 minutes residence time, almost complete hydrolysis was observed for hemicelluloses of various wood and herbaceous biomass materials (Mok and Antal Jr., 1992). Over 90% of hemicellulose was solubilized at 227°C for 135 s from almond shells (Pou-Ilinas et al., 1990).

3.3.3 Lignin

Lignin is the most complex compound in lignocellulosic feedstocks and is highly resistant to degradation. Chemical structure of lignin is cross-linked phenolic heteropolymers with p-hydroxyphenyl-propanoid units held together by C–C or C–O–C bonds. Trans-p-coumaryl alcohol, coniferyl alcohol, and sinapyl alcohol are the three monomeric building blocks of lignin (Bobleter, 1994). Lignin decomposes over a broad range of temperature as the scission of various oxygen functional groups occurs at different temperatures due to their thermal stabilities (Brebu and Vasile, 2010). Lignin degradation is shown to follow Arrhenius equation with variation in activation energy based on the method of isolating lignin (Schniewind, 1989). For instance, the activation energy for the lignin processed in sulfuric acid was 46 kJ/mol (Beall, 2007), whereas it was calculated to be 37 kJ/mol for Kraft-Pine lignin (Zhang et al., 2008). The similar activation energy (39.35 kJ/mol) was determined for guaiacol (model compound of lignin) decomposition in near-critical and supercritical water (Kanetake et al., 2007).

3.3.4 Proteins

Proteins are macromolecules made of amino acids and contribute to most of the nitrogen in the biomass. Most of this nitrogen gets incorporated in bio-oil during the HTL process, affecting the properties of bio-oil. The peptide bonds of proteins are more stable than α-1,4 and α-1,6 glycosidic bonds in glucose and starch, which leads to lower amino acids yield, even at optimal conditions (Rogalinski et al., 2008). The yields of amino acids in HTL are lower (10%) also due to subsequent degradation of amino acids in the hydrothermal environment (Peterson et al., 2008). Investigation of bovine serum albumin (BSA) as a model substance for protein showed highest amino acid concentration up to 310°C at 30 seconds residence time, after which degradation of amino acids started (Rogalinski et al., 2005). The degradation products were gaseous compounds (e.g., carbon dioxide, carbon monoxide, hydrogen, methane), low alkanes and alkenes, alcohols (up to C5), amides, aldehydes, and carboxylic acids. The two main reactions that occurred during hydrolysis of amino acids are deamination to produce ammonia and organic acids, and decarboxylation to produce carbonic acid and amines. The formation of long chain hydrocarbons and aromatic ring type structures is attributed to polymerization of these degradation products (Barreiro et al., 2013). The decomposition rate of amino acids can be described as a first-order kinetic reaction followed by Arrhenius equation with activation energy ranging from 148–166 kJ mol^{-1} for different amino acids (Sato et al., 2004). The degradation of amino acids in the hydrothermal environment is quick, for example, more than 70% glycine and alanine degrade in less than 20 and 30 seconds respectively at 350°C (Klingler et al., 2007). Reaction mechanism of alanine decomposition is shown in Fig. 7. At temperatures above 250°C, the amino acid decomposition rate has even been observed to exceed the hydrolysis rate (Rogalinski et al., 2005).

3.4 Elemental composition

Oxygen, carbon, and hydrogen content of the feedstock is an important parameter in deciding the quality of bio-oil produced, where low oxygen content, low moisture content, and high HHV are the most favorable properties. The composition of bio-oil is highly variable which affects the

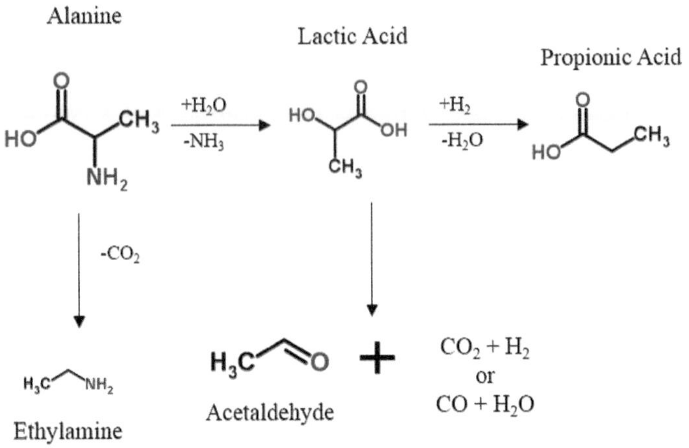

Fig. 7. Alanine decomposition reaction mechanism (adapted from Sato et al. 2004).

HHV of the product. Due to the increased carbon content in biocrude as compared with raw feedstock, the HHV of the biocrude increases. Table 2 summarizes and compares the HHV of bio-oil formed with their feedstock.

4. Effect of Process Variables on HTL Process

As discussed earlier, a variety of end products are formed at the end of hydrothermal decomposition of biomass. Bio-oil yield and quality of these end products are highly dependent on the process conditions such as operating temperature, residence time, pressure, heating rate, particle size, etc. This section of the chapter will discuss the effect of some of these important operating parameters on bio-oil yield.

4.1 Temperature

Temperature is an important factor in depolymerization of biomass which is a dominant reaction during initial stages of HTL (Fig. 3) and requires sufficient temperatures to overcome the activation energies of bond cessation. The optimum temperature depends on the type of biomass, however, in general, intermediate temperatures are shown to produce high bio-oil yields. Low temperatures (< 280°C) cause incomplete decomposition of biomass components and result in lower oil yields. Very high temperatures result in high operational cost as well as low oil yields because high temperatures support gas formation (Kosinkova et al., 2015; Zhong and Wei, 2004). For instance, an increase in temperature in HTL of *Spirulina platensis* up to 350°C at 60 minutes retention time produced the yield of bio-oil up to 39.9%, whereas a further increase in temperature to 380°C reduced the bio-oil yield to 36% (Jena et al., 2011a). It was observed

Table 2. Elemental composition and heating value of biocrude from various feedstocks.

Biomass	C (%)	H (%)	N (%)	O (%)	HHV (MJ/kg)	Biomass HHV	Reference
Spirulina	55–83	8.3–9.9	5.3–6.6	0.64–28.85	25.2–39.9		Jena et al. (2011a)
Spirulina	69.5–75.4	9.3–11.4	5.0–7.1	8.5–11.7	33.4–37.8		Ross et al. (2010)
Spirulina	72.7–75.4	9.2–10.8	4.6–7	8.7–10.9	34.8–36.8	21.2	Biller and Ross (2011)
Spirulina	68.9	8.9	6.5	14.9	33.2		Vardon et al. (2011)
Nannochloropsis	74.6–81.2	7.1–10.8	0.73–4.4	5.3–11.8	37–39	17.9	Brown et al. (2010)
Nannochloropsis	68.1–74.7	8.8–10.6	3.8–4.3	10.4–18.9	34.5–39	17.9	Biller and Ross (2011)
Chlorella	69.9–74	9.1–12.9	4.3–6.4	8.9–15.8	33.2–39.9		Ross et al. (2010)
Chlorella	70.7–73.6	8.6–10.7	4.9–5.3	10.7–14.8	33.2–37.1	23.2	Biller and Ross (2011)
Dunaliella tertiolecta	59.36	7.58	0.96	32.10	28.42		Zou et al. (2009)
Microcystis viridis	57.9–63.3	6.4–8.0	3.7–7.5	19.7–30	27–31		Yang et al. (2004)
Porphyridium	46.1–72.8	5.6–9.1	3.2–5.7	13.3	22.8–36.3	14.7	Biller and Ross (2011)
Mixed-culture algal biomass (AW)	59.1–71.4	7.58–8.98	2.5–4.92	15.4–30.3	25.8–33.3	12.9	Chen et al. (2014)
Swine manure	71.2	9.5	3.7	15.6	34.7		Vardon et al. (2011)
Anaerobic sludge	66.6	9.2	4.3	18.9	32		Vardon et al. (2011)
Bagasse	45–67	5.6–7.4	0.46–0.78	25–50	22–30	17.8	Kosinkova et al. (2015)
Pine wood	65.7–71.4	7.3–8	0	20.6–24.1	28.5–31.8	19.7	Xu and Etcheverry (2008)
Beech wood	76.7	7.1	0.1	16.1	34.9		Demirbaş (2000)

that the increase in temperature favored production of gases; gaseous products increased by 20% with an increase in temperature from 350 to 380°C. Another study with *Nanochloropsis* reported a maximum bio-oil yield of 43% at 350°C, and yield dropped with further increase in temperature due to increase in gaseous products (Brown et al., 2010). Similar results were observed by Chen et al. (2014) for liquefaction of mixed culture algae. The bio-oil yield was increased with a temperature rise from 260 to 300°C, however, a drop was observed with further temperature increase to 320°C (Chen et al., 2014). At very high temperatures, intermediate compounds are decomposed to gases and light volatile compounds are condensed into solid form (Jena et al., 2011a). Due to change in its ionic properties at temperatures above critical point (374°C, 22.1 MPa), water's catalytic power improves and these conditions favor the formation of gases (Akhtar and Amin, 2011; Jena et al., 2011a). Other than secondary decompositions, recombinations of free radical reactions at high temperatures cause char production and lower biocrude yields (Akhtar and Amin, 2011). Figure 8 shows the effect of operating temperature on bio-oil yields in various strains of microalgae (Anastasakis and Ross, 2011; Brown et al., 2010; Garcia Alba et al., 2011; Jena et al., 2011a; Shuping et al., 2010; Xiu et al., 2010). The yield is different due to the variation in chemical composition of each strain; however, the general trend of change in bio-oil yield with temperature is similar.

Similar trends have been observed from HTL of woody and lignocellulosic biomass. Karagoz et al. (2004) observed about 2.3 fold biocrude yield from sawdust HTL at 280°C compared to that at 180°C. A

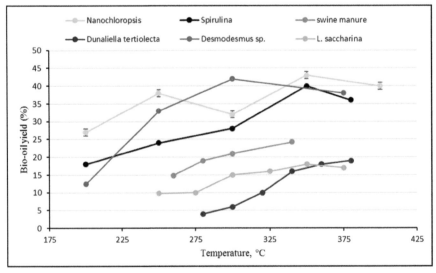

Fig. 8. Effect of temperature on bio-oil yields from HTL of various algal species.

shift of temperature from subcritical (200°C) to near supercritical (350°C) during HTL of Jack pine powder in the H_2 environment was observed to increase the oil yield from 19% to 44% with corresponding reduction in solid residues and gases (Xu and Etcheverry, 2008).

4.2 Pressure

Pressure is an important parameter during the HTL process as adequate pressure is necessary to maintain the single phase media at high temperatures. Single phase media avoids the large enthalpy energies required for phase change of solvents. Keeping pressure above the critical pressure of media, hydrolysis rates, and biomass dissolution can be controlled, which can boost favorable reaction pathways for liquid fuel production (Akhtar and Amin, 2011). Pressure also increases the solvent density which affects the solvent penetration into biomass and its conversion. At supercritical conditions, pressure does not have a significant effect on product yields because effect of pressure on water properties is negligible under these conditions.

4.3 Residence time

Residence time plays a critical role in the process efficiency and final bio-oil yield. Several researchers have investigated the effect of residence time on HTL of a variety of feedstocks and concluded that residence time directly controls the biomass conversion and final product composition. Similar to the temperature effect, biomass conversion increases continuously with the increase in reaction time. However, bio-oil yield increases up to a point and then starts decreasing for very long residence times. Karagoz et al. (2004) observed that during liquefaction of sawdust at low temperature (180°C), oil yield increased from 3.7% to 5.3% with an increase in residence time from 15 minutes to 60 minutes. However, during liquefaction at 250 and 280°C, oil yields at 60 minutes were observed about 16% and 22% lower respectively than those at 15 minutes. Long residence times at high temperatures may cause cracking of bio-oil and some of the intermediate products are converted to gases (Jena et al., 2011a). Due to the increase in condensation reactions, very long residence times also result in char formation. Jena et al. (2011) observed about 48% increase in gaseous products with a change in residence time from 60 minutes to 120 minutes. Yang et al. (2004) also observed lower oil yields for 60-minute residence than that at 30 minutes during liquefaction of *Microcystis viridis* at 300 and 340°C.

4.4 Solvents

Solvents are an alternative to water to aid in HTL. Various solvents have been investigated to increase the bio-oil yield with reduced oxygen content at mild operating conditions. Alcohols such as methanol and ethanol have lower critical temperatures and pressures than water, which could allow performing HTL at a low temperature and pressure, and results in energy efficient and economic feasible process. In addition, high molecular weight products from cellulose, hemicellulose, and lignin are expected to readily dissolve in these alcohols due to their low dielectric constant as compared to water. Yamazaki et al. (2006) investigated liquefaction of beech wood in various straight-chain alcohols in sub-critical and super-critical conditions. As much as 90% wood was found to liquefy at 350°C, regardless of the alcohol used. Cellulose was found to be more resilient to liquefaction in alcohols at 270°C but on increasing the temperature all the three components, cellulose, hemicellulose, and lignin were readily liquefied (Yamazaki et al., 2006). Miller et al. (1999) observed high conversions (~93%) of Kraft- and organosolv-derived lignin in KOH/ethanol at 290°C and the maximum value was reached in 10–15 minutes (Miller et al., 1999). Another study has investigated the effect of polar and non-polar solvents on the HTL of *Nanochloropsis* sp. and observed highest bio-oil yield (39%) with non-polar solvents, hexadecane, and decane. However, the highest amount of carbon observed in bio-oil (74–76%) was produced with polar solvents, chloroform, and dichloromethane (Valdez et al., 2011).

4.5 Catalyst

Catalysts are commonly used in HTL process to improve the bio-oil yield, increase the reaction rates, and reduce solids/char production. Both homogeneous and heterogeneous catalysts have been investigated for HTL. Homogeneous catalysts such as alkaline salts of Na and K are well used for algal biomass (Anastasakis and Ross, 2011; Duan and Savage, 2010; Shuping et al., 2010; Yang et al., 2004; Zhou et al., 2010) and lignocellulosic biomass (Karagöz et al., 2006; Song et al., 2004) and have shown to increase the bio-oil yield in case of multiple biomass (Table 3). One of the main catalytic effects of these alkaline salts is the acceleration of water gas shift reaction and favoring the formation of CO_2 and H_2. The H_2 gas produced acts as a reducing agent and thereby increases the heating value of the bio-oil (Toor et al., 2011). The catalytic activity of these salts are ranked as K_2CO_3 > KOH > Na_2CO_3 > NaOH for HTL of wood biomass, indicating potassium salts to be more effective than sodium salts. The activity was measured in terms of increase in liquid yield and reduction of solids/char (Karagöz et al., 2005). These catalysts are also believed to promote decarboxylation by inhibiting the dehydration of biomass. Increased decarboxylation of fatty acids was observed with homogeneous catalyst (KOH), where the breakdown of fatty acids increased from 2% to 32% (Watanabe et al., 2006). However, it is found that the bio-oil produced in the presence of homogeneous catalysts have increased oxygen content (Ross et al., 2010).

Table 3. Effect of some catalysts on bio-oil yield during hydrothermal liquefaction.

Biomass	Catalyst	Operating conditions	Bio-oil yield (%)	Ref.
Nanochlopsis sp.	No catalyst	350°C, 60 min Catalyst: Biomass loading 0.5; inert conditions	35	(Duan and Savage, 2010)
	Pd/C		57	
	Pt/C		48	
	Ru/C		52	
	Ni/SiO$_2$-Al$_2$O$_3$		53	
	CoMo/γ-Al$_2$O$_3$		54	
	zeolite		44	
Laminaria Saccharina	No catalyst		20	
	KOH	Catalyst: Biomass loading 0.5	12	(Anastasakis and Ross, 2011)
Enteromorpha prolifera	No catalyst	300°C, 30 min, 5% catalyst loading	20.4	(Zhou et al., 2010)
	Na$_2$CO$_3$		23	
D. tertiolecta cake	No catalyst	360°C, 50 min, 5% catalyst loading	20	(Shuping et al., 2010)
	Na$_2$CO$_3$		25.8	
Microcystis viridis	No catalyst	300°C, 30 min, 5% catalyst loading	23	(Yang et al., 2004)
	Na$_2$CO$_3$		30	
	No catalyst	340°C, 30 min, 5% catalyst loading	29	
	Na$_2$CO$_3$		33	
Corn stalks	No catalyst	1% catalyst loading	33.4	(Song et al., 2004)
	Na$_2$CO$_3$		47.2	
Wood	No catalyst	280°C, 15 min, 1% catalyst loading		(Karagöz et al., 2005)
	K$_2$CO$_3$			

Heterogeneous catalysts, on the other hand, are mostly used during hydrothermal gasification as these catalysts favor gas formation reducing the liquid production. Also, the use of heterogeneous catalysts bring up the issues of sintering, poisoning, fouling, or intra-particle diffusion limitations. The main products of catalytic hydrothermal gasification are methane and carbon dioxide with supported nickel catalysts and hydrogen and carbon dioxide with supported palladium and platinum catalysts (Toor et al., 2011). Some heterogeneous catalysts have also been investigated for HTL but due to their inclination towards gas formation, as opposed to liquid fuel production, they were not found to have much success. Christensen et al. (2014) investigated the effect of ZrO$_2$ during HTL of dried distiller's grains with solubles (DDGS) and observed that it did not have any significant impact on bio-oil yield, therefore was concluded to be a weak catalyst for HTL. Similarly, another heterogeneous catalyst, Raney-Ni and HZSM-5 type zeolite catalysts, also did not affect HTL of *Chlorella pyrenoidosa* (Zhang et al., 2013). Duan and Savage (2010) investigated 6 different heterogeneous catalysts (Table 3) under non-reducing and reducing conditions. Under the latter conditions (with H_2), no effect of catalyst was observed, however, with inert conditions, higher bio-oil yield was observed with the catalysts used but the increase was similar for all the catalysts (Duan and Savage, 2010). Contrary to this, Biller et al. (2011) found a slight increase in bio-oil yield but significant increase in the heating value and level of deoxygenation up to 10% was observed with heterogeneous catalysts used with microalgae (Biller et al., 2011).

As for bio-oil yields, catalysts have shown to improve the bio-oil properties by increasing the carbon content and HHV and reducing the oxygen content (Zhou et al., 2010).

Other than the important process parameters discussed above, some other parameters such as heating rate, particle size, initial pH, etc. also affect the bio-oil yield during the process. Most of the studies are performed at laboratory scale reactors that require large time for heating and do not specifically study the effect of heating rate. In general, higher heating rates during the HTL process are considered better as they improve the biomass decomposition and reduce char formation. Other than process yields, the product composition and quality are also highly affected by the process parameters, which is beyond the scope of this chapter. For additional details on the effect of process parameters, readers are referred to several good review papers on this topic (Akhtar and Amin, 2011; Guo et al., 2015b; Peterson et al., 2008).

5. Industrial Applications and Techno-Economic Feasibility

Although oxygen content of bio-oil produced from HTL is lower than that of pyrolysis bio-oil, it is still significantly high to necessitate its upgradation before it can be used with conventional diesel. Renewable diesel produced after upgradation of bio-oil can directly replace the petroleum diesel fuel. The purpose of upgrading is to reduce the oxygen content and viscosity and increase the heating value and stability. The upgradation of bio-oil can be done through hydro-deoxygenation, catalytic cracking of bio-oil, emulsification, and steam reforming. Bio-char is formed as a solid byproduct from HTL, which has some excellent properties as a soil conditioner. Bio-char increases the soil retention of nutrients and chemical for plants use and holds a high amount of carbon in it, thereby improving the soil and water quality (Yu et al., 2017). Soil fertility has been seen to improve with bio-char on tropical soils. This also prevents the leaching of nutrients into the groundwater. Bio-char also has good catalytic properties, which makes it usable in chemical industries. The aqueous phase obtained from HTL has high nutrient content in terms of nitrogen which makes it usable for the growth of microorganisms like microalgae and cyanobacteria. The reuse of these nutrients in the growth of microorganisms can reduce the biomass production cost and improve overall process economics. Various applications of bio-oil and the byproducts formed with HTL are the properties of the process that make it favorable for industrial consideration.

HTL uses wet feedstock and avoids the energy required for biomass drying, however, it is still an energy intensive process due to the requirement of very high temperature and pressures. The HTL process also eliminates the need for oil extraction, a cost-centered process, and accomplishes bio-oil production with high energy efficiency. Due to the high-pressure requirement, the capital cost for HTL process is expected to be higher than those from other thermochemical technologies. The overall process feasibility can be evaluated by looking at the complete process and considering the return expected on the final product and the byproducts. There are several factors affecting this analysis such as feedstock composition, process efficiencies, plant size, process inputs, and value of co-products. The techno-economic analysis is an important functional tool to assess the economic viability of any process at the industrial scale. Considering the huge commercial potential of HTL process, a few techno-economic studies have been conducted in recent years to estimate the process economics.

Zhu et al. (2014) performed techno-economic analysis for HTL and upgradation process for woody biomass at 2000 dry metric ton/day processing capacity using the current state of the art (base case), as well as considering potentially improved technologies (goal case). By considering the assumptions of improved technology and reducing organic loss to the water phase, oil production in goal was found to be about 43% cheaper ($4.44/GGE as compared to $2.52/GGE, respectively) (Zhu et al., 2014). Due to an additional hydrocracking process in the goal case, hydrogen and natural gas consumption were found to be higher than that of the base case. However, lower waste disposal cost compensates this to some extent. Hydrotreating and hydrocracking processes accounted for nearly half of the total capital cost. Assumptions of lower temperature and pressure of HTL reactors resulted in 10% lower installation cost for goal case, which indicates the scope of process improvements. Product yield and feedstock price were two most sensitive factors in deciding the selling price of oil. The selling price increased by 26% by doubling the feedstock cost. Ou et al. (2015) and Barlow et al. (2016) also reported that feedstock price is the most important parameter in deciding the production cost of oil during HTL of algae. Delrue et al. (2013) compared various technologies to convert algae into biodiesel and observed that due to high product yields the production cost of biodiesel using HTL pathway was 12% less than that of using wet lipid extraction pathway. However, the net energy ratio (ratio of energy input to energy output) was found to be 9% lower than that of wet lipid extraction pathway (Delrue et al., 2013). The net energy ratio is highly dependent on the biocrude yields. In case of techno-economic analysis of an integrated algal biorefinery, assumptions of optimum conditions to produce maximum crude oil resulted in net energy ratio of 0.33 compared to 1.65 in the base case (Barlow et al., 2016).

Summaries of some of the techno-economic studies on HTL of various feedstocks have been provided in Table 4.

Table 4. Bio-oil cost, as estimated by different studies, from hydrothermal liquefaction.

Biomass	Plant capacity, (dry metric ton/day)	Annual bio-oil production (MG/yr)	Bio-oil cost ($/gal)[a]	Reference
Woody biomass	2000	42.9	4.62	(Zhu et al., 2014)
Defatted microalgae	2000	51.3	2.62	(Ou et al., 2015)
Lipid extracted algae	608	26.9	2.20–7.55	(Zhu et al., 2013)
Microalgae	1340	70	4.96	(Jones et al., 2014)
Microalgae	-	10	9.41	(Davis et al., 2011)

[a] adjusted to inflation for 2016 price.

6. Conclusion

Hydrothermal liquefaction is a very promising technology for high moisture feedstock conversion. Wet biomass is processed at high temperatures and pressure in an aqueous medium that eliminates the need for drying and improves the energy efficiency and economics of the process. This process yields a high energy density bio-oil as the main product and some gaseous and aqueous byproducts. Bio-oil yield is highly dependent on the operating conditions, the temperature being the most critical parameter. Although there are several commercialization challenges for HTL, economic analysis shows a huge potential for this process. One of the hurdles to be crossed is to understand the chemistry of conversion of the whole biomass, which is a complex structure. Conversion kinetics of individual compounds are hugely useful in understanding the reaction mechanism. However, the interaction of these components in different biomass still needs understanding and investigation.

References

Akhtar, J. and N. A. S. Amin. 2011. A review on process conditions for optimum bio-oil yield in hydrothermal liquefaction of biomass. Renewable and Sustainable Energy Reviews 15(3): 1615–1624.

Anastasakis, K. and A. Ross. 2011. Hydrothermal liquefaction of the brown macro-alga Laminaria saccharina: effect of reaction conditions on product distribution and composition. Bioresource Technology 102(7): 4876–4883.

Barlow, J., R. C. Sims and J. C. Quinn. 2016. Techno-economic and life-cycle assessment of an attached growth algal biorefinery. Bioresource Technology 220: 360–368.

Barreiro, D. L., W. Prins, F. Ronsse and W. Brilman. 2013. Hydrothermal liquefaction (HTL) of microalgae for biofuel production: state of the art review and future prospects. Biomass and Bioenergy 53: 113–127.

Beall, F. 2007. Thermogravimetric analysis of wood lignin and hemicelluloses. Wood and Fiber Science 1(3): 215–226.

Becker, E.W. 1994. Microalgae: biotechnology and microbiology. Cambridge University Press.

Bhandari, N., D. G. Macdonald and N. N. Bakhshi. 1984. Kinetic studies of corn stover saccharification using sulphuric acid. Biotechnology and Bioengineering 26(4): 320–327.

Biller, P. and A. B. Ross. 2011. Potential yields and properties of oil from the hydrothermal liquefaction of microalgae with different biochemical content. Bioresource Technology 102(1): 215–225.

Biller, P., R. Riley and A. Ross. 2011. Catalytic hydrothermal processing of microalgae: decomposition and upgrading of lipids. Bioresource Technology 102(7): 4841–4848.

Biller, P., A. B. Ross, S. Skill, A. Lea-Langton, B. Balasundaram, C. Hall, R. Riley and C. Llewellyn. 2012. Nutrient recycling of aqueous phase for microalgae cultivation from the hydrothermal liquefaction process. Algal Research 1(1): 70–76.

Bobleter, O. 1994. Hydrothermal degradation of polymers derived from plants. Progress in Polymer Science 19(5): 797–841.

Brebu, M. and C. Vasile. 2010. Thermal degradation of lignin—a review. Cellulose Chemistry & Technology 44(9): 353.

Brennan, L. and P. Owende. 2010. Biofuels from microalgae—a review of technologies for production, processing, and extractions of biofuels and co-products. Renewable and Sustainable Energy Reviews 14(2): 557–577.

Brown, T. M., P. Duan, and P. E. Savage. 2010. Hydrothermal liquefaction and gasification of Nannochloropsis sp. Energy & Fuels 24(6): 3639–3646.

Cantero, D. A., M. D. Bermejo and M. J. Cocero. 2013. Kinetic analysis of cellulose depolymerization reactions in near critical water. The Journal of Supercritical Fluids 75: 48–57.

Changi, S. M., J. L. Faeth, N. Mo and P. E. Savage. 2015. Hydrothermal reactions of biomolecules relevant for microalgae liquefaction. Industrial & Engineering Chemistry Research 54(47): 11733–11758.

Chen, W.-T., Y. Zhang, J. Zhang, G. Yu, L. C. Schideman, P. Zhang and M. Minarick. 2014. Hydrothermal liquefaction of mixed-culture algal biomass from wastewater treatment system into bio-crude oil. Bioresource Technology 152: 130–139.

Christensen, P. S., G. Peng, F. Vogel and B. B. Iversen. 2014. Hydrothermal liquefaction of the microalgae Phaeodactylum tricornutum: impact of reaction conditions on product and elemental distribution. Energy & Fuel 28: 5792–5803.

Davis, R., A. Aden and P. T. Pienkos. 2011. Techno-economic analysis of autotrophic microalgae for fuel production. Applied Energy 88(10): 3524–3531.

Delrue, F., Y. Li-Beisson, P.-A. Setier, C. Sahut, A. Roubaud, A.-K. Froment and G. Peltier. 2013. Comparison of various microalgae liquid biofuel production pathways based on energetic, economic and environmental criteria. Bioresource Technology 136: 205–212.

Demirbaş, A. 2000. Mechanisms of liquefaction and pyrolysis reactions of biomass. Energy Conversion and Management 41(6): 633–646.

Duan, P. and P. E. Savage. 2010. Hydrothermal liquefaction of a microalga with heterogeneous catalysts. Industrial & Engineering Chemistry Research 50(1): 52–61.

Elliott, D. C., P. Biller, A. B. Ross, A. J. Schmidt and S. B. Jones. 2015. Hydrothermal liquefaction of biomass: developments from batch to continuous process. Bioresource Technology 178: 147–156.

Fu, J., X. Lu and P. E. Savage. 2010. Catalytic hydrothermal deoxygenation of palmitic acid. Energy & Environmental Science 3(3): 311–317.

Garcia Alba, L., C. Torri, C. Samorì, J. van der Spek, D. Fabbri, S. R. Kersten and D. W. Brilman. 2011. Hydrothermal treatment (HTT) of microalgae: evaluation of the process as conversion method in an algae biorefinery concept. Energy & Fuels 26(1): 642–657.

Guo, M., W. Song and J. Buhain. 2015a. Bioenergy and biofuels: History, status, and perspective. Renewable and Sustainable Energy Reviews 42: 712–725.

Guo, Y., T. Yeh, W. Song, D. Xu and S. Wang. 2015b. A review of bio-oil production from hydrothermal liquefaction of algae. Renewable and Sustainable Energy Reviews 48: 776–790.

Holliday, R. L., J. W. King and G. R. List. 1997. Hydrolysis of vegetable oils in sub- and supercritical water. Industrial & Engineering Chemistry Research 36(3): 932–935.

Jena, U., K. Das and J. Kastner. 2011a. Effect of operating conditions of thermochemical liquefaction on biocrude production from Spirulina platensis. Bioresource Technology 102(10): 6221–6229.

Jena, U., N. Vaidyanathan, S. Chinnasamy and K. Das. 2011b. Evaluation of microalgae cultivation using recovered aqueous co-product from thermochemical liquefaction of algal biomass. Bioresource Technology 102(3): 3380–3387.

Jones, S., Y. Zhu, D. Anderson, R. Hallen, D. C. Elliott, A. J. Schmidt, K. Albrecht, T. Hart, M. Butcher and C. Drennan. 2014. Process design and economics for the conversion of algal biomass to hydrocarbons: whole algae hydrothermal liquefaction and upgrading. US Department of Energy Bioenergy Technologies Office.

Kabyemela, B. M., T. Adschiri, R. M. Malaluan and K. Arai. 1997. Kinetics of glucose epimerization and decomposition in subcritical and supercritical water. Industrial & Engineering Chemistry Research 36(5): 1552–1558.

Kabyemela, B. M., T. Adschiri, R. M. Malaluan and K. Arai. 1999. Glucose and fructose decomposition in subcritical and supercritical water: detailed reaction pathway, mechanisms, and kinetics. Industrial & Engineering Chemistry Research 38(8): 2888–2895.

Kanetake, T., M. Sasaki and M. Goto. 2007. Decomposition of a lignin model compound under hydrothermal conditions. Chemical Engineering & Technology 30(8): 1113–1122.

Karagöz, S., T. Bhaskar, A. Muto, Y. Sakata and M. A. Uddin. 2004. Low temperature hydrothermal treatment of biomass: effect of reaction parameters on products and boiling point distributions. Energy& Fuels 18: 234–41.

Karagöz, S., T. Bhaskar, A. Muto, Y. Sakata, T. Oshiki and T. Kishimoto. 2005. Low-temperature catalytic hydrothermal treatment of wood biomass: analysis of liquid products. Chemical Engineering Journal 108(1): 127–137.

Karagöz, S., T. Bhaskar, A. Muto and Y. Sakata. 2006. Hydrothermal upgrading of biomass: effect of K_2CO_3 concentration and biomass/water ratio on products distribution. Bioresource Technology 97(1): 90–98.

King, J., R. Holliday and G. List. 1999. Hydrolysis of soybean oil. in a subcritical water flow reactor. Green Chemistry 1(6): 261–264.

Klingler, D., J. Berg and H. Vogel. 2007. Hydrothermal reactions of alanine and glycine in sub-and supercritical water. The Journal of Supercritical Fluids 43(1): 112–119.

Kosinkova, J., J. A. Ramirez, J. Nguyen, Z. Ristovski, R. Brown, C. S. Lin and T. J. Rainey. 2015. Hydrothermal liquefaction of bagasse using ethanol and black liquor as solvents. Biofuels, Bioproducts and Biorefining 9(6): 630–638.

Kumar, A., J. B. Cameron and P. C. Flynn. 2004. Pipeline transport of biomass. Proceedings of the Twenty-Fifth Symposium on Biotechnology for Fuels and Chemicals Held May 4–7, 2003, in Breckenridge, CO. Springer, pp. 27–39.

Kumar, D. and G. S. Murthy. 2012. Life cycle assessment of energy and GHG emissions during ethanol production from grass straws using various pretreatment processes. The International Journal of Life Cycle Assessment 17(4): 388–401.

Kumar, S. and R. B. Gupta. 2008. Hydrolysis of microcrystalline cellulose in subcritical and supercritical water in a continuous flow reactor. Industrial & Engineering Chemistry Research 47(23): 9321–9329.

Lehr, V., M. Sarlea, L. Ott and H. Vogel. 2007. Catalytic dehydration of biomass-derived polyols in sub- and supercritical water. Catalysis Today 121(1): 121–129.

Li, F., L. Liu, Y. An, W. He, N. J. Themelis and G. Li. 2016. Hydrothermal liquefaction of three kinds of starches into reducing sugars. Journal of Cleaner Production 112: 1049–1054.

Miller, J., L. Evans, A. Littlewolf and D. Trudell. 1999. Batch microreactor studies of lignin and lignin model compound depolymerization by bases in alcohol solvents. Fuel 78(11): 1363–1366.

Mok, W. S. L. and M. J. Antal Jr. 1992. Uncatalyzed solvolysis of whole biomass hemicellulose by hot compressed liquid water. Industrial & Engineering Chemistry Research 31(4): 1157–1161.

Nagamori, M. and T. Funazukuri. 2004. Glucose production by hydrolysis of starch under hydrothermal conditions. Journal of Chemical Technology and Biotechnology 79(3): 229–233.

Ou, L., R. Thilakaratne, R. C. Brown and M. M. Wright. 2015. Techno-economic analysis of transportation fuels from defatted microalgae via hydrothermal liquefaction and hydroprocessing. Biomass and Bioenergy 72: 45–54.

Peterson, A. A., F. Vogel, R. P. Lachance, M. Fröling, M. J. Antal Jr. and J. W. Tester. 2008. Thermochemical biofuel production in hydrothermal media: a review of sub- and supercritical water technologies. Energy & Environmental Science 1(1): 32–65.

Pou-llinas, J., J. Canellas, H. Driguez and M. R. Vignon. 1990. Steam pretreatment of almond shells for xylose production. Carbohydrate Research 207(1): 126–130.

Rogalinski, T., S. Herrmann and G. Brunner. 2005. Production of amino acids from bovine serum albumin by continuous sub-critical water hydrolysis. The Journal of Supercritical Fluids 36(1): 49–58.

Rogalinski, T., K. Liu, T. Albrecht and G. Brunner. 2008. Hydrolysis kinetics of biopolymers in subcritical water. The Journal of Supercritical Fluids 46(3): 335–341.

Ross, A., P. Biller, M. Kubacki, H. Li, A. Lea-Langton and J. Jones. 2010. Hydrothermal processing of microalgae using alkali and organic acids. Fuel 89(9): 2234–2243.

Sasaki, M., T. Adschiri and K. Arai. 2004. Kinetics of cellulose conversion at 25 MPa in sub- and supercritical water. AIChE Journal 50: 192–202.

Sato, N., A. T. Quitain, K. Kang, H. Daimon and K. Fujie. 2004. Reaction kinetics of amino acid decomposition in high-temperature and high-pressure water. Industrial & Engineering Chemistry Research 43(13): 3217–3222.

Schniewind, A. P. 1989. Concise encyclopedia of wood & wood-based materials. Pergamon Press; MIT Press.

Shuping, Z., W. Yulong, Y. Mingde, I. Kaleem, L. Chun and J. Tong. 2010. Production and characterization of bio-oil from hydrothermal liquefaction of microalgae Dunaliella tertiolecta cake. Energy 35(12): 5406–5411.

Song, C., H. Hu, S. Zhu, G. Wang and G. Chen. 2004. Nonisothermal catalytic liquefaction of corn stalk in subcritical and supercritical water. Energy & Fuels 18(1): 90–96.

Spatari, S., Y. Zhang and H. L. MacLean. 2005. Life cycle assessment of switchgrass- and corn stover-derived ethanol-fueled automobiles. Environmental Science & Technology 39(24): 9750–9758.

Tier, I. 2010. California Renewable Diesel Multimedia Evaluation.

Toor, S. S., L. Rosendahl and A. Rudolf. 2011. Hydrothermal liquefaction of biomass: a review of subcritical water technologies. Energy 36(5): 2328–2342.

Valdez, P. J., J. G. Dickinson and P. E. Savage. 2011. Characterization of product fractions from hydrothermal liquefaction of Nannochloropsis sp. and the influence of solvents. Energy & Fuels 25(7): 3235–3243.

Vardon, D. R., B. Sharma, J. Scott, G. Yu, Z. Wang, L. Schideman, Y. Zhang and T. J. Strathmann. 2011. Chemical properties of biocrude oil from the hydrothermal liquefaction of Spirulina algae, swine manure, and digested anaerobic sludge. Bioresource Technology 102(17): 8295–8303.

Watanabe, M., T. Iida and H. Inomata. 2006. Decomposition of a long chain saturated fatty acid with some additives in hot compressed water. Energy Conversion and Management 47(18): 3344–3350.

Xiu, S., A. Shahbazi, V. Shirley and D. Cheng. 2010. Hydrothermal pyrolysis of swine manure to bio-oil: effects of operating parameters on products yield and characterization of bio-oil. Journal of Analytical and Applied Pyrolysis 88(1): 73–79.

Xu, C. and T. Etcheverry. 2008. Hydro-liquefaction of woody biomass in sub- and super-critical ethanol with iron-based catalysts. Fuel 87(3): 335–345.

Yamazaki, J., E. Minami and S. Saka. 2006. Liquefaction of beech wood in various supercritical alcohols. Journal of Wood Science 52(6): 527–532.

Yang, Y., C. Feng, Y. Inamori and T. Maekawa. 2004. Analysis of energy conversion characteristics in liquefaction of algae. Resources, Conservation and Recycling 43(1): 21–33.

Yu, K. L., P. L. Show, H. C. Ong, T. C. Ling, J. C. Lan, W. Chen and J. Chang. 2017. Microalgae from wastewater treatment to biochar—Feedstock preparation and conversion technologies. Energy Conversion and Management 150: 1–13.

Yin, S. and Z. Tan. 2012. Hydrothermal liquefaction of cellulose to bio-oil under acidic, neutral and alkaline conditions. Applied Energy 92: 234–239.

Yu, G., Y. Zhang, L. Schideman, T. Funk and Z. Wang. 2011. Distributions of carbon and nitrogen in the products from hydrothermal liquefaction of low-lipid microalgae. Energy & Environmental Science 4(11): 4587–4595.

Yu, Y., Z. M. Shafie and H. Wu. 2013. Cellobiose decomposition in hot-compressed water: importance of isomerization reactions. Industrial & Engineering Chemistry Research 52(47): 17006–17014.

Zhang, B., H.-J. Huang and S. Ramaswamy. 2008. Reaction kinetics of the hydrothermal treatment of lignin. Applied Biochemistry and Biotechnology 147(1-3): 119–131.

Zhang, B., M. von Keitz and K. Valentas. 2009. Thermochemical liquefaction of high-diversity grassland perennials. Journal of Analytical and Applied Pyrolysis 84(1): 18–24.

Zhang, J., W.-T. Chen, P. Zhang, Z. Luo and Y. Zhang. 2013. Hydrothermal liquefaction of Chlorella pyrenoidosa in sub- and supercritical ethanol with heterogeneous catalysts. Bioresource Technology 133: 389–397.

Zhong, C. and X. Wei. 2004. A comparative experimental study on the liquefaction of wood. Energy 29(11): 1731–1741.

Zhou, D., L. Zhang, S. Zhang, H. Fu and J. Chen. 2010. Hydrothermal liquefaction of macroalgae Enteromorpha prolifera to bio-oil. Energy & Fuels 24(7): 4054–4061.

Zhou, Y., L. Schideman, G. Yu and Y. Zhang. 2013. A synergistic combination of algal wastewater treatment and hydrothermal biofuel production maximized by nutrient and carbon recycling. Energy & Environmental Science 6(12): 3765–3779.

Zhu, Y., K. O. Albrecht, D. C. Elliott, R. T. Hallen and S. B. Jones. 2013. Development of hydrothermal liquefaction and upgrading technologies for lipid-extracted algae conversion to liquid fuels. Algal Research 2(4): 455–464.

Zhu, Y., M. J. Biddy, S. B. Jones, D. C. Elliott and A. J. Schmidt. 2014. Techno-economic analysis of liquid fuel production from woody biomass via hydrothermal liquefaction (HTL) and upgrading. Applied Energy 129: 384–394.

Zou, S., Y. Wu, M. Yang, C. Li and J. Tong. 2009. Thermochemical catalytic liquefaction of the marine microalgae Dunaliella tertiolecta and characterization of bio-oils. Energy & Fuels 23(7): 3753–3758.

Chemical Preprocessing

CHAPTER 12

Chemical Preprocessing of Feedstocks for Improved Handling and Conversion to Biofuels

John Earl Aston

1. Introduction

1.1 Overview

Biomass feedstocks typically contain inorganic chemical species referred to as "ash" that represent non-convertible material, cause equipment fouling, and can act as inhibitors in conversion or biopower processes. These inorganic species are derived from both exogenous (soil) and physiological sources. The constituents of soil ash have been found to reduce the effectiveness of dilute-acid pretreatment for biochemical conversions by affecting the buffering capacity and neutralization (Johnson et al., 2013; Carpenter et al., 2014). In addition, during the biochemical conversion of dilute acid hydrolysates, ash from the original feedstock and that released during pretreatment and hydrolysis decreases microbial activity and ethanol yield (Johnson et al., 2013; Zha et al., 2014; Palmqvist and Hahn-Hägerdal, 2000; Casey et al., 2013), and must be replaced by convertible material on a cost basis.

Idaho National Laboratory, P.O. Box 1625, Idaho Falls, Idaho 83415.
  Email: john.aston@inl.gov

Ash, particularly that from physiological sources, in biomass contains alkaline earth and alkali metals such as calcium, magnesium, sodium, and potassium (Scott et al., 2001; Mohan et al., 2006; Patwardhan et al., 2010). Even trace quantities of these metals can affect the rate, decomposition mechanism, product yield, and degradation temperature of thermochemical conversions (Carpenter et al., 2014; Bridgewater, 2012; Liden et al., 1988). Several studies have compared the influence of individual ash species on the performance of thermochemical conversions (Patwardhan et al., 2010; Mourant et al., 2011). Table 1 summarizes the impact of compositional attributes on different conversion processes.

Because ash is distributed variably by type in biomass feedstocks, different preprocessing tools are required to remove different types of ash. In addition, since individual ash species behave differently in each conversion pathway, the preprocessing technologies should be selected based on the targeted end use. As an example, chemical preprocessing using low-severity hydrothermal or dilute acid leaching is excellent for removing alkaline earth metals and alkali metals for thermochemical conversions (Chen et al., 2011; Deng et al., 2013; Aston et al., 2016a). Conversely, alkaline extractions may affect structural changes in biomass that reduce energy requirements of densification and allow for the removal of silica, a significant ash component in herbaceous feedstocks and deleterious to biochemical conversions due to buffering effects and equipment wear. Various non-aqueous extractions may be used with different effects depending on the pH, polarity, and separation of the solvent. This chapter describes the consideration and application of chemical preprocessing to improve overall feedstock quality and handling for conversion applications.

Although such chemical preprocessing methods can reduce ash and improve feedstock quality, they also require capital and maintenance investments to purchase equipment, acquire and dispose of chemicals, and in the case of aqueous preprocessing, to dry the feedstock if necessary. The study by Thompson et al. (2016) has suggested that chemically preprocessing only high-ash fractions of mechanically generated feedstock fractions can result in sufficient overall feedstock quality upon recombination with the untreated feedstock, while minimizing reactor, chemical, disposal, and drying costs by allowing the low-ash feedstock fractions to bypass the chemical preprocessing step. Using this approach, generated feedstock fractions that may have been mechanically and chemically preprocessed can be combined with feedstocks that have not been preprocessed in any fashion via the formulation of feedstock blends. Thus, formulations can be developed to achieve least-cost feedstock blend comprised of various feedstock fractions that meet quality requirements for a given conversion pathway. Formulation combined with mechanical and chemical treatments can lower feedstock costs and increase feedstock quantities since they

Table 1. Impact of compositional attributes on pretreatment, fermentation, pyrolysis, and hydrothermal liquefaction (Li et al., 2016).

Chemical property	Impact on pretreatment and hydrolysis
Acetyl group	Increases biomass recalcitrance and reduces cellulase accessibility.
	Pretreatment generates *in situ* source of acetic acid, cause enzyme inhibition, and sugar yield reduction.
Lignin content	Affects biomass recalcitrance, pretreatment severity, enzyme activity, sugar yield and operation cost.
Ash	Not fully clarified. Requires solid/liquid separation and disposal with additional costs.
Impact on microbial fermentation	
Acetic acid	Affects the ethanol yield. Requires conditioning process with additional cost.
Phenolics	Highly inhibitory with low molecular weight phenolic compounds that are most toxic. Partition into biological membranes causes loss of integrity and affects cell growth. Require detoxification step and recovery process as solid fuels or co-products.
Salts and ionic compounds	Inhibit cell growth and sugar consumptions, ethanol productivities, not significantly ethanol yields but require long fermentation time instead, which is not economic for industrial ethanol production.
Impact on pyrolysis and upgrading	
Ash/ash composition	Affects pyrolysis oil yields; increases the pretreatment, conversion, and transportation costs. Traces quantities that can affect the product quality, composition, and rates of pyrolysis. Ash compositions affect char composition and disposal. Soluble ash affects acidity/quality. Upgrading: catalytic effect on vapor cracking. Causes phase separation, poor mixing. Difficulty in handling, storage, and processing.
Total alkali metals	Pyrolysis: affect the conversion pathway, product distribution, and bio-oil yields. Increase lignin degradation and char production. Exist in pyrolysis vapor and require upstream scrubbing system or a guard bed of catalyst before the upgrading reactor with the added cost to reduce the deactivation and coke formation. Upgrading: small concentrations can catalyze significant chemical changes during storage and upgrading. Can deposit on the catalyst surface, poison active sites, and cause deactivation. Cause deposition of solids, slag formation, erosion, and corrosion to boilers, engine, and turbines.
Impact on hydrothermal liquefaction bio-crude oil	
Ash	Affects solid residue yield, bio-crude yield, and product distribution. Covers the surface of organic matter, hampers mass transfer, and thermal chemical reaction. Requires solid/liquid separation step following hydrothermal liquefaction to have solids-free and mineral-free bio-crude. Higher ash content and lower lignin content lead to a higher conversion rate and yield. Difficulties in upgrading process. Causes phase separation, poor mixing.
Alkaline and alkaline earth metals	Catalytic effect. Effect on heavy oil yield depends on lignin content—the lower the lignin content, the weaker the catalytic effect.

allow additional lower costs and lower quality feedstocks to be used. This represents an adaptation from the animal feed industry where feedstock ingredients are combined with an assortment of end uses in a manner that takes advantage of available resources while still meeting quality requirements.

2. Applications of Chemical Preprocessing

Biofuel feedstocks are not typically densified (beyond bailing for herbaceous feedstocks) prior to delivery of the biorefinery. Once at the refinery, feedstock intended for biochemical conversion is dried, stored, size-reduced, and delivered to a chemical pretreatment reactor. In thermochemical conversion supply systems, materials (typically forest products or residues) are collected, chipped, and then transported to a conversion facility. Additional size reduction via grinding or chipping may occur at the facility. Before delivery to the biorefinery or conversion facility, it may be advantageous to apply other preprocessing technologies, including chemical methods, to improve the handling or quality of the feedstock. This may be done at a delocalized location to generate more uniform feedstocks that can be delivered to multiple facilities, which in turn may receive material from multiple facilities to improve feedstock supply chain economics (Seungdo and Dale, 2016). Several chemical preprocessing methods that have been investigated are described here.

2.1 Chemical preprocessing to improve densification and recalcitrance

2.1.1 Alkaline preprocessing

Dilute alkali preprocessing can result in the alkaline hydrolysis of the ester bonds that cross-links between lignin and xylan. The extent of this hydrolysis is heavily dependent on temperature, with significant hydrolysis per hydroxyl molecule occurring over 160°C. The result is reduced substrate recalcitrance for dilute acid pretreatments and subsequent biochemical conversions. In addition, such hydrolytic reaction results in lignin solubilization. Some hemicellulose solubilization may also occur depending on the severity of the conditions (pH and temperature). Solubilized lignin and hemicelluloses may be recovered via precipitation or used as a substrate for hydrocarbon production via bioreforming catalytic processes (Thompson et al., 2016). An example of lignin degradation that may occur is shown below in Fig. 1.

Alkaline preprocessing can often be divided into two of the more common applications: treatment with mineral alkali (e.g., NaOH, KOH or $CaOH_2$); or treatment with aqueous ammonia (NH_3). Dilute alkali processes are generally very effective in the pretreatment of herbaceous

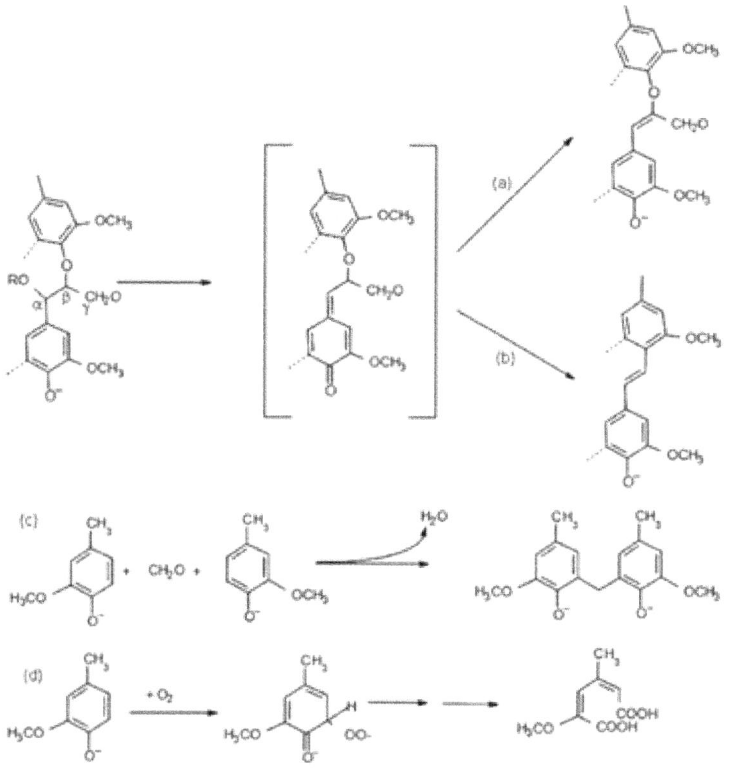

Fig. 1. Lignin degradation reactions in alkaline conditions involving- and -aryl ether linkages (a) cleavage of -aryl ether linkage (b) cleavage of CH_2O group (c) example of a possible condensation reaction. (d) Example of alkaline oxygen degradation of lignin (Adapted from Sanchez et al., 2011).

crops, including agricultural residues, as it results in the dissolution of lignin and silica from the feedstock, both of which are desirable outcomes for biochemical conversion preprocessing. While the silica is typically not considered to be of significant value, the recovery of the dissolved lignin is often required as it has value as a boiler fuel. Both lignin and silica can be recovered through acid precipitation; however, the lignin is of higher quality and value if it can be precipitated separately from the silica. Some previous works using the black liquor resulting from rice straw substrate in bioethanol production has shown that this can be accomplished (Minu et al., 2012). Additionally, some recent work also demonstrated a two-step process where silica and lignin were precipitated separately. At an overall level, the recovery of organics as high as 98% had been measured. The researchers reported that the precipitation of silica and organics were sufficiently separated by pH to enable the isolated precipitation of each constituent.

The work by Aston (2016b) showed that the collection of the precipitate formed at about pH 5 removed most of the silica from the solution. Removal of the precipitate from solution after further acidification to pH 3 allowed for the collection of most of the lignin in solution. The collected lignin fraction contained less than 2.5 wt% silica. As an added benefit, the energy to grind the alkaline preprocessed corn stover was 13.8% less than the unprocessed corn stover, and a 23.1% decrease in required pelleting energy was also observed. Finally, the density and pellet durability were statistically similar in the processed and unprocessed materials. The density of the alkaline treated materials was slightly higher than that of the raw materials (Aston et al., 2016b).

2.1.2 Ammonia preprocessing

Dilute aqueous ammonia-based processes are effective for improving the enzyme digestibility of low-lignin feedstocks, such as corn stover, switchgrass, and miscanthus. As such, it is a useful preprocessing technology for biochemical conversions. Typically, biomass is soaked in aqueous NH_3 for several days at ambient conditions (Kim and Lee, 2007). After soaking, the remaining solids are separated by filtering or pressing and washed to remove solubilized lignin. The soaking in aqueous ammonia pretreatment can result in up to 75% solubilization of the lignin, despite measured retention of nearly 100% of the glucan and 85% of the xylan (Kim and Lee, 2007).

2.1.3 Solvent preprocessing

Organosolv processes use organic solvents to remove lignin from lignocellulosic biomass. In addition to lignin solubilization, hemicellulose may also be removed through a subsequent acid-catalyzed hydrolysis. This combination results in improved recalcitrance and enzymatic digestibility of the remaining cellulose. Common solvents for the process include ethylene glycol, methanol, ethanol, and acetone. Temperatures used for the process can be as high as 200°C, but lower temperatures may be sufficient depending on the amount of catalyst used (Sun and Cheng, 2002). The effectiveness of such applications varies with the solvent type, solid loadings, and temperatures used.

2.2 Chemical preprocessing to remove ash

2.2.1 Acidic preprocessing

Acidic preprocessing can improve feedstock recalcitrance and cellulose accessibility by hydrolyzing ether linkages in the hemicellulose and cellulose fractions in biomass feedstock (Wyman et al., 2009). Requirements include heat, water, and lower pH to ensure proton availability. It is worth noting that hot water alone, with no acid catalyst, can lower the pH of a leaching medium due to the hydrolysis of acyl groups from the lignin which generates acetic acid (Yoon, 1998).

In addition to affecting some amount of hydrolysis, dilute acid leaching is also very effective for removing physiological ash components, particularly the alkaline earth and alkali metals that are deleterious to thermochemical conversions. As such, it may be possible to expand the types of feedstock available for such energy production pathways. For example, woody biomass is often used for thermochemical conversions such as pyrolysis and hydrothermal liquefaction (HTL) because the low ash content allows for high yields with less catalytic poisoning, slagging, and equipment fouling (Kenney et al., 2013; Carpenter et al., 2014; Bridgewater, 2012; Carrier et al., 2013). However, one of the most prominent and widely available lignocellulosic feedstocks is corn stover which typically possesses relatively high amounts of inorganic contaminants that serve to add expense to the logistics, processing, and conversion (Foust et al., 2009) of corn stover and other high-ash, herbaceous feedstocks. Unfortunately, many of the ash species found in corn stover that adversely affect thermochemical conversions are physiological, and therefore would require a chemical preprocessing step to be removed.

To this end, some recent work has shown that at five wt% solids, dilute acid leaching efficiently removed 97.3% of the alkali metals and alkaline earth metals which can negatively affect degradation pathways during pyrolysis (Aston et al., 2016a). Also, up to 98.4% of the chlorine and 88.8% of the phosphorus, which can cause equipment corrosion and foul upgrading catalysts, respectively, were removed. At 25°C in the absence of acid, only 6.8% of the alkali metals and alkaline earth metals were removed; however, 88.0% of chloride was removed. This lower temperature could be ideal for HTL applications as some alkaline earth and alkali metals catalyze beneficial reactions and improve yield, whereas chlorine would still have a corrosive effect (Aston et al., 2016a). Finally, the ratio of alkaline/acidic ash species, which has been suggested to proportionately relate to slagging in biopower applications (Teixeira et al., 2012), was reduced with water washing and dilute acid leaching. The initial molar alkali/acid ratio of the ash species present in the untreated corn stover was 0.38 (significant slagging risk (Teixeira et al., 2012)). At 5 wt% solids, this ratio was decreased to 0.18 (moderate slagging risk) at 0 wt% acid and 90°C, and was decreased to 0.07, 0.08, and 0.06 at 0.5 wt% acid at 25°C, 50°C and 90°C, respectively (little or no slagging risk) (Aston et al., 2016a). It should be noted that a subsequent water wash is required to remove the sodium introduced with the alkali.

2.2.2 SPORL pretreatments

Sulfite pretreatment to overcome recalcitrance of lignocellulose (SPORL) is a technology that has been shown to significantly increase enzymatic saccharification of normally recalcitrant woody material (Zhang et al., 2012a; Shuai et al., 2010). SPORL is applied over a relatively short period of time using a solution of a sulfite salt at 160–180°C and pH 2–4 for 30 min. The pH may be adjusted using sulfuric acid. Following this application, the material is size-reduced for conversion. Some size reduction and additional reduction of recalcitrance may occur during a pressure flash at the conclusion of the SPORL treatment. SPORL treatments result in more easily digestible substrates via the solubilization of lignin and hemicellulose. In addition, when compared to traditional acid treatments, SPORL occurs at relatively high pH which results in lower amounts of produced fermentation inhibitors. Again, if it is considered necessary to reduce the introduced sulfur, then an additional water wash step must be used.

2.2.3 Ionic liquid pretreatments

Ionic liquids (IL) are salts, typically composed of a small anion and a large organic cation, which exist as liquids at room temperature and have very low vapor pressures (Gellerstadt, 2009). Interest in the use of ILs has increased due to the inherent ability to tune the chemistry of the pretreatment solution by adjusting the type and concentrations of ILs. ILs have been shown to dissolve a wide variety of biomass types, in some cases into specific fractions that can be used for different conversion pathways or applications (van Rantwijk et al., 2003; Caes et al., 2011; Vancov et al., 2012). ILs have specifically been observed to dissolve and decrystallize cellulose (Wang et al., 2011) and have been used in conjunction with acid catalysts (Dadi et al., 2007) and NH_3 treatments (Zhang et al., 2012b) to achieve nearly maximal yields of cellulose to glucose.

2.2.4 Supercritical CO_2 pretreatments

A supercritical fluid is a material in a state above its critical temperature and pressure. Supercritical fluids possess near-liquid density and gas-like transport properties of diffusivity and viscosity. As such, they are an interesting media to facilitate mass transfer. Kim and Hong (2001) found that the treatment of supercritical CO_2 and steam provided an improvement in overall yield of sugar from wheat straw. Recent work with high pressure (200 bar) CO_2-H_2O pretreatment of several different herbaceous biomass types resulted in biphasic mixtures of an H_2O-rich liquid phase and a CO_2-rich supercritical phase (Kim and Hong, 2001). This can result in significant particle size reduction, drying, and removal of any ash and extractives that are water or CO_2 soluble during the flash removal of the supercritical fluid via a pressure reduction (Luterbacher et al., 2012).

3. Properties Governing Chemical Preprocessing

3.1 Properties related to ash removal

The solubility of ash and ash species in neutral, acidic, or alkaline conditions vary with the complexation of the species. For example, monovalent sodium and potassium salts are quite soluble. However, divalent calcium and magnesium ions are only slightly soluble when associated with divalent cations such as oxalate or carbonate. In some cases, the addition of protons will lower the pH and can protonate the divalent cation to make soluble oxalic acid and relatively soluble calcium sulfate. This can also be applied to trivalent iron and aluminum salts. In addition, protons can attack organic-bound carboxylate groups to form fixed carboxylic acid and soluble sulfate salts.

Among the common ash species, silica is a special case because it forms a network of cross-linked oxygen and silicon atoms. It will dissolve in acid but is especially affected by alkaline sodium hydroxide solutions that cause solid silica to form soluble sodium silicate, Na_2SiO_3.

3.2 Diffusion considerations from biomass particles

After some or all of each ash the component is dissolved in a leaching solution, diffusion transports the ions out of the biomass particle to the bulk liquid phase. This diffusion is expected to be primarily through the many straight, parallel tubules that characterize plant tissue other than bark. Measurements show that diffusion coefficients are 5–20 times higher in the direction of the tubules than across them, while the diffusion coefficients through the tubules are about 2–5 times less than reported in liquid solution without impediments (Saltberg et al., 2006). These barriers in biomass can be simple collisions with the tubule walls that hinder progress or interactions of the diffusing ion with bound oppositely-charged moieties, such as carboxylate groups, that attract and immobilize the diffusing ion briefly.

The distance that the ions must diffuse is a function of particle size and geometry. In general, the length from the center of the particle to the surface along the shortest dimension of the particle sets the representative distance that ions must diffuse. For disc-shaped particles, the half-thickness is this length. For cylindrical particles of biomass, such as needles or shredded construction and demolition wastes, the radius is the length. In each case, the difference in diffusion coefficients associated with tubule orientation must be considered. For example, the overall diffusion coefficient of a cylindrical piece of biomass with half-length twice its radius should be more influenced by the diffusion coefficient down the half-length because ions will diffuse in that direction 5–10 times faster than along the radius. The direction with the largest ratio of diffusion coefficient to length determines the direction in which to measure the diffusion length. This direction need not be along one of the principal axes of the particle shape. For instance, a chipper that cuts pieces with the grain at 45 degrees to the large flat surfaces of the chip should use that grain direction to determine the diffusion length. In this example, that length would be the chip's half-thickness times the square root of two to account for the 45-degree slant.

In a one-dimensional rectangular system, diffusion of molecules is described by the following differential equation, simplified from Flick's Equations in conjunction with appropriate boundary conditions:

$$dC/dt = -D \, (dC/dx)$$

Here, C is the liquid phase concentration of the diffusion species, t is time, D is the diffusion coefficient or diffusivity, and x is a measure of distance. The left-side derivative expresses the accumulation or loss of the diffusing species, while the right side shows that the local concentration gradient and a physical constant set the rate of diffusive transport.

While it is possible to write and solve the differential equations describing time-dependent multidirectional diffusion in rectangular, cylindrical, and spherical geometries, the uncertainties in the necessary parameters—which ash species are present, the initial concentration and distribution of each, the appropriate diffusion constants for the particular biomass type being considered, the amount of bound carboxylate that interacts with both diffusing ions and added acid, the distribution of particle sizes in the material to be leached, the variability of all these factors with season of harvest—can lead to illusory accuracy. In the design of a practical leaching process, a substantial amount of design conservatism might be added to the solution of these equations.

It is nonetheless possible to simply approximate how long the process should take. The exact solutions for transient diffusion in slabs, cylinders, or spheres depend on the same dimensionless measure of time multiplied by a geometry-dependent factor with a value of order one. This dimensionless time, also called the Fourier number and abbreviated Fo, is calculated as $\alpha t / L^2$ where α is the diffusion coefficient (cm^2/s) for the ion(s) of interest, t is time (seconds), and L is the characteristic diffusion length (cm). At a dimensionless time of value one, the diffusion front penetrated the particle to a depth L to produce a substantial but not complete reduction of the concentration of diffusible material. At a dimensionless time of two or three, the changes in concentration at depth L will be largely complete. Labeling the characteristic time for these changes as t, it can be calculated as

$\alpha t / L^2 = 1$

$t = L^2 / \alpha$

where time can be modified by adjusting either L or α. The characteristic length is readily reduced by chipping or grinding the biomass to smaller particle sizes. Reducing the average size by a factor of two reduces the characteristic time by a factor of four. However, such grinding is an energy-intensive operation especially as the target particle size gets smaller. Smaller particles also increase the risks of plugging, bridging, or especially dust explosions.

Changing the diffusion coefficient α is more difficult. Its value depends on the ion of interest, so being more specific about the objective of ash reduction might lead to using a larger, faster value for α. For instance, calcium and magnesium salts have diffusivities near 1.2×10^{-5} cm^2/s, while sodium is 1.6×10^{-5} cm^2/s and potassium 1.9×10^{-5} cm^2/s. The exact value of diffusivity varies with the anion that must accompany these cations. These positive cations cannot outrun their anions without creating a spontaneous separation of charges; therefore, both ions must diffuse together. With any of these cations, a heavier, larger, and therefore slower anion such as sulfate would lead to a lower joint diffusion coefficient compared to, say, chloride. For example, $MgCl_2$ has a diffusion coefficient of 1.1×10^{-5} cm^2/s while $MgSO_4$ is 0.7×10^{-5} cm^2/s (Lobo, 1993).

The larger geometric factors of cylinders and spheres indicate that ash in them can be leached significantly faster than ash in flat slab-like chips. However, more time and work input will be required to size-reduce biomass to cylindrical or spherical shapes with radii comparable to the half-thickness of coin-shaped chips. The flat slab result with its geometric factor near one reinforces the discussion about the use of Fo as a characteristic time for this mass transfer process. The relationships governing the diffusion from different geometrical shapes is shown in Table 2.

Table 2. Asymptotic behavior of remaining fraction of ash to be leached. Fo = $\alpha t / L^2$.

Geometry	Fraction remaining	Value of geometric factor
Flat slab	$8/\pi^2 * \exp[-8/\pi^2 * Fo]$	$8/\pi^2 = 0.811$
Long cylinder	$4/b^2 * \exp[-b^2 * Fo]$	$b = 2.405, b^2 = 5.78$
Sphere	$6/\pi^2 * \exp[-\pi^2 * Fo]$	$\pi^2 = 9.87$

3.3 Removal of poorly soluble salts

Although leaching methods may be effective for highly soluble compounds such as sodium carbonate; the leaching fluid might not initially dissolve all the poorly soluble salts that are present, meaning that only a fraction of the salt becomes mobile. As some of the slightly soluble material diffuses out of the biomass pores, more of the undissolved material will be able to dissolve. Because the loss of material by diffusion occurs first near the mouth of the pore, at the surface of the particle, the undissolved material there is the first to be dissolved. A front exists between the inner region with undissolved salt and the outer region where all of that poorly soluble salt is not only dissolved but reduced to less than saturation concentration by diffusion out of the pore. This front gradually moves into the biomass particle, progressing more slowly with time as the distance to the particle surface increases, thereby decreasing the concentration gradient in the region where diffusion is occurring.

3.4 Temperature effects

Temperature can also affect the rate of diffusion. Diffusivity in the liquid phase is proportional to absolute temperature (Hsu and Bird, 1960). Raising the leaching temperature from ambient to just below boiling (from 25 to 90°C) is a modest 22% increase in absolute temperature and therefore in the diffusion coefficient and diffusion rate. Raising the temperature might prove helpful in a second way with low-solubility species. By increasing their solubility, hence, concentration in the biomass pores, a higher temperature can increase the concentration gradient that is the driving force for diffusion out of the biomass.

The rate at which leaching occurs at various temperatures can be theoretically determined by fitting an exponential function from zero leaching at time zero to an asymptotic approach to the leaching fraction at some nth time interval. At time n, the maximum extent of leaching is reached and the additional removal of ash asymptotes with time. Based on experimental data for a given system and feedstock, a leaching value of λ (lamda) at some time (i) can be approximated from the observed maximum extent of leaching λmax (lamda-max) at a time n if the rate constant is known. Conversely, the rate constant can be approximated from experimentally observed extents of leaching.:

$$\lambda = \lambda_{max} \times [1 - \exp(-kt)]$$
$$k = -1/(2\ hr) \times ln[1 - (\lambda_{i\ hr} / \lambda_{n\ hr})]$$

Most of the values for this rate constant are near 0.4 hr^{-1} or higher. The inverse of this rate constant is mathematically the characteristic time for the leaching expression above. The resulting values of about 2.5 hour or less agree well with the value obtained from theory (Hu et al., 2017).

4. Integrated Preprocessing Approaches

4.1 Considerations of decentralized feedstock preprocessing facilities

Feedstock preprocessing sites that are not coupled with biorefineries or conversion facilities can serve as preprocessing depots that receive a broad range of feedstocks, including both high- and low-quality. As such, the overall amount of material available to a given feedstock depot would be increased over a scenario where only traditional feedstocks are used. Preprocessing capabilities at such sites may then be used to apply appropriate technologies to feedstocks to generate feedstock blends that could be used to generate supply for the local biorefineries and conversion facilities (Thompson et al., 2013). Feedstock blending and formulation are already used in coal power generation, grain, animal feeds, and biopower (Boavida et al., 2004; Shih and Frey, 1995; Reddy and Krishna, 2009). Coupling such an approach with preprocessing methods may contribute to the cost-effective production and delivery of feedstocks that are of sufficient quality for a given conversion pathway (Thompson et al., 2013). To this end, work by Thompson et al. (2016) demonstrated an exercise where the removal of total and elemental ash by preprocessing, the cost of preprocessing and available feedstock, and feedstock cost were evaluated to identify least-cost blends for thermochemical, biochemical, and combustion applications. The exercise was based on assumptions aligned with projected throughputs of depot scale processing centers (Searcy and Hess, 2010; Nguyen et al., 2014).

Unlike biochemical conversions and combustion, much tighter ash specifications are suggested for thermochemical conversions, specifically the different pyrolysis reactions. A recent Pacific Northwest National Laboratory process design report examining fast pyrolysis of lignocellulosic biomass had acceptable bio-oil yields using a feedstock with 0.9% total ash (Jones et al., 2013). A separate study found that bio-oil yields were reduced from 62% for pine (0.1% ash) to 44.7% for rapeseed straw (6.1% ash) (Oasmaa et al., 2010). This suggests that chemical preprocessing methods must be applied to feedstocks other than clean pine before they are delivered to a thermochemical conversion facility. However, it is likely that the application of a chemical preprocessing method to an entire feedstock would be cost prohibitive due to chemical and drying costs (aqueous leaching and extraction), chemical and chemical recovery costs (solvent extractions), or compression costs (supercritical fluid extraction).

The development and integration of feedstock preprocessing methodologies with feedstock formulation and blending is one potential solution to this problem. This approach may effectively use the available feedstocks for any geographical area and/or market and preprocess and blend them in a way that produces a least-cost blend that will meet quality specifications for a given conversion application. To this end, fractions of feedstocks traditionally considered as unsuitable for thermochemical conversions may be integrated into blendstocks for different types of pyrolysis.

4.2 Application of integrated preprocessing to thermochemical conversions

It was demonstrated by Thompson et al., 2013 that air classification can be used to separate corn stover and grass clippings into light fractions that contain relatively high amounts of ash and are suitable for combustion. The opposite fraction is relatively low ash, however, of particular interest to thermochemical conversion since it still contains relatively high amounts of alkali metals and alkaline earth metals which affect pyrolysis kinetics, yield, and change the reaction pathways and decomposition mechanisms as a function of temperature (Wyman et al., 2009; Yoon, 1998). In the work done by Aston et al., 2016a, potassium, sodium, calcium, and magnesium combined to represent 24.5 ± 1.4 wt% of the ash in multi-pass corn stover used in this study or approximatley 1.9% (19,000 ppm) of the total feedstock. Clean pine, a substrate widely considered to be suitable for pyrolysis, typically contains 500 to 2,000 ppm of alkali metals and alkaline earth metals (Dyer and Ragauskas, 2005; Sannigrahi et al., 2010).

In a similar fashion, the work done by Thompson et al., 2016 combining a dilute-acid leaching preprocessing step with the mechanical fractionation of high-ash fractions may be a viable approach to generating fractions of herbaceous biomass suitable for thermochemical applications while keeping drying costs relatively low by significantly reducing the total amount of solids that require aqueous treatment. For example, the application of 1.0 wt% sulfuric acid at 90°C to multi-pass corn stover resulted in a 97.3% removal of roughly 19,000 ppm of alkaline metals and alkaline earth metals which would yield a final concentration of approximately 500 ppm (Aston et al., 2017). However, because clean pine normally contains between 500 and 2,000 ppm of alkaline earth and alkali metals (Dyer and Ragauskas, 2005; Sannigrahi et al., 2010), it is likely that only a portion of the corn stover would require dilute acid leaching before being recombined with unleached material and supplying a possible blendstock for pyrolysis. It should be noted that unlike their inhibitory effects in pyrolysis, these ash species serve as natural catalysts in HTL (Pan et al., 2006). Therefore, a chemical preprocessing step to improve feedstock quality for pyrolysis may make it less suitable for HTL.

4.3 Application of integrated preprocessing to biochemical conversions

Herbaceous and low-quality feedstocks such as corn stover, switchgrass, and grass clippings can be combined for biochemical conversion, possibly at a lower cost than a higher quality feedstock alone such as single pass corn stover. Although ash is not widely considered to cause significant problems in biochemical conversions, it does cause abrasiveness in the feedstock handling systems. The ash is also inert, thus reducing the yield on total mass into a reactor. A suggested specification for total ash in the feedstock is 5 percent (Bonner et al., 2014).

Using such a specification, many research studies have shown that least cost blends consisting of a 75:25 mixture of stover/switchgrass by weight may be feasible. Using such methods, blends can be constructed by combining various raw feedstock types and feedstock fractions generated via different preprocessing methods. As a result, the feedstock delivered to a biorefinery or conversion facility can be blended for quality and cost based on conversion type and equipment requirements.

4.4 Application of integrated preprocessing to combustion

With respect to combustion, the use of mechanical preprocessing and chemical preprocessing technologies have been shown to reduce the acidity of the inorganic species of a given feedstock. Mechanical technologies such as size separation and air classification have been shown to segregate entrained soil particles from cleaner material (Lacey et al., 2015), and chemical by the removal of alkaline earth and alkaline metals via dilute acid or hot water leaching (Aston et al., 2016b). Using high-ash fractions such as mechanically fractionated corn stover or typically high ash grass clippings in such a manner would lend some ability to recover value from less suitable feedstock fractions via combustion.

5. Summary

There are several experimentally tested chemical preprocessing technologies to remove ash from biomass feedstocks including alkaline preprocessing, ammonia preprocessing, solvent preprocessing, acidic preprocessing, SPORL pretreatments, ionic liquid pretreatments, and supercritical CO_2 pretreatments. These chemical preprocessing applications vary significantly in efficacy and cost. As such, the feedstock conversion specifications and any potential to combine with other mitigation strategies, including mechanical preprocessing and blending, are important considerations for designing an overall ash mitigation strategy. As a result of the varied physical and chemical nature of biomass feedstocks and various end-point conversion technologies, the preprocessing technologies that may be necessary to improve feedstock quality and industry sustainability will necessarily be varied. Ultimately, this will require a nuanced understanding of preprocessing effects on feedstock handling, storage characteristics, pretreatment behavior, and conversion yield.

References

Aston, J. A., J. A. Lacey, V. S. Thompson and D. N. Thompson. 2016a. Alkaline deacetylation of corn stover: Effects on lignin, silica and required densification energies. In Idaho Falls, Idaho, USA: Idaho National Laboratory, Manuscript in Preparation.

Aston, J. E., D. N. Thompson and T. L. Westover. 2016b. Performance assessment of dilute-acid leaching to improve corn stover quality for thermochemical conversion. Accepted with Revisions, Fuel.

Aston, J. E. and D. N. Thompson. 2017. Application of varied alkaline preprocessing conditions and sequential dilute acid pretreatment to remove inorganics from high-ash multipass corn stover. Manuscript in preparation.

Boavida, D., P. Abelha, I. Gulyurtlu, B. Valentim and M. J. L. Sousa. 2004. A study on coal blending for reducing NO_x and N_2O levels during fluidized bed combustion. Clean Air 5: 175–191.

Bonner, I. J., W. A. Smith, J. J. Einerson and K. L. Kenney. 2014. Impact of harvest equipment on ash variability of baled corn stover biomass for bioenergy. Bioenergy Research 7: 845–855.

Bridgwater, A. V. 2012. Review of fast pyrolysis of biomass and product upgrading. Biomass Bioenergy 38: 68–94.

Caes, B. R., J. B. Binder, J. J. Blank and R. T. Raines. 2011. Separable fluorous ionic liquids for the dissolution and saccharification of cellulose. Green Chem. 13: 2719–2722.

Carpenter, D., T. L. Westover, S. Czernik and W. Jablonski. 2014. Biomass feedstocks for renewable fuel production: a review of the impacts of feedstock and pretreatment on the yield and product distribution of fast pyrolysis bio-oils and vapors. Green Chemistry 16: 384–406.

Carrier, M., J. E. Joubert, S. Danje, T. Hugo, J. Gorgens and J. Knoetze. 2013. Impact of the lignocellulosic material on fast pyrolysis yields and product quality. Bioresource Technol. 150: 129–38.

Casey, E., N. S. Mosier, J. Adamec, Z. Stockdale, N. Ho and M. Sedlak. 2013. Effect of salts on the Co-fermentation of glucose and xylose by a genetically engineered strain of Saccharomyces cerevisiae. Biotechnol. Biofuels 6.

Chen, H., T. Namioka and K. Yoshikawa. 2011. Characteristics of tar, NOx precursors and their absorption performance with different scrubbing solvents during the pyrolysis of sewage sludge. Appl. Energ. 88: 5032–4501.

Dadi, A. P., C. A. Schall and S. Varanasi. 2007. Mitigation of cellulose recalcitrance to enzymatic hydrolysis by ionic liquid pretreatment. Appl. Biochem. Biotechnol. 137: 407–421.

Deng, L., T. Zhang and D. F. Che. 2013. Effect of water washing on fuel properties, pyrolysis and combustion characteristics, and ash fusibility of biomass. Fuel Process Technol. 106: 712–720.

Dyer, T. J. and A. J. Ragauskas. 2005. Deconvoluting chromophore formation and removal during kraft pulping: influence of metal cations. International Symposium on wood, fibre and pulping chemistry, Auckland, New Zealand, May 2005.

Foust, T., A. Aden, A. Dutta and S. Phillips. 2009. An economic and environmental comparison of a biochemical and a thermochemical lignocellulosic ethanol conversion processes. Cellulose 16: 547–65.

Gellerstedt, G. 2009. Chemistry of chemical pulping. pp. 91–120. *In*: Ek M., G. Gellerstedt and G. Henriksson (eds.). Pulp and Paper Chemistry and Technology: Pulping Chemistry and Technology. Walter de Gruyter, Berlin, Germany.

Hu, H., T. L. Westover, R. Cherry, J. E. Aston, J. A. Lacey and D. N. Thompson. 2017. Process simulation and cost analysis for removing inorganics from wood chips using combined mechanical and chemical preprocessing. BioEnergy Research 10: 237–247.

Hsu, H. W. and R. B. Bird. 1960. Multicomponent diffusion problems. AIChE Journal 6: 516–524.

Johnsson, L., B. Alriksson and N. O. Nilvebrant. 2013. Bioconversion of lignocellulose: inhibitors and detoxification. Biotechnol. Biofuels 6: 16.

Jones, S., P. Meyer, L. Snowden-Swan, A. Padmaperuma, E. Tan, A. Dutta, J. Jacobson and K. Cafferty. 2013. Process design and economics for the conversion of lignocellulosic biomass to hydrocarbon fuels: Fast pyrolysis and hydrotreating bio-oil pathway. *In*. National Renewable Energy Laboratory, Pacific Northwest National Laboratory.

Kenney, K. L., W. A. Smith, G. L. Gresham and T. L. Westover. 2013. Understanding biomass feedstock variability. Biofuels 4: 111–27.

Kim, K. H. and J. Hong. 2001. Supercritical CO_2 pretreatment of lignocellulose enhances enzymatic cellulose hydrolysis. Bioresour. Technol. 77: 139–144.

Kim, T. H. and Y. Y. Lee. 2007. Pretreatment of corn stover by soaking in aqueous ammonia at moderate temperature. Appl. Biochem. Biotechnol. 136: 81–92.

Lacey, J. A., J. E. Aston, T. L. Westover, R. S. Cherry and D. N. Thompson. 2015. Removal of introduced inorganic content from chipped forest residues via air classification. Fuels 160: 265–273.

Li, C., J. E. Aston, J. A. Lacey, V. S. Thompson and D. N. Thompson. 2016. Impact of feedstock quality and variation on biochemical and thermochemical conversion. In Press, Renew. Sust. Energ. Rev.

Liden, A. G., F. Berruti and D. S. Scott. 1988. A kenetic model for the production of liquids form the flash pyrolysis of biomass. Chemical Engineering Communications 65: 207–21.

Lobo, C. and M. D. Cohen. 1993. Hydration of type K expansive cement paste and the effect of silica fume: II. Pore solution analysis and proposed hydration mechanism. Cement and Concrete Research 23: 104–114.

Luterbacher, J. S., J. W. Tester and L. P. Walker. 2012. Two-temperature stage biphasic CO_2-H_2O pretreatment of lignocellulosic biomass at high solid loadings. Biotechnol. Bioeng. 109: 1499–1507.

Minu, K., K. K. Jiby and V. V. N. Kishore. 2012. Isolation and purification of lignin and silica from the black liquor generated during the production of bioethanol from rice straw. Biomass & Bioenergy 39: 210–217.

Mohan, D., C. U. Pittman and P. H. Steele. 2006. Pyrolysis of wood/biomass for bio-oil: a critical review. Energ. Fuel 20: 848–889.

Mourant, D., Z. Wang, M. He, X. S. Wang, M. Garcia-Perez and K. Ling. 2011. Mallee wood fast pyrolysis: Effects of alkali and alkaline earth metallic species on the yield and composition of bio-oil. Fuel 90: 2915–2922.

Nguyen, L., K. G. Cafferty, E. M. Searcy and S. Spatari. 2014. Uncertainties in life cycle greenhouse gas emissions from advanced biomass feedstock logistics supply chains in kansas. Energies 7: 7125–7146.

Oasmaa, A., Y. Solantausta, V. Arpiainen, E. Kuoppala and K. Sipila. 2010. Fast pyrolysis bio-oils from wood and agricultural residues. Energy & Fuels 24: 1380–1388.

Palmqvist, E. and B. Hahn-Hägerdal. 2000. Fermentation of lignocellulosic hydrolysates. I: inhibition and detoxification. Bioresource Technol. 74: 17–24.

Pan, X. J., N. Gilkes and J. N. Saddler. 2006. Effect of acetyl groups on enzymatic hydrolysis of cellulosic substrates. Holzforschung 60: 398–401.

Patwardhan, P. R., J. A. Satrio, R. C. Brown and B. H. Shanks. 2010. Influence of inorganic salts on the primary pyrolysis products of cellulose. Bioresource Technol. 101: 4646–4655.

Reddy, D. V. and N. Krishna. 2009. Precision animal nutrition: A tool for economic and eco-friendly animal production in ruminants. Livest. Res. Rural Dev. 21: 36.

Saltberg, A., H. Brelid and H. Theliander. 2006. Removal of metal ions from wood chips during acidic leaching: Comparison between Scandinavian softwood, birch and eucalyptus. Nord. Pulp Pap. Res. J. 21: 507–512.

Sanchez, O., R. Sierra and C. J. Almeciga-Diaz. 2011. Delignification Process of Agro-Industrial Wastes an Alternative to Obtain Fermentable Carbohydrates for Producing Fuel 10.5772/22381.

Sannigrahi, P., A. J. Ragauskas and G. A. Tuskan. 2010. Poplar as a feedstock for biofuels: A review of compositional characteristics. Biofuels, Bioproducts and Biorefining 4: 209–226.

Scott, D. S., L. Paterson, J. Piskorz and D. Radlein. 2001. Pretreatment of poplar wood for fast pyrolysis: rate of cation removal. J. Anal. Appl. Pyrol. 57: 169–76.

Searcy, E. M. and J. R. Hess. 2010. Uniform-Format Feedstock Supply System Design for Lignocellulosic Biomass: A Commodity-Scale Design to Produce an Infrastructure-Compatible Biocrude From Lignocellulosic Biomass. In Idaho Falls, Idaho, USA: Idaho National Laboratory.

Seungdo, K. and B. E. Dale. 2016. A distributed cellulosic biorefinery system in the US Midwest based on corn stover. Biofuels, Bioprod., Biorefin. 10: 819–832.

Shih, J. S. and H. C. Frey. 1995. Coal blending optimization under uncertainty. European Journal of Operational Research 83: 452–465.

Shuai, L., Q. Yang and J. Y. Zhu. 2010. Comparative study of SPORL and dilute-acid pretreatments of spruce for cellulosic ethanol production. Bioresour. Technol. 101: 3106–3114.

Sun, Y. and J. Y. Cheng. 2002. Hydrolysis of lignocellulosic materials for ethanol production: a review. Bioresour. Technol. 83: 1–11.

Teixeira, P., H. Lopes, I. Gulyurtlu, N. Lapa and P. Abelha. 2012. Evaluation of slagging and fouling tendency during biomass co-firing with coal in a fluidized bed. Biomass and Bioenergy 39: 192–203.

Thompson, D. N., T. Campbell, B. Bals, T. Runge, F. Teymouri and L. Ovard. 2013. Application of low-severity pretreatment chemistries for commoditization of lignocellulosic feedstock. Biofuels 4: 323–340.

Thompson, D. N., T. Campbell, B. Bals, T. Runge, T. Farzaneh and L. P. Ovard. 2013. Chemical preconversion: application of low-severity pretreatment chemistries for commoditization of lignocellulosic feedstock. Biofuels 4: 323–340.

Thompson, V. S., J. A. Lacey, D. Hartley, M. A. Jindra, J. E. Aston and D. N. Thompson. 2016. Application of air classification and formulation to manage feedstock cost, quality and availability for bioenergy. Fuel 180: 497–505.

van Rantwijk, F., R. M. Lau and R. A. Sheldon. 2003. Biocatalytic transformations in ionic liquids. Trends Biotechnol. 21: 131–138.

Vancov, T., A. S. Alston, T. Brown and S. McIntosh. 2012. Use of ionic liquids in converting lignocellulosic material to biofuels. Renewable Energy 45: 1–6.

Wang, Y., M. Radosevich, D. Hayes and N. Labbe. 2011. Compatible ionic liquid-cellulases system for hydrolysis of lignocellulosic biomass. Biotechnol. Bioeng. 108: 1042–1048.

Wyman, C. E., B. E. Dale and R. T. Elander. 2009. Comparative sugar recovery and fermentation data following pretreatment of poplar wood by leading technologies. Biotechnol. Prog. 25: 333–339.

Yoon, H. H. 1998. Pretreatment of lignocellulosic biomass by autohydrolysis and aqueous ammonia percolation. Korean J. Chem. Eng. 15: 631–636.

Zha, Y., J. Westerhuis, B. Muilwijk, K. Overkamp, B. Nijmeijer and L. Coulier. 2014. Identifying inhibitory compounds in lignocellulosic biomass hydrolysates using an exometabolomics approach. BMC Biotechnology 14: 22.

Zhang, C., J. Y. Zhu, R. Gleisner and J. Sessions. 2012a. Fractionation of forest residues of Douglas-fir for fermentable sugar production by SPORL pretreatment. Bioenerg. Res. 5: 978–988.

Zhang, Z. H., W. Q. Wang and X. Y Liu. 2012b. Kinetic study of acid-catalyzed cellulose hydrolysis in 1-butyl-3-methylimidazolium chloride. Bioresour. Technol. 112: 151–155.

CHAPTER 13

Acid Preprocessing Treatments
Benefit for Bioconversion of Biomass for Liquid Fuels and Bioproduct Production

*Nick Nagle** and *Erik Kuhn*

1. Introduction

Acid feedstock preprocessing includes a variety of unit operations that occur upstream of conversion that results in value-added feedstock intermediates with enhanced conversion performance and efficiency. Acid preprocessing encompasses a wide range of methods, including ensiled feedstocks treated with biologically produced organic acids, acid catalyst impregnation of feedstocks prior to pretreatment, and more recently reduced-severity preprocessing/pretreatment using carboxylic acids. Acid preprocessing improves feedstock quality and enzymatic digestibility by reducing non-structural components, such as ash, increasing catalyst penetration prior to pretreatment, and debranching xylan chains prior to downstream bioconversion. A major advancement in pretreatment processes was the transition from using acid concentrations of 10%–30% (w/w) to 0.5%–2% (w/w) dilute sulfuric acid while increasing glucose yields which was made possible by advancements in enzyme technology and reduction in price (Jung and Kim, 2015). Incorporating chemical and physical preprocessing into bioconversion processes, coupled with multiple improvements in

National Bioenergy Center, National Renewable Energy Laboratory, 15013 Denver West Parkway Golden, CO, 80401.
* Corresponding author: nick.nagle@nrel.gov

enzyme biotechnology, blurs the distinction between what constitutes an acid pretreatment process from the evolving lower-severity preprocessing treatments. These treatments allow greater flexibility at both the farm gate and biorefinery, while reducing costs for the conversion process.

For example, acid catalyst impregnation, which has additional benefits such as reducing feedstock ash and extractive content, while increasing catalyst penetration into the biomass pores, improving biomass conversion but envisioning a supply chain where acid impregnation is geographically decoupled to the biofuels conversion plant would likely prove infeasible. Shipping acid impregnated biomass to a conversion facility at 2,000 MT/ day would be difficult due to feedstock handling and material corrosion issues and this would cause duplication with wastewater treatment and other utilities that would also be necessary for the conversion facility. A geographically decoupled preprocessing facility that produces multiple grades of densified feedstock to reduce downstream shipping costs may be feasible, but extensive LCA and TEA studies would be required to ensure sustainability and economic viability before such an investment is made. Whether or not preprocessing is performed at the conversion facility, from a technical perspective feedstock preprocessing can improve conversion performance by mitigating the impacts of feedstock variability and the associated impacts on conversion yield in the downstream unit operations. Here we explore the examples of acid treatments to reduce biomass recalcitrance and enhance feedstock performance.

Native or natural lignocellulosic biomass can be classified into two categories: herbaceous, which encompasses agricultural residuals, energy crops, and forage materials; and woody feedstocks, consisting of wood waste, bark, trees, and shrubs. The chemical composition of these materials varies with geographic regions, plant species, environmental conditions, and cultivation practices (Monono et al., 2013; Hu et al., 2010; Limayem and Ricke, 2012). Either together or separately, storage and logistics can increase variation in composition, moisture, particle size, and distribution. The ranges of individual chemical components of major sources of terrestrial feedstocks are shown in Table 1.

The variable composition in biomass feedstocks affects both yield and biorefinery operation. For example, increased lignin content increases biomass recalcitrance, requiring higher severity pretreatment conditions to achieve the same conversion yield and volume targets. High ash content can neutralize acid catalysts, requiring increased acid loading affecting downstream processes and potentially increasing salt concentration and salinity in the fermentation process.

Biomass structure is heterogeneous, representing macro, meso, and micro scales. Corn stover, a potential feedstock available in large quantities (site billion ton study) (Fig. 1), includes husks, leaves, sheaths, stalks, and

Table 1. Compositional analysis results from representative lignocellulosic biomasses (% w/w).

Biomass	Cellulose	Hemicellulose	Lignin
Hardwood (poplar)	50–53	26–29	15–16
Softwood (Pine)	45–50	25–35	25–35
Wheat straw	35–39	23–30	12–16
Corn cob	33–41	31–36	6–16
Corn stalk	35–40	16–35	7–18
Rice straw	29–35	23–26	17–19
Rice husks	29–36	12–29	15–25
Sorghum straw	32–35	24–27	15–21
Switchgrass	35–40	25–30	15–22

(Cai et al., 2017; Jørgensen et al., 2007).

cobs (Viamajala et al., 2006); however, over 60% of the dry mass is in the node and internode tissue supporting the plant and providing transport for nutrients and water. The interior of these structures house a vast array of vascular tissues, including the xylem and phloem, as well as void spaces and fissures that allow transport between the cells. Viamajala et al. demonstrated uptake by using selective dyes, proving that the majority of transport in corn stover is associated with vascular bundles and, to a lesser extent, in pits, cell junctions, and fractures in the cell walls. The exterior waxy rind is dense and water resistant, minimizing water loss and serving as a barrier to catalyst transport. Entrapped air within the biomass structure is also a major barrier to catalyst impregnation (Buxton et al., 2003).

Size reduction by milling or grinding serves to open biomass structure and is also an effective preprocessing strategy that can improve catalyst uptake and access of enzymes to polymeric carbohydrates. However, size reduction can collapse the pores, vascular tissues, and pits impeding catalyst transport into the biomass tissues leading to non-uniform pretreatment and ultimately poor enzymatic hydrolysis conversion. At the micro-scale, the degree of cellulose polymerization (DP), cellulose crystallinity, pore volume, lignin content and structure, and syringyl/guaiacyl (S/G) ratio impact enzymatic hydrolysis through non-specific binding of the enzymes on the biomass. The presence of hemicellulose in biomass, often cross-linked to cellulose, can also restrict access of the cellulase enzymes to cellulose reducing enzymatic hydrolysis (McCann and Carpita, 2015).

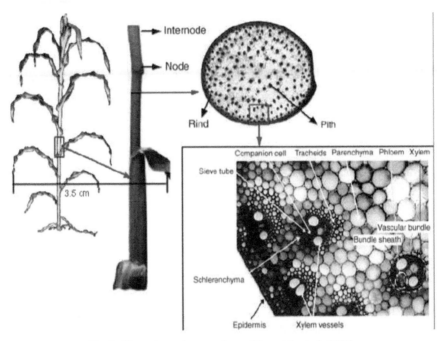

Fig. 1. Tissue types in corn stover (Viamajala et al., 2006).

2. Biological Preconversion

For more than 3,000 years, acid processing, better known as ensiling, of forage crops has been widely used at a variety of scales (Pakarinen et al., 2011). Organic acids, predominately lactic acid, produced by anaerobic lactic acid bacteria solubilize non-structural biomass components, remove the acetyl linkage from the lignocellulosic backbone in forage crops while inhibiting microbial degradation by lowering both the pH and oxygen content of the intra- and extra-particle liquid. These conditions reduce spoilage and dry matter (DM) losses and increase the digestibility of the silage in ruminate livestock. More recently, ensiling or wet storage has been evaluated as a storage process for biofuel feedstocks. Potential benefits include reduced fire risk, increased on-site storage at the biorefinery, and compatibility with existing agricultural schemes for transportation, processing, and storage.

Additional benefits may result in reducing bioconversion requirements (i.e., lower pretreatment severity may be achieved through feedstock ensiling). Several reports of increased digestibility in ensiled switchgrass, canary grass, corn stover, triticale straw, barley, and whole crop silage, for both biogas and biofuel production (Chen et al., 2007; Thomsen et al., 2008; Herrmann et al., 2011). Biomethane production from ensiled maize, sorghum, forage rye, and triticale showed an increase in biogas production from 11% to 16% as compared to fresh non-ensiled feedstock controls. However, the increased methane yields may not offset dry matter losses resulting from extended storage periods (Oleskowicz-Popiel et al., 2011).

The reduced pretreatment severity by employing ensiling as an acid preprocessing treatment has been demonstrated in herbaceous feedstocks for ethanol production. Glucose release during enzymatic hydrolysis of ensiled maize, rye, and clover grass increased 51.5%, 36.5%, and 41.9%, respectively. Ethanol production of the ensiled feedstocks was higher than the non-ensiled feedstocks but still less than 50% of the theoretical yield. Both the ensiled and non-ensiled feedstocks were then hydrothermally treated (HTT), a process using only hot water at elevated temperatures below 250°C to pretreat biomass prior to fermentation. Ethanol yield in the HTT ensiled feedstocks were 72.0%, 80.7%, and 75.7% of the theoretical yield based on the C6 sugar contents in the maize, rye, and clover grass, respectively. Ethanol yields from ensiled hydrothermal pretreated feedstocks were higher than non-ensiled controls.

High levels of glucose release from enzymatic hydrolysis were also observed after hydrothermal treatment (HTT) of ensiled wheat straw as compared to the experimental controls (Ambye-Jensen et al., 2013). Glucose yields from the HTT treated ensiled wheat straw were 1.8 times the yield of non-ensiled HTT treated wheat straw at 180°C or 170°C after enzymatic hydrolysis of the solid fractions. The overall enzymatic hydrolysis yields of HTT ensiled wheat straw at 180°C were 78% (w/w) and 87% (w/w) for glucose and xylose, respectively. While ensiling alone was not a sufficient stand-alone pretreatment, ensiling grass combined with HTT increased the enzymatic biomass convertibility and decreased the required temperature of the HTT from 190°C to 180–170°C.

Digman et al., 2010 investigated a full-scale, on-farm hybrid approach combining both storage and acid preprocessing. Design variables included feedstock type, method of application, either on-line, treated or thoroughly mixed (TTM) feedstock and two application rates for sulfuric acid 28 g or 54 g/kg dry matter (DM) for both feedstocks. Switchgrass and reed canarygrass were both treated with 18 N sulfuric acid applied during bagging. The on-line application acid was sprayed using multiple nozzles at low pressure onto the feedstocks as it was bagged based on the acid mass flow rate and drove frequency of the bagger. The weight of the material and the amount of acid used was monitored using load cells. The thoroughly mixed (TTM) feedstocks were prepared using a triolet feed mixer (TMR). Acid was sprayed on the feedstocks during mixing for 10 minutes matching the acid loading used for the on-line application. The mixed treated feedstocks were then bagged and stored with the on-line prepared feedstocks anaerobically at ambient temperature, pressure, and stored for 90 days (Digman et al., 2010). Density, uniformity, pH distribution, composition, and enzymatic cellulose conversion were measured at the end of the study.

The results from the compositional analysis (Table 2) show similar glucose, xylose, and lignin concentrations of the fresh, in-line, and mixed samples for both switchgrass and reed canary grass. pH measurements showed the uniform acid distribution for the on-line application compared to the TTM materials, which demonstrated non-uniform acid distribution. The concentration of organic acids was low, presumably related to inhibition of microbial activity by sulfuric acid. Cellulose conversion to ethanol using simultaneous saccharification and fermentation (SSF) was the highest at 56% in the on-line treatment for reed canary grass at an acid loading of 50 g (kg DM)$^{-1}$. No differences in cellulose conversion were observed for switchgrass at either application rate or method. The previous results of cellulose conversion at 100 g (kg DM)$^{-1}$ acid loading reached 46%–83% for switchgrass and reed canary grass, respectively. Sulfuric acid treatment preserved both biomass substrates minimizing dry matter losses under on-farm conditions during anaerobic incubation at ambient temperature and pressure. The ethanol yields, while low, were obtained under realistic storage and environmental conditions suggesting that farm gate preprocessing may lower bioconversion conditions at the biorefinery.

Table 2. Post cell wall composition in feedstocks after sulfuric acid application of 50 g (kg DM)$^{-1}$.

	Glucose g (kg DM)$^{-1}$	Xylose	Arabinose	Lignin
Reed Canary grass				
Fresh	342	229	33	159
On-line	330	223	27	156
Mixed	330	223	30	154
Switchgrass				
Fresh	349	244	34	163
On-line	344	249	31	154
Mixed	344	249	32	155

3. Prehydrolysis of Biomass

In 1880, Franz Heinz demonstrated that biomass hydrolysis using concentrated sulfuric acid conducted at low temperatures and ambient pressures fractionated biomass into soluble sugar streams (Bergius, 1937). Later, in 1881, some researchers evaluated the efficacy of other mineral acids to produce sugars from woody feedstocks (Harris and Lang, 1950).

While concentrated acid was effective at deconstructing biomass, the high concentrations (10% to 30% w/w) required acid recycling and neutralization and proved to be costly for fuel production. Sulfuric acid has been the most common acid used for concentrated and dilute-acid pretreatment; however, other acids such as nitric, phosphoric, and hydrochloric acid have been explored as well.

A two-stage acid pretreatment process was developed to increase both cellulose conversion and sugar recovery while reducing inhibitor formation. In the first stage, known as prehydrolysis, a potential preprocessing method, labile hemicellulose, either xylose or mannose is removed using a low severity pretreatment condition. This can be accomplished by either water or an acid of lower concentration or by reducing reaction temperature and/or reaction time. The pretreated solids are then washed to recover the sugars and pretreated at a high severity condition after reapplying catalyst using either a higher catalyst concentration or by increasing time and temperature in the second stage. This approach allows more process flexibility, increases hemicelluloses yields using lower acid consumption as compared to concentrated acid pretreatment (Torget et al., 1996a; Torget et al., 1996b). However, cellulose conversion to glucose was fundamentally limited by kinetic consideration to yields no greater than 60%.

More recently, improvements in enzymes have led to reductions in acid concentration during prehydrolysis from 6% to 0.5% (w/w). The effect of lower acid concentrations is lower catalyst cost and fewer chemicals needed for neutralizing the acid. However, dilute-acid pretreatment conducted at an elevated time and temperature still produce potent inhibitors such as furfural, hydroxymethylfurfural, and other sugar degradation products, which can inhibit or reduce fermentation yields (Jung and Kim, 2015; Limayem and Ricke, 2012). Costly acid-resistant metals or claddings are still required for the process equipment used for dilute-acid biomass pretreatment (Nitsos et al., 2013).

Acid prehydrolysis is also used to remove the hemicellulose fraction in soft- and hard-woods prior to biomass pulping. The value prior to pulping focuses on a biorefinery approach for forestry resources, potentially increasing the revenue of paper mills (Stoklosa and Hodge, 2014). Hemicellulose is removed from the woody biomass by autohydrolysis where acetic acid is released from the acetyl groups naturally occurring in the woody biomass. The resulting hemicellulose liquor contains primarily oligomeric sugars and requires hydrolysis of the oligomers using enzymatic hydrolysis to produce monomeric sugars for fermentation into ethanol or other high-value chemicals (Tunc et al., 2013; Tunc et al., 2008). The remaining cellulose fraction is then pulped using Kraft or thermomechanical processing. Removing the hemicellulose fraction reduces energy requirements of pulping in both the processes. However, the pulp strength after prehydrolysis is affected and, in some instances, lowered (Helmerius et al., 2010). The low-severity approach using reduced acid loading offers the potential to use existing paper digesters and in-plant equipment, providing opportunities for a forestry-based biorefinery (Christopher et al., 2012).

4. Acid Leaching and Impregnation

4.1 Mineral acids

Washing or leaching biomass with either hot water or dilute acid can remove non-structural ash reducing additional wear and corrosion of biorefinery processing equipment. For thermochemical operation reducing both structural and non-structural ash is a primary consideration to prevent fouling, catalyst poisoning, and reduced product yields. Aston et al., 2016 demonstrated 97% of alkali and alkali metal reduction in corn stover by washing with dilute sulfuric acid at 25°C, compared to a 6.8% reduction using water alone. More importantly, the ratio of alkali/acid ash species was reduced from 0.38 to 0.06, significantly reducing the risk of slagging in combustion applications (Aston et al., 2016). Thompson et al., 2003 evaluated the impact of time, temperature, chemical type, and chemical concentration on ash removal from wheat straw (Thompson et al., 2003). Sulfuric acid at 0.1% (w/w) to 5.0% (w/w), sodium hydroxide at 0.1% (w/w) to 1.0% (w/w) and deionized water were used with reaction temperatures of 25°C, 37°C, and 50°C for 0.5 h to 24 h. Dilute acid was more effective at removing K and Cl, while dilute alkali was more effective at removing Si. The mass loss increased with increasing temperature of the acid washes, but no loss in cellulose or hemicellulose was detected. This low severity approaches, along with other forms of chemical treatments, demonstrate the potential for ash removal using dilute acids or alkali with minimal feedstock alterations (Thompson et al., 2013).

Catalyst impregnation has a different purpose than acid leaching, although similar chemicals are used. Impregnation using acid or alkali can reduce catalyst loading into pretreatment, improve process yields, and reduce inhibitor production. Impregnation before pretreatment allows for a uniform distribution of the catalyst and improves reaction kinetics. Impregnation at higher acid or alkali concentrations, at elevated temperature (40°C to 60°C), and with longer reaction times are more effective than leaching or washing biomass. Pre-impregnation of biomass using low concentrations of sulfuric acid 1.0% (w/w), before pretreatment has been demonstrated (Tucker et al., 2003; Linde et al., 2006; Weiss et al., 2009) or sulfur dioxide (SO_2) (Shi et al., 2011; Olsen et al., 2015) at low temperature for herbaceous and woody biomass. The concentration of components in the biomass, such as ash and non-structural soluble components are reduced and removed from the feedstock by separating the impregnated liquor from the biomass. This produces an enriched solid feedstock with a higher proportion of lignocellulosic structural sugars and lower ash content. In this approach, milled biomass is soaked in an acid solution in either a recalculating bath system (Tucker et al., 2003) or in a jacket-heated paddle reactor (Weiss et al., 2009). The former approach is industrially relevant, but the latter would have limitations for commercial-scale acid impregnation because high-solids paddle mixers have size limitations and high energy consumption due to the large size of the agitator shaft. Briefly, to ensure effective pretreatment, the impregnation time and temperature conditions were selected to maximize acid penetration into the biomass leading to higher xylose yields using this method (Linde et al., 2006). Corn stover was prepared at a solid loading of 8% (w/w) using a low sulfuric acid concentration of 1.0% (w/w). Time and temperature were held at 60 ± 5°C for two hours. The acid-impregnated biomass (AIB) was drained and then pressed or dewatered using a compression screw press. The resulting solids content ranged from 45% to 50% (w/w). The composition of the AIB compared to the initial corn stover composition is shown in Table 3. Glucan, xylan, and lignin were enriched by selectively removing extractives, proteins, and sucrose. Ash content was decreased by approximately 45% while increasing glucan content by 13.9%. Sucrose, a fermentable sugar extracted during the acid impregnation process, is degraded during pretreatment at this reaction time and temperature. The AIB resulting from

Table 3. Compositional analysis of native milled corn stover and after impregnation with 1 % (w/w) sulfuric acid.

Component	Initial corn stover (%)	Acid impregnated biomass (%)
Glucan	34.0 ± 0.8	39.3 ± 0.8
Xylan	22.0 ± 0.6	24.7 ± 0.6
Galactan	1.6 ± 0.1	1.5 ± 0.3
Arabinan	3.1 ± 0.1	3.0 ± 0.4
Sucrose	4.0 ± 0.4	0
Lignin	12.4 ± 0.1	18.0 ± 0.3
Acetyl	2.9 ± 0.1	3.9 ± 0.9
Ash	7.1 ± 1.0	3.9 ± 0.9
Protein	1.6 ± 0.2	ND*
Extractives	8.3 ± 0.4	0
Total	96.8 ± 0.6	93.1 ± 0.8

Standard deviation based on triplicate analysis.
*ND not detected.

this treatment has enriched sugar content, and reduced ash and other non-productive components, improving pretreatment and enzymatic hydrolysis yields (Weiss et al., 2009).

Tucker et al. (2003) conducted a steam gun pretreatment using AIB corn stover, followed by enzymatic hydrolysis of the pretreated solids. Over 93% of the initial xylan was recovered as monomeric xylose at conditions of 1.0% sulfuric acid (w/w), 190°C, and 3 minutes residence time. Ethanol fermentation of washed solids from corn stover, pretreated using an enzyme loading of 15 FPU/g of cellulose, gave ethanol yields in excess of 85%. Weiss et al. (2009) were able to achieve similar yields for soluble xylose, monomeric and oligomeric, of 86% and 81% monomeric using the identical steam pretreatment reactor with corn stover AIB.

4.2 Sulfur dioxide (SO_2)

SO_2 has been used for pretreating woody feedstocks because it rapidly impregnates biomass tissue, providing uniform distribution of the catalyst. Several advantages of SO_2 impregnation include: solid/liquid separation is not required (Shi et al., 2011), reducing water handling and recovery. For the smaller batch pretreatment studies, the moist feedstock is impregnated with gaseous SO_2 by introducing the gas into a bag with the feedstock at a concentration of 0.0 to 0.1 g SO_2 per gram of dried feedstock, depending on the target final acid's concentration. The bag is sealed, and the SO_2 impregnates the feedstock for 12–15 hours at room temperature. The feedstock is then removed and pretreated usually employing a steam-based pretreatment process (Olsen et al., 2015). Using a response surface design, Olsen et al. (2015) evaluated the impact of wood chip size and initial moisture content on the steam explosion of SO_2 impregnated lodgepole pine chips. Glucan solubilization ranged from 17% to 91%, suggesting a near-complete glucan hydrolysis of the wood chips. The effect of chip size was not a significant factor; however, initial moisture content in the chip prior to pretreatment was significant, and modeling results suggested that a moisture content of 50% had a significant effect on the composition of the insoluble solid fraction after pretreatment. The recovery of hemicellulose-derived sugars was the highest (over 80%) at a moisture content of 47.5% (w/w). In a related study, Shi et al. evaluated the impact of reaction time and temperature with either 1% (w/w) sulfuric acid or 5% (w/w) SO_2 loading on pretreatment of switchgrass. The total glucose and xylose yield from enzymatic hydrolysis with 48.6 mg enzyme/g glucan for pretreatment at 40 minutes and 140°C were 86% and 87% with sulfuric acid and SO_2, respectively. Higher xylose yields were obtained at shorter residence times, while glucose yields were the highest at longer residence times (Shi et al., 2011).

5. Combinatory Approaches to Enhance Acid Preprocessing

Combining additional processing, either physical or chemical, to a dilute-acid pretreatment can reduce pretreatment severity requirements, transitioning from a traditional acid pretreatment process to a lower severity approach for biomass hydrolysis. By coupling steam explosion with sulfuric or SO_2 impregnated biomass, hydrolysis is further enhanced often using lower acid loading compared to non-steam exploded pretreatment (Tucker et al., 2003). Steam explosion is typically initiated at a temperature of 160°–260°C (corresponding pressure, 0.69–4.83 MPa) for several seconds to a few minutes before the material is discharged through a narrow orifice (Weiss et al., 2009). The shearing of biomass increases surface area and decreases particle size and crystallinity index, rendering enzymatic hydrolysis more effective (Liu et al., 2013).

Chemical preprocessing, such as sulfite pretreatment to overcome recalcitrance of lignocellulosic (SPORL), is an effective process for softwood chips or other forms of recalcitrant woody biomass (Zhu et al., 2010). In this process, wood chips first react with a solution of sodium bisulfite (or calcium, magnesium, or other bisulfite) at 160°–190°C and pH 2–5 (for about 10–30 minutes in batch operations). The pretreated wood chips are then fiberized through mechanical milling (using a disk refiner) to generate a fibrous substrate for ethanol production. The pretreatment process reduces the energy required for size reduction and removes lignin producing a highly-digestible substrate for enzymatic hydrolysis. Some yeast strains were inhibited during the entire slurry SSF of the SPORL-treated lodgepole pine; however, the other yeast strains achieved 90.1% and 73.5% theoretical ethanol yields (Zhou et al., 2014).

Deacetylating feedstocks before acid pretreatment can reduce the amount of acid needed for pretreatment while lowering pretreatment severity. Although acid preprocessing has been discussed for the bulk of this chapter, mild sodium hydroxide preprocessing can reduce feedstock variability and increase conversion pretreatment and enzymatic hydrolysis performance and efficiency. Low concentrations of sodium hydroxide (< 1.0% w/w) debranch xylan chains and solubilize up to 85% of acetyl groups producing acetic acid, a potent fermentation inhibitor. The process also solubilizes 10%–15% of lignin (Chen et al., 2012a). Feedstock deacetylation followed by either dilute sulfuric-acid (Chen et al., 2012b) or oxalic acid (Kundu and Lee, 2015) pretreatment demonstrated higher hemicellulose yields and reduced inhibitor production and acid consumption. Deacetylated corn stover was pretreated at 160°C, 10 min, with 0.26% H_2SO_4 (combined severity of 1.15 to 1.20) and the combined sugar yield from pretreatment and enzymatic hydrolysis was 77% achieving a total (monomeric plus oligomic) xylose concentration of 100 g/L (Shekiro et al., 2014). The additional benefits obtained from treating deacetylated biomass with organic acids, such as oxalic and maleic acid, include the reduced need for acid-resistant equipment and lower chemical usage for neutralization as compared to mineral acid pretreatment and demonstrating high selectivity for cellulose substrates. These double carbocyclic acids are similar to the catalytic core of cellulases and cellulose-binding molecules serving as proton donors, mimicking the carboxylic group of amino acids residues on the cellulase active site (Jung and Kim, 2015). Pretreatment using these catalysts has been shown to be less toxic to microorganisms and to produce fewer sugar degradation products. Combined with deacetylation, this offers an additional pathway for biomass hydrolysis. Kundu et al., 2014, pretreated deacetylated yellow poplar with 50 mM oxalic acid at 150°C for 20 min. The pretreated slurry was fermented, without detoxification, using the SSF process producing a final ethanol yield of 0.47 g ethanol/g sugar.

6. Conclusion

Acid preprocessing of primarily herbaceous feedstocks covers a wide spectrum of use and application of acid treatments to improve feedstock reactivity prior to pretreatment. These processes include *in situ* biologically produced organic acids for feedstock ensiling, pre-impregnated biomass with an acid catalyst, and prehydrolysis of woody biomass for future forest biorefineries to provide both valuable co-products and reduced energy and processing inputs to the pulping process. More recently, the use of dicarboxylic acids to mimic the enzymatic hydrolysis of biomass has been used to reduce pretreatment requirements potentially reducing the need for acid-resistant materials for pretreatment. While many of these treatments combine elements of pretreatment, preprocessing, and prehydrolysis processes, improvements in enzymes and enzyme technology will reduce the pretreatment severity required for high sugar yields shifting bioconversion towards lower severity pretreatment/preprocessing strategies that enhance and improve the biomass substrate for enzymatic hydrolysis. These trends, while exciting for the potential they offer, further blur the distinction between acid preprocessing and acid pretreatment, potentially decoupling feedstock processing from the biorefinery to the farm gate.

Acknowledgements

This work was authored by Alliance for Sustainable Energy, LLC, the manager and operator of the National Renewable Energy Laboratory for the U.S. Department of Energy (DOE) under Contract No. DE-AC36-08GO28308. Funding provided by U.S. Department of Energy Office of Energy Efficiency and Renewable Energy Bioenergy Technologies Office. The views expressed in the article do not necessarily represent the views of the DOE or the U.S. Government. The U.S. Government retains and the publisher, by accepting the article for publication, acknowledges that the U.S. Government retains a nonexclusive, paid-up, irrevocable, worldwide license to publish or reproduce the published form of this work, or allow others to do so, for U.S. Government purposes.

References

Ambye-Jensen, Morten, Sune Tjalfe Thomsen, Zsófia Kádár and Anne S. Meyer. 2013. Ensiling of wheat straw decreases the required temperature in hydrothermal pretreatment. Biotechnology for Biofuels 6 (August): 116. doi:10.1186/1754-6834-6-116.

Aston, John, E., David N. Thompson and Tyler L. Westover. 2016. Performance assessment of dilute-acid leaching to improve corn stover quality for thermochemical conversion. Fuel 186 (December): 311–19. doi:10.1016/j.fuel.2016.08.056.

Bergius, Friedrich. 1937. Conversion of wood to Carbohydrates. Industrial & Engineering Chemistry 29(3): 247–53. doi:10.1021/ie50327a002.

Buxton, Dwayne, R., Richard E. Muck, Joseph H. Harrison, J. M. Wilkinson, K. K. Bolsen and C. J. Lin. 2003. History of silage. pp. 1–30. *In*: Silage Science and Technology: Agronomy Monograph, American Society of Agronomy, Crop Science Society of America, Soil Science Society of America. doi:10.2134/agronmonogr42.c1.

Cai, Junmeng, Yifeng He, Xi Yu, Scott W. Banks, Yang Yang, Xingguang Zhang, Yang Yu, Ronghou Liu and Anthony V. Bridgwater. 2017. Review of physicochemical properties and analytical characterization of lignocellulosic biomass. Renewable and Sustainable Energy Reviews 76 (September): 309–22. doi:10.1016/j.rser.2017.03.072.

Chen, Xiaowen, Joseph Shekiro, Mary Ann Franden, Wei Wang, Min Zhang, Erik Kuhn, David K. Johnson and Melvin P. Tucker. 2012a. The impacts of deacetylation prior to dilute acid pretreatment on the bioethanol process. Biotechnology for Biofuels 5: 8. doi:10.1186/1754-6834-5-8.

Chen, Xiaowen, Joseph Shekiro, Rick Elander and Melvin Tucker. 2012b. Improved xylan hydrolysis of corn stover by deacetylation with high solids dilute acid pretreatment. Industrial & Engineering Chemistry Research 51(1): 70–76. doi:10.1021/ie201493g.

Chen, Ye, Ratna R. Sharma-Shivappa and Chengci Chen. 2007. Ensiling agricultural residues for bioethanol production. Applied Biochemistry and Biotechnology 143(1): 80–92. doi:10.1007/s12010-007-0030-7.

Christopher, Lew, James H. Clark and George A. Kraus. 2012. Integrated Forest Biorefineries. Royal Society of Chemistry.

Digman, Matthew Francis, Kevin J. Shinners, Richard E. Muck and Bruce S. Dien. 2010. Full-scale on-farm pretreatment of perennial grasses with dilute acid for fuel ethanol production. BioEnergy Research 3(4): 335–41. doi:10.1007/s12155-010-9092-4.

Harris, E. and B. Lang. 1950. The decomposition of wood by acids. In Wood Saccharification, 390–413. http://www.unece.lsu.edu/biofuels/documents/2003–2006/bf03_024.pdf.

Helmerius, Jonas, Jonas Vinblad von Walter, Ulrika Rova, Kris A. Berglund and David B. Hodge. 2010. Impact of hemicellulose pre-extraction for bioconversion on birch kraft pulp properties. Bioresource Technology 101(15): 5996–6005. doi:10.1016/j.biortech.2010.03.029.

Herrmann, Christiane, Monika Heiermann and Christine Idler. 2011. Effects of ensiling, silage additives and storage period on methane formation of biogas crops. Bioresource Technology 102(8): 5153–61. doi:10.1016/j.biortech.2011.01.012.

Hu, Zhoujian, Robert Sykes, Mark F. Davis, E. Charles Brummer and Arthur J. Ragauskas. 2010. Chemical profiles of switchgrass. Bioresource Technology 101(9): 3253–57. doi:10.1016/j.biortech.2009.12.033.

Jørgensen, Henning, Jan Bach Kristensen and Claus Felby. 2007. Enzymatic conversion of lignocellulose into fermentable sugars: challenges and opportunities. Biofuels, Bioproducts and Biorefining 1(2): 119–34. doi:10.1002/bbb.4.

Jung, Young Hoon and Kyoung Heon Kim. 2015. Chapter 3—Acidic pretreatment. pp. 27–50. *In*: Ashok Pandey, Sangeeta Negi, Parameswaran Binod, and Christian Larroche (eds.). Pretreatment of Biomass. Amsterdam: Elsevier. doi:10.1016/B978-0-12-800080-9.00003-7.

Kundu, Chandan and Jae-Won Lee. 2015. Optimization conditions for oxalic acid pretreatment of deacetylated yellow poplar for ethanol production. Journal of Industrial and Engineering Chemistry 32 (December): 298–304. doi:10.1016/j.jiec.2015.09.001.

Limayem, Alya and C. Steven Ricke. 2012. Lignocellulosic biomass for bioethanol production: current perspectives, potential issues and future prospects. Progress in Energy and Combustion Science 38 (4): 449–67. doi:10.1016/j.pecs.2012.03.002.

Linde, Marie, Mast Galb and Guido Zacchi. 2006. Steam pretreatment of acid-sprayed and acid-soaked barley straw for production of ethanol. Applied Biochemistry and Biotechnology 130(1-3): 546–62. doi:10.1385/ABAB:130:1:546.

Liu, Zhi-Hua, Lei Qin, Feng Pang, Ming-Jie Jin, Bing-Zhi Li, Yong Kang, Bruce E. Dale and Ying-Jin Yuan. 2013. Effects of biomass particle size on steam explosion pretreatment performance for improving the enzyme digestibility of corn stover. Industrial Crops and Products 44 (January): 176–84. doi:10.1016/j.indcrop.2012.11.009.

McCann, Maureen C. and Nicholas C. Carpita. 2015. Biomass recalcitrance: a multi-scale, multi-factor, and conversion-specific property. Journal of Experimental Botany 66(14): 4109–18. doi:10.1093/jxb/erv267.

Monono, Ewumbua M., Paul E. Nyren, Marisol T. Berti and Scott W. Pryor. 2013. Variability in Biomass Yield, Chemical Composition, and ethanol potential of individual and mixed herbaceous biomass species grown in North Dakota. Industrial Crops and Products 41 (January): 331–39. doi:10.1016/j.indcrop.2012.04.051.

Nitsos, Christos K., Chrysa M. Mihailof, Konstantinos A. Matis, Angelos A. Lappas and Kostas S. Triantafyllidis. 2013. Chapter 7—The role of catalytic pretreatment in biomass valorization toward fuels and chemicals. pp. 217–60. *In*: The Role of Catalysis for the Sustainable Production of Bio-Fuels and Bio-Chemicals. Amsterdam: Elsevier. doi:10.1016/B978-0-444-56330-9.00007-3.

Oleskowicz-Popiel, Piotr, Anne Belinda Thomsen and Jens Ejbye Schmidt. 2011. Ensiling—wet-storage method for lignocellulosic biomass for bioethanol production. Biomass and Bioenergy 35(5): 2087–92. doi:10.1016/j.biombioe.2011.02.003.

Olsen, Colin, Valdeir Arantes and Jack Saddler. 2015. Optimization of chip size and moisture content to obtain high, combined sugar recovery after sulfur dioxide-catalyzed steam pretreatment of softwood and enzymatic hydrolysis of the cellulosic component. Bioresource Technology 187(July): 288–98. doi:10.1016/j.biortech.2015.03.084.

Pakarinen, Annukka, Pekka Maijala, Seija Jaakkola, Frederick L. Stoddard, Maritta Kymäläinen and Liisa Viikari. 2011. Evaluation of preservation methods for improving biogas production and enzymatic conversion yields of annual crops. Biotechnology for Biofuels 4(July): 20. doi:10.1186/1754-6834-4-20.

Shi, Jian, Mirvat A. Ebrik and Charles E. Wyman. 2011. Sugar yields from dilute sulfuric acid and sulfur dioxide pretreatments and subsequent enzymatic hydrolysis of switchgrass. Bioresource Technology 102(19): 8930–38. doi:10.1016/j.biortech.2011.07.042.

Stoklosa, Ryan J. and David B. Hodge. 2014. Chapter 4—Integration of (Hemi)-cellulosic biofuels technologies with chemical pulp production. pp. 73–100. *In*: Biorefineries, Amsterdam: Elsevier. doi:10.1016/B978-0-444-59498-3.00004-X.

Thompson, David N., Peter G. Shaw and Jeffrey A. Lacey. 2003. Post-harvest processing methods for reduction of silica and alkali metals in wheat straw. Applied Biochemistry and Biotechnology 105(1-3): 205–18. doi:10.1385/ABAB:105:1-3:205.

Thompson, David N., Timothy Campbell, Bryan Bals, Troy Runge, Farzaneh Teymouri and Leslie Park Ovard. 2013. Chemical preconversion: application of low-severity pretreatment chemistries for commoditization of lignocellulosic feedstock. Biofuels 4(3): 323–40. doi:10.4155/bfs.13.15.

Thomsen, M. H., J. B. Holm-Nielsen, P. Oleskowicz-Popiel and A. B. Thomsen. 2008. Pretreatment of whole-crop harvested, ensiled maize for ethanol production. Applied Biochemistry and Biotechnology 148(1-3): 23–33. doi:10.1007/s12010-008-8134-2.

Torget, Robert, Christos Hatzis, Tammy Kay Hayward, Teh-An Hsu and George P. Philippidis. 1996b. Optimization of reverse-flow, two-temperature, dilute-acid pretreatment to enhance biomass conversion to ethanol. Applied Biochemistry and Biotechnology 57–58(1): 85. doi:10.1007/BF02941691.

Torget, Robert, Christos Hatzis, Tammy Kay Hayward, Teh-An Hsu and George P. Philippidis. 1996. Optimization of reverse-flow, two-temperature, dilute-acid pretreatment to enhance biomass conversion to ethanol. Applied Biochemistry and Biotechnology 57–58(1): 85. doi:10.1007/BF02941691.

Tucker, Melvin P., Kyoung H. Kim, Mildred M. Newman and Quang A. Nguyen. 2003. Effects of temperature and moisture on dilute-acid steam explosion pretreatment of corn stover and cellulase enzyme digestibility. Applied Biochemistry and Biotechnology 105(1-3): 165–77. doi:10.1385/ABAB:105:1-3:165.

Tunc, M. Sefik and Adriaan R. P. van Heiningen. 2008. Hemicellulose extraction of mixed southern hardwood with water at 150°C: Effect of Time. Industrial & Engineering Chemistry Research 47(18): 7031–37. doi:10.1021/ie8007105.

Tunc, Mehmet Sefik, Juben Chheda, der Heide Evert van, Jerry Morris and Heiningen Adriaan van. 2013. Pretreatment of hardwood chips via autohydrolysis supported by acetic and formic acid. Holzforschung 68(4): 401–409. doi:10.1515/hf-2013-0102.

Viamajala, Sridhar, Michael J. Selig, Todd B. Vinant, Melvin P. Tucker, Michael E. Himmel, James D. McMillan and Stephen R. Decker. 2006. Catalyst transport in corn stover internodes. Applied Biochemistry and Biotechnology 130(1-3): 509–27. doi:10.1385/ABAB:130:1:509.

Weiss, Noah D., Nicholas J. Nagle, Melvin P. Tucker and Richard T. Elander. 2009. High xylose yields from dilute acid pretreatment of corn stover under process-relevant conditions. Applied Biochemistry and Biotechnology 155(1-3): 115–25. doi:10.1007/s12010-008-8490-y.

Zhou, Haifeng, Tianqing Lan, Bruce S. Dien, Ronald E. Hector and J. Y. Zhu. 2014. Comparisons of five saccharomyces cerevisiae strains for ethanol production from sporl-pretreated lodgepole pine. Biotechnology Progress 30(5): 1076–83. doi:10.1002/btpr.1937.

Zhu, J. Y., Xuejun Pan and Ronald S. Zalesny. 2010. Pretreatment of woody biomass for biofuel production: energy efficiency, technologies, and recalcitrance. Applied Microbiology and Biotechnology 87(3): 847–57. doi:10.1007/s00253-010-2654-8.

CHAPTER 14

Compositional and Structural Modification of Lignocellulosic Biomass for Biofuel Production by Alkaline Treatment

Kingsley L. Iroba, Majid Soleimani and *Lope G. Tabil**

1. Introduction

There is a general consensus that global climate change is caused by forced warming resulting from greenhouse gas (GHG) emissions, which is mainly from the combustion of non-renewable fossil fuels (Bush et al., 2014). As part of the strategies to mitigate the above effect, there is a need to reduce the carbon footprint and enhance the sustainability of energy supply. Lignocellulosic biomass is one alternative that is a readily available and renewable source of energy (Demirbas et al., 2009), with annual production of approximately 200 billion tonnes worldwide (Zhang, 2008). Lignocellulosic biomass has long been recognized as a potentially sustainable source of mixed simple sugars for fermentation to biofuels and production of other biomaterials. Production of bioethanol from cellulosic biomass will help to reduce the augmentation of atmospheric CO_2 concentrations and create a cleaner environment. It will also avoid the competition between food

Chemical and Biological Engineering Department, University of Saskatchewan, 57 Campus Drive, Saskatoon, SK, S7N 5A9, Canada.
* Corresponding author: lope.tabil@usask.ca

and energy, create domestic job opportunities, and generate revenue for governments both at federal and provincial/state or local levels.

Lignocellulosic materials are regarded as an alternative energy source for bioethanol production to reduce our reliance on fossil fuels. Lignocellulosic substances consist mainly of cellulose, hemicellulose, and lignin. The very nature of lignocellulosic biomass presents resistance and recalcitrance to biological and chemical degradation during enzymatic hydrolysis and the subsequent fermentation process (Iroba et al., 2013b). This leads to a very low conversion rate, which makes the process economically not feasible. It also results in little or no accessibility and digestibility of the energy-based compounds in biomass (Iroba and Tabil, 2013a; Iroba et al., 2013b). Cellulose and hemicellulose are structural carbohydrate polymers that can be hydrolyzed to fermentable monomers for ethanol biosynthesis. Lignin is an aromatic polymer with a complex structure forming a barrier around the carbohydrates and limiting their exposure to the biocatalytic agents. Production of cellulose-based ethanol and bio-based products through biochemical reactions are dependent on the quality of substrate, in addition to other factors, namely process parameters, biocatalysts, and many others. Lignin removal from the lignocellulosic matrix using biological, chemical, and physical pretreatments could increase the exposure of cellulose to the biocatalyst, and consequently, improve carbohydrates recovery and the performance of the biochemical reaction in the next stage of biochemical applications (Wyman, 1999).

The main purpose of this chapter is to discuss compositional and structural modification of biomass by the application of alkaline pretreatment method for biochemical conversion. With this knowledge from the perspective of biomass, biomass delignification, biomass liquefaction, kinetics of delignification, and mechanisms of alkaline modification of lignocellulose will also be discussed. To have a better and detailed understanding of alkaline pretreatment, different alkaline pretreatment methods, advantages and limitations of alkaline pretreatment, lignin valorization, and comparison of alkaline pretreatment with other biomass pretreatment methods will also be presented.

2. The Complex Nature of Lignocellulosic Biomass

Lignocellulosic biomass is a composite of cellulose, hemicellulose, and lignin. Lignin (which is normally generated in the stalk during crop maturity) is a cell wall cross-linked by the heterogeneous aromatic polymer. Lignin is a major component of lignocellulosic crops, which are abundantly available on earth. It is the second most abundant biopolymer on earth after cellulose. Lignin is the woody substance which gives plants the firmness that they use for strength, for defense against pathogens, and to transport water in their tissues (Linger et al., 2014). It is the largest source of bioaromatics (benzene-derived bio-based chemicals) in the world and a major source of energy. Lignin can be used in the production of biomaterials and biochemicals (bioproducts). Lignin is a biopolymer made up of aromatic building blocks which are cross-linked by carbon and ether linkages. It is produced as a waste stream in many industrial processes, such as in the production of bioethanol from lignocellulosic biomass and during the pulping of wood for cellulose production. Currently, lignin is primarily used as a source of energy in a process plant. Lignin binds hemicellulose and cellulose with cellulose positioned at the inner core of the structure (Fan et al., 2006). Hemicellulose is a carbohydrate polymer composed of 5- and 6-carbon sugars and has a random, branched, and amorphous structure with little strength. Cellulose is a linear polysaccharide of D-glucose residues linked by β-1, 4-glycosidic bonds with a high degree of polymerization (Ramesh and Singh, 1993). The sugars within cellulose and hemicellulose can be accessed for cellulosic bioethanol production by ethanologenic microorganisms (Ramesh and Singh, 1993). However, the composite nature of the molecule, particularly the lignin fraction, presents resistance to biological and chemical degradation during enzymatic hydrolysis/saccharification and the subsequent fermentation process. This leads to a very low conversion rate, which reduces economic feasibility (Söderström et al., 2003; Fan et al., 2006; Chandra et al., 2007; Taherzadeh and Karimi, 2008). Sokhansanj et al. (2006) studied the production and distribution of cereal straw (wheat, barley, oats, and flax) in the Canadian prairies over a period of 10 years (1994–2003). They reported that the provinces of Alberta, Saskatchewan, and Manitoba collectively produced roughly 37 million tonnes of wheat, barley, oat, and flax straw annually. Lignocellulosic agricultural biomass residues represent an abundant, inexpensive, and readily available source of renewable lignocellulosic biomass (Lynd et al., 1999; Liu et al., 2005). To overcome the aforementioned challenges, pretreatment strategies are employed to deconstruct/disrupt the lignocellulosic matrix, and possibly decrease the crystalline nature of cellulose, so that the cellulose and hemicellulose can be more efficiently hydrolyzed by enzymes (Taherzadeh and Karimi, 2008). A series of stages is involved in biochemical cellulosic ethanol production:

1) size reduction via grinding to increase the surface area for reactions (Fan et al., 2006);
2) pretreatment using different methods to break down the lignified structure to increase the enzymatic digestibility of cellulose and hemicellulose (Taherzadeh and Karimi, 2008);
3) hydrolysis/saccharification which converts the complex carbohydrates (cellulose and hemicellulose) into simple sugars such as pentoses (xylose and arabinose) and hexoses (glucose, galactose, mannose) (Van Wyk, 1997; Kumar et al., 2009);

4) fermentation of the simple sugars to produce cellulosic bioethanol which is further separated via distillation (Chung et al., 2005).

Several technologies have been developed in the past years for the conversion process and pretreatment to enhance the release of the energy present in lignocellulosic biomass. Pretreatment methods can be physical, chemical, physicochemical, and biological processes. Some of the pretreatment methods include steam explosion, biological, wet oxidation, acid hydrolysis, microwave-chemical, supercritical CO_2 explosion (SC-CO_2), and alkaline such as sodium hydroxide, calcium hydroxide, and ammonia fibre explosion (AFEX) pretreatment (Montane et al., 1998; Okano et al., 2005; Palonen et al., 2004; Saha et al., 2005; Zhu et al., 2006; Kim and Hong, 2001; Alizadeh et al., 2005; Iroba and Tabil, 2013a; Iroba et al., 2013b). The above studies have shown that there are different shortcomings associated with each of this pretreatment method. Such shortcomings include long residence time, incomplete breakdown of the lignin and carbohydrate matrix, generation of inhibiting compounds, high capital costs and low rate of hydrolysis. A detailed and specific description of these shortcomings is discussed in Table 3. Recently, the clear goal of the bioenergy community is to make pretreatment of biomass cost-competitive with respect to the other bioenergy sources that are economically viable, easy to scale up, and free from by-products (with toxic impact on the fermentation biocatalyst) that subsequently inhibit microbial ethanol fermentation.

3. Alkaline Pretreatment of Biomass

Pretreatment methods can be physical, chemical, physicochemical, and biological processes. Among the promising chemical pretreatment technologies, alkaline pretreatment has received much attention. Alkaline removal of lignin is a method that was originally developed and extensively used in the pulp and paper industry. In this method, an alkaline solution alone or in combination with other process enhancers is used to reduce or remove lignin from the structure of lignocellulose (Savy and Piccolo, 2014). Application of pretreatment such as alkaline technique on lignocellulosic biomass is essential to enhance the accessibility and digestibility characteristics of the energy potentials (cellulose and hemicellulose) leading to the production of simple sugars higher than 20% of the theoretical maximum (Fan et al., 2006). The accessibility and digestibility characteristics of lignocellulosic biomass can be enhanced by modifying the structure of cellulose-hemicellulose-lignin matrix via application of preprocessing and pretreatment methods (Iroba et al., 2013b; Sun and Cheng, 2002). It is postulated that the interaction of alkali and lignocellulosic biomass causes saponification of intermolecular ester bonds linkages within the biomass (Sun and Cheng, 2002; Feist et al., 1970). One of the major effects of alkaline pretreatment is the delignification of lignocellulosic biomass, which increases the remaining carbohydrates reactivity. The lignin contents of the biomass influence the effect of alkaline pretreatment (Fan et al., 1987). Alkaline pretreatment also removes acetyl and different kinds of uronic acid substitutions on hemicellulose, which lowers the extent of enzymatic hydrolysis of cellulose and hemicellulose (Chang and Holtzapple, 2000). Saponification of intermolecular ester bonds linkages within the biomass also promotes the swelling of lignocellulosic biomass beyond water-swollen dimensions and favoring increased enzymatic and microbiological penetration into the cell-wall fine structure (Iroba et al., 2013b; Sun and Cheng, 2002; Feist et al., 1970). The removal of such linkages increases the porosity of lignocellulosic biomass, leading to an increase in internal surface area (Iroba et al., 2013b; Nlewem and Thrash, 2010; Fan et al., 1987), which subsequently helps to separate the structural linkages between lignin and the complex carbohydrates, disrupts lignin structure and reduces cellulose crystallinity (Iroba et al., 2013b; Fan et al., 1987) as shown in Fig. 1 below:

Karp et al. (2014) reported that alkaline pretreatment has the capability of partial depolymerization and disintegration of lignin in a complex

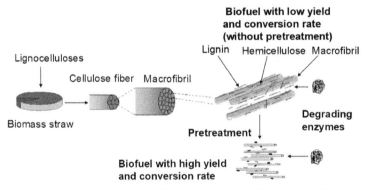

Fig. 1. Effect of pretreatment on accessibility or biodegradability of lignocellulosic biomass (adapted from Mohammad and Karimi, 2008).

mechanically structured biomass. Sodium, potassium, calcium, and ammonium hydroxides are acceptable alkaline pretreatment agents (Kumar et al., 2009). This pretreatment method requires the use of some bases/ alkaline solutions to remove lignin and a portion of the hemicellulose, and successfully improve the accessibility and digestibility of cellulose by enzymes (Jeoh et al., 2007; Mohammad and Karimi, 2008). Alkaline pretreatment enhancing the saccharification process can be done at low temperatures but at relatively long period of time with a high concentration of the alkaline solution, this will be explained in detail later in this chapter (Kumar et al., 2009; Mohammad and Karimi, 2008). It has been applied as a pretreatment technique in biogas production (Mohammad and Karimi, 2008). The effect of alkaline pretreatment is dependent on the lignin content of the biomass materials (Kumar et al., 2009; Kim and Holtzapple, 2006a). Alkaline processes result in less sugar degradation than the acid pretreatment. Most of the caustic salts from the process can be recovered and/or regenerated (Kumar et al., 2009). Alkaline pretreatment is more effective on agricultural biomass residues than on wood materials due to the lower degree of polymerization (DP) of the former group of materials (Mohammad and Karimi, 2008; Xu et al., 2007). A drawback of alkaline pretreatment is the increase of ash content of the treated biomass compared to the original material. This drawback increases as the alkaline concentration increases (Rai and Mudgal, 1987).

3.1 Biomass delignification

Alkaline delignification is an effective method of lignin dissolution and removal. Using this process, a substantial portion of hemicellulose is also extracted from the lignocellulosic matrix via solubilization, which is done through saponification of ester bonds existing in the hemicellulose structure or between hemicellulose and lignin. This would lead to the lignin modification, biomass swelling, decreased cellulose crystallinity and degree of polymerization, and increased porosity and internal surface area of the solid fraction obtained after alkaline pretreatment (Iroba et al., 2013b; Zheng et al., 2009).

Comparing the effectiveness of sodium hydroxide, sulfuric acid, ozone, and hydrogen peroxide in cotton stalk delignification, Silverstein et al. (2007) reported that sodium hydroxide pretreatment was the most effective method of delignification (65.6% lignin removal at 2% NaOH conc. and 121°C, 15 psi pressure, in 90 min) resulting in a 61% cellulose enzymatic conversion (Silverstein et al., 2007). However, sulfuric acid pretreatment was reportedly more effective in xylan removal (95.23% xylan reduction at 2% acid conc. and 121°C, 0.1 MPa pressure, in 90 min). In an investigation by Park and Kim (2012) who used alkaline percolation process, a 12-fold increase in enzymatic cellulose digestibility was reported for eucalyptus. They also reported a doubled delignification performance in percolation method as compared to the soaking method of pretreatment. Furthermore, they also found a more improved enzyme digestibility using sodium hydroxide solution compared to the other alkaline agents, namely potassium hydroxide, ammonia, and sodium carbonate.

Among the chemical pretreatments methods, alkaline hydrolysis is the most effective process of breaking the cross-links between lignin, hemicellulose, and cellulose with less disintegration of hemicellulose (Gaspar et al., 2007). For example, an advantage with soaking in aqueous ammonia at low temperature is the efficient removal of lignin and at the same time minimizing hemicellulose removal and retaining this fraction alongside cellulose for a biochemical reaction. Therefore, the retained fractions can be hydrolyzed to fermentable carbohydrates. In an investigation by Kim et al. (2008) on barley biomass treated with 15% aqueous ammonia at 75°C for 48 h, a lignin solubilization of 66% and the consequent saccharification yields of respectively, 83% and 63% were obtained for glucan and xylan. Lime pretreatment has shown to be an efficient method to reduce the lignin content of biomass with the capability of retaining the major part of the carbohydrates in the solid fraction (Sierra et al., 2009).

Oxidizing substances such as hydrogen peroxide, ozone, and oxygen can improve lignin removal. Lignin could be converted into simple molecules like carboxylic acid and components due to the high reactivity of the oxidizing agents with the aromatic ring under severe conditions. However, these molecules released during the reaction might work as microbial inhibitors in ethanol fermentation process. Furthermore, hemicellulose might be converted into chemical components called degradation products which are no longer suitable or convertible to sugars. However, under controlled conditions, alkaline peroxide efficiently removes lignin with a minimal amount of biochemical inhibitors generated during the reaction (Saha and Cotta, 2006). In wet oxidation (WO), oxygen or air is applied as the oxidative agent in an alkaline aqueous system at elevated pressure (5–20 MPa) and temperature (150°C–30°C) (Jorgensen et al., 2007). According to Klinke et al. (2002), a delignification of 65% is achievable using alkaline (sodium carbonate) wet oxidation on wheat biomass. They reported a high cellulose recovery (96%) and a fairly high enzymatic convertibility (67%) under the optimal condition of wet oxidation at 195°C, 12 bar oxygen, and 6.5 g/l sodium carbonate. The extract from 100 g straw consisted of 16 g hemicellulose, 11 g low-molecular-weight carboxylic acids, 0.48 g monomeric phenols, and 0.01 g furoic acid. Acetic acid and formic acid were the main inhibitors in the extract. Overall, wet alkaline oxidation generates fewer amounts of inhibitors during the process as compared to the neutral and acidic conditions (Martin et al., 2006). High hydrolysis rates and ethanol fermentation yields have been reported using wet oxidation pretreatment. Peterson et al. (2007) reported 66%, 70%, and 52% of the theoretical ethanol yield, respectively from winter rye, oilseed rape, and faba bean straw when a wet oxidation pretreatment at 195°C and 12 bar for 15 min using 2 g/l sodium carbonate was employed.

The other method of delignification is ozonolysis in which ozone is sparged into the biomass suspension for a specified period at normal pressure and temperature, to solubilize lignin and possibly hemicellulose. Although the primary target is lignin, some loss of carbohydrates is expected during ozonolysis. In a comparative study by Kaur et al. (2012), alkaline pretreatment (sodium hydroxide 1–4% (w/v) at 121°C in a residence time of 30 to 90 min) was compared with ozone pretreatment (45 mg/l of ozone for 150 min) to pretreat cotton stalks. In the ozone treatment, lignin and xylan were reduced by 42% and 23%, respectively, while glucan increased by almost 24%. However, alkaline pretreatment with higher efficiency using 4% sodium hydroxide for 60 min resulted in a decreased content of lignin (47%) and xylan (35%) and increased glucan content (83%).

3.2 Mechanisms of alkaline modification of lignocellulose

Among the three main components of lignocellulose (cellulose, hemicellulose, and lignin), the main goal of alkaline processing is to modify or depolymerize lignin. Lignin as a macromolecule is always linked to hemicellulose, and it is the result of cross-linking of its building block which is phenylpropane. The main building blocks of lignin in plant biomass are coniferyl alcohol, sinapyl alcohol, and p-coumaryl alcohol (Fig. 2). The typical linkages in the structure of lignin are phenylpropane β-aryl ether, phenylpropane α-aryl ether, phenyl coumaran, biphenyl, and diaryl ether. The most important functional groups that impact the reactivity of lignin are a phenolic hydroxyl group, methoxyl group, and aldehyde group.

Alkaline modification of lignin mainly occurs with the cleavage of aryl ether bonds with the mechanisms shown in Fig. 3. In an alkaline medium with the presence of only hydroxyl nucleophilic group, α-O-4 is very easy to cleave and is possible even at low temperatures. However, it does not react when the hydroxyl group is esterified or in the presence of sulfhydryl ion. On the other hand, β-O-4 does not react with hydroxyl group alone, and the linkage, in this case, would be converted to a vinyl ether structure which is stable and difficult to cleave, and sometimes would be associated with the release of the terminal aldehyde group generating formaldehyde (HCHO). However, in an alkaline medium in the presence of sulfhydryl ion, β-O-4 linkage could be cleaved very fast because of the very high nucleophilic strength of sulfhydryl ion.

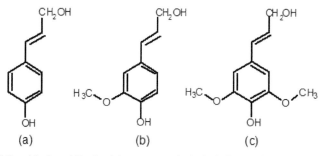

Fig. 2. Building blocks of lignin; (a) p-coumaryl alcohol, (b) coniferyl alcohol, (c) sinapyl alcohol (structures are resketched).

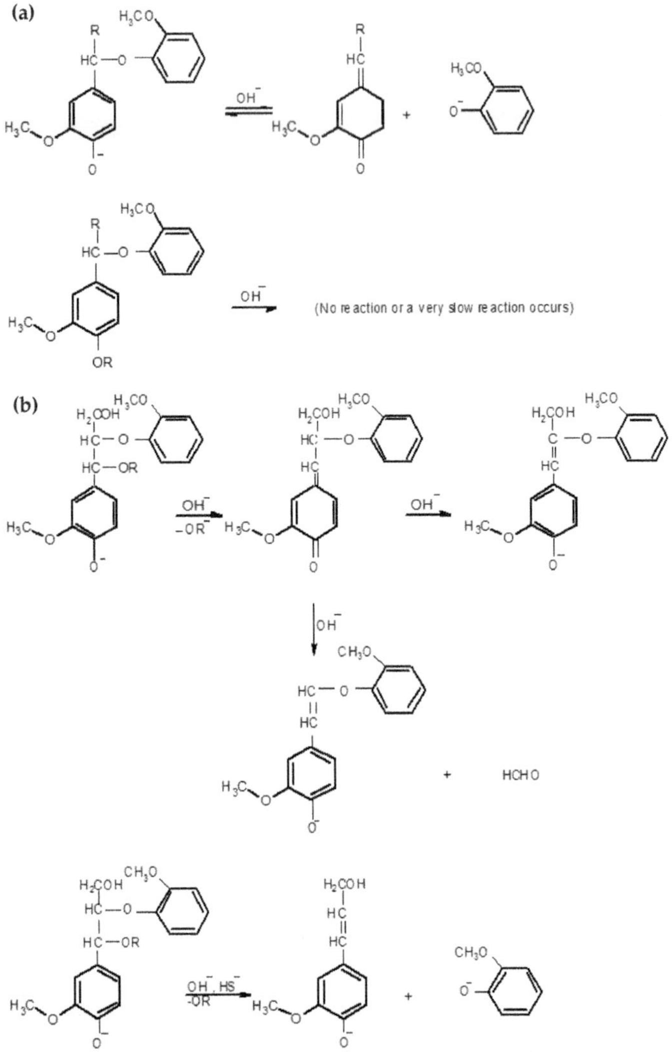

Fig. 3. Lignin reactions under alkaline condition; (a) possibility or impossibility of cleavage reaction at α-aryl ether position, and (b) possibility or impossibility of cleavage reaction at β-aryl ether position (structures are resketched).

3.3 Kinetics of delignification

A number of studies have been conducted to investigate the kinetics of the delignification reaction in different types of biomass. Whitmire and Maiti (2002) used two xylanase enzymes in different molecular weights (67 and 20 kD) for lignin removal in softwood. They reported that the susceptible lignin to xylanase catalyst is mainly a function of catalyst molecular size and pore size distribution in the pulp. The results indicated that the catalyst with lower molecular size resulted in a higher lignin removal (48 wt%) as compared to the catalyst with a higher molecular size which resulted in 30 wt% lignin removal.

Soleimani et al. (2015) conducted a kinetic study on alkaline delignification of an intact (untreated oat hulls) biomass and the byproduct of acid-catalyzed hydrolysis all from oat hull, in temperatures up to 100°C and residence time of up to 150 min. A lignin removal of over 80% for both types of biomass (intact and prehydrolyzed biomass) at higher temperature was obtained. They also concluded that delignification kinetics of both substances could be expressed by two-phase models describing lignin dissolution in bulk and terminal phases. Application of acid-catalyzed hydrolysis accelerated the reaction compared to the intact biomass, and the capacity of lignin dissolution increased, especially at the lower range of temperature (30°C and 65°C) as shown in Fig. 4. The kinetic data indicated that by increasing the temperature from 30°C to 100°C, removal of lignin would be shifting from residual phase to the bulk phase that causes acceleration of the process.

Kim and Holtzapple (2006a) also found that a two-stage kinetic model could best describe the delignification of corn stover biomass using lime-pretreatment. According to their results, increased process temperature

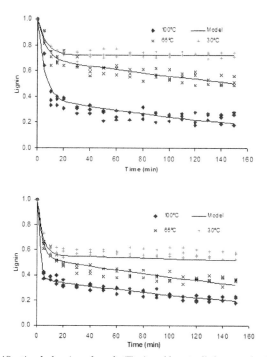

Fig. 4. The delignification behavior of crude (Top) and hemicellulose-prehydrolyzed (Bottom) biomass graphed by temperature in a two-stage kinetic model (Soleimani et al., 2015).

resulted in an improved delignification performance in bulk stage as compared to that of the residual stage.

The influence of microbial pretreatment on the kinetics of kraft delignification of wood chips was investigated by Mendonca et al. (2002). They reported that initial phase of delignification was accelerated and shortened by the microbial pretreatment due to the depolymerization of lignin during the biotreatment. Therefore, a combination of biological or chemical pretreatments with chemical delignification would be an option to be applied to integrated processes in biofuel or biorefinery industry to attain more efficient transformations with cost-effective products.

Other than batch reactors that have been used commercially and have been extensively used by researchers for delignification process, continuous reactors also have been used for conducting this process for kinetic studies to some extent. In a study by Lucia et al. (2002) on kraft delignification and oxygen delignification, it has been shown that high kappa number (bleachable or reactive) pulp is easier to delignify as compared to low kappa number pulp. This is because of the higher content of reactive phenolic hydroxyl groups and lower level of condensed lignin in high kappa pulp. To overcome the difficulty of lignin reduction in a low kappa pulp, an oxygenated alkaline solution was utilized in a flow-through reactor allowing the solution to be fresh and concentration to be constant (Ji and Heiningen, 2007). Jafari et al. (2014) conducted an experimental kinetic study on oxygen delignification of a high kappa (kappa = 65) pulp in a continuous flow-through reactor. The results obtained for the product (low enough kappa number and acceptable viscosity) by the continuous system was comparable with the results that are obtained in a conventional batch system. At an overall level, on comparing delignification for pulp and paper industry with that for the biofuel industry, it was found that delignification in the latter needs to be performed under milder conditions to be able to retain pentose sugars, to minimize energy input and cost, and to minimize generation of fermentation inhibitors.

3.4 Biomass liquefaction

Hydrothermal liquefaction is a biofuel production technology with the capability of converting lignocellulosic or other types of biomass to bio-oil (Zhou et al., 2011). Hydrothermal liquefaction can be performed under neutral, acidic, or alkaline conditions. Various alkaline substances have been used as catalysts for alkaline liquefaction of plant-based materials, including sodium carbonate, potassium carbonate, potassium hydroxide, sodium hydroxide, and calcium hydroxide. Yin and Tan (2012) conducted liquefaction reactions at different pH levels (3, 7, and 14-provided by sodium hydroxide), and temperature and residence time in the range of 275–320°C and up to 30 min, respectively. They reported that under neutral and acidic conditions, the bio-oil mainly composed of 5-hydroxymethyl furfural (HMF), however, the bio-oil from liquefaction under alkaline condition mainly composed of C_{2-5} carboxylic acids. Under the alkaline condition, lignocellulosic biomass would be mainly liquefied to bio-oil, although the original reaction medium is aqueous. Bio-oil is a complex liquid made of a wide range of chemical compositions. The typical components of bio-oil are glycol aldehyde dimers, anhydroglucose, 1, 3-dihydroxyacetone dimers, 5-HMF, furfural, polyols, organic acids, hydrocarbons, and phenolic compounds (Zhou et al., 2011). In terms of the effect of process parameters in an alkaline medium (using 0.83% sodium carbonate) on reaction rate, it has been shown that cellulose degradation starts at temperatures of less than 533 K, and its decomposition is accelerated at a temperature range of 533 K to 573 K. The generation of bio-oil was reported to occur at temperatures of over 533 K with maximum yields at temperature range of 593 K to 613 K (Minowa et al., 1997). In a similar investigation by Karagoz et al. (2004) who conducted liquefaction using hydrothermal and alkaline catalyzed processes, it was reported that at a lower temperature (180°C for 15 min) of a hydrothermal process, 26.7 wt% of sawdust was convertible with a total oil yield of 3.7 wt%. Total oil yield was increased to 7.6 wt% and 8.5 wt% by increasing process temperature to 250°C (for 15 min) and 280°C (for 15 min), respectively. By conducting the liquefaction in the presence of calcium hydroxide at 280°C (for 15 min), a higher oil yield of 9.3 wt% and gas yield of 11.9% was obtained. Therefore, they concluded that a higher oil yield but lower water-soluble products are achievable in alkaline catalyzed liquefaction compared to the hydrothermal process.

Using temperatures over 300°C in liquefaction would increase the possibility of carbonization. Application of lower temperatures in liquefaction with a proper selection of catalyst would prevent or lower biomass carbonization and especially the use of alkaline catalysts would increase generation of phenol and benzenediol derivatives with decreased amounts of solid byproducts such as char and tar. The efficiency of alkaline catalysts for liquid phase formation comes down in the following order: potassium carbonate, potassium hydroxide, sodium carbonate, and sodium hydroxide (Zhou et al., 2011).

4. Sodium Hydroxide Pretreatment

Sodium hydroxide as an alkaline can effectively enhance lignocellulose digestibility by increasing internal surface area, decreasing the degree of polymerization and the crystallinity of cellulose, and separating structural linkages between lignin and carbohydrates (Fan et al., 1987). The digestibility of NaOH-treated hardwood increased with the decrease of lignin content (Millet et al., 1975). Sodium hydroxide pretreatment was also effective for enhancing the digestibility of wheat straw (Bjerre et al.,

1996). Zhang et al. (2008) studied the pretreatment of rice straw using 2% NaOH to remove lignin. It was demonstrated that this pretreatment increased cellulose by 54.83% and decreased lignin by 36.24%, which could improve and facilitate the enzymatic hydrolysis process. Analysis of wheat straws indicated that hemicellulose, lignin, and silica were solubilized by NaOH pretreatment. The digestibility of different structural polysaccharides was higher for NaOH-pretreated straw than the native straw. However, at lower concentrations of alkaline solution, cellulose showed resistance to solubilization but not at higher levels (above 7% w/w) (Zhang et al., 2008). Alkaline (sodium hydroxide) pretreatment of coastal bermudagrass for the recovery of enhanced reducing sugars was investigated by Wang et al. (2008). The effect of NaOH pretreatment at 50°C, 80°C and 100°C and 121°C using 0.5%, 0.75%, 1%, 2% and 3% (w/v) NaOH for 15, 30, 60 and 90 minutes was evaluated. The total reducing sugars, glucose, and xylose were analyzed. The optimal NaOH pretreatment conditions at 121°C for glucose and xylose production are 15 minutes and 0.75% NaOH. The highest yield of reducing sugars reached up to approximately 86% of the theoretical maximum for NaOH pretreatment. Wang et al. (2008) reported that sodium hydroxide is more efficient than lime at 121°C for improving the yield of reducing sugars.

Some recent studies in bioethanol production have shown that very mild alkaline pretreatment of corn stover, prior to an acid pretreatment, fractionates acetate and lignin into the aqueous phase (Karp et al., 2015; Chen et al., 2012). However, the main disadvantage to the development of this alkaline technology is that apart from removing lignin and acetate from the solid phase, high severity alkaline pretreatment is known to also solubilize polysaccharides, especially a significant amount of hemicellulose and other simple sugars (Chen et al., 2011). It is advantageous to alleviate the solubilization of hemicellulose during this alkaline pretreatment process to enhance the yield of digestible carbohydrates left in the solid phase. This can be challenging in an alkaline process, because increasing treatment severity results in the loss of hemicellulose (Karp et al., 2014). The pulp and paper industry also experiences similar difficulty where high retention of hemicellulose is advantageous resulting in higher pulp yields (Genco et al., 1990). To realize this, several additives have been developed to boost the retention of hemicellulose. One of such is the addition of sodium sulfide to a caustic NaOH solution in the kraft pulping process. Sodium sulfide forms NaOH in solution which acts as a strong nucleophile increasing the fragmentation of lignin (Genco et al., 1990).

5. Calcium Hydroxide Pretreatment

Kumar et al. (2009) stated that calcium hydroxide (slake lime) is an environmentally acceptable and suitable pretreatment agent and is the least expensive per kilogram of hydroxide. It is possible to recover the "calcium from an aqueous reaction system as insoluble calcium carbonate, by neutralizing it with "low-cost carbon dioxide; the calcium hydroxide after that can be regenerated using already discovered and well-accepted lime kiln technology (Kumar et al., 2009). Alkaline pretreatment plays a very important role in making cellulose accessible to enzyme attack. The removal of lignin enhances the effectiveness of enzyme by removing the non-productive adsorption sites, thereby, increasing the accessibility and digestibility of cellulose and hemicellulose (Kumar et al., 2009).

Kim and Holtzapple (2006b) investigated the pretreatment of corn stover with excess calcium hydroxide (0.5 g $Ca(OH)_2$/g raw biomass) in non-oxidative and oxidative conditions. The enzymatic degradability and "digestibility of lime-treated corn stover was affected by the change of structural features such as acetylation, lignification, and crystallinity resulting from the treatment" (Kim and Holtzapple, 2006b). Agricultural residues were treated at 100°C for one to two hours with lime (CaO) to increase the rate and degree of dry matter digestibility. Gandi et al. (1997) concluded that the lime treatment roughly doubled the digestibility making it an efficient technique to improve and upgrade the digestibility of agricultural biomass residues.

Studies have been carried out on lime pretreatment of switchgrass at 100°C for 2 h (Chang et al., 1997), wheat straw at 85°C for 3 h (Chang et al., 1998), corn stover at 100°C for 13 h (Karr and Holtzapple, 1998; 2000), and poplar wood at 150°C for 6 h with 14-atm oxygen (Chang et al., 2001). The results reported by the above authors follow the same trend, increasing the accessibility of the cellulose and hemicellulose. Chang and Holtzapple (2000) study indicated that adding air/oxygen to the reaction system can significantly improve the delignification of the biomass. Chang et al. (2001) performed oxidative lime pretreatment of poplar wood at 150°C for 6 h with 78% removal of lignin and 71% improvement of the glucose yield from enzymatic hydrolysis. Lime (0.5 g lime/g raw biomass) was used to pretreat corn stover in non-oxidative and oxidative conditions at 25°C, 35°C, 45°C, and 55°C. The optimal condition was found to be 55°C for 4 weeks with aeration (Kim and Holtzapple, 2005). Low reagent cost and safety, and the recovery of lime from water as insoluble calcium carbonate by reaction with carbon dioxide benefit the lime pretreatment method (Playne, 1984; Chang et al., 1997).

Wang et al. (2008) investigated the alkaline pretreatment of coastal bermudagrass for bioethanol production using lime (calcium hydroxide) to enhance reducing sugars recovery. The authors used lime (0.1 g $Ca(OH)_2$/g raw biomass) as a pretreatment at room temperature, 50°C, 80°C, and 121°C. Their results indicated that increasing the temperature decreased the optimal pretreatment time at the same lime loading. The reducing sugars production under optimal pretreatment time was enhanced by 8% of theoretical maximum from room temperature to 80°C.

6. Ammonia Fiber Explosion (AFEX)

This process involves the explosion of the biomass feedstocks at a relatively low temperature to prevent sugar decomposition (Alizadeh et al., 2005). AFEX is an alkaline physicochemical pretreatment mechanism. The lignocellulosic biomass is placed in liquid ammonia at a temperature range of 90–100°C for a time duration of about 5–30 min, followed by a decrease in pressure (Alizadeh et al., 2005). The required parameters in the AFEX technique are ammonia and water loading, time, temperature, and blow down pressure (Alizadeh et al., 2005; Teymouri et al., 2004). The yield of this pretreatment method is only a pretreated solid material, unlike other alkaline pretreatments methods that generate slurry which can be separated into fractions of solid and liquid (Mosier et al., 2005). The lignin present in the lignocellulosic biomass can effectively be modified while the cellulose and hemicellulose portions may remain unbroken. The AFEX process can substantially enhance the enzymatic hydrolysis if the optimum conditions are established. The optimum conditions for AFEX are dependent on the lignocellulosic biomass (Alizadeh et al., 2005). Teymouri et al. (2004) studied the effect of pretreatment using AFEX on corn stover where the highest glucan and xylan conversions and ethanol yields from AFEX-treated corn stover were achieved. It was pointed out that enzymatic hydrolysis of corn stover treated under optimal AFEX conditions (90°C, 60% (dry weight basis, 5 min, and ammonia-corn stover mass ratio of 1:1) yielded almost 98% glucan conversion and 80% xylan conversion against 29 and 16% for untreated corn stover, respectively (Teymouri et al., 2004). This method of pretreatment has one major advantage; it does not result in the formation of inhibitors as by-products/sugar monomers, unlike other alkaline pretreatment methods that produce such inhibitors, as furans (Teymouri et al., 2004).

However, there are some disadvantages associated with the AFEX pretreatment method. It is more efficient on lignocellulosic biomass with lower lignin concentrations, as it does not substantially solubilize hemicellulose unlike other types of pretreatment processes (e.g., dilute-acid pretreatment) (Teymouri et al., 2004). Finally, it involves a difficult recycling of the feed ammonia as a reusable gas stream after the pretreatment to decrease the pretreatment cost and to keep the environment safe at the same time (Sun and Cheng, 2002; Wyman, 1996). Since ammonia is expensive, not being able to reuse it totally has a negative effect on the process economy (Zheng et al., 1995). Furthermore, the corrosive and toxic nature of ammonia would necessitate considering the appropriate design of equipment and operations (Sun and Cheng, 2002; Teymouri et al., 2004; Wyman, 1996).

7. Combined Microwave and Alkaline Pretreatment

Microwaves fill up a transitional portion in the electromagnetic (EM) spectrum between radio-frequency and infrared radiation. This portion of EM corresponds to a frequency range of 300 MHz to 300 GHz, with higher frequency range used for telecommunication and radar transmissions (Ryynanen, 1995). To avoid interference, almost all domestic and industrial microwaves operate typically either at 900 MHz or 2.45 GHz (Punidadas et al., 2003; Ryynanen, 1995). Microwave (MW) is part of the EM spectrum that results in dielectric heating of materials via induced molecular vibration and friction resulting from dipole rotation or ionic polarization (Ramaswamy and Tang, 2008). The heating process is based on volumetric heat generation. This implies that heat transfer is from inward to outward unlike the conventional heating system. MW ovens can also be used for commercial purposes including rapid extraction, blanching, drying, sterilization, pasteurization, selective heating, disinfestations, tempering, an enhanced reaction kinetics, among others (Ramaswamy and Tang, 2008). Microwave radiation has been applied in different fields of science. It has been successfully applied in the fields of organic chemistry by Larhed et al. (2002) in accelerating organic transformations at a reduced reaction time. To widen the scope and science behind MW radiation, it has been used in the lignocellulosic biomass pretreatment. The microwave-alkaline pretreatment is a more efficient pretreatment method than the traditional heating alkaline pretreatment due to the effective acceleration of the reactivity of cellulosic material (Zhu et al., 2005). Several researchers have applied this method on lignocellulosic biomass concluding that it helps in breaking down lignin, while improving its accessibility and digestibility to hydrolytic enzymes. Zhu et al. (2006) investigated the effects of three different microwave-chemical pretreatment processes on rice straw (microwave-alkali, microwave-acid-alkali, and microwave-acid-alkali-H_2O_2). The results show that rice straw pretreated with microwave-acid-alkali-H_2O_2 had the highest rate of hydrolysis and glucose content in the hydrolysate. Donepudi and Muthukumarappan (2009) studied the effect of microwave power level and processing time with different levels of concentrations of acid and alkaline on the yield of glucose and xylose from corn stover. It was reported that increasing the microwave processing time from 0 to 5 minutes led to a remarkable increase in glucose and xylose yields by 58.7 and 149.6%, respectively. The effect of microwave and alkali pretreatment on rice straw and its enzymatic hydrolysis was investigated by Zhu et al. (2005). The result was compared with the alkali-alone pretreated process. The results indicated that rice straw subjected to microwave-alkali pretreatment had more glucose content in the hydrolysate and higher hydrolysis rate than the method that involved alkali pretreatment alone. Pretreatment with higher microwave power with shorter processing time and lower microwave power

Table 1. Chemical composition (% dry basis) of untreated and microwave pre-treated of wheat and barley straw at power 713 W (Kashaninejad and Tabil, 2011).

Wheat straw	Protein	Lignin	Ash	Starch	Cellulose	Hemicellulose
Untreated	1.99[b]	8.33[a]	6.33[f]	1.11[d]	44.99[b]	27.96[a]
Microwave-distilled water	2.24[a]	8.01[c]	8.87[e]	1.48[b]	39.69[d]	22.62[b]
Microwave-NaOH (1%)	1.41[e]	7.82[d]	17.32[b]	1.89[a]	35.82[e]	12.32[d]
Microwave-NaOH (2%)	1.36[f]	7.09[f]	34.77[a]	0.27[f]	34.77[f]	4.06[f]
Microwave-Ca(OH)$_2$ (1%)	1.85[c]	8.11[b]	12.24[d]	0.69[e]	45.66[a]	14.94[c]
Microwave-Ca(OH)$_2$ (2%)	1.52[d]	7.55[e]	15.89[c]	1.31[c]	42.56[c]	11.10[e]
Barley straw						
Untreated	1.61[d]	11.95[a]	6.03[d]	0.79[c]	46.93[a]	27.40[a]
Microwave-distilled water	2.01[a]	8.85[b]	6.28[d]	1.08[b]	45.25[b]	27.21[a]
Microwave-NaOH (1%)	1.80[b]	6.65[e]	16.96[b]	0.60[e]	40.81[c]	8.74[c]
Microwave-NaOH (2%)	1.62[d]	4.52[f]	41.43[a]	0.54[f]	35.22[d]	5.46[d]
Microwave-Ca(OH)$_2$ (1%)	1.81[b]	7.27[d]	13.21[c]	0.72[d]	41.01[c]	15.00[b]
Microwave-Ca(OH)$_2$ (2%)	1.68[c]	8.16[c]	16.73[b]	1.19[a]	41.24[c]	8.97[c]

Superscript letters indicate that means with the same letters designation in a column are not significantly different at P = 0.05.

with longer pretreatment time have a fairly similar effect on the weight loss and composition at the same energy consumption (Zhu et al., 2005).

Kashaninejad and Tabil (2011) investigated the effect of microwave and microwave-chemical pretreatments on wheat and barley straw grind with hammer mill screen with an opening size of 1.58 mm at 713 W microwave power level. The data from this study is presented in Table 1.

Kashaninejad and Tabil (2011) reported that the chemical composition of untreated wheat and barley straw was almost identical except for the lignin content. The lignin content of barley straw (11.95%) was higher than wheat straw (8.33%), although the chemical composition of cereal straw varies with variety, location, and cultural practices employed in growing the crop. Table 1 also shows the remarkable changes in chemical composition of biomass samples after microwave and microwave alkali pretreatments, particularly in hemicellulose, lignin, cellulose, and ash contents. It was evident that biomass samples pretreated by microwave-alkali technique have lower lignin, hemicellulose, and cellulose than samples pretreated with microwave distilled water or untreated samples. Moreover, degradation and depolymerization of biomass lignin to smaller phenolic components is another influence of microwave alkali pre-treatment. Typically, an increase in alkali concentration from 1 to 2% (w/v) resulted in significantly higher solubility of lignin and hemicellulose in both biomass samples. Microwave-NaOH pre-treatment was more effective than microwave-Ca(OH)$_2$ to hydrolyze and dissolve lignocellulosic components. For both feedstocks, pretreatment with NaOH resulted in significantly lower lignin, hemicellulose, and cellulose over pretreatment with Ca(OH)$_2$. These results are not surprising because the effects of microwave-based processes depend on the polar characteristics or dipole moments of the system—NaOH solution has much higher dipole moments than Ca(OH)$_2$. Higher ash content was also recorded in biomass samples pretreated by microwave-alkali treatment compared to untreated or microwave distilled water treated samples, and it increased remarkably with an increasing alkali concentration. The release of lignin accompanied by the disruption of the biomass structure and change in the crystallinity of cellulose are the main advantages of microwave-alkali pretreatment, but the increase in the ash content of pretreated wheat and barley straw samples to 34.77 and 41.43% after pretreatment by 2% NaOH is an issue to be addressed. Further experiments in the laboratory showed that almost all excess ash content is easily discharged by deionized water washing, the results of which are not presented here (Kashaninejad and Tabil, 2011). However, MW is known for non-uniform heat distribution which could cause thermal runaway during the pretreatment process (Ramaswamy and Tang, 2008; Punidadas et al., 2003; Ryynanen, 1995).

8. Combined Radio-Frequency and Alkaline Pretreatment

Pre-hydrolysis pretreatment of lignocellulosic biomass, using radio-frequency (RF) with a low concentration of NaOH solution as a catalyst, can breakdown the cementing lignin matrix, and subsequently enhance the accessibility and digestibility of the energy potentials (cellulose and hemicellulose), and improve the enzymatic hydrolysis rates of lignocellulosic biomass for biofuel production. Therefore, Iroba et al. (2013b) investigated alkaline (NaOH) pretreatment on barley straw, using radio-frequency-based dielectric heating (1.5 kW and 27.12 MHz laboratory dryer, Strayfield, Theale, Reading). Three levels of temperature (70, 80, and 90°C), five levels of biomass:NaOH solution ratio (1:4, 1:5, 1:6, 1:7, and 1:8), 1 h soaking time, screen size of 1.6 mm, 1% w/v NaOH concentration, and 20 min residence time were used for the pretreatment.

Iroba et al. (2013b) reported that the addition of NaOH to lignocellulosic biomass during the RF pretreatment causes swelling of particles within the mixture. This swelling is a result of the saponification of intermolecular ester bonds linkages within the biomass. It enhances the penetration of enzymatic and microbiological activities into the fine cell-wall structure of lignocellulosic biomass. Subsequently, it assists in the separation of the structural linkages between lignin and the polysaccharides (cellulose and hemicellulose), increases the porosity of lignocellulosic biomass, and results in an increase in internal surface area. Lignocellulosic biomass absorbs more NaOH than water, because of the hydrophobic nature of lignin, which acts as an external cross-link binder on the biomass matrix and shields the hydrophilic structural polysaccharides.

The effect of the alkaline pretreatment was evaluated through chemical composition analysis of the pretreated and non-treated biomass samples. It was reported that the use of NaOH solution and the biomass:NaOH solution ratio played a vital role in the breakdown of the lignified matrix. Table 2 shows that the preserved cellulose from the raw sample (non-treated) is higher than that from the RF alkaline pretreated samples because of the initial degradation of the sugars during the pretreatment process. The same observation applies to hemicellulose. This implies that there is a

Table 2. Acid insoluble lignin, furfural, cellulose, hemicellulose, and ash contents (% dry basis) of radio frequency alkaline pretreated and non-treated barley straw grind (Iroba et al., 2013b).

Chemical compositions	Temp (°C)	NT	Biomass:NaOH solution ratio				
			1:4	1:5	1:6	1:7	1:8
Acid Insoluble Lignin (%)	24	20.12	-	-	21.57	17.99	-
	70	-	19.56	17.8	17.91	17.82	18.53
	80	-	19.22	18.22	18.64	18.10	18.86
	90	-	20.66	17.64	18.38	18.05	19.27
Average Furfural (%)	24	4.43	-	-	3.14	4.89	-
	70	-	3.51	3.85	4.29	5.34	3.66
	80	-	3.37	3.75	3.54	4.52	3.78
	90	-	4.29	6.09	3.30	4.52	3.56
Cellulose (%)	24	42.5	-	-	30.93	25.78	-
	70	-	22.25	24.21	28.25	27.39	26.73
	80	-	22.37	21.07	31.94	26.08	18.44
	90	-	26.93	24.65	30.37	21.33	22.68
Hemicellulose (%)	24	29.98	-	-	29.12	23.40	-
	70	-	23.00	21.63	23.18	24.27	26.76
	80	-	22.14	21.38	26.14	26.75	18.32
	90	-	26.24	21.05	22.36	22.27	19.00
Ash Content (%)	24	6.03	-	-	11.94	13.17	-
	70	-	9.51	10.93	11.57	13.25	14.11
	80	-	9.79	11.15	12.02	13.61	14.73
	90	-	9.76	10.78	11.43	13.42	13.83

Temp = Temperature; NT = non-treated biomass; 1% NaOH concentration was used.

trade-off between the breakdown of the biomass matrix/creating pores in the lignin and enhancing the accessibility and digestibility of the cellulose and hemicellulose. The more the biomass matrix (lignin bond) is broken down, the more the components of interest (cellulose and hemicellulose) were degraded. Varga et al. (2002) reported that 1%, 5%, and 10% NaOH treatments reduced hemicellulose in the solid fraction by 65.9%, 79.2%, and 88.2%, respectively; it also decreased cellulose by 30.8%, 41.9%, and 53.3%, respectively.

Iroba et al. (2013b) reported that the treatment combinations using the biomass to NaOH solution ratio of 1:6 at all temperatures (70, 80, and 90°C) resulted in optimum yields of cellulose and hemicellulose. At optimal conditions, pretreatment enhanced hemicellulose yields by 75–97%, compared to enhancements of 67–75% yield for cellulose. Hemicellulose had a higher yield because of its random, branched, and amorphous structure with little strength unlike the crystalline, highly ordered cellulose with a high degree of polymerization that requires more severe pretreatment conditions. Radiofrequency-assisted alkaline pretreatment resulted in lower acid insoluble lignin.

The disruption and deconstruction of the lignified matrix are also associated with the dipole interaction, flip-flop rotation, and friction generated between the electromagnetic charges from the radio-frequency and the ions and molecules from the NaOH solution and the biomass. Based on the obtained data, Iroba et al. (2013b) predicted that this radio-frequency alkaline pretreatment would decrease the required amount and cost of enzymes by up to 64% compared to using non-treated biomass.

Alkaline pretreatment enhances accessibility and digestibility of the cellulose and hemicellulose. However, it increases the ash content. The ash content increased from about 60 to 140% depends on the ratio of biomass to NaOH solution. The ratio of biomass to NaOH solution and temperature have significant effects ($P < 0.05$) on the ash content. Ash content is a measure of the mineral content and other inorganic matter, structural or extractable in biomass (Sluiter et al., 2008). The lower the ratio of biomass to the NaOH solution, the higher is the ash content, due to the high concentration of mineral content of sodium in the NaOH solution at a lower ratio. This problem of increased ash content can be addressed by washing the pretreated samples. Washing the pretreated samples reduced the ash content to 2.63–4.37% in which about 25–55% depending on the ratio of biomass to the NaOH solution. There was no statistical difference observed for furfural (Iroba et al., 2013b).

It was concluded that in comparison with other pretreatment technologies such as steam explosion, radio-frequency alkaline pretreatment involves lower temperatures and pressures. Radio-frequency heating generates uniform heat distribution (Fig. 5) with no local hot spots or

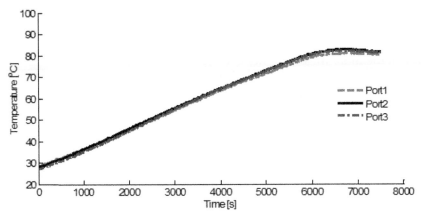

Fig. 5. The temperature profile of biomass-NaOH solution (1% w/v) at 80°C, 1.6 mm hammer   screen size, and biomass:NaOH solution ratio of 1:8 using the blown glass reactor. Ports 1–3 represent the radial temperature profile taken from the three ports of the reactor (Iroba et al., 2013b).

thermal runaway, unlike the microwave alkaline pretreatment. However, radio-frequency alkaline pretreatment involves longer pretreatment time (about 140–280% longer, depending on the ratio) mainly due to the low power level (1.5 kW) of the radio-frequency setup used in the study (Iroba et al., 2013b).

The radio-frequency alkaline pretreatment combinations (ratio 1:6 of biomass to NaOH solution at 24°C–90°C) that resulted in an optimum yield of cellulose and hemicellulose were selected and then enzymatically digested with a combined mixture of cellulase and β-glucosidase enzymes at 50°C for 96 h on a shaking incubator at 250 revs/min. The glucose in the hydrolyzed samples was subsequently quantified. The results obtained confirmed the effectiveness of the radio-frequency alkaline pretreatment process. The average available percentage glucose yield that was released during the enzymatic hydrolysis for bioethanol production ranged from 78 to 96%, depending on the treatment combination. On the other hand, non-treated sample has available average percentage glucose yield of just below 12% (Iroba et al., 2013b).

9. Limitations of Alkaline Pretreatment

The alkaline hydrolysis breaks down the cellulose-hemicellulose-lignin matrix into smaller amorphous molecules (Sun et al., 2004). Alkaline pretreatments are usually effective. However, they have some disadvantages which should not be ignored. The effect of alkaline pretreatment depends on the lignin content present in the biomass material. Some studies have concluded that alkaline pretreatment is more effective on agricultural biomass residues than on wood materials (Mohammad and Karimi, 2008; Xu et al., 2007). There are some disadvantages associated with the AFEX pretreatment method. It is more efficient on the lignocellulosic biomass that has a small amount of lignin, and does not substantially solubilize hemicellulose unlike other types of pretreatment processes (e.g., dilute-acid pretreatment) (Teymouri et al., 2004). It requires the use of specialized equipment that is corrosion resistant, it also requires extensive washing of the equipment and subsequently resulting to difficulty in the disposal of the chemical wastes (Sun et al., 2004). Using AFEX process, there is difficulty in the recycling of the feed ammonia as a reusable gas stream after the pretreatment to decrease the pretreatment cost and to keep the environment safe at the same time (Sun and Cheng, 2002; Wyman, 1996). Since ammonia is expensive, not being able to totally reuse ammonia has a negative effect on the process economy (Zheng et al., 1995). Furthermore, the corrosive and toxic nature of ammonia may result in design and operational difficulty (Sun and Cheng, 2002; Teymouri et al., 2004; Wyman, 1996). Alkaline pretreatment can be done at low temperatures but at relatively long period of time with a high concentration of the alkaline solution (Kumar et al., 2009; Mohammad and Karimi, 2008). The summary of the merits and demerits of the various pretreatment methods on lignocellulosic biomass is given below.

With the state of technology for current approaches to produce bioethanol from lignocellulosic biomass, many of these technologies will still require important investment in scale-up and process integration to be applicable for lignocellulosic biomass conversion to biofuels. A considerable finding of techno-economic and life-cycle analyses on biofuel production will be necessary for the long-term economic and environmental sustainability of biorefineries.

10. Lignin Valorization

Lignin is readily available and is a good source of aromatics. In biofuel production, lignin is typically considered to be a hindrance to cost-effectively obtaining carbohydrates. Current lignocellulose biomass conversion to biofuels requires the breakdown of lignin to liberate sugars that can be converted into advanced fuels. Despite being the most energy-dense polymer in the plant cell walls, the lignocellulose biomass conversion process results in a significant amount of lignin waste product that could be utilized for other byproducts (Linger et al., 2014). The residual lignin is currently primarily burned for process heat because it is challenging to depolymerize and upgrade into fuels or chemicals (Linger et al., 2014; Ragauskas et al., 2014; Chundawat et al., 2011; Tuck et al., 2012; Himmel et al., 2007). Using lignin for value-added applications would allow better

Table 3. Summary of merits and demerits of various processes used for the pretreatment of lignocellulosic biomass (adapted from Fan et al. (2006); Mohammad and Karimi (2008); Iroba and Tabil (2013a).

Pretreatment process	Advantages	Limitations
Alkaline pretreatment	Lesser sugar degradation, removes lignin; increases accessible surface area,utilizes lower temperatures and pressures, most caustic salts can be recovered and/or regenerated	Long residence times required; there are chemical requirements, dependent on the lignin content of the lignocellulosic biomass
Steam explosion	Results in lignin transformation; cost-effective, increases accessible surface area, decreases degrees of polymerization, usually rapid treatment rate, among the effective and promising techniques for industrial applications	Degradation of a portion of the xylan; incomplete disruption and breakdown of the lignin, carbohydrate matrix; generation of inhibiting compounds to microorganisms, chemicals are required, typically needs harsh conditions
Ammonia fiber explosion	Increases accessible surface area, eliminates lignin and hemicellulose to an extent; does not generate inhibitors for downstream processes, promising processes for industrial applications	Not efficient for biomass materials with high lignin content, there are chemical requirements
CO_2 explosion	Increases accessible surface area; no formation of inhibitory/toxic compounds, non-flammability, readily available at low cost, and environmental acceptability	High capital cost for high-pressure equipment, depends on the type of LB
Acid hydrolysis	Hydrolyzes hemicellulose to xylose and other simple sugars; changes lignin structure	High cost of corrosion resistant equipment; formation of inhibitory substances, there are chemical requirements
Mechanical comminution	Decreases cellulose crystallinity, increases the accessible surface area and pore size, causes a decrease in degrees of polymerization, chemicals are generally not required for these process	Power consumption is generally higher than resident biomass energy
Biological	Degrades lignin; low energy requirements, no chemical requirement, environmentally benign, decreases the degree of polymerization of cellulose	Rate of hydrolysis is very low, very low treatment rate, not considered for commercial application, degrades hemicelluloses

use of biomass and make biorefining more economically attractive. In the context of maintaining high sugar yields for the production of biofuels and biochemicals from cellulose and hemicellulose components, there is a need to also obtain high value from lignin in industrial-scale biorefineries which will require significant research and development. It could be a potential additive or an ingredient for the production of adhesives, resins, detergents, plastics (PVC, polyolefins, polyurethanes, polyesters) and aromatic biochemicals (Gosselink, 2015; Singh and Simmons, 2014). This will include technologies that are typically more advanced than the current lignin valorization.

Singh and Simmons (2014) are presently discovering new depolymerization procedures for lignin, which convert this polymer into simple monomers that can be used for the production of high-value chemicals from a renewable source. The depolymerization techniques currently being examined are also overcoming the recalcitrant nature of lignin in lignocellulosic biomass conversion. These chemical-based methods avoid the use of specialized equipment and the need for thermo-chemical processes which require combinations of high pressure and temperatures, diluted acids or alkalis, with high energy inputs. Ionic liquid solvents are used to extract lignin from biomass and depolymerize the macromolecules at lower temperatures. The ionic liquids derived from biomass conversion have also been used to depolymerize lignin, which greatly enhances the renewable value of this method (Singh and Simmons, 2014). Some of the high-value biochemicals derived from renewable lignin depolymerization process include benzene, toluene, xylene, styrene, biphenols, cyclohexane, syringaldehyde, vanillin and vanillic acid (Singh and Simmons, 2014). Developing biobased chemicals will increase the profitability of second-generation biofuels production (Gosselink, 2015). Linger et al. (2014) demonstrated that the use of aromatic catabolic pathways enable an approach to valorize lignin by overcoming its inherent heterogeneity to produce biofuels, biochemicals, and biomaterials.

11. Summary

It is evident that alkaline pretreatment fractionates lignin leaving low molecular mass cellulose and hemicellulose. Alkaline pretreatment techniques create reaction sites for enzymatic hydrolysis with a corresponding higher glucose yield relative to the non-treated biomass. The alkaline pretreated samples require low enzyme loading with a corresponding higher glucose yield, unlike the non-treated biomass. The glucose yield indicates the amount of readily available glucose in the biomass that can be released during the enzymatic hydrolysis which will be available for bioethanol production. Therefore, it can be concluded that to improve the yield of lignocellulosic biomass for bioethanol production, it is necessary to perform pretreatment before the enzymatic hydrolysis step.

References

Alizadeh, H., F. Teymouri, T. I. Gilbert and B. E. Dale. 2005. Pretreatment of switchgrass by ammonia fiber explosion (AFEX). Applied Biochemistry and Biotechnology 124: 1133–1141.

Bjerre, A. B., A. B. Olesen and T. Fernqvist. 1996. Pretreatment of wheat straw using combined wet oxidation and alkaline hydrolysis resulting in convertible cellulose and hemicellulose. Biotechnology and Bioengineering 49: 568–577.

Bush, E. J., J. W. Loder, T. S. James, L. D. Mortsch and S. J. Cohen. 2014. An overview of Canada's changing climate. pp. 23–64. *In:* Warren, F. J. and D. S. Lemmen (eds.). Canada in a Changing Climate: Sector Perspectives on Impacts and Adaptation. Government of Canada, Ottawa, ON, http://www.nrcan.gc.ca/sites/www.nrcan.gc.ca/files/earthsciences/pdf/assess/2014/pdf/Chapter2 Overview_Eng.pdf. (2016/04/04).

Chandra, R. P., R. Bura, W. E. Mabee, A. Berlin, X. Pan and J. N. Saddler. 2007. Substrate pretreatment: The key to effective enzymatic hydrolysis of lignocellulosics? Advances in Biochemical Engineering/Biotechnology 108: 67–93.

Chang, V. S., B. Burr and M. T. Holtzapple. 1997. Lime pretreatment of switchgrass. Applied Biochemistry and Biotechnology 63-65: 3–19.

Chang, V. S., M. Nagwani and M. T. Holtzapple. 1998. Lime pretreatment of crop residues bagasse and wheat straw. Applied Biochemistry and Biotechnology 74: 135–159.

Chang, V. S. and M. T. Holtzapple. 2000. Fundamental factors affecting biomass enzymatic reactivity. Applied Biochemistry and Biotechnology 84: 5–37.

Chang, V. S., M. Nagwani, C. H. Kim and M. T. Holtzapple. 2001. Oxidative lime pretreatment of high-lignin biomass. Applied Biochemistry and Biotechnology 94: 1–28.

Chen, X. W., J. Shekiro, M. A. Franden, W. Wang, M. Zhang, E. Kuhn, D. K. Johnson and M. P. Tucker. 2012. The impacts of deacetylation prior to dilute acid pretreatment on the bioethanol process. Biotechnology for Biofuels 5(8): 1–14.

Chundawat, S. P. S., G. T. Beckham, M. E. Himmel and B. E. Dale. 2011. Deconstruction of lignocellulosic biomass to fuels and chemicals. Annual Review of Chemical and Biomolecular Engineering 2: 121–145.

Chung, Y. C., A. Bakalinsky and M. H. Penner. 2005. Enzymatic saccharification and fermentation of xylose-optimized dilute acid-treated lignocellulosics. Applied Biochemistry and Biotechnology 121-124: 947–961.

Demirbas, A., M. Fatih, M. Balat and H. Balat. 2009. Potential contribution of biomass to the sustainable energy development. Energy Conversion and Manage 50(7): 1746–1760.

Donepudi, A. and K. Muthukumarappan. 2009. Effect of microwave pretreatment on sugar recovery from corn stover. Transactions of ASAE, Paper Number: 097057. Reno, Nevada. June 21–June 24, doi:10.13031/2013.27442.

Fan, L. T., M. M. Gharpuray and Y. H. Lee. 1987. Cellulose Hydrolysis Biotechnology Monographs. Springer, Berlin, p. 57.

Fan, L. T., L. Yong-Hyun and M. M Gharpuray. 2006. The nature of lignocellulosics and their pretreatments for enzymatic hydrolysis. Advances in Biochemical Engineering/Biotechnology 23: 157–187.

Feist, W. C., A. J. Baker and H. Tarkow. 1970. Alkali requirements for improving digestibility of hardwoods by rumen micro-organisms. Journal of Animal Science 30: 832–835.

Gandi, J., M. T. Holtzapple, A. Ferrer, F. M. Byers, N. D. Turner, M. Nagwani and S. Chang. 1997. Lime treatment of agricultural residues to improve rumen digestibility. Animal Feed Science and Technology 68: 195–211.

Gaspar, M., G. Kalman and K. Reczey. 2007. Corn fiber as a raw material for hemicellulose and ethanol production. Process Biochemistry 42: 1135–1139.

Genco, M. J., N. Busayasakul, K. H. Medhora and W. Robbin. 1990. Hemicellulose retention during kraft pulping. Tappi Journal 73: 223–233.

Gosselink, R. 2015. Lignin valorization towards materials, chemicals and energy. 2nd Lund symposium on lignin and hemicellulose valorisation, Nov. 3–4, 2015, Palaestra, Lund, Sweden.

Himmel, M. E., S. Y. Ding, D. K. Johnson, W. S. Adney, M. R. Nimlos, J. W. Brady and T. D. Foust. 2007. Biomass recalcitrance: Engineering plants and enzymes for biofuels production. Science 315(5813): 804–807.

Iroba, K. L. and L. G. Tabil. 2013a. Lignocellulosic biomass: Feedstock characteristics, pretreatment methods and pre-processing for biofuel and bioproduct applications, U.S. and Canadian perspective. pp. 61–98. *In*: Zhang, B. and Y. Wang (eds.). Biomass Processing, Conversion and Biorefinery. New York: Nova Science Publishers, Inc.

Iroba, K. L., L. G. Tabil, T. Dumonceaux and O. D. Baik. 2013b. Effect of alkaline pretreatment on chemical composition of lignocellulosic biomass using radio frequency heating. Biosystems Engineering 116: 385–398.

Jafari, V., H. Sixta and A. Heininen. 2014. Kinetics of oxygen delignification of high-kappa pulp in a continuous flow-through reactor. Industrial and Engineering Chemistry Research 53: 8385–8394.

Jeoh, T., C. I. Ishizawa, M. F. Davis, M. E. Himmel, W. S. Adney and D. K. Johnson. 2007. Cellulase digestibility of pretreated biomass is limited by cellulose accessibility. Biotechnology and Bioengineering 98(1): 112–122.

Ji, Y. and A. Heininen. 2007. A new CSTR for oxygen delignification mechanism and kinetics study. Pulp and Paper Canada 108: 38–42.

Jorgensen, H., J. B. Kristensen and C. Felby. 2007. Enzymatic conversion of lignocellulose into fermentable sugars: Challenges and opportunities. Biofuels Bioproducts and Biorefining 1: 119–134.

Karagoz, S., T. Bhaskar, A. Muto, Y. Sakata and M. A. Uddin. 2004. Low-temperature hydrothermal treatment of biomass: effect of reaction parameters on products and boiling point distributions. Energy and Fuels 18: 234–241.

Karp, E. M., B. S. Donohoe, M. H. O'Brien, P. N. Ciesielski, A. Mittal, M. J. Biddy and G. T. Beckham. 2014. Alkaline pretreatment of corn stover: bench-scale fractionation and stream characterization. ACS Sustainable Chemistry & Engineering 2: 1481–1491.

Karp, E. M., M. G. Resch, B. S. Donohoe, P. N. Ciesielski, M. H. O'Brien, J. E. Nill, A. Mittal, M. J. Biddy and G. T. Beckham. 2015. Alkaline pretreatment of switchgrass. ACS Sustainable Chemistry & Engineering 3: 1479–1491.

Karr, W. E. and M. T. Holtzapple. 1998. The multiple benefits of adding non-ionic surfactant during the enzymatic hydrolysis of corn stover. Biotechnology and Bioengineering 59: 419–427.

Karr, W. E. and M. T. Holtzapple. 2000. Using lime pretreatment to facilitate the enzymatic hydrolysis of corn stover. Biomass and Bioenergy 18: 189–199.

Kashaninejad, M. and L. G. Tabil. 2011. Effect of microwave-chemical pre-treatment on compression characteristics of biomass grinds. Biosystems Engineering 108: 36–45.

Kaur, U., H. S. Oberoi, V. K. Bhargav, R. Sharma-Shivappac and S. S. Dhaliwal. 2012. Ethanol production from alkali- and ozone-treated cotton stalks using thermotolerant Pichia kudriavzevii HOP-1. Industrial Crops and Products 37: 219–226.

Kim, K. H. and J. Hong. 2001. Supercritical CO_2 pretreatment of lignocellulose enhances enzymatic cellulose hydrolysis. Bioresource Technology 77: 139–144.

Kim, S. and M. T. Holtzapple. 2005. Lime pretreatment and enzymatic hydrolysis of corn stover. Bioresource Technology 96: 1994–2006.

Kim, S. and M. T. Holtzapple. 2006a. Delignification kinetics of corn stover in lime pre-treatment. Bioresource Technology 97: 778–785.

Kim, S. and M. T. Holtzapple. 2006b. Effect of structural features on enzyme digestibility of corn stover. Bioresource Technology 97: 583–591.

Kim, T. H., F. Taylor and K. B. Hicks. 2008. Bioethanol production from barley hull using SAA (soaking in aqueous ammonia) pretreatment. Bioresource Technology 99: 5694–5702.

Klinke, H. B., B. K. Ahring, A. S. Schmidt and A. B. Thomsen. 2002. Characterization of degradation products from alkaline wet oxidation of wheat straw. Bioresource Technology 82: 15–26.

Kumar, P., D. M. Barrett, M. J. Delwiche and P. Stroeve. 2009. Methods for pretreatment of lignocellulosic biomass for efficient hydrolysis and biofuel production. Industrial and Engineering Chemistry Research 48: 3713–3729.

Larhed, M., C. Moberg and A. Hallberg. 2002. Microwave-accelerated homogenous catalysis in organic chemistry. Accounts of Chemical Research 35: 717–727.

Linger, J. G., D. R. Vardon, M. T. Guarnieri, E. M. Karp, G. B. Hunsinger, M. A. Franden, C. W. Johnson, G. Chupka, T. J. Strathmann, P. T. Pienkos and G. T. Beckham. 2014. Lignin valorization through integrated biological funneling and chemical catalysis PNAS 111(33): 12013–12018. www.pnas.org/cgi/doi/10.1073/pnas.1410657111.

Liu, R., H. Yu and Y. Huang. 2005. Structure and morphology of cellulose in wheat straw. Cellulose 12: 25–34.

Lucia, L. A., A. J. Ragauskas and F. S. Chakar. 2002. Comparative evaluation of oxygen delignification processes for low- and high-lignin content softwood kraft pulps. Industrial and Engineering Chemistry Research 41: 5171–5180.

Lynd, L. R., C. E. Wyman and T. U. Gerngross. 1999. Biocommodity engineering. Biotechnology Progress 15: 777–793.

Martin, C., Y. Gonzalez, T. Fernandez and A. B. Thomsen. 2006. Investigation of cellulose convertibility and ethanolic fermentation of sugarcane bagasse pretreated by wet oxidation and steam explosion. Journal of Chemical Technology and Biotechnology 81: 1669–1677.

Mendoca, R., A. Guerra and A. Ferraz. 2002. Delignification of pinus taeda wood chips treated with ceriporiopsis subvermispora for preparing high-yield kraft pulp. Journal of Chemical Technology and Biotechnology 77: 411–418.

Millett, M. A., A. J. Baker and L. D. Satter. 1975. Pretreatments to enhance chemical, enzymatic and microbiological attack of cellulosic materials. Biotechnology and Bioengineering Symposium 5: 193–219.

Minowa, T., F. Zhen, T. Ogi and G. Varhegyi. 1997. Liquefaction of cellulose in hot compressed water using sodium carbonate: Products distribution at different reaction temperatures. Journal of Chemical Engineering of Japan 30: 186–190.

Mohammad, J. T. and K. Karimi. 2008. Pretreatment of lignocellulosic wastes to improve ethanol and biogas production. International Journal of Molecular Sciences 9: 1621–1651.

Montane, D., X. Farriol, J. Salvado, P. Jollez and E. Chornet. 1998. Application of steam explosion to the fractionation and rapid vapor-phase alkaline pulping of wheat straw. Biomass and Bioenergy 14(3): 261–276.

Mosier, N., C. Wyman, B. Dale, R. Elander, Y. Y. Lee, M. Holtzapple and M. Ladisch. 2005. Features of promising technologies for pretreatment of lignocellulosic biomass. Bioresource Technology 96: 673–686.

Nlewem, K. C. and M. E. Thrash. 2010. Comparison of different pretreatment methods based on residual lignin effect on the enzymatic hydrolysis of switchgrass. Bioresource Technology 101: 5426–5430.

Okano, K., M. Kitagawa, Y. Sasaki and T. Watanabe. 2005. Conversion of Japanese red cedar (Cryptomeria japonica) into a feed for ruminants by white-rot basidiomycetes. Animal Feed Science and Technology 120: 235–243.

Palonen, H., A. B. Thomsen, M. Tenkanen, A. S. Schmidt and L. Viikari. 2004. Evaluation of wet oxidation pretreatment for enzymatic hydrolysis of softwood. Applied Biochemistry and Biotechnology 117(1): 1–17.

Park, Y. C. and J. S. Kim. 2012. Comparison of various alkaline pretreatment methods of lignocellulosic biomass. Energy 47: 31–35.

Peterson, A., M. H. Thomsen, H. Hauggaard-Nielsen and A. B. Thomsen. 2007. Potential bioethanol and biogas production using lignocellulosic biomass from winter rye, oilseed rape and faba bean. Biomass and Bioenergy 31: 812–819.

Playne, M. J. 1984. Increased digestibility of bagasse by pretreatment with alkalis and steam explosion. Biotechnology and Bioengineering 26(5): 426–433.

Punidadas, R., D. Chantal, K. Tatiana, H. S. Ramaswamy and G. B. Awuah. 2003. Radio frequency heating of foods: principles, applications and related properties—a review. Critical Reviews in Food Science and Nutrition 43(6): 587–606.

Ragauskas, A. J., G. T. Beckham, M. J. Biddy, R. Chandra, F. Chen, M. F. Davis, B. H. Davison, R. A. Dixon, P. Gilna, M. Keller, P. Langan, A. K. Naskar, J. N. Saddler, T. J. Tschaplinski, G. A. Tuskan and C. E. Wyman. 2014. Lignin valorization: Improving lignin processing in the biorefinery. Science 344(6185): 1246843-1–1246843-10.

Rai, S. N. and V. D. Mudgal. 1987. Effect of sodium hydroxide and steam pressure treatment on the utilization of wheat straw by rumen microorganisms. Biological Wastes 21: 203–212.

Ramaswamy, H. and J. Tang. 2008. Microwave and radio frequency heating. Food Science and Technology International 14(5): 423–427.

Ramaswamy, H. S. and M. A. Tung. 1981. Thermophysical properties of apples in relation to freezing. Journal of Food Science 46: 724–728.

Ramesh, C. K. and A. Singh. 1993. Lignocellulose biotechnology: current and future prospects. Critical Reviews in Biotechnology 13(2): 151–172.

Ryynanen, S. 1995. The electromagnetic properties of food materials: A review of the basic principles. Journal of Food Engineering 26: 409–429.

Saha, B. C., L. B. Iten, M. A. Cotta and Y. V. Wu. 2005. Dilute acid pretreatment, enzymatic saccharification and fermentation of wheat straw to ethanol. Process Biochemistry 40: 3693–3700.

Saha, B. C. and M. A. Cotta. 2006. Ethanol production from alkaline peroxide pretreated enzymatically saccharified wheat straw. Biotechnology Progress 22: 449–453.

Savy, D. and A. Piccolo. 2014. Physical chemical characteristics of lignins separated from biomasses for second-generation ethanol. Biomass and Bioenergy 62: 58–67.

Sierra, R., C. B. Granda and M. Holtzapple. 2009. Short-term lime pretreatment of poplar wood. Biotechnology Progress 25: 323–332.

Silverstein, R. A., Y. Chen, R. R. Sharma-Shivappa, M. D. Boyette and J. Osborne. 2007. A comparison of chemical pretreatment methods for improving saccharification of cotton stalks. Bioresource Technology 98: 3000–3011.

Singh, S. and B. Simmons. 2014. Biofuels: Increasing the Value of Lignin. Discovering Effective Methods of Depolymerizing Lignin Will Improve Economics of Biorefineries and Create a Renewable Resource for Chemicals. Sandia National laboratories, 7011 East Avenue Livermore, CA.

Sluiter, A., B. Hames, R. Ruiz, C. Scarlata, J. Sluiter and D. Templeton. 2008. Determination of ash in biomass, Laboratory Analytical Procedure (LAP). National Renewable Energy Laboratory Jan. Technical Report No.: NREL/TP-510-42622. Golden, CO: NREL.

Söderström, J., L. Pilcher, M. Galbe and G. Zacchi. 2003. Two-step steam pretreatment of softwood by dilute H_2SO_4 impregnation for ethanol production. Biomass and Bioenergy 24(6): 475–486.

Sokhansanj, S., S. Mani, M. Stumborg, R. Samson and J. Fenton. 2006. Production and distribution of cereal straw on the Canadian prairies. Canadian Biosystems Engineering 48: 3.39–3.46.

Soleimani, M., L. G. Tabil and C. Niu. 2015. Delignification of intact biomass and cellulosic coproduct of acid-catalyzed hydrolysis. AIChE Journal 61: 1783–1791.

Sun, X. F., F. Xu, R. C. Sun, Y. X. Wang, P. Fowler and M. S. Baird. 2004. Characteristics of degraded lignins obtained from steam exploded wheat straw. Polymer Degradation and Stability 86: 245–256.

Sun, Y. and J. J. Cheng. 2002. Hydrolysis of lignocellulosic materials for ethanol production. Bioresource Technology 83(1): 1–11.

Taherzadeh, M. J. and K. Karimi. 2008. Pretreatment of lignocellulosic wastes to improve ethanol and biogas production: a review. International Journal of Molecular Science 9: 1621–1651.

Teymouri, F., L. Laureano-Pérez, H. Alizadeh and B. E. Dale. 2004. Ammonia fiber explosion treatment of corn stover. Applied Biochemistry and Biotechnology 115(1-3): 951–963.

Tuck, C. O., E. Pérez, I. T. Horváth, R. A. Sheldon and M. Poliakoff. 2012. Valorization of biomass: Deriving more value from waste. Science 337(6095): 695–699.

Van Wyk, J. P. H. 1997. Cellulose hydrolysis and cellulase adsorption after pretreatment of cellulose materials. Biotechnology Techniques 11(6): 443–445.

Varga, E., Z. Szengyel and K. Reczey. 2002. Chemical pretreatments of corn stover for enhancing enzymatic digestibility. Applied Biochemistry and Biotechnology 98(1-3): 73–87.

Wang, Z., D. R. Keshwani, A. P. Redding and J. J. Cheng. 2008. Alkaline Pretreatment of Coastal Bermudagrass for Bioethanol Production. ASABE Annual International Meeting, Rhode Island Convention Center Providence, Rhode Island, June 29–July 2, Paper Number: 084013.

Whitmire, D. and B. Maiti. 2002. Xylanase effects on pulp delignification. Chemical Engineering Communications 189: 608–622.

Wyman, C. E. 1996. Handbook on Bioethanol: Production and Utilization; Applied Energy Technology Series. Taylor & Francis: Washington DC, USA.

Wyman, C. E. 1999. Biomass ethanol: technical progress, opportunities, and commercial challenges. Annual Review of Energy and Environment 24: 189–226.

Xu, Z., Q. Wang, Z. Jiang, X. Yang and Y. Ji. 2007. Enzymatic hydrolysis of pretreated soybean straw. Biomass and Bioenergy 31: 162–167.

Yin, S. and Z. Tan. 2012. Hydrothermal liquefaction of cellulose to bio-oil under acidic, neutral and alkaline conditions. Applied Energy 92: 234–239.

Zhang, Y. H. P. 2008. Reviving the carbohydrate economy via multiproduct lignocellulose biorefineries. Journal of Industrial Microbiology and Biotechnology 35: 367–375.

Zheng, Y., H. M. Lin, J. Wen, N. Cao, X. Yu and G. T. Tsao. 1995. Supercritical carbon dioxide explosion as a pretreatment for cellulose hydrolysis. Biotechnology Letters 17: 845–850.

Zheng, Y., Z. Pan and R. Zhang. 2009. Overview of biomass pretreatment for cellulosic ethanol production. International Journal of Agricultural and Biological Engineering 2: 51–68.

Zhou, C. H., X. Xia, C. X. Lin, D. S. Tong and J. Beltramini. 2011. Catalytic conversion of lignocellulosic biomass to fine chemicals and fuels. Chemical Society Reviews 40: 5588–5617.

Zhu, S., Y. Wu, Z. Yu, J. Liao and Y. Zhang. 2005. Pretreatment by microwave/alkali of rice straw and its enzymic hydrolysis. Process Biochemistry 40: 3082–3086.

Zhu, S., Y. Wu, Z. Yu, C. Wang, F. Yu, S. Jin, Y. Ding, R. Chi, J. Liao and Y. Zhang. 2006. Comparison of three microwave/chemical pretreatment processes for enzymatic hydrolysis of rice straw. Biosystems Engineering 93(3): 279–283.

CHAPTER 15

Impacts, Challenges, and Economics of Ionic Liquid Pretreatment of Biomass

Jipeng Yan, Ling Liang, Todd R. Pray and *Ning Sun**

1. Introduction

Lignocellulosic biomass mainly consists of cellulose, hemicelluloses, and lignin, which has different reactivity during chemical and physical pretreatment. Cellulose is built up by 10,000 glucose units through β (1→4)-glucosidic bonds and has a highly crystalline and less ordered amorphous region (Sjöström, 1993). The strong hydrogen bonds and fibrous structure of cellulose enable its high resistance to pretreatments. Compared to cellulose, hemicelluloses have a lower degree of polymerization (less than 200) and short branched chain connections, which can be relatively easily broken down to release fermentable sugars. Lignin, an amorphous heteropolymer, is mainly built up by phenylpropanoid units linked by β-O-4, β→5, α-O-4, β-β, 4-O-5, and 5→5 bonds (Sjöström, 1993), which has a chemical degradation resistance between cellulose (most resistant) and hemicelluloses (least resistant) (Liu, 2010). In the process of physical and chemical pretreatment, lignin prevents agents from passing through the cell wall to protect cellulose and hemicelluloses from degradation. Therefore,

Advanced Biofuels Process Demonstration Unit, Biological Systems and Engineering Division, Lawrence Berkeley National Laboratory, Emeryville, CA, United States.
Emails: jipengyan@lbl.gov; lliang@lbl.gov; tpray@lbl.gov
* Corresponding author: nsun@lbl.gov

in order to obtain fermentable sugars from lignocellulose, an efficient pretreatment method is required to overcome the natural recalcitrance of biomass contributed by the cellulose crystallinity (Beckham et al., 2011), the presence of lignin (Pu et al., 2007), functional groups on hemicelluloses (Brodeur et al., 2011), and interwoven linkages among these major components (Sathitsuksanoh et al., 2013).

Conversion of biomass into advanced biofuels and value-added chemicals offers an opportunity to reduce dependence on fossil fuels and subsequently reduce carbon emissions. Due to the sustainability of plant-based biomass, the expanding bioenergy industry has the potential to contribute to the development of the rural economy and energy security (Lynd et al., 2005). Although biomass resources have promising merits, biomass recalcitrance has been recognized as the grand challenge for efficient utilization (Dekker and Wallis, 1983; Lu et al., 2012). Therefore, different biomass pretreatment methods, such as physical (Zhu and Pan, 2010), irradiation (Wasikiewicz et al., 2005), alkaline (Zhao et al., 2008), acid (Hinman et al., 1992; Grethlein and Converse, 1991), steam explosion (Duff and Murray, 1996), sulfite (Zhu et al., 2009), hot-water extraction (Yan et al., 2015; Yan et al., 2013; 2016), and IL pretreatment (Li et al., 2010a), are developed to overcome the challenges by breaking inter- or intra-molecular linkages among lignin, hemicellulose, and cellulose. Among various pretreatment technologies, IL pretreatment has been receiving significant attention as a potential "green" process that enables fractionation of a wide range of feedstocks due to its high processing efficiency (Li et al., 2010a; Sun et al., 2011; Sun et al., 2013b; Sun et al., 2015; Xu et al., 2012; Ragauskas et al., 2014; Perez-Pimienta et al., 2015).

Figure 1 shows two common approaches for IL-based bioprocess. Figure 1a is the conventional approach including pretreatment, washing and ILs recovery, saccharification, and fermentation. The washing and ILs recovery step can help to improve IL usage efficiency and avoid inhibition effect in the downstream process (Ouellet et al., 2011). However, a significant amount of water consumption and wastewater treatment cause issues for process scale-up and economy. More recently, the one-pot process was developed as shown in Fig. 1b, where pretreatment and saccharification are performed in a single vessel consequently without washing and followed directly by sugar extraction and ILs recovery. Approximately, 2–15 fold reductions in water usage were observed as compared to the conventional approach (Shi et al., 2013a). Xu et al. (2016a) reported an overall 40 percent reduction in the cost of cellulosic ethanol production and a reduction in local burdens on water resources and waste management infrastructure with the help of a high-gravity and one-pot process design (Xu et al., 2016a).

Typically, ILs refer to salts consisting of anions and cations, which melt under 100°C and have a very low vapor pressure at room temperature. The

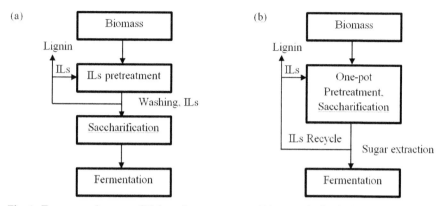

Fig. 1. Two general routes of IL-based pretreatment of biomass for biofuels production (Shi et al., 2013a; Xu et al., 2016a).

thermodynamic and physicochemical properties of ILs can be controlled by selection of anions and cations. Therefore, the specific combination of the anion and cation could be developed to meet some targeted properties, thus called "designer solvent" (Freemantle, 1998). Table 1 lists a summary of ILs used in biomass pretreatment. Among a large group of ILs, imidazolium-based ILs are well-studied to efficiently breakdown various biomass through solvation of the major components in the plant cell wall (Sun et al., 2009; Sun et al., 2014; Xu et al., 2012). However, imidazolium-based ILs are expensive and toxic to microorganisms which are the major hurdle for process development and commercialization. Another group of ILs built up with cholinium cations and amino acid anions were developed as "bionic liquids", and reported to selectively remove lignin in rice straw and significantly improve saccharification rates and sugar yields (Hou et al., 2012). Compared to imidazolium base ILs, bionic liquids are more compatible with enzymes and microbes and are relatively low-cost. Although the detailed mechanism of IL pretreatment is still not clear, two main reasons contribute to the increase of biomass reactivity: (1) dissolution of cellulose and reduction of its crystallinity (Li et al., 2010a; Swatloski et al., 2002); and (2) selective delignification (Sun et al., 2014; Hou et al., 2012). The physical and chemical modifications of lignin and polysaccharides during pretreatment have a great impact on the following step of enzymatic hydrolysis, which includes enzyme loading, hydrolysis rate, enzymatic digestibility, and product recovery and purification (Torr et al., 2012; Zhang, 2008).

In this chapter, we discuss the impact of IL pretreatment on the major biomass components for the benefit of enzymatic conversion. Both technical and economic challenges of IL pretreatment were discussed to provide some perspectives on future process development and optimization.

Table 1. Summary of ILs used in biomass pretreatment.

Chemical name	Structure of ILs		Reference
1-allyl-3-methylimidazolium chloride		Cl⁻	(Wang et al., 2014)
1-butyl-3-methylimidazolium bromide		Br⁻	(Xu et al., 2014)
1-butyl-3-methylimidazolium chloride		Cl⁻	(Dadi et al., 2006; Nguyen et al., 2010; Zhao et al. 2009; Li et al., 2010b; Fu et al., 2010; Kim et al., 2010)
1-butyl-3-methylimidazolium hexafluorophosphate			(Li et al., 2010b)

Name	Cation	Anion	Reference
1-butyl-3-methylimidazolium acetate			(Zhao et al., 2009)
1-benzyl-3-methylimidazolium chloride		Cl⁻	(Lee et al., 2009)
1-butyl-1-methylpyrrolidinium chloride		Cl⁻	(Li et al., 2010b)
1-butyl-3-methylimidazolium methylsulfate			(Brandt et al., 2011)
N,N-dimethylethanolammonium formate			(Fu et al., 2010)

Table 1 contd....

...Table 1 contd.

Chemical name	Structure of ILs	Reference
N,N-dimethylethanolammonium acetate		(Fu et al., 2010; Li et al., 2010b)
N,N-dimethylethanolammonium glycolate		(Fu et al., 2010)
N,N-dimethylethanolammonium succinate		(Fu et al., 2010)
1-ethyl-3-methylimidazole acetate		(Shill et al., 2011; Samayam and Schall, 2010; Nguyen et al., 2010; Zhao et al., 2009; Fu et al., 2010; Wang et al., 2011; Lee et al., 2009)
1-ethyl-3-methylimidazolium dimethyl phosphate		(Li et al., 2010b)

Name	Cation	Anion	References
1-ethyl-3-methylimidazolium diethyl phosphate			(Li et al., 2009; Kamiya et al., 2008; Li et al., 2010b)
1-ethyl-3-methylimidazolium chloride		Cl⁻	(Binder and Raines, 2010; Nguyen et al., 2010)
1-ethyl-3-methylimidazolium hydrogen sulfate			(Nguyen et al., 2010)
Monoethylammoniumhy hydrogen sulfate			(George et al., 2015)
Diethylammonium hydrogen sulfate			(George et al., 2015)

Table 1 contd. ....

...*Table 1 contd.*

Chemical name	Structure of ILs		Reference
Triethylammonium hydrogen sulfate			(George et al., 2015)
Monoethanolammonium hydrogen sulfate			(George et al., 2015)
Diethanolammonium hydrogen sulfate			(George et al., 2015)
Triethanolammonium hydrogen sulfate			(George et al., 2015)
Diisopropylammonium hydrogen sulfate			(George et al., 2015)
1,3-dimethylimidazolium methyl sulfate			(Lee et al., 2009)

Name	Cation	Anion	Reference
1,3-dimethylimidazolium dimethyl phosphite			(Li et al., 2010b)
Cholinium acetate			(Sun et al., 2014; Xu et al., 2016a)
Cholinium aspartate			(Xu et al., 2016a)
Cholinium glycine			(Hou et al., 2012)
Cholinium lysine			(Hou et al., 2012; Sun et al., 2014; Xu et al., 2016a)
Cholinium alanine			(Hou et al., 2012)

Table 1 contd. ….

...Table 1 contd.

Chemical name	Structure of ILs		Reference
Cholinium serine			(Hou et al., 2012)
Cholinium threonine			(Hou et al., 2012)
Cholinium methionine			(Hou et al., 2012)
Cholinium phenylalanine			(Hou et al., 2012)

2. How Does Ionic Liquid Work?

In order to evaluate the impact of IL pretreatment on the chemical composition of biomass for the benefit of enzymatic conversion, we need to understand the interactions between ILs and biomass in the process. The performance of the IL pretreatment is affected by key factors, including biomass type and solid loads, process conditions (temperature, residence time, and agitation), rector configuration, and recovery technologies (Singh and Simmons, 2013). In general, a higher temperature and incubation time lead to more lignin removal (Li et al., 2010a; Sun et al., 2009). To understand how ILs interacts with biomass, confocal Raman microscopy and confocal fluorescence microscopy were employed which provided new insight into the molecular-level behaviors (Sun et al., 2013a). During IL-pretreatment, cell wall swelling occurs primarily in the secondary plant cell walls and the compound middle lamella was merely affected. Lignin in the secondary cell walls dissolves rapidly, while cellulose dissolves at high speed regardless of distribution in the cell walls. The dissolution ability of lignin and cellulose varied among different type of cells. Tracheids showed much faster lignin and cellulose dissolution rate than parenchyma cells. Sclerenchyma cells have intermediate lignin dissolution rate and comparable cellulose dissolution rate to tracheids (Sun et al., 2013a). The removal of lignin and the change of cellulose morphology through IL pretreatment enable a larger surface area of the biomass and higher accessibility to enzymes leading to higher sugar yields in the following enzymatic hydrolysis process (Li et al., 2010a; Dadi et al., 2007; Lee et al., 2009; An et al., 2015; Yang and Fang, 2015; Asakawa et al., 2016).

Dissolution of cellulose in ILs is mainly due to the interruption of inter- and intra-molecular hydrogen bonding and formation of new hydrogen bonds between the anion and carbohydrate hydroxyl groups (Youngs et al., 2007). The dissolution of lignin is driven by the interactions of hydrogen bonding and π–π stacking with ILs cations (Janesko, 2011). The solvation ability of ILs is governed by their polarity, which can be quantified by the Kamlet-Taft parameters using solvatochromic dyes such as 4-nitroaniline and N, N-diethyl-4-nitroaniline and UV-Vis spectroscopy (Kamlet and Taft, 1976; Taft and Kamlet, 1976; Yokoyama et al., 1976; Lungwitz et al., 2010). In the Kamlet-Taft system, three solvent parameters, including hydrogen bond acidity (α), hydrogen bond basicity (β), as well as solvent polarizability (π^*) are defined, which present as a measure of the IL as a hydrogen bond donor, a hydrogen bond acceptor, and a measure of the dipolarity/polarizability of the ILs, respectively. Since the β value quantifies an IL's ability to accept a hydrogen bond, its magnitude is primarily determined by the anion. ILs with higher β values, and more recently reported, ILs with larger differences between β and α, i.e., net basicity (β-α), tend to dissolve cellulose more efficiently (Hauru et al., 2012). Upon precipitation with anti-solvent, the regenerated cellulose often has reduced crystallinity and is easily digested by cellulase (Doherty et al., 2010). Hydrogen bond basicity not only affects an IL's capacity to dissolve and/or swell lignocellulose, but also acts as an excellent predictor of biomass pretreatment efficacy (Doherty et al., 2010) ILs with higher β values significantly remove lignin, reduce cellulose crystallinity, and result in higher glucose yields after enzymatic saccharification (Doherty et al., 2010).

2.1 Impact of IL pretreatment on cellulose

The hydroxyl groups in cellulose form hydrogen bonds with the anions in ILs, which leads to the dissolution of cellulose (Remsing et al., 2006). Both experimental studies using nuclear magnetic resonance (NMR) (Remsing et al., 2006) and theoretical studies employing molecular dynamics simulations (Liu et al., 2011) were reported on the interactions between cellulose and ILs. Liu and coworkers investigated the interactions that govern the dissolution and regeneration of cellulose in a complex system including 1-ethyl-3-methylimidazolium acetate, water, and cellulose using molecular dynamics simulations, which proposed that the addition of water changes the structure of the IL and bonding between the IL and cellulose leading to the precipitation of cellulose. Cellulose crystal structure is changed from cellulose I to cellulose II with decreased crystallinity after solvation by the ILs (Liu et al., 2011).

Although likely to be a major factor in the ability of salt to dissolve cellulose, the potential for hydrogen bonding between hydroxyl groups and the anion is not the only factor; the cation appears to have an influence as well, although this is not yet entirely understood. For instance, cellulose is reported to be virtually insoluble in 1-alkyl-3-methylimidazolium chlorides ([CnC1Im]Cl, with 'n' being the number of carbons in the linear alkyl chain) when the alkyl chain is propyl (ca. 0.5 wt% solubility) or decyl (Vitz et al., 2009). Moreover, within the series of $[C_nC_1Im]Cl$ ILs, those with an even number of carbons in their alkyl chain substituent exhibit higher capacities for cellulose dissolution as compared to those having an odd number, for relatively short alkyl chains (pentyl and shorter) (Erdmenger et al., 2007). 1-allyl-3-methylimidazolium chloride ([AC$_1$Im]Cl) is reported to be more efficient than $[C_4C_1Im]Cl$ (Zhang et al., 2005), although this may be a result of the decreased viscosity of the IL produced by the double bond in the side chain, which could facilitate cellulose dissolution. Additionally, the smaller size of the [Amim]$^+$ cation and its more polarizable nature might contribute to greater interactions with cellulose.

2.2 Impact of ILs pretreatment on lignin

The covalent linkages are not only within each polymer but also in the lignin-carbohydrate complex (LCC) which decreases the solubility of lignocellulose and the substrate's enzyme accessibility (Lee et al., 2009; Mosier et al., 2005) and cause challenges in the process of pretreatment and saccharification. Also, lignin can also bind with cellulase and restrict swelling of cellulose which leads to a lower enzymatic hydrolysis efficiency (Kumar et al., 2012). Therefore, one of the major functions of ILs is to remove lignin and expose cellulose and hemicellulose (Mosier et al., 2005). Many studies have shown that IL pretreatment can efficiently remove lignin (up to 90 percentage) and inform the biomass (Li et al., 2010a; Dadi et al., 2007; Sun et al., 2009; Lee et al., 2009; Tan et al., 2009; An et al., 2015; Asakawa et al., 2016). To investigate the changes of chemical compositions, ATR-FTIR and Raman spectroscopy were employed (Li et al., 2011). For ATR-FTIR analysis, six characterized chemical bonds between lignin and polysaccharides were chosen and compared to explicate the changes, which included the bands at 1510 cm^{-1} (aromatic skeletal from lignin), 1329 cm^{-1} (syringyl and guaiacyl condensed lignin), 1235 cm^{-1} (C-O stretching in lignin & hemicellulose), 1375 cm^{-1} (C-H deformation in cellulose & hemicellulose), 1745 cm^{-1} (carbonyl C=O stretching), and 900 cm^{-1} (anti-symmetric out of plane ring stretch of amorphous cellulose) (Li et al., 2011). For Raman spectroscopy, four brands were monitored to explain delignification at a molecular level, which was at 1170 cm^{-1} (phenol), 1266 cm^{-1} (guaiacyl ring breathing), 1600 cm^{-1} (aromatic ring stretch), and 1620 cm^{-1} (ring-conjugated C=C bond of coniferaldehyde) (Chu et al., 2010). After IL-pretreatment, the changes of these brands are used as an indicator of breakage/modification of the aromatic ring. Pu et al. (2007) investigated the solubility of lignin obtained from kraft pulp in 1-hexyl-3-methylimidazolium trifluoromethanesulfonate ([C$_6$mim] [CF$_3$SO$_3$]), 1,3-dimethylimidazolium methylsulfate ([C$_1$mim][MeSO$_4$]), and 1-butyl-3-methylimidazolium methylsulfate ([C$_4$mim][MeSO$_4$]) (Pu et al., 2007). The solubility of lignin varied among the ILs containing [C$_4$mim]$^+$, which indicated that the solubility of lignin was governed by the properties of the anions. Ji et al. (2012) conducted mechanism study of modeled lignin 1-(4-methoxyphenyl)-2-methoxyethanol (LigOH) dissolution and regeneration of 1-allyl-3-methylimidazolium chloride (AmimCl) (Ji et al., 2012). It was observed that the dissolution of lignin was realized by forming of hydrogen bonds between AminCl and LigOH, which were weakened by the addition of water and led to the precipitation of lignin.

2.3 Impact of IL pretreatment on enzymatic conversion

Enzymatic saccharification of cellulose and hemicellulose including hydrolysis kinetics and sugar yields is highly related to biomass crystallinity and delignification efficiency (Lee et al., 2009; Dadi et al., 2007; Kumar et al., 2009; Zhang and Lynd, 2004; Xu et al., 2016b). As discussed in the former sections, IL pretreatment leads to the dissolution of carbohydrates and/or lignin of biomass (Li et al., 2010a). The surface area of biomass is significantly increased while cellulose crystallinity is decreased (Li et al., 2010a; Dadi et al., 2007; Lee et al., 2009; Tan et al., 2009; Asakawa et al., 2016; Xu et al., 2016b; Hou et al., 2015). Figure 2 shows the comparison of SEM micrographs before and after IL pretreatment. A clear increase of surface area can be observed from Fig. 2. Under IL pretreatment, intra- and inter-molecular hydrogen bonding is removed, which causes the increase of surface area and binding sites in the resulting cellulose fibers (Li et al., 2010a). The physical and chemical modification of biomass significantly increases enzyme accessibility, conversion efficiency, and initial hydrolysis rate (Li et al., 2010a; Dadi et al., 2007; Mosier et al., 2005). Investigations of the effect of IL pretreatment on enzymatic hydrolysis were conducted using various feedstocks.

Li et al. (2011) reported the influence of physicochemical changes on enzymatic digestibility of [C_2mim][OAc] and ammonia fiber expansion (AFEX) pretreated corn stover (Li et al., 2011). Compared to AFEX, [C_2mim][OAc] pretreatment can significantly disrupt the native crystal structure of corn stover. More than 70 percent of the theoretical sugar yield was obtained after 48 hours of pretreatment using commercial enzyme cocktails' hydrolysis on both pretreated biomass. In comparison to AFEX, lower enzyme loading and shorter hydrolysis time were required for [C_2mim][OAc] pretreated corn stover to achieve similar sugar yields.

Arora et al. investigated the effect of residence time and temperature on IL pretreatment of switchgrass using [C_2mim][OAc], which included biomass delignification, xylan and glucan depolymerization, porosity, surface area, sugar conversion kinetics, and sugar yields. A strong

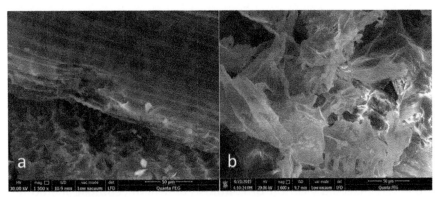

Fig. 2. SEM micrographs of untreated rice straw (a) and [Dmim][(OCH$_3$)$_2$PO$_2$]-treated sample (b) (Xu et al., 2016b).

correlation between pretreatment condition and biomass modification were observed. The results showed that an increase of pretreatment temperature accelerates delignification. Softening or melting of lignin is the main reason for the increase in enzymatic hydrolysis kinetics of the pretreated biomass. After [C_2mim][OAc] pretreatment, the surface area of pretreated switchgrass increased by a factor of ~30 fold with an average pore size of 10–15 nm, which greatly increased the surface accessibility for cellulolytic enzymes and improved the enzymatic hydrolysis.

Besides agricultural residuals and grasses, the application of ILs on processing woody biomass was also reported. Torr et al. (2012) investigated the impact of [C_2mim][OAc] pretreatment on the chemistry, and enzymatic digestibility of *Pinus radiate* compression wood under pretreatment conditions of 120°C and 155°C for 3 h (Torr et al., 2012). The results showed that the structure and composition of *Pinus radiate* was modified, including large removal of hemicellulose, loss of ether linkage, the formation of condensed structures in the lignin, decrease of cellulose crystallinity, and possible depolymerization of polysaccharide chains, which significantly improve the enzymatic digestibility of the cellulose. Compared with the untreated biomass, the glucose yield was increased by 11 to 30 times and achieved 90% by the end of 24-hour saccharification. Recently, Torr et al. (2016) reported a study on the relevance of enzymatic digestibility and biomass features, including accessible surface area and porosity, cellulose crystallinity, and lignin content using two ILs ([C_2mim][OAc] and [C_2mim] Cl) and pine wood (Torr et al., 2016). It was concluded that the increased surface area and porosity are closely correlated with the improvement of enzymatic digestibility.

Moreover, municipal solid waste (MSW) was investigated as a potential feedstock for biofuels production using imidazolium-based IL pretreatment with reduced feedstock cost and improved sugar conversion performance. During the IL-pretreatment, blending MSW with another type of biomass feedstock, such as corn stover, can help to decrease the viscosity and increase the pretreatment efficiency (Sun et al., 2015). After IL pretreatment and enzymatic saccharification, up to 84 percent glucose and 75 percent xylose were released from the MSW blends (Sun et al., 2015).

3. Challenges to the Process

3.1 Process scale-up challenges

Although the combination of IL pretreatment and enzymatic hydrolysis has shown its advantages, including feedstock agnostic, high sugar yields, fast kinetics, and extensive delignification, significant challenges remain prior to the large-scale demonstration. The scale-up of IL-based biomass pretreatment is still at an early stage with several possibilities of IL structures and process configurations unidentified. The major challenges include the high cost of ILs, lack of knowledge about process scale-up, as well as evaluation of different process and reactor configurations. Technically, pretreatment scale-up requires that the ILs have excellent thermal stability, recyclability, and low toxicity. Besides, the operating difficulties of the current IL process scale-up still exist in the material handling and product recovery.

Pretreatment process is always aimed at high solid loadings to save the cost of the solvent. With solid load above ten percentage (w/w), the viscosity of slurry increases significantly which challenges torque burden of the mixing impeller. Most of the reported solid loads are in the range of 5–30 percent by weight. Based on the experience at Advanced Biofuels Process Demonstration Unit (ABPDU), high solid loading (over 30 percent) makes mixing and material handling very challenging. The reactor must be equipped with sufficient driving force to enable stirring and efficient mass/heat transfer, which is critical to ensure homogeneous slurry and to prevent overheating of the biomass. Compared with a turbine impeller which is commonly used for liquid mixing, an anchor or helicalimpeller is more suitable for higher solid loadings. Flow breaker could be added to further enhance the mixing. Also, reactors, tanks and, transport/delivery devices should be able to tolerate the pH of the ILs, which might be highly caustic.

In order to overcome these challenges, research has been conducted in the areas including the reduction of the cost and usage of ILs and enzymes (Binder and Raines, 2010; Sun et al., 2013b), development of the efficient IL recycling technologies (Dibble et al., 2011), optimization of enzyme cocktails (Dimarogona et al., 2012; Park et al., 2012), increase of solid loading during mixing and/or saccharification (Cruz et al., 2013), employment of integrated systems approaches such as "one-pot" (Shi et al., 2013b; Xu et al., 2016a), and evaluation of process scale-up (Li et al., 2017; Liang et al., 2017). A coupon testing has been carried out in a recent scale-up study on IL-acidolysis process, where the chloride-based ILs, 1-butyl-3-methylimidazolium chloride ($[C_4C_1Im]Cl$), and 1-ethyl-3-methylimidazolium chloride ($[C_2C_1Im]Cl$), were used at pretreatment step followed by acid (hydrochloride acid) hydrolysis to produce sugar monomers (Li et al., 2017). The results indicate that the corrosion of the acidic ILs $[C_4C_1Im]Cl$, and $[C_2C_1Im]Cl$ to metals are very limited with corrosion rate between 0 and 1 mile per year (mpy) and $[C_2C_1Im]Cl$ was less corrosive compared to $[C_4C_1Im]Cl$ (Li et al., 2017). While slight corrosion may happen during high-temperature reactions (140°C for pretreatment and 105°C for acidolysis), the overall corrosion impact is not a severe concern for the Hastelloy C276 Parr reactor (Li et al., 2017). As reported in the literature, a common IL-based biomass pretreatment was conducted in the temperature range of 50°C–160°C (Vasheghani Farahani et al., 2016; Elgharbawy et al., 2016; Mahmood et al.,

2016). Heating/cooling during the biomass conversion process is one of the most energy consuming factors, especially for large scale. The optimum temperature usually takes into account the types of ILs, target sugar yields, the heating efficiency of reaction vessels, stirring rate, and so on.

Furthermore, one of the general properties of ILs is their high electrical conductivity which opened up their applications as electrolytes in batteries, double-layer capacitors or solar cells. However, the high conductivity of ILs may cause electrical hazards in biomass pretreatment operation. From a safety standpoint, IL leakage to bare electrical wires can cause short circuit or fire. In key operating areas with possible IL contact, chemical-resistant, heat-resistant and waterproof wires should be used to ensure the safety.

3.2 IL inhibition of biochemical conversion process

The toxicity of most of ILs remains unknown or undisclosed publicly. It has been reported that some ILs inhibit the subsequent enzymatic saccharification and fermentation (Ouellet et al., 2011; Nemestóthy et al., 2017). The enzymes are easily deactivated in the presence of certain ILs (Elgharbawy et al., 2016). Similarly, ILs may restrain microbial growth and the production of secondary metabolites in fermentation (Ouellet et al., 2011). Therefore, identification of the ILs that can be tolerated by downstream processes has become a new approach to IL process development (Xu et al., 2016a). IL-tolerance also becomes a key factor for genetic engineering of the strains to produce advanced biofuels for biochemicals (Dickinson et al., 2016). So far, reports of potential ILs that can achieve high sugar yields with low IL/biomass ratio in bench scale (1–10 L) biomass pretreatment are still limited. The removal of excess ILs from pretreated biomass becomes particularly important. It can be achieved through solid-liquid separation followed by extensive washing/dilution of pretreated biomass (Li et al., 2013).

3.3 The economics of IL process

The high cost of ILs has been one of the major impediments to IL utilization in the cellulosic biorefinery (George et al., 2015). Identification of both efficient and economically viable ILs is crucial for future research and process commercialization. Approximately, there are a thousand structures of ILs reported with only one-third of them being commercially available. Significant price swings (from $1/kg to $800/kg) exist in the IL market based on the IL structure and production scale (Klein-Marcuschamer et al., 2011). Taking into account the economics, a commercially viable IL has to be recyclable and reusable efficiently.

In IL manufacture, synthesis and purification are the main steps affecting the IL cost. For example, when ILs are synthesized through neutralization of an organic amine with a mineral acid, which does not require complicated purification, the cost will be far less. Solvent properties can also be well-controlled by selection of anion and cation for a specific task. As discussed above, cholinium and amino acids based ILs are getting more attention in biomass pretreatment (Sun et al., 2014). Lignin and hemicellulose-derived compounds have also been used as potential raw materials for IL synthesis lately. These polymers represent inexpensive and abundant streams from a variety of biomass processing industries including textiles, pulping/paper, and biofuels (Socha et al., 2014).

The techno-economic analysis (TEA) models of making biofuels from lignocellulosic materials at a commercial scale are instructive tools to help researchers evaluate the economic impacts of the process and to identify the process challenges and bottlenecks (Klein-Marcuschamer et al., 2010; Klein-Marcuschamer et al., 2011; Klein-Marcuschamer et al., 2013; Klein-Marcuschamer et al., 2012).

4. Conclusion

Certain ILs are highly efficient solvents or catalysts for biomass pretreatment to overcome the natural recalcitrance of lignocellulosic biomass contributed by cellulose crystallinity, the presence of lignin, functional groups on hemicelluloses, and interwoven linkages among these major components. The physical and chemical modifications of lignin and polysaccharides during pretreatment have a great impact on the following step of enzymatic hydrolysis, which includes enzyme loading, hydrolysis rate, enzymatic digestibility, and product recovery. The partial solubilization of the major components of biomass significantly increases enzyme accessibility and efficiency.

The combination of IL pretreatment and enzymatic hydrolysis has shown its advantages including feedstock agnostic, high sugar yields, fast kinetics, and extensive delignification. However, there are still challenges to be addressed, such as high cost of ILs and lack of knowledge of process scale-up, and evaluation of different process configurations. More efforts are needed to increase the IL process' efficiency and economic viability including an increase of solid loading for both pretreatment and saccharification, reduction of the cost and usage of IL and enzyme, and development of efficient IL recycling technologies and integrated processes.

References

An, Y.-X., M.-H. Zong, H. Wu and N. Li. 2015. Pretreatment of lignocellulosic biomass with renewable cholinium ionic liquids: Biomass fractionation, enzymatic digestion and ionic liquid reuse. Bioresource Technology 192: 165–171. doi:http://dx.doi.org/10.1016/j.biortech.2015.05.064.

Asakawa, A., T. Oka, C. Sasaki, C. Asada and Y. Nakamura. 2016. Cholinium ionic liquid/cosolvent pretreatment for enhancing enzymatic saccharification of sugarcane bagasse. Industrial Crops and Products 86: 113–119. doi:http://dx.doi.org/10.1016/j. indcrop.2016.03.046.

Beckham, G. T., J. F. Matthews, B. Peters, Y. J. Bomble, M. E. Himmel and M. F. Crowley. 2011. Molecular-level origins of biomass recalcitrance: Decrystallization free energies for four common cellulose polymorphs. The Journal of Physical Chemistry B 115(14): 4118–4127. doi:10.1021/jp1106394.

Binder, J. B. and R. T. Raines. 2010. Fermentable sugars by chemical hydrolysis of biomass. Proc. Natl. Acad. Sci. U.S.A. 107(10): 4516–4521. doi:10.1073/pnas.0912073107.

Brandt, A., M. J. Ray, T. Q. To, D. J. Leak, R. J. Murphy and T. Welton. 2011. Ionic liquid pretreatment of lignocellulosic biomass with ionic liquid-water mixtures. Green Chem. 13(9): 2489–2499.

Brodeur, G., E. Yau, K. Badal, J. Collier, K. B. Ramachandran and S. Ramakrishnan. 2011. Chemical and physicochemical pretreatment of lignocellulosic biomass: a review. Enzyme Research 2011: 17. doi:10.4061/2011/787532.

Chu, L.-Q., R. Masyuko, J. V. Sweedler and P. W. Bohn. 2010. Base-induced delignification of miscanthus x giganteus studied by three-dimensional confocal raman imaging. Bioresource Technology 101(13): 4919–4925. doi:http://dx.doi.org/10.1016/j. biortech.2009.10.096.

Cruz, A. G., C. Scullin, C. Mu, G. Cheng, V. Stavila, P. Varanasi, D. Xu, J. Mentel, Y.-D. Chuang, B. A. Simmons and S. Singh. 2013. Impact of high biomass loading on ionic liquid pretreatment. Biotechnology for Biofuels 6(1): 1–10. doi:10.1186/1754-6834-6-52.

Dadi, A. P., S. Varanasi and C. A. Schall. 2006. Enhancement of cellulose saccharification kinetics using an ionic liquid pretreatment step. Biotechnol. Bioeng. 95(5): 904–910.

Dadi, A. P., C. A. Schall and S. Varanasi. 2007. Mitigation of cellulose recalcitrance to enzymatic hydrolysis by ionic liquid pretreatment. Appl. Biochem. Biotechnol. 137(1): 407–421. doi:10.1007/s12010-007-9068-9.

Dekker, R. F. H. and A. F. A. Wallis. 1983. Autohydrolysis-explosion as pretreatment for the enzymic saccharification of sunflower seed hulls. Biotechnology Letters 5(5): 311–316. doi:10.1007/bf01141131.

Dibble, D. C., C. Li, L. Sun, A. George, A. Cheng, Ö. P. Çetinkol, P. Benke, B. M. Holmes, S. Singh and B. A. Simmons. 2011. A facile method for the recovery of ionic liquid and lignin from biomass pretreatment. Green Chemistry 13(11): 3255. doi:10.1039/c1gc15111h.

Dickinson, Q., S. Bottoms, L. Hinchman, S. McIlwain, S. Li, C. L. Myers, C. Boone, J. J. Coon, A. Hebert, T. K. Sato, R. Landick and J. S. Piotrowski. 2016. Mechanism of imidazolium ionic liquids toxicity in Saccharomyces cerevisiae and rational engineering of a tolerant, xylose-fermenting strain. Microbial Cell Factories 15(1): 1–13. doi:10.1186/s12934-016-0417-7.

Dimarogona, M., E. Topakas, L. Olsson and P. Christakopoulos. 2012. Lignin boosts the cellulase performance of a GH-61 enzyme from Sporotrichum thermophile. Bioresour. Technol. 110: 480–487. doi:10.1016/j.biortech.2012.01.116.

Doherty, T. V., M. Mora-Pale, S. E. Foley, R. J. Linhardt and J. S. Dordick. 2010. Ionic liquid solvent properties as predictors of lignocellulose pretreatment efficacy. Green Chemistry 12(11): 1967–1975. doi:10.1039/C0GC00206B.

Duff, S. J. B. and W. D. Murray. 1996. Bioconversion of forest products industry waste cellulosics to fuel ethanol: a review. Bioresource Technology 55(1): 1–33. doi:10.1016/0960-8524(95)00122-0.

Elgharbawy, A. A., M. Z. Alam, M. Moniruzzaman and M. Goto. 2016. Ionic liquid pretreatment as emerging approaches for enhanced enzymatic hydrolysis of lignocellulosic biomass. Biochemical Engineering Journal 109: 252–267. doi:http://dx.doi.org/10.1016/j. bej.2016.01.021.

Erdmenger, T., C. Haensch, R. Hoogenboom and U. S. Schubert. 2007. Homogeneous tritylation of cellulose in 1-butyl-3-methylimidazolium chloride. Macromolecular Bioscience 7(4): 440–445. doi:10.1002/mabi.200600253.

Freemantle, M. 1998. Designer solvents. Chemical & Engineering News Archive 76(13): 32–37. doi:10.1021/cen-v076n013.p032.

Fu, D., G. Mazza and Y. Tamaki. 2010. Lignin extraction from straw by ionic liquids and enzymatic hydrolysis of the cellulosic residues. J. Agric. Food Chem. 58(5): 2915–2922. doi:10.1021/jf903616y.

George, A., A. Brandt, K. Tran, S. M. S. N. S. Zahari, D. Klein-Marcuschamer, N. Sun, N. Sathitsuksanoh, J. Shi, V. Stavila, R. Parthasarathi, S. Singh, B. M. Holmes, T. Welton, B. A. Simmons and J. P. Hallett. 2015. Design of low-cost ionic liquids for lignocellulosic biomass pretreatment. Green Chemistry 17(3): 1728–1734. doi:10.1039/C4GC01208A.

Grethlein, H. E. and A. O. Converse. 1991. Common aspects of acid prehydrolysis and steam explosion for pretreating wood. Bioresource Technology 36(1): 77–82. doi:10.1016/0960-8524(91)90101-O.

Hauru, L. K. J., M. Hummel, A. W. T. King, I. Kilpeläinen and H. Sixta. 2012. Role of solvent parameters in the regeneration of cellulose from ionic liquid solutions. Biomacromolecules 13(9): 2896–2905. doi:10.1021/bm300912y.

Hinman, N. D., D. J. Schell, C. J. Riley, P. W. Bergeron and P. J. Walter. 1992. Preliminary estimate of the cost of ethanol-production for ssf technology. Appl. Biochem. Biotechnol. 34-5: 639–649. doi:10.1007/bf02920584.

Hou, X.-D., T. J. Smith, N. Li and M.-H. Zong. 2012. Novel renewable ionic liquids as highly effective solvents for pretreatment of rice straw biomass by selective removal of lignin. Biotechnol. Bioeng. 109(10): 2484–2493. doi:10.1002/bit.24522.

Hou, X.-D., J. Xu, N. Li and M.-H. Zong. 2015. Effect of anion structures on cholinium ionic liquids pretreatment of rice straw and the subsequent enzymatic hydrolysis. Biotechnol. Bioeng. 112 (1): 65–73. doi:10.1002/bit.25335.

Janesko, B. G. 2011. Modeling interactions between lignocellulose and ionic liquids using DFT-D. Physical Chemistry Chemical Physics 13(23): 11393–11401. doi:10.1039/C1CP20072K.

Ji, W., Z. Ding, J. Liu, Q. Song, X. Xia, H. Gao, H. Wang and W. Gu. 2012. Mechanism of lignin dissolution and regeneration in ionic liquid. Energy & Fuels 26(10): 6393–6403. doi:10.1021/ef301231a.

Kamiya, N., Y. Matsushita, M. Hanaki, K. Nakashima, M. Narita, M. Goto and H. Takahashi. 2008. Enzymatic *in situ* saccharification of cellulose in aqueous-ionic liquid media. Biotechnol. Lett. 30(6): 1037–1040. doi:10.1007/s10529-008-9638-0.

Kamlet, M. J. and R. W. Taft. 1976. The solvatochromic comparison method. I. The beta-scale of solvent hydrogen-bond acceptor (HBA) basicities. Journal of the American Chemical Society 98(2): 377–383. doi:10.1021/ja00418a009.

Kim, S. J., A. A. Dwiatmoko, J. W. Choi, Y. W. Suh, D. J. Suh and M. Oh. 2010. Cellulose pretreatment with 1-n-butyl-3-methylimidazolium chloride for solid acid-catalyzed hydrolysis. Biores. Technol. 101(21): 8273–8279.

Klein-Marcuschamer, D., P. Oleskowicz-Popiel, B. A. Simmons and H. W. Blanch. 2010. Technoeconomic analysis of biofuels: a wiki-based platform for lignocellulosic biorefineries. Biomass and Bioenergy 34(12): 1914–1921. doi:10.1016/j.biombioe.2010.07.033.

Klein-Marcuschamer, D., B. A. Simmons and H. W. Blanch. 2011. Techno-economic analysis of a lignocellulosic ethanol biorefinery with ionic liquid pre-treatment. Biofuels, Bioproducts and Biorefining 5: 562–569.

Klein-Marcuschamer, D., P. Oleskowicz-Popiel, B. A. Simmons and H. W. Blanch. 2012. The challenge of enzyme cost in the production of lignocellulosic biofuels. Biotechnol. Bioeng. 109(4): 1083–1087. doi:10.1002/bit.24370.

Klein-Marcuschamer, D., C. Turner, M. Allen, P. Gray, R. G. Dietzgen, P. M. Gresshoff, B. Hankamer, K. Heimann, P. T. Scott, E. Stephens, R. Speight and L. K. Nielsen. 2013. Technoeconomic analysis of renewable aviation fuel from microalgae, Pongamia pinnata, and sugarcane. Biofuels, Bioproducts and Biorefining 7(4): 416–428. doi:10.1002/bbb.1404.

Kumar, L., V. Arantes, R. Chandra and J. Saddler. 2012. The lignin present in steam pretreated softwood binds enzymes and limits cellulose accessibility. Bioresource Technology 103(1): 201–208. doi:http://dx.doi.org/10.1016/j.biortech.2011.09.091.

Kumar, R., G. Mago, V. Balan and C. E. Wyman. 2009. Physical and chemical characterizations of corn stover and poplar solids resulting from leading pretreatment technologies. Bioresource Technology 100(17): 3948–3962. doi:http://dx.doi.org/10.1016/j.biortech.2009.01.075.

Lee, S. H., T. V. Doherty, R. J. Linhardt and J. S. Dordick. 2009. Ionic liquid-mediated selective extraction of lignin from wood leading to enhanced enzymatic cellulose hydrolysis. Biotechnol. Bioeng. 102(5): 1368–1376. doi:10.1002/bit.22179.

Li, C., G. Cheng, V. Balan, M. S. Kent, M. Ong, S. P. Chundawat, L. Sousa, Y. B. Melnichenko, B. E. Dale, B. A. Simmons and S. Singh. 2011. Influence of physico-chemical changes on enzymatic digestibility of ionic liquid and AFEX pretreated corn stover. Bioresour. Technol. 102(13): 6928–6936. doi:10.1016/j.biortech.2011.04.005.

Li, C., B. Knierim, C. Manisseri, R. Arora, H. V. Scheller, M. Auer, K. P. Vogel, B. A. Simmons and S. Singh. 2010a. Comparison of dilute acid and ionic liquid pretreatment of switchgrass: Biomass recalcitrance, delignification and enzymatic saccharification. Bioresource Technology 101: 4900–4906. doi:10.1016/j.biortech.2009.10.066.

Li, C., D. Tanjore, W. He, J. Wong, J. L. Gardner, K. L. Sale, B. A. Simmons and S. Singh. 2013. Scale-up and evaluation of high solid ionic liquid pretreatment and enzymatic hydrolysis of switchgrass. Biotechnology for Biofuels 6(1): 1–13. doi:10.1186/1754-6834-6-154.

Li, C., L. Liang, N. Sun, V. S. Thompson, F. Xu, A. Narani, Q. He, D. Tanjore, T. R. Pray, B. A. Simmons and S. Singh. 2017. Scale-up and process integration of sugar production by acidolysis of municipal solid waste/corn stover blends in ionic liquids. Biotechnology for Biofuels 10(1): 13. doi:10.1186/s13068-016-0694-8.

Li, Q., Y. He, M. Xian, G. Jun, X. Xu, J. Yang and L. Li. 2009. Improving enzymatic hydrolysis of wheat straw using ionic liquid 1-ethyl-3-methyl imidazolium diethyl phosphate pretreatment. Biores. Tech. 100(14): 3570–3575.

Li, Q., X. Jiang, Y. He, L. Li, M. Xian and J. Yang. 2010b. Evaluation of the biocompatibile ionic liquid 1-methyl-3-methylimidazolium dimethylphosphite pretreatment of corn cob for improved saccharification. Appl. Microbiol. Biotechnol. 87(1): 117–126. doi:10.1007/s00253-010-2484-8.

Liang, L., C. Li, F. Xu, Q. He, J. Yan, T. Luong, B. A. Simmons, T. R. Pray, S. Singh, V. S. Thompson and N. Sun. 2017. Conversion of cellulose rich municipal solid waste blends using ionic liquids: feedstock convertibility and process scale-up. RSC Advances 7(58): 36585–36593. doi:10.1039/C7RA06701A.

Liu, H. B., K. L. Sale, B. A. Simmons and S. Singh. 2011. Molecular dynamics study of polysaccharides in binary solvent mixtures of an ionic liquid and water. Journal of Physical Chemistry B 115: 10251–10258. doi:10.1021/jp111738q.

Liu, S. 2010. Woody biomass: niche position as a source of sustainable renewable chemicals and energy and kinetics of hot-water extraction/hydrolysis. Biotechnology Advances 28(5): 563–582. doi:10.1016/j.biotechadv.2010.05.006.

Lu, H., R. Hu, A. Ward, T. E. Amidon, B. Liang and S. Liu. 2012. Hot-water extraction and its effect on soda pulping of aspen woodchips. Biomass and Bioenergy 39(0): 5–13. doi:http://dx.doi.org/10.1016/j.biombioe.2011.01.054.

Lungwitz, R., V. Strehmel and S. Spange. 2010. The dipolarity/polarisability of 1-alkyl-3-methylimidazolium ionic liquids as function of anion structure and the alkyl chain length. New Journal of Chemistry 34(6): 1135–1140. doi:10.1039/B9NJ00751B.

Lynd, L. R., C. Wyman, M. Laser, D. College. 2005. Strategic Biorefinery Analysis: Analysis of Biorefineries. In 2005, p. 40.

Mahmood, H., M. Moniruzzaman, S. Yusup and H. M. Akil. 2016. Pretreatment of oil palm biomass with ionic liquids: a new approach for fabrication of green composite board. Journal of Cleaner Production 126: 677–685. doi:http://dx.doi.org/10.1016/j.jclepro.2016.02.138.

Mosier, N., C. Wyman, B. Dale, R. Elander, Y. Y. Lee, M. Holtzapple and M. Ladisch. 2005. Features of promising technologies for pretreatment of lignocellulosic biomass. Bioresource Technology 96(6): 673–686. doi:http://dx.doi.org/10.1016/j.biortech.2004.06.025.

Nemestóthy, N., G. Megyeri, P. Bakonyi, P. Lakatos, L. Koók, M. Polakovic, L. Gubicza and K. Bélafi-Bakó. 2017. Enzyme kinetics approach to assess biocatalyst inhibition and deactivation caused by [bmim][Cl] ionic liquid during cellulose hydrolysis. Bioresource Technology 229: 190–195. doi:https://doi.org/10.1016/j.biortech.2017.01.004.

Nguyen, T. D., K. R. Kim, S. J. Han, H. Y. Cho, J. W. Kim, S. M. Park, J. C. Park and S. J. Sim. 2010. Pretreatment of rice straw with ammonia and ionic liquid for lignocellulose conversion to fermentable sugars. Biores. Technol. 101(19): 7432–7438. doi:10.1016/j.biortech.2010.04.053.

Ouellet, M., S. Datta, D. C. Dibble, P. R. Tamrakar, P. I. Benke, C. Li, S. Singh, K. L. Sale, P. D. Adams, J. D. Keasling, B. A. Simmons, B. M. Holmes and A. Mukhopadhyay. 2011. Impact of ionic liquid pretreated plant biomass on Saccharomyces cerevisiae growth and biofuel production. Green Chemistry 13(10): 2743–2749. doi:10.1039/C1GC15327G.

Park, J. I., E. J. Steen, H. Burd, S. S. Evans, A. M. Redding-Johnson, T. Batth, P. I. Benke, P. D'haeseleer, N. Sun, K. L. Sale, J. D. Keasling, T. S. Lee, C. J. Petzold, A. Mukhopadhyay, S. W. Singer, B. A. Simmons and J. M. Gladden. 2012. A thermophilic ionic liquid-tolerant cellulase cocktail for the production of cellulosic biofuels. PLoS ONE 7: 1–10. doi:10.1371/journal.pone.0037010.

Perez-Pimienta, J. A., M. G. Lopez-Ortega, J. A. Chavez-Carvayar, P. Varanasi, V. Stavila, G. Cheng, S. Singh and B. A. Simmons. 2015. Characterization of agave bagasse as a function of ionic liquid pretreatment. Biomass and Bioenergy 75: 180–188. doi:10.1016/j.biombioe.2015.02.026.

Pu, Y., N. Jiang and A. J. Ragauskas. 2007. Ionic liquid as a green solvent for lignin. Journal of Wood Chemistry and Technology 27(1): 23–33. doi:10.1080/02773810701282330.

Ragauskas, A. J., G. T. Beckham, M. J. Biddy, R. Chandra, F. Chen, M. F. Davis, B. H. Davison, R. A. Dixon, P. Gilna, M. Keller, P. Langan, A. K. Naskar, J. N. Saddler, T. J. Tschaplinski, G. A. Tuskan and C. E. Wyman. 2014. Lignin valorization: improving lignin processing in the biorefinery. Science (New York, NY) 344: 1246843. doi:10.1126/science.1246843.

Remsing, R. C., R. P. Swatloski, R. D. Rogers and G. Moyna. 2006. Mechanism of cellulose dissolution in the ionic liquid 1-n-butyl-3-methylimidazolium chloride: a 13C and 35/37Cl NMR relaxation study on model systems. Chemical Communications (12): 1271–1273. doi:10.1039/B600586C.

Samayam, I. P. and C. A. Schall. 2010. Saccharification of ionic liquid pretreated biomass with commercial enzyme mixtures. Bioresource Technology 101(10): 3561–3566. doi:10.1016/j.biortech.2009.12.066.

Sathitsuksanoh, N., A. George and Y. H. P. Zhang. 2013. New lignocellulose pretreatments using cellulose solvents: a review. Journal of Chemical Technology & Biotechnology 88(2): 169–180. doi:10.1002/jctb.3959.

Shi, J., J. M. Gladden, N. Sathitsuksanoh, P. Kambam, L. Sandoval, D. Mitra, S. Zhang, A. George, S. W. Singer, B. A. Simmons and S. Singh. 2013a. One-pot ionic liquid pretreatment and saccharification of switchgrass. Green Chemistry 15: 2579–2589. doi:10.1039/c3gc40545a.

Shi, J., V. S. Thompson, N. A. Yancey, V. Stavila, B. A. Simmons and S. Singh. 2013b. Impact of mixed feedstocks and feedstock densification on ionic liquid pretreatment efficiency. Biofuels 4: 63–72. doi:10.4155/bfs.12.82.

Shill, K., S. Padmanabhan, Q. Xin, J. M. Prausnitz, D. S. Clark and H. W. Blanch. 2011. Ionic liquid pretreatment of cellulosic biomass: Enzymatic hydrolysis and ionic liquid recycle. Biotechnol. Bioeng. 108(3): 511–520. doi:10.1002/bit.23014.

Singh, S. and B. A. Simmons. 2013. Ionic liquid pretreatment: Mechanism, performance, and challenges. Aqueous Pretreatment of Plant Biomass for Biological and Chemical Conversion to Fuels and Chemicals: 223–238. doi:10.1002/9780470975831.ch11.

Sjöström, E. 1993. Wood Chemistry. Fundamentals and Applications, 2nd ed. Academic Press, San Diego. doi:http://dx.doi.org/10.1016/B978-0-08-092589-9.50005-X.

Socha, A. M., R. Parthasarathi, J. Shi, S. Pattathil, D. Whyte, M. Bergeron, A. George, K. Tran, V. Stavila, S. Venkatachalam, M. G. Hahn, B. A. Simmons and S. Singh. 2014. Efficient biomass pretreatment using ionic liquids derived from lignin and hemicellulose. Proceedings of the National Academy of Sciences 111(35): E3587–E3595. doi:10.1073/pnas.1405685111.

Sun, N., M. Rahman, Y. Qin, M. L. Maxim, H. Rodríguez and R. D. Rogers. 2009. Complete dissolution and partial delignification of wood in the ionic liquid 1-ethyl-3-methylimidazolium acetate. Green Chemistry 11: 646–655. doi:10.1039/b822702k.

Sun, N., W. Li, B. Stoner, X. Jiang, X. Lu and R. D. Rogers. 2011. Composite fibers spun directly from solutions of raw lignocellulosic biomass dissolved in ionic liquids. Green Chemistry 13: 1158. doi:10.1039/c1gc15033b.

Sun, L., C. Li, Z. Xue, B. A. Simmons and S. Singh. 2013a. Unveiling high-resolution, tissue specific dynamic changes in corn stover during ionic liquid pretreatment. RSC Adv. 3(6): 2017–2027. doi:10.1039/c2ra20706k.

Sun, N., H. Liu, N. Sathitsuksanoh, V. Stavila, M. Sawant, A. Bonito, K. Tran, A. George, K. L. Sale, S. Singh, B. A. Simmons and B. M. Holmes. 2013b. Production and extraction of sugars from switchgrass hydrolyzed in ionic liquids. Biotechnology for Biofuels 6: 39. doi:10.1186/1754-6834-6-39.

Sun, N., R. Parthasarathi, A. M. Socha, J. Shi, S. Zhang, V. Stavila, K. L. Sale, B. A. Simmons and S. Singh. 2014. Understanding pretreatment efficacy of four cholinium and imidazolium ionic liquids by chemistry and computation. Green Chemistry 16: 2546–2557. doi:10.1039/c3gc42401d.

Sun, N., F. Xu, N. Sathitsuksanoh, V. S. Thompson, K. Cafferty, C. Li, D. Tanjore, A. Narani, T. R. Pray, B. A. Simmons and S. Singh. 2015. Blending municipal solid waste with corn stover for sugar production using ionic liquid process. Bioresource Technology 186: 200–206. doi:10.1016/j.biortech.2015.02.087.

Swatloski, R., S. Spear, J. Holbrey and R. Rogers. 2002. Dissolution of cellulose with ionic liquids. J. Am. Chem. Soc. 124(18): 4974–4975.

Taft, R. W. and M. J. Kamlet. 1976. The solvatochromic comparison method. 2. The alpha-scale of solvent hydrogen-bond donor (HBD) acidities. Journal of the American Chemical Society 98(10): 2886–2894. doi:10.1021/ja00426a036.

Tan, S. S. Y., D. R. MacFarlane, J. Upfal, L. A. Edye, W. O. S. Doherty, A. F. Patti, J. M. Pringle and J. L. Scott. 2009. Extraction of lignin from lignocellulose at atmospheric pressure using alkylbenzenesulfonate ionic liquid. Green Chemistry 11(3): 339–345. doi:10.1039/B815310H.

Torr, K. M., K. T. Love, Ö. P. Çetinkol, L. A. Donaldson, A. George, B. M. Holmes and B. A. Simmons. 2012. The impact of ionic liquid pretreatment on the chemistry and enzymatic digestibility of Pinus radiata compression wood. Green Chemistry 14(3): 778. doi:10.1039/c2gc16362d.

Torr, K. M., K. T. Love, B. A. Simmons and S. J. Hill. 2016. Structural features affecting the enzymatic digestibility of pine wood pretreated with ionic liquids. Biotechnol. Bioeng. 113: 540–549. doi:10.1002/bit.25831.

Vasheghani Farahani, S., Y.-W. Kim and C. A. Schall. 2016. A coupled low temperature oxidative and ionic liquid pretreatment of lignocellulosic biomass. Catalysis Today 269: 2–8. doi:http://dx.doi.org/10.1016/j.cattod.2015.12.022.

Vitz, J., T. Erdmenger, C. Haensch and U. S. Schubert. 2009. Extended dissolution studies of cellulose in imidazolium based ionic liquids. Green Chemistry 11(3): 417–424. doi:10.1039/B818061J.

Wang, C., D. Yan, Q. Li, W. Sun and J. Xing. 2014. Ionic liquid pretreatment to increase succinic acid production from lignocellulosic biomass. Bioresource Technology 172: 283–289. doi:10.1016/j.biortech.2014.09.045.

Wang, Y., M. Radosevich, D. Hayes and N. Labbé. 2011. Compatible ionic liquid-cellulases system for hydrolysis of lignocellulosic biomass. Biotechnology and Bioengineering 108(5): 1042–1048. doi:10.1002/bit.23045.

Wasikiewicz, J. M., F. Yoshii, N. Nagasawa, R. A. Wach and H. Mitomo. 2005. Degradation of chitosan and sodium alginate by gamma radiation, sonochemical and ultraviolet methods. Radiat. Phys. Chem. 73(5): 287–295. doi:10.1016/j.radphyschem.2004.09.021.

Xu, F., Y.-C. Shi and D. Wang. 2012. Enhanced production of glucose and xylose with partial dissolution of corn stover in ionic liquid, 1-ethyl-3-methylimidazolium acetate. Bioresource Technology 114: 720–724. doi:10.1016/j.biortech.2012.03.023.

Xu, F., J. Sun, N. V. S. N. M. Konda, J. Shi, T. Dutta, C. D. Scown, B. A. Simmons and S. Singh. 2016a. Transforming biomass conversion with ionic liquids: process intensification and the development of a high-gravity, one-pot process for the production of cellulosic ethanol. Energy Environ. Sci. 9: 1042–1049. doi:10.1039/C5EE02940F.

Xu, J.-K., Y.-C. Sun and R.-C. Sun. 2014. Ionic liquid pretreatment of woody biomass to facilitate biorefinery: structural elucidation of alkali-soluble hemicelluloses. ACS Sustainable Chemistry & Engineering 2(4): 1035–1042. doi:10.1021/sc500040j.

Xu, J., X. Wang, X. Liu, J. Xia, T. Zhang and P. Xiong. 2016b. Enzymatic *in situ* saccharification of lignocellulosic biomass in ionic liquids using an ionic liquid-tolerant cellulases. Biomass and Bioenergy 93: 180–186. doi:http://dx.doi.org/10.1016/j.biombioe.2016.07.019.

Yan, J., N. Joshee and S. Liu. 2013. Kinetics of the hot-water extraction of paulownia elongata woodchips. Journal of Bioprocess Engineering and Biorefinery 2(1): 1–10. doi:10.1166/jbeb.2013.1041.

Yan, J., D. Kiemle and S. Liu. 2015. Quantification of xylooligomers in hot water wood extract by 1H–13C heteronuclear single quantum coherence NMR. Carbohydrate Polymers 117(0): 903–909. doi:http://dx.doi.org/10.1016/j.carbpol.2014.10.031.

Yan, J., N. Joshee and S. Liu. 2016. Utilization of hardwood in biorefinery: a kinetic interpretation of pilot-scale hot-water pretreatment of *Paulownia elongata* woodchips. Journal of Biobased Materials and Bioenergy 10: 1–10.

Yang, C.-Y. and T. J. Fang. 2015. Kinetics of enzymatic hydrolysis of rice straw by the pretreatment with a bio-based basic ionic liquid under ultrasound. Process Biochem. 50(4): 623–629. doi:http://dx.doi.org/10.1016/j.procbio.2015.01.013.

Yokoyama, T., R. W. Taft and M. J. Kamlet. 1976. The solvatochromic comparison method. 3. Hydrogen bonding by some 2-nitroaniline derivatives. Journal of the American Chemical Society 98(11): 3233–3237. doi:10.1021/ja00427a030.

Youngs, T. G. A., C. Hardacre and J. D. Holbrey. 2007. Glucose solvation by the ionic liquid 1,3-dimethylimidazolium chloride: a simulation study. The Journal of Physical Chemistry B 111(49): 13765–13774. doi:10.1021/jp076728k.

Zhang, H., J. Wu, J. Zhang and J. He. 2005. 1-allyl-3-methylimidazolium chloride room temperature ionic liquid: a new and powerful nonderivatizing solvent for cellulose. Macromolecules 38(20): 8272–8277. doi:10.1021/ma0505676.

Zhang, Y.-H. P. 2008. Reviving the carbohydrate economy via multi-product lignocellulose biorefineries. Journal of Industrial Microbiology & Biotechnology 35(5): 367–375. doi:10.1007/s10295-007-0293-6.

Zhang, Y.-H. P. and L. R. Lynd. 2004. Toward an aggregated understanding of enzymatic hydrolysis of cellulose: Noncomplexed cellulase systems. Biotechnol. Bioeng. 88(7): 797–824. doi:10.1002/bit.20282.

Zhao, H., G. Baker and J. Cowins. 2009. Fast enzymatic saccharification of switchgrass after pretreatment with ionic liquids. Biotechnol. Progr. doi:Doi: 10.1002/btpr.331.

Zhao, Y. L., Y. Wang, J. Y. Zhu, A. Ragauskas and Y. L. Deng. 2008. Enhanced enzymatic hydrolysis of spruce by alkaline pretreatment at low temperature. Biotechnol. Bioeng. 99(6): 1320–1328. doi:10.1002/bit.21712.

Zhu, J. Y., X. J. Pan, G. S. Wang and R. Gleisner. 2009. Sulfite pretreatment (SPORL) for robust enzymatic saccharification of spruce and red pine. Bioresource Technology 100(8): 2411–2418. doi:10.1016/j.biortech.2008.10.057.

Zhu, J. Y. and X. J. Pan. 2010. Woody biomass pretreatment for cellulosic ethanol production: Technology and energy consumption evaluation. Bioresource Technology 101(13): 4992–5002. doi:10.1016/j.biortech.2009.11.007.

CHAPTER 16

Ammonia Fiber Expansion and its Impact on Subsequent Densification and Enzymatic Conversion

*Bryan D. Bals** and *Timothy J. Campbell*

1. Introduction

Agricultural cellulosic feedstocks—including both dedicated energy crops such as miscanthus and switchgrass as well as agricultural residues such as corn stover and wheat straw—have long been widely under-utilized resources. Being composed primarily of complex sugars and high-energy lignin, these feedstocks could be used as building blocks for chemicals and fuels (Lange, 2007). Over the last decade, a considerable number of research studies have been performed on the conversion of these feedstocks into intermediates or final products, whether through the improvement of plant genetics, the use of novel technologies to chemically treat the biomass, and/ or the modification of fermentation organisms (Singh et al., 2017).

However, comparatively little research has been focused on the logistics of transporting these feedstocks. Such feedstocks are notoriously heterogeneous, of low density, difficult to safely and effectively store, and have poor handling characteristics (Hess et al., 2009). While processing

MBI, 3815 Technology Boulevard, Lansing, MI 48910, USA.
* Corresponding author: bals@mbi.org

engineers would prefer large volumes to be centrally processed at a conversion facility, taking advantage of economies of scale, the cost and logistics of transporting thousands of tons of bales per day to a single location quickly become insurmountable (Argo et al., 2013). A solution is to densify the biomass off-site at a pre-conversion facility, commonly called *depots* (Hess et al., 2009).

An intriguing possibility is to perform other preprocessing operations at these depots in addition to densification (Thompson et al., 2013). Cellulosic biomass requires a chemical or mechanical pretreatment prior to being enzymatically broken down into sugars (Mosier et al., 2005). Several pretreatment options are considered in the literature, but most of these require high temperatures, high pressures, expensive chemicals, complicated equipment, and/or produce multiple streams, all of which are more difficult to perform in a depot (Dale and Ong, 2012). In contrast, ammonia fiber expansion (AFEX™) treatment (Fig. 1) is currently being studied and designed specifically to be performed within a depot (Eranki et al., 2011).

Fig. 1. Representation of biomass preprocessing within a pre-conversion facility (depot). After its on-farm collection, the biomass is brought to the depot where it is first chemically treated before being physically compressed into pellets. The final pellets can be shipped out and are ready as-is for subsequent saccharification and fermentation.

2. AFEX Treatment

Ammonia fiber expansion (AFEX) treatment was first envisioned in the 1980s as an alternative to the ammoniation practices commonly performed on farms for the purpose of increasing nitrogen content and digestibility of forages for ruminants (Dale and Moreira, 1982). The treatment consists of contacting biomass with gaseous or liquid ammonia at a relatively high concentration (0.3–3.0 g ammonia per g dry biomass) within a pressurized vessel. The biomass contains some moisture (0.3–2.0 g water per g dry biomass) which is enough to react with the ammonia to produce ammonium hydroxide but not enough to create a separate liquid stream. The temperature of the reaction is relatively mild (50–150°C) and the pressure is dependent on these three variables—ammonia concentration, moisture content, and reaction temperature. After the reaction is completed, the pressure is released and the ammonia escapes as a gas. Thus, no liquid stream is present and no mass is lost in the cellulosic biomass. Thus, for every ton of biomass that enters the reactor, virtually one ton of biomass exits the reactor (Chundawat et al., 2013).

Several studies have focused on the numerous chemical changes within the biomass that occur during the AFEX process (Chundawat et al., 2010; Chundawat et al., 2011). Of primary interest is ammonia reacting with ester linkages within the cell wall which aids in separating the lignin from the hemicellulose. The primary product formed from this reaction is acetamide which is the most common nitrogenous compound found in AFEX-treated biomass (Chundawat et al., 2010). Ferulic and coumeric esters are also amidated during the process. Unlike with acid pretreatments, sugar depolymerization and degradation are minimal. Some hemicellulose (10–30%) is broken down to oligomeric form during the AFEX process but very little is converted to monomeric xylose or other sugars. Note that because no solubles stream is removed during AFEX treatment, the oligomeric hemicellulose is still present and available as a sugar source for enzymatic hydrolysis and subsequent microbial fermentation. It is possible that some sugars are lost to Maillard reactions with ammonia, producing imidazoles and other compounds. In one study on corn stover, less than 1 g of Maillard products were produced from AFEX per kg corn stover. Thus, sugar degradation is not a concern for AFEX treatment (Chundawat et al., 2010).

Several physical changes occur during AFEX treatment as well (Chundawat et al., 2011). Of particular interest is lignin, which is soluble in alkaline conditions. During the AFEX process, the ammonia is soaked into the cell wall, partially solubilizing the lignin. As the pressure of the reactor is released, the ammonia moves towards the surface of the biomass, carrying the lignin (as well as some hemicellulose) with it. Most of the ammonia then evaporates, which redeposits the solubilized lignin and hemicellulose onto the surface of the biomass (Donohoe et al., 2011). Lignin is not significantly depolymerized during this process, but several lignin branching points and lignin-carbohydrate linkages are removed (Singh et al., 2015). In addition to lignin and hemicellulose redeposition, micro-pores are generated throughout the cell wall due to both the pressure differential as well the relocation of lignin and/or hemicellulose. These micro-pores allow for enzymes to access cellulose more readily, leading to an increase in hydrolysis rate and yield (Chundawat et al., 2011). The final major physical change is swelling of cellulose. This too leads to greater enzyme access and the rapid conversion of biomass.

AFEX treatment has been tested on a wide variety of crop residues as well as dedicated energy crops. In general, it has been determined that AFEX treatment works best on members of the grass family, likely due to higher acetate content and other cell wall's structural differences. Garlock et al. (2012a) tested AFEX treatment on multiple samples of mixed species and determined that AFEX was not effective when the grass content of the mix was less than 60% by dry weight. Likewise, tests with poplar (Balan et al., 2009), alfalfa (Bals et al., 2010), palm empty fruit bunch fiber (Lau et al., 2010), and black locust (Garlock et al., 2012b) only saw modest gains in cellulose and hemicellulose conversion after AFEX treatment. Among grasses, the most important factor in assessing the amount of sugars obtained after AFEX treatment appears to be the lignin content. Grasses with high lignin content, such as sugarcane bagasse and miscanthus, require more severe conditions (higher ammonia loadings and higher temperatures) than grasses with lower lignin content such as corn stover and rice straw. Table 1 shows a list of AFEX conditions and sugar yields using a standard hydrolysis protocol.

Table 1. Conversion of glucan and xylan of several members of the grass family after AFEX treatment.

Biomass	Glucan conversion (%)[1]	Xylan conversion (%)
Switchgrass	95	83
Reed canary grass	87	80
Corn stover	95	60
Sugarcane bagasse	90	70
Rice straw	87	52
Miscanthus	85	70
Sweet sorghum	80	90

[1]All the data is for saccharification carried out at low solids loading and approximately 15 FPU Spezyme CP/g glucan (Genencor). Data adapted from Chundawat et al. (2013).

3. AFEX Reaction Design

Designing an AFEX reactor specifically for a detached preprocessing facility (i.e., a depot) presents several challenges (Campbell et al., 2013). Due to the smaller facility, the equipment must be relatively low cost (as a depot would not be able to capture the same economies of scale as a large facility) and low maintenance (due to less maintenance staff at a depot). In particular, the two major challenges are moving biomass against a pressure gradient and removing and concentrating ammonia. Initial industrial AFEX designs were based on the pulp industry, with large Pandia-style reactors bringing the biomass into the reactor via a screw conveyor (Eggeman and Elander, 2005). After the reaction, the reactor is flashed (sudden pressure drop) to remove some ammonia and then sent to a dryer to remove the residual ammonia. This ammonia is then recompressed via a compressor. While this design was seen as competitive for a commercial-scale cellulosic refinery on the order of 2000 tons of biomass processed per day, the costs did not scale well for smaller depots on the order of 100–400 tons/day (MBI, unpublished data). Furthermore, the screw conveyor and ammonia dryer were seen as high-maintenance pieces of equipment and thus undesirable as an approach for depots.

An alternative design, called the Packed Bed AFEX design (Campbell and Teymouri, 2015), was invented in 2011 at MBI (Michigan Biotechnology Institute, East Lansing, MI). In this approach, biomass is added to empty vertical reactors in a batch process. The vessel is then sealed and steam is added to heat and moisten the biomass. Ammonia is then added as a vapor which permeates through the biomass and reacts with the water present. While this reaction is occurring, a second reactor is loaded with biomass and steam is added. When the first vessel has completed its residence time, valves between the two reactors are opened, allowing approximately half of the ammonia to flow from the treated bed of biomass to the second reactor. Steam is then introduced to the top of the first reactor to strip ammonia from the biomass as a vapor, which is then channeled through a compressor and loaded to the top of the second vessel, allowing the second vessel to be treated. The first vessel can then be opened (as the pressure is at atmospheric once all ammonia is removed), the treated biomass removed, and fresh untreated biomass added to the reactor. Thus, the ammonia continually cycles back and forth between the two reactors (with a small amount of makeup ammonia added each time), while the biomass itself is added as a batch process. This flow scheme is shown in Fig. 2. Approximately 0.65 kg low-pressure steam is required per kg biomass to pre-steam and removes ammonia. Approximately 0.6 kg of ammonia are required per kg biomass, but only 0.02 kg ammonia/kg biomass is consumed.

While simple, there are a couple of important design features that must be incorporated to make this system work. The ammonia addition and removal rely on a certain amount of porosity in the biomass bed, and ammonia flow significantly decreases at biomass density of 125 kg/m^3 dry weight or higher (Campbell and Teymouri, 2015). The bottom of a bed of biomass can become compressed and lose porosity if too much weight is

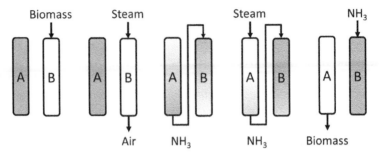

Fig. 2. Schematic showing the process flows during AFEX treatment of two beds of biomass. While biomass in Reactor A is being treated with ammonia, fresh biomass is loaded into Reactor B. Steam is then added, which forces out the excess air. When the residence time for Reactor A is complete, the valves between the reactors are opened and ammonia flows to Reactor B. Using a compressor (not shown) and steam, the remaining ammonia is removed from Reactor A and placed in Reactor B. Reactor A may now be unloaded, while makeup ammonia is added to Reactor B.

added above it. To avoid this compression, the biomass is added as sets of multiple baskets which slide in and out of the reactor, so that the baskets support the weight of the biomass rather than the biomass itself, eliminating compression. As an added benefit, these baskets are easy to add or remove from the reactor, reducing loading and unloading times and eliminating bridging. Secondly, the reactors must be vertical because ammonia vapor is denser than steam, and thus will be pushed out of the bottom of the reactor as a relatively pure stream when steam is added from the top. If a horizontal reactor is used, the ammonia and steam vapors will mix, leading to inefficient ammonia removal as the water must be removed from the mixed ammonia/water stream before reuse.

This design has several advantages that make it ideal for a depot setting. There are no moving parts within the reactor itself; it is an empty shell with a single quick-opening closure. Thus, it has a relatively low capital and maintenance cost compared to pulp and paper equipment. Likewise, the system is robust and does not require temperature control. All heat is added by the initial steaming and the reaction of ammonia and water; there is no other heating source. Because sugar degradation is minimal during the AFEX reaction, there is no need to finely control the reaction, and thus no need for highly-trained operators or complicated process control schemes. Because the ammonia is captured directly onto the next bed of biomass, there is no need for excess tanks or other equipments for ammonia's recovery. Thus, the only expensive and potentially high-maintenance piece of equipment is the ammonia compressor. Alternative, compressor-free approaches such as a quench system (condensing all ammonia that must be repressurized, pressurizing with a pump, and then vaporizing the liquid) are also possible but have not been tested at a pilot scale.

MBI first developed this packed bed system in a laboratory skid manifold with 10 cm diameter pipes that were 90 cm long. The maximum pressure rating on these pipes was 1.4 MPa, which was deemed insufficient to rapidly add ammonia or raise the temperature to levels used in other AFEX reactors. Despite these limitations, the sugar release during enzymatic hydrolysis of corn stover treated in the packed bed reactor matched that of previous laboratory-scale AFEX reactors (Bals, 2015). Likewise, ammonia's removal demonstrated that 95% of the ammonia can be removed at very high purity. Based on this success, MBI constructed a 45 cm diameter, 2.75 m tall pilot-scale AFEX reactor capable of 2 MPa operating pressures. The additional pressure allowed the ammonia to penetrate rapidly and increase the temperature to 140°C, sufficient to treat the biomass in a timely manner. Hydrolysis performance on the AFEX-treated material has been maintained and ammonia loading time decreased as compared to the laboratory-scale reactor. Likewise, ammonia recycling via steam stripping was also deemed successful, although some ammonia is removed as an ammonia/water mixture rather than as a vapor, and thus must be dewatered prior to recycling. To date, more than 600 cycles of AFEX treatment have been performed on corn stover and wheat straw in this reactor with no safety concerns or failed batches. An example of temperature and pressure profiles through a typical run in the pilot scale reactor is shown in Fig. 3.

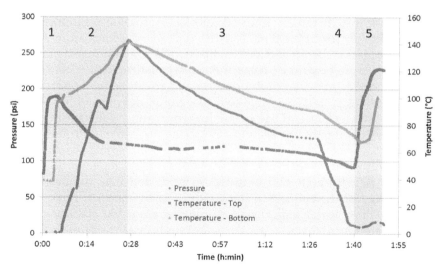

Fig. 3. Example of processing conditions during AFEX treatment in the pilot scale reactor. The different shadings refer to the different stages of pretreatment, namely pre steaming (1); ammonia addition (2); the soaking period (3); depressurization (4); and steam stripping (5). While the soaking residence time is shown as greater than one hour in this example, actual residence times are likely to be 20–30 minutes. Different temperatures are present at the top and bottom of the reactor, and so both are shown. Data is from MBI's pilot-scale reactor.

4. Suitability of AFEX-Treated Biomass for Densification

AFEX-treated biomass is highly suitable for densification. As stated previously, the output of the AFEX process is a solid material rather than slurry. Thus, there is no need for solid-liquid separation after treatment. Because steam is used to remove ammonia, however, the biomass (approximately 40% moisture) exiting the reactor is moist, and will likely need drying to ~15% moisture prior to densification. Conventional rotary drum dryers are suitable for this operation. Residual ammonia not removed during the AFEX process would also be removed in the dryer which means an ammonia scrubbing system would also need to be included in the exhaust.

More importantly, the lignin that was redistributed to the surface of the biomass is tacky, and thus acts as a natural adhesive during the densification process (Dale et al., 2014). Thus, no additional binder is needed, eliminating the cost of steam or syrup addition. In addition, this natural adhesive reduces the energy required to produce the pellets, which likely leads to increased throughput. A smaller length to diameter ratio in the die used for densification may also be possible, which would reduce capital cost and decrease energy use in the pelletizer.

5. Properties of Densified AFEX-Treated Biomass

The primary purpose of densification is to increase the bulk density of biomass, which ranges from 40–100 kg/m³ for loose biomass and 150–200 kg/m³ for baled biomass. The density of pellets from AFEX-treated biomass is slightly higher than that of pelletized untreated material. These tests were performed using a flat-die pelletizer, which consists of biomass added from the top by gravity and rollers above a horizontal die. In these tests, MBI was able to achieve bulk densities of 575 kg/m³ for corn stover pellets and 544 kg/m³ for wheat straw pellets using a flat die pelletizer compared to 444 and 519 kg/m³, respectively, for untreated biomass (Campbell et al., 2013). These results were similar to the loose bulk density of AFEX-treated corn stover obtained by Hoover et al. (2014), also in a flat die pelletizer, with loose bulk densities ranging from 590–630 kg/m³ depending on the conditions. In contrast, Karki et al. (2015) obtained lower bulk densities, ranging from 400–500 kg/m³, for AFEX-treated corn stover, switchgrass, and cordgrass briquettes produced using a ComPAKco densification method. The lower bulk density appears to be such due to the shape of the briquettes rather than the inefficiency of conversion. The density of individual briquettes ranged from 1400–1600 kg/m³ as compared to 1000–1300 kg/m³ for AFEX pellets obtained by Hoover et al. In comparison, corn grain has a bulk density of 720 kg/m³ or 610 kg/m³ for dry bulk density. Thus, it's likely that AFEX pellets will remain slightly less dense than corn grain.

Table 2. Bulk density and durability of multiple types of densified AFEX-treated biomass.

Biomass	Type	Density	Durability	Source
Untreated (Raw) Biomass				
Corn stover	Pellet	444 kg/m³	96.6%	Campbell et al., 2013
Wheat straw	Pellet	519 kg/m³	94.3%	Campbell et al., 2013
AFEX Treated Biomass				
Corn stover	Pellet	575 kg/m³	99.2%	Campbell et al., 2013
Wheat straw	Pellet	544 kg/m³	99.1%	Campbell et al., 2013
Corn stover – 4 mm	Pellet	631 kg/m³	99.1%	Hoover et al., 2014
Corn stover – 6 mm	Pellet	634 kg/m³	98.1%	Hoover et al., 2014
Switchgrass	Briquette	434 kg/m³	91.1%	Karki et al., 2015
Corn stover	Briquette	430 kg/m³	92.6%	Karki et al., 2015
Cordgrass	Briquette	425 kg/m³	87.1%	Karki et al., 2015

Likewise, pellet durability was significantly improved via AFEX treatment compared to the untreated, pelletized material. High durability is important for transportation, ensuring that the pellets maintain their shape throughout transport and handling which prevents losses. Pellet durability was 94% for untreated wheat straw and 97% for untreated corn stover but AFEX pellets of both feedstocks were in excess of 99% durable (Campbell et al., 2013). Similar results were observed by Hoover et al. (2014) for AFEX-treated corn stover pellets, although Karki et al. (2015) observed lower durability (80–95%) for the AFEX briquettes. This is likely due to the shape of the briquettes, having square edges compared to the round pellets.

Another important property is the stability of the pellets with respect to contamination. When AFEX pellets are produced at high moisture and are not dried afterward, they are susceptible to fungal growth (MBI, internal data). However, at low moistures (< 20%), AFEX pellets have been shown to be stable for multiple years, maintaining the same hydrolysis and fermentation yields even after being stored for extended periods at room temperature (MBI, unpublished data). Furthermore, it appears that the AFEX process may also reduce the ability of the pellets to absorb water. Bonner et al. (2015) placed both untreated and AFEX-treated corn stover pellets in a moisture adsorption chamber at 95% humidity for three days. The untreated pellets showed pellet deformation, observable mold growth, and increased weight due to moisture addition. In contrast, AFEX pellets maintained their shape and showed no observable contamination. It is likely that this is due to the lignin redeposition during the AFEX process, which may make the pellet surfaces more hydrophobic. Thus, it is expected that, if dried properly, AFEX pellets can be stored in real-world conditions for several months, a key factor in depot operations.

Based on this information, AFEX pellets appear to meet the criteria needed for biomass logistics, handling, and storage. While their density is not quite as high as corn grain, it is still 10x greater than raw undensified biomass. Likewise, the high durability suggests they will not be damaged during transport, and the low adsorption of atmospheric moisture suggests that the pellets can be stored indefinitely. In brief, the resulting pellets are able to be handled in a manner similar to corn grain. Given the large infrastructure already in place for grain, and given that several large biorefineries already exist that handle grain, it is expected that AFEX pellets could be handled and transported to biorefineries from AFEX depots in a relatively simple manner.

6. Impact of Densification on Hydrolysis of AFEX-Treated Biomass

While densification at a depot will decrease the transportation costs, it is necessary to ensure that densification will not negatively impact the subsequent saccharification and fermentation at a refinery. There is a concern, for example, that densification can compact the pores inside the biomass restricting enzyme access. It is known that densification reduces the water absorption properties of biomass by collapsing pores which has been linked to reduced sugar hydrolysis (Luo et al., 2011).

Despite these concerns, pelletized AFEX-treated corn stover offered a marginal improvement in hydrolysis performance over unpelletized AFEX-treated corn stover in shake flasks (Bals et al., 2013). This was likely due in part to the decreased particle size because of the grinding that occurs within the pelletizer. Also, it is possible that due to the heat generated within the pelletizer some additional pretreatment beyond physical grinding occurs during the process.

While the benefit is small within a shake flask, significant improvements in hydrolysis performance are observed when moving to a stirred tank reactor (Sarks et al., 2016). It is speculated that this improvement may be due in part to superior mixing. Because the pellets break down slowly over time and absorb less water than unpelletized material, there is more free liquid during the initial stages of liquefaction when the biomass is in the pellet form. This allows for enzymes to move more freely throughout the system. This has also been demonstrated to allow for higher solid loading during the hydrolysis stage. Likewise, the stirred tank reactors had pH control which was unavailable for the shake flasks. Given the slow dissolution of pellets, it is likely that, without controlled pH, the pH was not at optimal conditions during the liquefaction in shake flasks. This may be a secondary reason for the large improvement from shake flasks to stirred tank reactors. At 22% solid loading, pelletized AFEX-treated corn stover produced a yield of 90% glucose and 70% xylose within 30 hours of hydrolysis at an enzyme loading of 7 mg/g glucan (Sarks et al., 2016).

Hoover et al. (2014) also performed hydrolysis at high solid loading (20% total solids) on AFEX-treated corn stover that was pelletized at different conditions. When the biomass was first milled to 4 mm particle size, the pelletized material performed significantly better than the unpelletized biomass at 79% glucose conversion compared to 72% at 168 hours. This was true regardless of the die speed of the pelletizer (40–60 Hz). In contrast, when larger particle size was used during the pelletizing process, no improvement in hydrolysis yield was observed. Furthermore, pre-heating the biomass prior to pelletization decreased yields at 72 hours as compared to the untreated material, suggesting an adverse impact of pelletization. This further suggests that it will be necessary to optimize the pelletization process when proceeding to commercial scale to ensure no adverse downstream impacts. It also suggests that further optimization with densification can improve hydrolysis yields.

In contrast, Rijal et al. (2014) tested hydrolysis and fermentation on briquetted AFEX biomass and did not see any improvement as compared to undensified AFEX-treated biomass. In fact, for prairie cordgrass, densification reduced sugar yields. These hydrolysis tests were performed at a low solid loading rather than the high solid loading described previously for pellets. Thus, the advantages of briquetting may not have been observed at this concentration. Furthermore, although the authors do not speculate on the reason for this discrepancy, briquetting may hinder sugar production in ways pelletization does not. It may also be a peculiarity unique to cordgrass as this adverse impact is not observed in either switchgrass or corn stover. The cordgrass contained a much higher concentration of soluble as compared to the other two types of biomass, which may be more susceptible to damage during the densification process.

Sundaram and Muthukumarappan (2016a) also attempted to discern the impact of various physical properties on the hydrolysis results of AFEX-treated corn stover, switchgrass, and prairie cordgrass pellets. In contrast to Hoover et al. (2014), no significant difference in sugar yields was observed when pelletized at temperatures ranging from 75°C–125°C. These pellets were made using a single screw extruder rather than a flat die pelletizer, however, which may be the reason for the different results. Likewise, no impact of particle size on sugar yields was observed. In contrast, untreated material that was pelletized at different conditions did have different sugar yields, with more extreme pelleting conditions producing more digestible pellets. The authors state that these results suggest that densification can be performed under mild conditions, with no external heating, and with minimal additional grinding. If so, then this suggests that pelletization costs can be reduced relative to pelletizing native, untreated material. Sundaram and Muthukumarappan (2016b) also demonstrated that pellets of different species (e.g., corn stover and switchgrass) can be blended and hydrolyzed together with no adverse impact on yields.

Comparatively, little research has been performed on the fermentation of hydrolyzed AFEX pellets. There should be no concerns regarding the mixability of the slurry after hydrolysis; the conditions should be nearly identical for pellets and unpelleted biomass. The only concern would be if pelletization produced any inhibitory compounds in the biomass. Sarks et al. (2016) performed fermentation using *Zymomonas mobilis* after hydrolysis. Complete glucose utilization and nearly complete xylose utilization was observed with high conversion to ethanol. Furthermore, a seed train for the fermentation organism was demonstrated that used hydrolysates of AFEX pellets as the sugar source. It, therefore, appears unlikely that any inhibitory products were formed during pelletization.

Thus, there does not appear to be any adverse impact of pelletization on the subsequent hydrolysis and fermentation of AFEX-treated biomass. In contrast, slight improvements in sugar yields are likely. Furthermore, pelletization at different conditions may further improve these yields and all the studies suggest that low energy inputs during pelletization appear to be viable. However, experiences with briquetted prairie cordgrass suggest that these improvements may not be consistent across all types of biomass and all methods of densification.

7. Design of an AFEX Depot

Because of the ability to rapidly densify biomass that has been AFEX treated, the AFEX technology is currently being developed solely for the depot concept rather than as a part of a centralized biorefinery. The current packed-bed design of AFEX is suitable for a 100–200 ton/day depot rather than a 2000+ ton/day biorefinery. While no firm engineering design has been developed for this size, there are currently no major concerns. All equipment in the depot can be easily constructed at this size and most of the equipment is already standard pieces being built by multiple companies. While the treatment occurs at a relatively high pressure (~2 MPa), it is highly unlikely that an overpressure situation can occur given that there is no external heating. Ammonia release is the largest safety concern, although no safety incidences have occurred in the pilot facility. Likewise, residual ammonia off-gassing could also be a safety concern but any residual ammonia would be removed in the dryer. Thus, a scrubber would be necessary. Transferring biomass into and out of the reactor can be performed using baskets, and transferring biomass into and out of the baskets appears to be a relatively simple task.

As currently envisioned, the AFEX depot is expected to be a relatively simple operation, bringing in the biomass bales and then selling dried AFEX pellets. The "upstream" operations, prior to AFEX treatment, include only a bale grinder and mixer. The grinder—likely covered to prevent dust—would grind the bales to a 3–5 cm particle size. This milled material would then pass to a mixer, where it would be mixed with water to the proper moisture content. Process control over this operation may be difficult, as the moisture in the incoming bales may not be clear. However, although the process will not be as energy-efficient, the effectiveness of the process is not hampered by any additional moisture. Mixing water and biomass can take place within commercial ribbon blenders and would require a residence time of only a few minutes. After the biomass is at the correct moisture, it is added to a basket and compressed to approximately 2X its loose bulk density using a simple piston.

The baskets filled with biomass are loaded into and out of the AFEX reactors using a piston. Afterwards, they are dumped into a rotary drum dryer. This dryer reduces the moisture of the biomass to approximately 15–18%, which is sufficient for subsequent pelletization. After drying, a hammer mill is used to mill the biomass to approximately 0.6 cm particle size. While not an absolute necessity, it was deemed more efficient than attempting to pelletize the larger, 3–5 cm particle size biomass. The milled biomass is fed directly to a ring-die pelletizer which reduces the moisture to approximately 12–15% due to the heat of friction. The pellets are cooled in a pellet cooler before being stored for shipment. Because the moisture content is low, the pellets will be stable and can be stored indefinitely. Any storage method suitable for grain, such as a silo or a bin, would also be suitable for the pellets. A drawing of this depot, showing the relative size of the operation, is shown in Fig. 4.

Depots would require water, electricity, and gas utilities, although the load for each of these would be relatively small. Depots would also require all the relevant ammonia permits and safety considerations. It is possible that depots could be co-located with other rural operations in order to share resources. One example might be grain elevators which would not only share utilities but also transportation infrastructure, including rail. Another option may be large anaerobic digesters or large cattle feeding operations which may have access to similar resources.

Of particular concern is determining the right size of a depot. In a practical sense, commercially available quick-opening hatches for a pressure vessel can be obtained at sizes up to 1.5 m diameter, which limits the size of the AFEX reactors. At this size, each reactor can produce approximately 25 metric dry tons of AFEX-treated biomass per day or 50 tons for a pair of reactors. Thus, increasing the size of the depot would require increasing the number of reactors. Because the compressor is only in use for a short time period during the entire cycle, a single compressor can be used to transfer ammonia in multiple pairs of reactors. Hence, it is preferable to

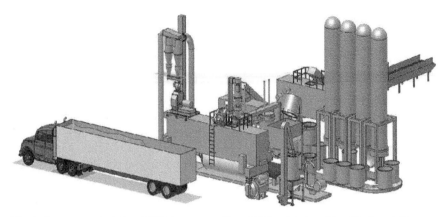

Fig. 4. Representation of an AFEX depot operating at 100 metric tons/day. The four tall towers at the right are the four AFEX reactors with baskets being carried around on a carousel to be loaded with biomass, brought into the AFEX reactors, and dumped into a dryer. A semi-truck is shown for scale. Drawing by Corrie Nichol, Idaho National Laboratory.

have multiple pairs of reactors to share the expense of the compressor. Furthermore, larger depot sizes would reduce the per-ton cost of labor and capital, which is a significant cost concern given the relatively low throughputs in a depot. In contrast, increasing the size of a depot decreases its logistical advantage.

In its analyses, MBI determined that depots of 100–200 dry tons per day of pellet production offered the optimal size in terms of balancing the cost of logistics vs. the fixed costs (MBI, unpublished data). At this size, it is still relatively simple to store small supplies of biomass bales and move them to the depot, as well as collect biomass from local communities. A depot of this size is able to produce enough revenue to justify the costs as well as being able to provide a large enough supply to entice a biofuel producer. In areas with high corn production, like Iowa, this would correspond to approximately 15–20 km radius of collection even at modest collection rates or roughly equivalent to one depot per county.

Lamers et al. (2015) calculated the cost of producing AFEX pellets to be $62.50 per metric ton, not inclusive of the cost of the biomass itself. In comparison, a conventional pelletization depot would produce untreated pellets at $47.80 per ton. Thus, the incremental cost of AFEX treatment is approximately $15 per ton. Lamers et al. (2015) also discussed a novel approach to pelletization at high moisture which reduced the cost of the depot by $17.00 per ton. While this approach has not been tested with AFEX treated biomass, it is possible that the cost of AFEX pellets can be reduced to $45 per metric ton.

8. Depot Deployment

One of the critical barriers to commercializing depots is the lack of a final customer. A single depot cannot supply all of the biomass needs for a biorefinery, and thus the biorefinery would require taking biomass from other sources until a depot network is developed. Therefore, the refinery would require bale handling and grinding, pretreatment facilities, etc., and so would not be designed to take advantage of the logistical improvements and simplicity of design that would entail if it is used solely for AFEX pellets. Because of this, the inherent advantages of pelletization would be limited, thus placing the AFEX pellets at a cost disadvantage compared to buying bales off the open market. Furthermore, since AFEX treatment is unlikely to be scaled to thousands of tons per day at a single location, any refinery that simultaneously accepts AFEX pellets from early depots as well as accepts incoming bales would likely be using two separate pretreatments. This may make it difficult if not impossible to optimize downstream processes, particularly in the use of enzymes.

Commercializing depots is a difficult task. Without a guaranteed market, the investment costs are too high to justify the depot, but a single depot cannot support the market. In order to break this paradox, the pellets must be suitable for an alternate market, preferably one that is already developed. For AFEX pellets, the cattle feed market is being considered as an alternative.

Ammoniation of biomass has been performed to improve the nutritive qualities of agricultural residues for decades (Horton, 1978; Morris and Mowat, 1980). Besides increasing the digestibility of fiber, the ammonia that reacts with the biomass can serve as a source of nitrogen to the animals as the micro-organisms within the cattle's rumen can convert the nitrogen into amino acids. Traditionally, ammoniation is performed by adding a small amount (1–3% of the dry weight of the biomass) to biomass under a tarp and allowing for a residence time of several weeks. This approach, performed by individual farmers, only slightly increases the digestibility of the agricultural residues and is often used during winters when green fodder is unavailable.

AFEX treatment, then, can be considered a more extreme approach to this traditional ammoniation but the primary results—increased fiber digestibility and increased nitrogen content—are the same. An approach to test the digestibility of AFEX-treated biomass for animal feed is to incubate the biomass with fluid obtained from a cattle's rumen. One can measure either the fiber remaining after the fermentation (Tilley and Terry, 1963) or the total gas produced from the fermentation. AFEX treatment was shown to dramatically increase both the rate and extent of fiber digestion for multiple types of biomass (Bals et al., 2010; Scott et al., 2011). In general, AFEX treatment was effective for members of the grass family that were not already highly digestible. For corn stover, AFEX treatment increased the extent of fiber digestibility by 75% and the rate by 25%. The increase in digestibility for AFEX treatment was also higher than that for the traditional ammoniation. Likewise, AFEX treatment increased gas production by 123% compared to untreated corn stover for a full diet (MBI, unpublished data).

Based on these results, AFEX-treated pellets should provide more energy to cattle than ammoniated biomass and could conceivably be used as an energy source for ruminant animals. To test this, Michigan State University performed a feeding trial on 12 Holstein steers in which 30% of their diet was replaced with AFEX-treated corn stover pellets and monitored the weight gain over a period of 160 days (Blummel et al., 2014). The key result of this trial was that the cattle ate the pellets, gained weight, and remained healthy throughout the trial. While the cattle on the AFEX diet gained less weight than the cattle on the control diet (160 kg vs. 200 kg), this is unsurprising given that the control diet was high in both corn and distiller's grains. In addition, carcass quality was similar to that of the cattle on a control diet with the primary difference being less fat in the AFEX diet. Based on these results, it appears that AFEX pellets can be used to support cattle growth, although not as effectively as corn grain. Further research is required to fully understand the value of AFEX pellets as cattle feed.

If there is value in AFEX pellets as cattle feed, then this would instantly produce a long-term, stable market for AFEX pellets. Beef cattle feeding operations are prevalent throughout the US Corn Belt, particularly in the western portion, and could be local markets for pellets. Assuming each animal consumes 4 kg dry weight pellets per day, then a depot would require 50,000 cattle to maintain a 200 ton per day production. However, regulatory approval may be required before AFEX pellets can be used for this market.

9. Future Work

AFEX processing has currently been demonstrated at the pilot scale using both corn stover (Sarks et al., 2016) and wheat straw (MBI, unpublished data). The next scale to test would be a demonstration scale, with a nominal capacity of at least 0.5 tons biomass treated per hour. This scale would prove not only the ammonia recovery and pretreatment effectiveness at a near commercial level but would also allow for demonstration of biomass handling. In particular, an automated method of basket loading and unloading would need to be designed for this scale. Furthermore, commercial ring-dye pelletizers can be tested at this scale as well as allowing for large-scale testing of AFEX pellets as either animal feed or for fermentation. After the demonstration at this scale, the AFEX depot design could likely be developed for commercial purposes assuming markets for the products.

Besides the final scale-up, there are other potential improvements to the AFEX depot concept. Currently, it is envisioned that the AFEX biomass, which is removed from the reactor at ~45% moisture, must be dried to ~15–20% moisture prior to pelletization. Current work at the Idaho National Laboratory has shown untreated corn stover at 30% moisture can be pelletized and dried after pelletization, with the resulting energy cost being lower than if dried first (Tumuluru, 2015). It remains to be seen if this approach can be used on AFEX-treated biomass as well.

In addition, while most pieces of equipment in an AFEX depot are relatively simple, low cost, and low maintenance, the ammonia compressor could be expensive. An alternative approach to the ammonia transfer would be to use a system similar to ammonia refrigeration units. Such a system may be more energy intensive, but would be relatively low capital. To date, no tests have been performed on the ammonia quench system for an AFEX reactor. Such a system would need to be designed and tested prior to determining which approach would be more suitable for an AFEX depot.

10. Summary

AFEX treatment can be effectively performed in a depot setting to produce pellets that are ready-made for saccharification and fermentation within a biorefinery. A low-cost, low-maintenance pretreatment reactor has been designed and demonstrated at the pilot scale. Pellets produced from this pilot reactor have been proven to be equal to or superior to unpelletized AFEX treated biomass when subjected to enzymatic hydrolysis and subsequent fermentation to ethanol, with no further pretreatment or preprocessing required. Furthermore, these AFEX pellets can potentially be sold as a cattle feed, thereby opening up another market and incentivizing the development of a biomass market. Given these advantages, AFEX shows strong potential as a pre-processing operation within a depot setting.

11. Acknowledgments

We would like to thank Farzaneh Teymouri, Chandra Nielson, Josh Videto, Tony Rinard, and Janette Moore for their support in developing the AFEX project. We would also like to thank Corrie Nichol for his permission to use his drawing of an AFEX depot. AFEX is a trademark of MBI.

References

Argo, A. M., E. C. D. Tan, D. Inman, M. H. Langholtz, L. M. Eaton, J. J. Jacobson et al. 2013. Investigation of biochemical biorefinery sizing and environmental sustainability impacts for conventional bale system and advanced uniform biomass logistics designs. Biofuels Bioprod. Biorefin. 7: 282–302.
Balan, V., L. C. Sousa, S. P. S. Chundawat, D. Marshall, L. N. Sharma, C. K. Chambliss and B. E. Dale. 2009. Enzymatic digestibility and pretreatment degradation products of AFEX-treated hardwoods (*Populus nigra*). Biotechnol. Progress 25: 365–375.

Bals, B. and H. Murnen, M. Allen and B. Dale. 2010.Ammonia fiber expansion (AFEX) treatment of eleven different forages: improvements to fiber digestibility *in vitro*. Anim. Feed Sci. Technol. 155: 147–155.

Bals, B. D., C. Gunawan, J. Moore, F. Teymouri and B. E. Dale. 2013. Enzymatic hydrolysis of pelletized AFEX™-treated corn stover at high solid loadings. Biotechnol. Bioeng. 111: 264–271.

Bals, B. D. 2015. Pilot scale production of fermentable sugars from corn stover via distributed preprocessing in AFEX™ depots. Presented at 37th Symposium on Biobased Fuels and Chemicals, San Diego, CA, April 27, 2015.

Blummel, M., B. Steele and B. E. Dale. 2014. Opportunities from second-generation biofuel technologies for upgrading lignocellulosic biomass for livestock feed. CAB Reviews 9: 41.

Bonner, I. J., D. N. Thompson, F. Teymouri, T. Campbell, B. Bals and J. S. Tumuluru. 2015. Impact of sequential ammonia fiber expansion (AFEX) pretreatment and pelletization on the moisture sorption properties of corn stover. Drying Technol. 33: 1768–1778.

Campbell, T. J., F. Teymouri, B. Bals, J. Glassbrook, C. D. Nielson and J. Videto. 2013. A packed bed ammonia fiber expansion reactor system for pretreatment of agricultural residues at regional depots. Biofuels 4: 23–34.

Campbell, T. J. and F. Teymouri. 2015. Process for treating biomass.US Patent # 9102964 B2, August 11, 2015.

Chundawat, S. P. S., R. Vismeh, L. N. Sharma, J. F. Humpula, L. C. Sousa and C. K. Chambliss. 2010. Multifaceted characterization of cell wall decomposition products formed during ammonia fiber expansion (AFEX) and dilute acid based pretreatments. Bioresour. Technol. 101: 8429–8438.

Chundawat, S. P. S., B. S. Donohoe, L. C. Sousa, T. Elder, U. P. Agarwal and F. Lu. 2011. Multi-scale visualization and characterization of lignocellulosic plant cell wall deconstruction during thermochemical pretreatment. Energy Environ. Sci. 4: 973–984.

Chundawat, S. P. S., B. Bals, T. Campbell, L. Sousa, D. Gao and M. Jin. 2013. Primer on ammonia fiber expansion pretreatment. pp. 169–200. *In*: Wyman, C. (ed.). Aqueous Pretreatment of Plant Biomass for Biological and Chemical Conversion to Fuels and Chemicals. Chichester: John Wiley & Sons, Ltd.

Dale, B. E. and M. J. Moreira. 1982. A freeze-explosion technique for increasing cellulose hydrolysis. Biotechnol. Bioeng. Symposium 12: 31–44.

Dale, B. E. and R. G. Ong. 2012. Energy, wealth, and human development: why and how biomass pretreatment research must improve. Biotechnol. Prog. 28: 893–898.

Dale, B. E., B. Ritchie and D. Marshall. 2014. Pretreated densified biomass products. US Patent # 8673031 B2, March 18, 2014.

Donohoe, B. S., T. B. Vinzant, R. T. Elander, V. R. Pallapolu, Y. Y. Lee, R. J. Garlock, V. Balan and B. E. Dale. 2011. Surface and ultrastructural characterization of raw and pretreated switchgrass. Bioresour. Technol. 102: 11097–11104.

Eggeman, T. and R. T. Elander. 2005. Process and economic analysis of pretreatment technologies. Bioresour. Technol. 96: 2019–2025.

Eranki, P. L., B. D. Bals and B. E. Dale. 2011. Advanced regional biomass processing depots: a key to the logistical challenges of the cellulosic biofuel industry. Biofuels Bioprod. Biorefin. 5: 621–630.

Garlock, R. J., B. Bals, P. Jasrotia, V. Balan and B. E. Dale. 2012a. Influence of variable species composition of the saccharification of AFEXTM pretreated biomass from unmanaged fields in comparison to corn stover. Biomass & Bioenergy 37: 49–59.

Garlock, R. J., Y. S. Wong, V. Balan and B. E. Dale. 2012b. AFEX pretreatment and enzymatic conversion of black locust (*Robinia pseudoacacia* L.) to soluble sugars. Bioenergy Res. 5: 306–318.

Hess, J. R., C. T. Wright, K. L. Kenney and E. M. Searcy. 2009. Uniform-format solid feedstock supply system: A commodity-scale design to produce an infrastructure-compatible bulk solid from lignocellulosic biomass. Idaho National Laboratory Report INL/EXT-09-15423.

Hoover, A. N., J. S. Tumuluru, F. Teymouri, J. Moore and G. Gresham. 2014. Effect of pelleting process variables on physical properties and sugar yields of ammonia fiber expansion pretreated corn stover. Bioresour. Technol. 164: 128–135.

Horton, G. M. J. 1978. The intake and digestibility of ammoniated cereal straws by cattle. Can. J. Anim. Sci. 58: 471–478.

Karki, B., K. Muthukumarappan, Y. Wang, B. Dale, V. Balan, W. R. Gibbons and C. Karunanity. 2015. Physical characteristics of AFEX-pretreated and densified switchgrass, prairie cordgrass, and corn stover. Biomass & Bioenergy 78: 164–174.

Lamers, P., M. S. Roni, J. S. Tumuluru, J. J. Jacobson, K. G. Cafferty, J. K. Hansen, K. Kenney, F. Teymouri and B. Bals. 2015. Techno-economic analysis of decentralized biomass processing depots. Bioresour. Technol. 194: 205–213.

Lange, J. P. 2007. Lignocellulose conversion: an introduction to chemistry, process and economics. Biofuels Bioprod. Biorefin. 1: 39–48.

Lau, M. J., M. W. Lau, C. Gunawan and B. E. Dale. 2010. Ammonia fiber expansion (AFEX) pretreatment, enzymatic hydrolysis, and fermentation on empty palm fruit bunch fiber (EPFBF) for cellulosic ethanol production. Appl. Biochem. Biotechnol. 162: 1847–1857.

Luo, X. L., J. Y. Zhu, R. Gleisner and H.Y . Zhan. 2011. Effects of wet-pressing-induced fiber hornification on enzymatic saccharification of lignocelluloses. Cellulose 18: 1055–1062.

Mosier, N., C. Wyman, B. Dale, R. Elander, Y. Y. Lee, M. Holtzapple and M. Ladisch. 2005. Features of promising technologies for pretreatment of lignocellulosic biomass. Bioresour. Technol. 96: 673–686.

Morris, P. J. and D. N. Mowat. 1980. Nutritive value of ground and/or ammoniated corn stover. Can. J. Anim. Sci. 60: 327–336.

Rijal, B., G. Biersbach, W. R. Gibbons and S. W. Pryor. 2014. Effect of initial particle size and densification on AFEX-pretreated biomass for ethanol production. Appl. Biochem. Biotechnol. 174: 845–854.

Sarks, C., B. D. Bals, J. Wynn, F. Teymouri, S. Schwegmann and K. Sanders. 2016. Scaling up and benchmarking of ethanol production from pelletized pilot scale AFEX-treated corn stover using *Zymomonas mobilis* 8b. Biofuels 7: 3.

Scott, S. L., R. S. Mbifo, J. Chiquette, P. Savoie and G. Turcotte. 2011. Rumen disappearance kinetics and chemical characterization of by-products from cellulosic ethanol production. Anim. Feed Sci. Technol. 165: 151–160.

Singh, R. S., A. Pandey, E. Gnansounou (eds.). 2017. Biofuels: Production and Future Perspectives. Boca Raton: CRC Press.

Singh, S., G. Cheng, N. Sathitsuksanoh, D. Wu, P. Varanasi, A. George, V. Balan, X. Gao, R. Kumar, B. E. Dale, C. E. Wyman and B. A. Simmons. 2015. Comparison of different biomass pretreatment techniques and their impact on chemistry and structure. Front. Energy Res. 2: 62.

Sundaram, V. and K. Muthukumarappan. 2016a. Impact of AFEXTM pretreatment and extrusion pelleting on pellet physical properties and sugar recovery from corn stover, prairie cordgrass, and switchgrass. Appl. Biochem. Biotechnol. DOI 10.1007/s12010-016-1988-9.

Sundaram, V. and K. Muthukumarappan. 2016b. Influence of AFEX™ pretreated corn stover and switchgrass blending on the compaction characteristics and sugar yields of the pellets. Ind. Crops Prod. 83: 537–544.

Thompson, D. N., T. Campbell, B. Bals, T. Runge, F. Teymouri and L. P. Ovard. 2013. Chemical preconversion: application of low-severity pretreatment chemistries for commoditization of lignocellulosic feedstock. Biofuels 4: 323–340.

Tilley, J. M. A. and R. A. Terry. 1963. A two-stage technique for the *in vitro* digestion of forage crops. Grass Forage Sci. 18: 104-111.

Tumuluru, J. S. 2015. High moisture corn stover pelleting in a flat die pellet mill fitted with a 6 mm die: physical properties and specific energy consumption. Energy Sci. Eng. 3: 327–341.

Index

Printed and bound by CPI Group (UK) Ltd, Croydon, CR0 4YY

24/10/2024

01778304-0019